Engineering Applications of Computational Methods

Volume 16

The book series Engineering Applications of Computational Methods addresses the numerous applications of mathematical theory and latest computational or numerical methods in various fields of engineering. It emphasizes the practical application of these methods, with possible aspects in programming. New and developing computational methods using big data, machine learning and AI are discussed in this book series, and could be applied to engineering fields, such as manufacturing, industrial engineering, control engineering, civil engineering, energy engineering and material engineering.

The book series Engineering Applications of Computational Methods aims to introduce important computational methods adopted in different engineering projects to researchers and engineers. The individual book volumes in the series are thematic. The goal of each volume is to give readers a comprehensive overview of how the computational methods in a certain engineering area can be used. As a collection, the series provides valuable resources to a wide audience in academia, the engineering research community, industry and anyone else who are looking to expand their knowledge of computational methods.

This book series is indexed in both the **Scopus** and **Compendex** databases.

Akeel A. Shah · Puiki Leung · Qian Xu ·
Pang-Chieh Sui · Wei Xing

New Paradigms in Flow Battery Modelling

Springer

Akeel A. Shah
Key Laboratory of Low-grade Energy
Utilization Technologies and Systems,
MOE
Chongqing University
Chongqing, China

Puiki Leung
Key Laboratory of Low-grade Energy
Utilization Technologies and Systems,
MOE
Chongqing University
Chongqing, China

Qian Xu
Institute for Energy Research
Jiangsu University
Zhenjiang, Jiangsu, China

Pang-Chieh Sui
School of Automotive Engineering
Wuhan University of Technology
Wuhan, Hubei, China

Wei Xing
School of Mathematics and Statistics
University of Sheffield
Sheffield, United Kingdom

ISSN 2662-3366 ISSN 2662-3374 (electronic)
Engineering Applications of Computational Methods
ISBN 978-981-99-2523-0 ISBN 978-981-99-2524-7 (eBook)
https://doi.org/10.1007/978-981-99-2524-7

This Springer imprint is published by the registered company Springer Nature Singapore Pte Ltd.
The registered company address is: 152 Beach Road, #21-01/04 Gateway East, Singapore 189721, Singapore

Contents

Chapter 1
Introduction

1.1 Motivation and Outline of this Book

Redox flow batteries (RFBs) were designed for large-and medium-scale energy storage and have historically been used as backup and standalone power systems. With renewable energy generation taking on an increasing share of electrical power production, the role of RFBs is expected to grow in importance. Indeed, they may well become critical to maintaining a continual supply of electricity to homes and businesses.

Despite major advances in the technology in the last two decades, there are still many hurdles that developers must overcome in order to deploy RFBs on a much wider scale. These include the fundamental development of new systems and materials, and the engineering of flow batteries on scales larger than is currently achievable, optimising and extending the lifetimes of existing systems, increasing the power and energy densities, and integration with other technologies.

Modelling and simulation, the focus of this book, can play a crucial role in overcoming these challenges, as well as in the operation and maintenance of flow batteries. Modelling and simulation for electrochemical systems is well established, especially for hydrogen and solid-oxide fuel cells and lithium-ion batteries. In recent years, traditional modelling approaches, largely based on continuum mechanics, have been augmented with a number of powerful complementary and sometimes alternative methods. These include methods for screening and designing new materials, such as electronic-structure calculations and molecular dynamics; methods for the detailed study of mass and charge transport at electrode/electrolyte interfaces, such as the lattice-Boltzmann and pore-network models; methods for the study of phase changes, microstructure formation, deposition and dendrite formation, such as phase-field models and the volume-of-fluid method; and alternatives to physics-based approaches for problems such as optimisation and end-of-life prediction, primarily based on machine learning.

A. A. Shah et al., *New Paradigms in Flow Battery Modelling*,
Engineering Applications of Computational Methods 16,
https://doi.org/10.1007/978-981-99-2524-7_1

The aforementioned methods can provide insights into problems that defy analysis using the traditional modelling approaches, and it is our intention in this book to introduce these methods to the reader, providing a thorough and detailed description of their application to flow batteries. At the present time, the vast majority of the applications are to be found in the fuel cell and lithium-ion battery literature. Another goal of this book, therefore, is to promote the use of these methods for RFBs. With an expected growth in interest in RFB technology, this is an ideal time to present this material, together with our thoughts and perspectives. We have deliberately laden the presentation with technical content so that readers are able to use this book as both a guide to the potential uses of the methods as well as to their implementation.

In this chapter we summarise the different types of energy storage technologies available or in development, so that we may place RFBs in a broader context. The relative advantages and disadvantages compared to other technologies, together with the potential applications of RFBs will be made clear. In Chap. 2 we will provide a more through introduction to RFB technologies, including the different types, their fundamental operation, components and materials, applications and open challenges in their development.

In Chap. 3 we introduce various physics-based methods for modelling scientific and engineering problems, outlining potential applications to the development of flow batteries as we present the methods. The subsequent two Chaps. 4 and 5 present case studies in the use of macroscopic, mesoscopic and multi-scale methods to the analysis of all-vanadium and iron-vanadium flow battery systems. Also provided in Chap. 3 is an outline of machine learning methods and surrogate models, introducing the reader to the basic terminology and to the basic framework used in most machine learning tasks.

In Chap. 6 we explain machine learning (including deep learning) methods in detail, and provide case studies in their application to flow battery systems, including quantitative-structure-activity-relationship models and machine learning-assisted macroscopic and quantum-mechanical models for all-vanadium and organic RFBs, respectively. In Chap. 7 we introduce alternative modelling approaches that combine data-driven and physics-based methods. Specifically, we outline reduced-order and multi-fidelity models. We also introduce time series methods, which are indispensable in the study of battery degradation and failure. We end with a case study in the use of multi-fidelity methods for electrochemical systems.

We note that numerical methods (meaning implementations of models) are largely ignored. In the main text we focus on the specification of models, including their physical or mathematical underpinnings. For the sake of completeness, and being aware that some readers may not be familiar with some of the methods we discuss in the main text, basic-level outlines of selected numerical techniques are included in Appendices A–D. These appendices cover basic linear systems, time-stepping methods, methods for solving partial differential equations and gradient-based optimisation techniques, respectively. They should provide sufficient understanding to follow most of the discussions related to implementations in the main text.

1.2 Why Energy Storage?

Global demand for electrical energy continues to grow in line with industrial developments and population growth. Power plants must be flexible enough to meet a constantly fluctuating demand and provide additional power in the case of disruption to supply from other sources. Energy storage currently plays a crucial role in ensuring demand is met and is likely to play a much more prominent role in the future, in a variety of settings.

In many developed countries, such as the United States, the capacity factor of electrical power generating sources is as low as 40%, so that less than half of the peak capacity is produced. Excess power generation and transmission facilities are therefore underused every year. Energy storage is used extensively in electric power industries. Insufficient energy storage may lead to problems, such as increased volatility, reduced reliability and threatened security.

There are many forms of energy storage [1–3], which can be distinguished along the lines of the mode in which energy is stored, the scale of the storage, and the timescales over which the energy can be delivered. Perhaps the most common and well-known applications currently are for portable electronics (mobile phones, laptops, etc.) in the form of batteries, now almost exclusively Li-ion, as well as the lead-acid batteries found primarily in conventional vehicles.

Batteries have a long history, dating back to the nineteenth century. Major breakthroughs were made in the second half of the twentieth century with nickel-and lithium-based batteries. In particular, the development of the first lithium-ion battery in 1985 by Asahi Chemical in Japan, and the lithium-polymer battery by Sony and Asahi Kasei in 1997 enabled the revolution in portable electronic devices that followed soon after [4].

Currently, the major growth areas are battery-powered vehicles and grid-scale energy storage for integrating intermittent power into grids. Before moving onto describing the different types of storage, we list below some well-known existing and emerging or potential (future) applications.

1. At the utilities level
    - Meeting peak demand in electricity
    - As a flexible means of frequency and voltage regulation for non-matched supply and demand
    - System operators must have reserve capacity (spinning or non-spinning) that can potentially be replaced by energy storage technologies
2. As backup power
    - Supply power during outages, for businesses and especially for emergency services such as hospitals

3. Electronics and transport
   - Rechargeable batteries for consumer portable electronics, especially mobile phones and laptops
   - Battery powered electric vehicles, either for hybrid power propulsion or for full electric propulsion
4. Distributed and smart grids
   - To enable further penetration of variable renewable energy technologies such as wind and photovoltaics into grids, with low-or zero-carbon emissions
   - Store excess energy from renewable or non-renewable sources in various forms when there is a surplus of electrical power and uses this stored energy when generation cannot meet the demand for electrical and thermal power

Traditionally, utilities used grid-scale energy storage for peak shaving and load levelling in thermal-cycle power plants [5]. Cheap electricity generated during off-peak hours was stored as potential energy by driving pumps to transport water to a high-lying reservoir. During peak hours of demand a turbine-generator was used to convert the potential energy to electricity. This form of energy storage led to more stable (steady) operation of the plant and afforded savings by curtailing the generation of electricity from more costly fuels during peak hours.

With the advent of new technologies, utilities have found further potential uses for energy storage, especially as the share of intermittent generation increases. Energy storage is in fact becoming a critical need, rather than simply a desirable objective to lower costs [6]. The two main ways in which the need becomes critical are (a) matching supply and demand and (b) stabilising the grid, especially with a large presence of wind power generation. It can also be used for spinning, non-spinning, and supplemental reserves (which are frequently mandated), as well as transmission congestion relief.

1.3 Types of Energy Storage

Categorisation of energy storage technologies is usually on the basis of the physical mechanism by which the energy is stored, as illustrated in Fig. 1.1. In general, energy can be stored in *electrical, thermal, mechanical, chemical* or *electrochemical* forms, with the electrical and electrochemical forms often placed together under 'electrical', or the chemical and electrochemical combined into 'chemical'.

1.3.1 Chemical Energy Storage

Chemical energy storage refers to the storage of energy within the chemical bonds of a fuel, such as methane CH_4, ethanol, methanol and hydrogen H_2 [7]. Almost

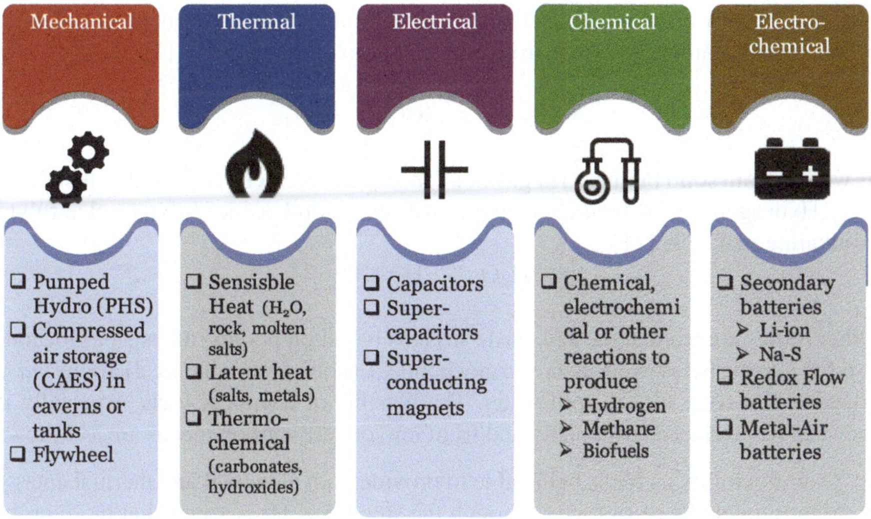

Fig. 1.1 An illustration of the categorisation of energy storage technologies in terms of the physical mechanisms by which the energy is stored

exclusively these fuels are *hydrocarbons*, with H_2 being the main exception. They can be used for a variety of purposes, but are primarily used for generating electricity *via* a conventional thermal means in a Rankine cycle. The fuel is combusted and used to produce high pressure steam by boiling water in a large vessel. The steam is subsequently used to drive a steam turbine, which is connected to an electrical generator. Various sources of fuel can be used for traditional steam-powered generation, almost all of them hydrocarbon based, e.g., coal and CH_4 (natural gas). Nuclear fuel and waste (agricultural, domestic, industrial) containing hydrocarbons are also mainstream fuels, as well as solar energy *via* concentrating lenses.

H_2 is an excellent energy carrier but is rarely used alone for direct combustion. It can be injected into fuel lines at up to 20% to enhance combustion [8] or it can be used in a H_2 polymer electrolyte membrane (PEMFC) or solid-oxide fuel cell (SOFC) to produce electricity directly. It can also be used to produce CH_4 or ammonia, the latter of which is required for fertilisers.

The heat produced from combustion-based electricity generation can also be used for heating purposes, such as in a district heating system. Storing heat, whether from a power plant or some other industrial process, is another form of energy storage that can be achieved by several means.

One of the main forms of sustainable chemical energy storage currently under investigation is the production of *synthetic natural gas* (SNG) and H_2 from excess electricity generated by variable renewable energy (VRE) sources, primarily wind turbines and solar photovoltaics (PV), during times of low demand and excess production of electrical power. The excess power is used for electrolysis of water followed by methanation with carbon dioxide CO_2. This is an known as *power-to-gas*

(P2G) [9]. SNG or synthetic methane can be produced by other means. It can be produced from biomass *via* anaerobic digestion involving glucose [10]

$$2C_6H_{12}O_6 \rightarrow 3CO_2 + 3CH_4$$

as well as from solid biomass *via* gasification and direct electrochemical reduction of CO_2. Hydrogen can be generated *via* electrolysis in solid-oxide, polymer-electrolyte or alkaline fuel cells [11]

$$2H_2O \rightarrow 2H_2 + O_2$$

Other routes are from algae and waste streams based on photosynthesis or biocatalysed electrolysis, as well as a thermochemical (non-electrochemical) route *via* a concentrated solar flux to split water. The benefits of chemical fuels, especially if they can be produced sustainably and from low-or zero-carbon means are as follows

1. Their flexibility as fuels, being able to provide both electrical and thermal energy
2. Mature methods of storage (although the storage of H_2 is somewhat more problematic than that of SNG)
3. Low-cost and mature means of utilisation to produce electrical and thermal energy
4. Lack of degradation over time, leading to long storage times
5. Capability of producing energy at grid (up to GW) scale

On the other hand, such fuels emit CO_2, which (without entering into the debate on anthropogenic contributions to climate change) conflicts with short-and medium-term policy goals worldwide. Nevertheless, even in Europe, in which CO_2 reductions have been pursued particularly aggressively, there is significant investment in the production and storage of synthetic fuels [12], at least as an intermediate solution to 'net-zero'.

1.3.2 Electrical Energy Storage

Electrical energy storage is ideal for situations in which a high-power output is required over a short duration [13]. Chief among these technologies is *supercapacitors*, also known as *ultracapacitors*, which can deliver short bursts of high power, but are unable to provide power over several hours. They are, therefore, characterised by high-power densities and low energy densities (or capacities). Capacitors have been used in circuit boards on a small scale for several decades, while supercapacitors, which are of a much higher capacity and have lower voltage limits, are still in a developmental stage [13].

Super-capacitors are comprised of two electrodes that are separated by an ion-permeable membrane, together with an electrolyte for conducting ions and allowing their passage between the electrodes [13]. When a voltage is applied, ions in the electrolyte form electric double layers with a polarity opposite to that of the electrodes. A positively polarised electrode will have a layer of negatively charge ions present at the electrode/electrolyte interface, together with a layer of positively charged ions that adsorb onto the negative layer to balance the charge.

Super-capacitors bridge the gap between batteries and conventional electrolytic capacitors, which store electrical energy in an electric field established between two electrodes separated by a dielectric, by virtue of charge separation [14]. The charging times are much shorter than those of batteries, as is the delivery of electrical energy. Moreover, they are able to withstand repeated charge-discharge cycles without significant degradation, in contrast to batteries.

The nature of supercapacitors makes them ideal for applications that demand rapid and repeated charge-discharge cycling, as opposed to long-term storage and delivery of energy. They have potential applications for regenerative braking, backup power, backup for static random-access memory and voltage/frequency stabilisation in grids to maintain safe operating limits [14]. They can also be used to extend the range and cycle life of batteries in hybrid electric vehicles (HEV). They can potentially provide a power boost to make the engine more responsive during acceleration and replace the battery for stop-start operation, as well as power the steering and brakes in the case of battery failure.

There are several variants of supercapacitors currently under development [13]. All supercapacitors store energy *via* an electrostatic double-layer capacitance, together with electrochemical pseudocapacitance (see Fig. 1.2). The former is achieved by the principle of electrostatic charge separation within a Helmholtz double layer, while pseudocapacitance is an electrochemical phenomenon in which electrical energy is stored by virtue of reduction and oxidation reactions involving charge-transfer reactions at electrode/electrolyte interfaces.

Electrostatic double-layer capacitors based primarily on carbon electrodes store energy through the double-layer effect, while electrochemical pseudocapacitors based on metal oxides or conducting polymers additionally possess a high degree of electrochemical pseudocapacitance. Hybrid capacitors, such as the lithium-ion and

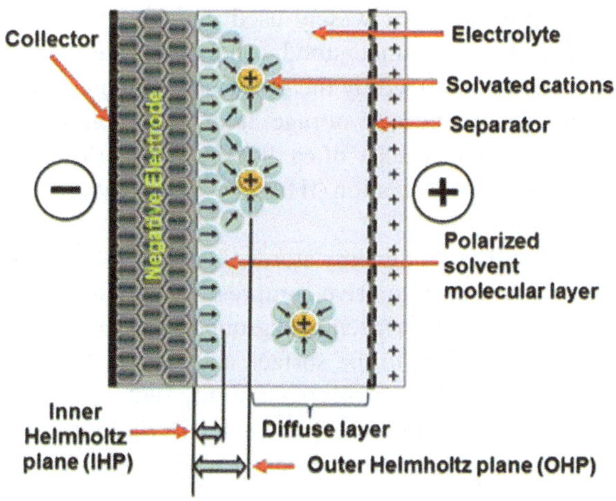

Fig. 1.2 Illustration of the working principles of a double-layer capacitor

zinc-ion capacitors [15], on the other hand, use one electrode favouring double-layer capacitance and the other favouring electrochemical capacitance.

The other main form of electrical energy storage is the *superconducting magnet*, which stores energy in a magnetic field induced by passing a DC current through a superconductor [16]. The conductor is kept at cryogenic temperatures (at which it becomes a superconductor) with virtually no resistance. Superconducting magnets consist of four parts: a superconducting coil with a magnet, a power conditioning system, a cryogenic system and a control unit. They have high efficiencies, in common with supercapacitors, high-power densities and they undergo fast charging. However, they also possess low capacities and involve very high capital costs. The very short durations (secs) of power delivery and the low energy densities limit their applications to those relevant to supercapacitors.

1.3.3 Mechanical Energy Storage

Mechanical energy storage comes in three main forms: *pumped hydro*, *compressed air* and *flywheels*. The first of these is a well-established technology with a long history. The principle is very simple: water is pumped uphill from a low-lying reservoir to a high-lying reservoir using pumps that are powered electrically. When electricity is required, the water is allowed to flow from the high-to low-lying reservoir through a turbine, which is connected to an electrical generator.

The technology is very energy efficient and low cost to operate (per kWh). On the other hand, it involves high capital costs and its major drawback is the requirement for the existence or creation of two reservoirs of water sufficiently separated in height, a very specific geological condition. Plants also take up to 10 years to build, with capital costs not recoverable potentially for decades. Moreover, they can potentially have serious environmental impacts, e.g., creating a dam in a river to make a reservoir, or changing the water temperature *via* pumping.

Pumped-hydro is widely used in Japan, the USA, South America, China, and Europe. It is well understood, reliable and safe. Its importance as a form of energy storage is underlined by the fact that it remains the only established technology for utility-scale electricity storage. It has been commercially deployed since 1890, with a current total capacity of ca. 9000 GWh worldwide according to The International Hydropower Association (IHA), dwarfing that of any other form of grid-scale storage technology.

Compressed air energy storage (CAES) is also based on a simple concept: excess electricity is used to drive compressors that pump air into a geological formation or (man-made) subsurface/aboveground storage vessel [17]. When demand increases, the air is released to the surface and heated with natural gas to drive a specialised combustion turbine involving a highly fuel-efficient process. Geological formations include depleted gas reservoirs, salt caverns, constructed rock caverns, gas fields and aquifers. The main advantages of CAES systems are their simplicity, the ease with which the components can be produced and sourced, their negligible losses and their scalability [17].

Man-made vessels, on the other hand, are high cost and currently small scale (< 10 MW), not being viable for practical application at the present time [18]. CAES based on natural formations suffers from the same problems as pumped hydro: it requires specific geological and geographical conditions, and it involves very high capital costs [19]. Its low volumetric energy density (a few kWh m^{-3} [20]) means that large volumes are required for storing the compressed air. Very few systems are in operation. Although a number have been planned in the last decade, most have not come to fruition, either being cancelled or delayed until further notice. Two recent successes are

1. A system in Goderich, Ontario with 10MWh storage for the Ontario Grid, built by Hydrostor in 2019 [21]
2. A 400 MWh installation in Zhangjiakou, China, achieving 70.4% efficiency [22]

Flywheels, sometimes referred to as 'kinetic batteries', are large rotating masses attached to a motor and generator for storing electricity [23]. The motor accelerates the flywheel using excess electricity and conversely the flywheel drives the generator to produce electrical power when required. Their main advantage is the very fast response times, making them suitable for voltage and frequency stabilisation. Their use in grids is already established, with a number of installations around the world. The duration of electrical power delivery is on the order of minutes to an hour, and efficiencies exceed 90%, with long life spans and low maintenance. They involve high capital costs and have low energy capacities and densities but possess high-power densities.

First-generation flywheels were based on steel, rotating on mechanical bearings. Newer generations use carbon-fibre composites and are able to store more energy. Moreover, fiction can be reduced by employing magnetic bearings [24] rather than traditional mechanical bearings.

While their use in vehicles for providing bursts of power and reducing fuel consumption is under investigation, the most prominent future application is for short-term spinning reserve for grid frequency regulation and for balancing rapid fluctuations in supply and demand. Examples include a 20 MW plant using 200 flywheels in New York, USA, opened in 2011 by Beacon Power [25], along with a similar plant at Hazle Township, Pennsylvania, USA opened in 2014. A flywheel system consisting of multiple flywheels on magnetic bearings was developed by NRStor in 2014 and is located in Ontario, Canada [26].

1.3.4 Thermal Energy Storage

As the name suggests, thermal energy storage (TES) is used to store energy in the form of heat and can be broadly categorised as follows [26].

1. *Sensible heat* technologies (H_2O, rock, and molten salts)
2. *Latent heat* technologies (salts and metals)
3. *Thermochemical* systems (carbonates and hydroxides).

The different technologies in this category operate under different temperature ranges, with systems also developed for cooling purposes. The sources of energy can include nuclear reactors, solar irradiance, waste heat, and geothermal energy [26]. When the source energy is low-grade thermal energy, TES can achieve round-trip efficiencies between 50% and 100%, but when the source energy is high-grade electrical energy, the efficiencies drop below 50%, compared to 80%–100% for battery energy storage.

This makes thermal energy storage more suitable for thermal power plants, especially concentrating solar and nuclear, in contrast to wind [27]. Thermal energy storage, on the other hand, typically has a longer cycle life than batteries and employs readily-available, abundant materials with low toxicity. In fact, the use of thermal energy storage with concentrated solar plants is increasing, with around half of the existing plants operating in tandem with thermal energy storage. The figure is more than 70% for those under construction or in the design stage [28]. The thermal energy stored is flexible, in that it can be used for electricity production or heating/cooling purposes.

Sensible heat storage involves materials that store heat within their specific heat capacity, elevating their temperature during the heating phase ('charging') without undergoing a phase change. The capacity of heat stored is determined by the density, volume, specific heat and temperature increment of the material involved. The storage materials can include water, thermal oils, rocks and for higher temperatures, molten salts such as a mixture of KNO_3, $NaNO_3$ and $NaNO_2$ [29], as well as liquid metals such as Na [30].

Latent heat storage usually relies on salt or metal materials changing phase during a constant-temperature process from solid to liquid (or the reverse) [26]. The materials are encapsulated in packed beds of spheres and the working fluid is passed over the packed bed to heat the material, or extract heat for thermal or electrical power. Solid-to-solid systems can also be used; although the specific latent heat is lower for these materials, they avoid issues related to leakage, as well as the requirement for encapsulation of the material. Liquid-to-gas systems possess the highest latent heats of phase change [31] but present enormous challenges in terms of the volume change, which explains why they are generally not adopted.

There is a vast array of suitable materials, which are summarised below [32].

1. *Organic materials* such as paraffins $CH_3{-}(CH_2)_{(n-2)}{-}CH_3$ and fatty acids R−COOH (R = alkyl group), polyethylene glycol $H{-}(O{-}CH_2{-}CH_2)_n{-}H$ and alcohols such as D-Mannitol $C_6H_{14}O_6$
2. *Salts* such as hydroxides, sulphates, chlorides, carbonates such as $CaCO_3$, and (especially) nitrates, such as $NaNO_3$
3. *Metals and alloys* such as Cu, Zn and Al, and alloys such as Zn-Mg

Latent heat storage is one to two orders of magnitude larger than sensible heat storage in terms of capacity. Its main drawback is the low thermal conductivity of the materials used, e.g., salts have a thermal conductivity in the range 0.5 and 1 W $m^{-1}K^{-1}$, with values for organic materials even lower. Moreover, these materials are flammable and the inorganic materials can corrode metal containers.

Thermochemical storage [33] dissociates various materials for either storing or releasing heat (e.g., from concentrated solar), as in the calcium carbonate reaction below

$$CaCO_3(s) + \Delta H \rightleftarrows 3CaO(s) + CO_2(g)$$

They operate at a mid-range temperature between 200 and 400 °C. When storing heat ('charging') the material dissociates with heating, releasing water vapour. The products are stored separately and can be stored for very long periods of time. During the 'discharging' phase, the products are brought into contact to undergo the reverse reaction and release heat in the process. Thermochemical storage possesses the highest energy density of all thermal storage technologies. Typical materials are magnesium sulphate $MgSO_4{\cdot}7H_2O$, calcium chloride $CaCl_2{\cdot}2H_2O$, lithium sulphate $Li_2SO_4{\cdot}H_2O$ and magnesium hydroxide $Mg(OH)_2$ [34, 35].

In all forms of thermal energy storage there is a need for more stable materials, lower cost materials, simpler reactor designs and process intensification before they can be adopted on a much wider scale.

1.3.5 Electrochemical Energy Storage

Electrochemical energy storage is the most familiar type of storage, given its presence in everyday life *via* mobile phones and laptops, and increasingly in vehicles as the source of power for propulsion, not to mention its longstanding use in the form of lead-acid batteries in conventional vehicles.

Batteries store and release electrical energy directly *via* complementary *reduction* and *oxidation* (or 'redox') reactions in separated electrodes. There are vast arrays of different chemistries (active redox couples), electrolytes, electrodes, separators, operating environments (acidic to alkaline), cell configurations and geometries available. The capacities, energy densities and power densities differ markedly across batteries. Their primary applications as single cells or stacks (series or parallel connected) are small- to medium-scale, from the aforementioned consumer electronics to backup power. Although various types of batteries have been introduced, the most widely used are lead-acid, nickel-cadmium, lithium-ion, metal-air and flow batteries. Below we discuss each of these technologies.

Lead Acid Batteries

Lead-acid batteries were the first type of rechargeable battery and have been in use for over a century. They use lead dioxide and metallic lead as the positive and negative electrode materials, respectively, in sulfuric acid electrolytes. Compared to other batteries, lead-acid batteries have advantages in terms of cost, high current densities, high cell voltages, and safety. However, they have relatively low specific energies (30–50 Wh kg^{-1}) and short cycle lives (< 500 deep cycles). They remain one of the most widely used rechargeable batteries and represent around 60% of installed battery power.

The most prominent applications are in motor vehicles to initiate the combustion process, and as a backup or standalone power systems, especially when cost and safety are the priorities. Most of the modern lead-acid batteries are valve-regulated (VRLA), in which gels and absorbed glass-mats are used rather than immersion in a liquid electrolyte made of sulfuric acid. This method is effective in replenishing the water that is consumed during operation, particularly during overcharge.

Nickel-cadmium and Nickel-metal Hydride Batteries

Nickel-cadmium and nickel-metal hydride batteries are the most extensively used batteries among those based on nickel electrodes (others are nickel-iron and nickel-zinc). Nickel-cadmium cells use nickel hydroxide and metallic cadmium as the positive and negative electrodes, respectively, in hydroxide electrolytes. Nickel-metal hydride batteries use a hydrogen-absorbing alloy, exhibiting capacities that are several times those of the nickel-cadmium family. They are also more environmentally friendly, given the growing concern regarding the toxicity of cadmium.

On the other hand, the self-discharge rate is higher and they are less tolerant to overcharge than nickel-cadmium cells. A concentrated potassium hydroxide electrolyte is often used for low temperature operation, while sodium hydroxide is used for systems operating at higher temperatures. These nickel-based batteries have long cycle lives, an overcharge capability, high discharge rates and can be operated at low temperatures.

Similar to VRLA batteries, sealed nickel-cadmium cells often consist of a pressure vessel that enables evolved hydrogen and oxygen to recombine. Depending on the cell structure, typical specific energies are in the range of 40–60 Wh kg^{-1} and cycle lives are up to several thousands cycles for vented cells. This type of battery is typically produced with capacities ranging from 10 mA h–20 A h, while vented stand-by units can possess capacities of over 1000 A h. The relatively high cost and the toxicity of cadmium has limited applications in recent decades.

Lithium-Ion Batteries

Lithium-ion batteries are the most common rechargeable batteries in consumer electronic applications due to their high energy densities, high cell voltages, long cycle lives and low weight when compared to other battery systems. The lithium-ion battery consists of intercalation electrodes in organic solvent electrolytes. Most commercial lithium-ion batteries do not involve metallic lithium and use metal-oxide intercalation electrodes with polymer gel electrolytes. The most common positive and negative electrode materials are lithium cobalt oxide ($LiCoO_2$) and graphitic carbon, respectively, which offer relatively high energy densities. Other lithium metal oxides, such as lithium iron phosphate ($LiFePO_4$), lithium manganese oxide (spinel $LiMn_2O_4$ or Li_2MnO_3-based lithium rich layered materials LMR-NMC), and lithium nickel manganese cobalt oxide ($LiNiMnCoO_2$ or NMC) may offer improved cycle lives or higher rate capabilities.

The positive and negative active masses are pasted on aluminium and copper current collectors, respectively. Microporous polymer sheets serve as separators for the two electrodes. Typical cell configurations are coin-cell, as well as cylindrical and prismatic. Compared to other batteries, lithium-ion batteries have advantages in terms of their specific energies (150–200 Wh kg^{-1}), high cell voltages (> 3.6 V), reasonable cycle lives (> 500 cycles) and relatively low self-discharge (< 10% per month). However, these batteries are not particularly low cost and require controlled charging processes to prevent overheating, fire hazards and explosions.

Metal-air Batteries

Metal-air batteries, which have a long history, have higher energy densities than most battery systems. The use of oxygen from atmospheric air in the positive electrode enables high specific energies at low cost. Electrolytes either based on aqueous or nonaqueous solvents and bifunctional air electrodes are the primary components. Metals, such as zinc, lithium, magnesium and aluminium are typical negative active materials. The positive electrode is air-breathing, with open porous architectures that enable a continuous oxygen supply. Aqueous zinc-air batteries are the most studied systems and have been commercialised as button cells and larger batteries for grid-scale energy storage. They offer a much lower fire risk than their nonaqueous counterparts and have practical specific energies between 350 and 500 Wh kg^{-1}. Despite their relatively low fabrication cost and high specific energies, most metal-air batteries tend to have relatively low-rate capability and suffer from poor energy efficiencies of < 60%, limited by the inefficiencies of the air electrode. The cycle lives of these batteries are also not satisfactory due to the lack of suitable catalysts with high stabilities.

Redox Flow Batteries

RFBs store and release electrical energy by making use of reversible electrochemical reactions involving active materials dissolved in liquid electrolytes, which flow through the electrochemical cells/stacks during the charge and discharge processes. The decoupling of power and energy is a key distinction of RFBs. The system energy is based on the volume of electrolytes and/or the concentrations of the usable active species, while the power rating is determined by the size and number of cells, which has been scaled-up to tens of MW. The negative and positive electrolytes are

(a) (b)

Fig. 1.3 **a** The Dalian Flow Battery Energy Storage Peak-shaving Power Station; **b** power modules

recirculated through pumps. Their mixing is avoided by using ion-exchange membranes or separators inside the cells.

The main advantages of RFBs are their safety, cycle lives (up to 10000 cycles), scalability (modular design) at reasonable cost, and deep discharge capability without damage to the batteries. However, they have relatively low specific energies (20–40 Wh kg^{-1}) and require additional components (pumps, sensors and external tanks). Conventional RFBs, such as the all-vanadium and iron-chromium batteries, store energy entirely in their electrolyte in the form of redox species. Hybrid flow batteries, such as zinc-bromine and zinc-iron, involve at least one phase change (solid or gaseous), typically metal electrodeposition. Such systems no longer fully decouple power and energy. Although the all-vanadium RFB is the most developed system, the costs of the electrolyte ($ 80 $(kW\ h)^{-1}$) and cell stack ($ > 150 $(kW\ h)^{-1}$) remain relatively high. Hence, recent efforts have focused on exploring lower cost active materials, as well as methods to increase the power density.

There have been a growing number of installations of redox flow batteries in recent years. The world's largest became operational in October 2022 and is based in Dalian, China, developed by the Dalian Institute of Chemical Physics and Dalian Rongke Power (see Fig. 1.3). It is expected to supply power for up to 200,000 residents per day, with an initial capacity of 400 MWh and a 100 MW output, serving as a power bank and assisting in the penetration of wind and solar into the grid. It is designed to be scaled-up to 200 MW with an 800-MWh capacity.

1.4 Summary

A comparison of different energy storage systems in terms of their power and energy ratings, costs and current status is provided in Table 1.1. Figure 1.4 illustrates the capacities and discharge times of the various technologies, and summarises their potential and existing applications. As is evident from this figure, the likely appli-

Table 1.1 Comparison of storage technologies

Technology	Power rating	Discharge duration	Response time	Efficiency	Parasitic losses	Lifetime	Capital cost			O & M cost		Status
							Power-related cost ($/kW)	Energy-related cost ($/kWh)	BOP ($/kW-y)	Fixed ($/kW-y)	Variable ($/kWh)	
Pumped-hydro	100–4000 MW	4–12 h	Sec.-min.	0.7–0.85	Evaporation	30 yrs	600	0–20	included	3.8	0.38	Commercial products
CAES (in reservoirs)	100–300MW	6–20 h	Sec.-min.	0.64	–	30 yrs	425–480	10-Mar	50	1.42	0.01	Commercial products
CAES (in vessels)	50–100MW	1–4 h	Sec.-min.	0.57	–	30 yrs	517	50	40	3.77	0.27	Concept
Flywheels (low speed)	1650 kW	3–120 s	1 cycle	0.9	1%	20 yrs	300	200–300	80	–	–	Commercial products
Flywheels (high speed)	750 kW	1 h	1 cycle	0.93	3%	20 yrs	350	500–25,000	1,000	7.5	0.4	Prototype in testing
Super-capacitors	100 kW	1 min	1/4 cycle	0.95	–	10,000 cycles	300	82,000	10,000	5.55	0.5	Some commercial products
SMES (Micro)	10 kW–10 MW	1 s–1 min	1/4 cycle	0.95	4%	30 yrs	300	72,000	10,000	26	2	Commercial products
SMES	10–10 MW	1–30 min	1/4 cycle	0.95	1%	30 yrs	300	2,000	1,500	8	0.5	Design concept
Lead-acid battery	50 MW	1 min–8 h	1/4 cycle	0.85	Small	5–10 yrs	200–300	175–250	50	1.55	1	Commercial products
NaS battery	10 MW	8 h	N.a.	0.75–0.86	5kW/kWh	5 yrs	259	245	40	N.a.	N.a.	In development
Zn-Br hybrid flow battery	1 MW	4 h	1/4 cycle	0.75	Small	2,000 cycles	1,500	200	included	N.a.	N.a.	In test/commercial units
All-V redox flow battery	3 MW	10 h	N.a	0.70–0.85	N.a	10 yrs	N.a.	175–190	N.a.	N.a.	N.a.	In test
Polysulphide Br flow battery	15 MW	20 h	N.a.	0.60–0.75	N.a.	2,000 cycles	1,200	75–190	N.a.	N.a.	N.a.	In test
Hydrogen (Fuel Cell)	250 kW	As needed	1/4 cycle	0.34–0.40	N.a	10–20 yrs	1,100–2,600	15-Feb	N.a.	10	1	In test
Hydrogen (Engine)	2 MW	As needed	Seconds	0.29–0.33	N.a.	10–20 yrs	950–1,850	15-Feb	N.a.	0.7	0.77	Available for demonstration

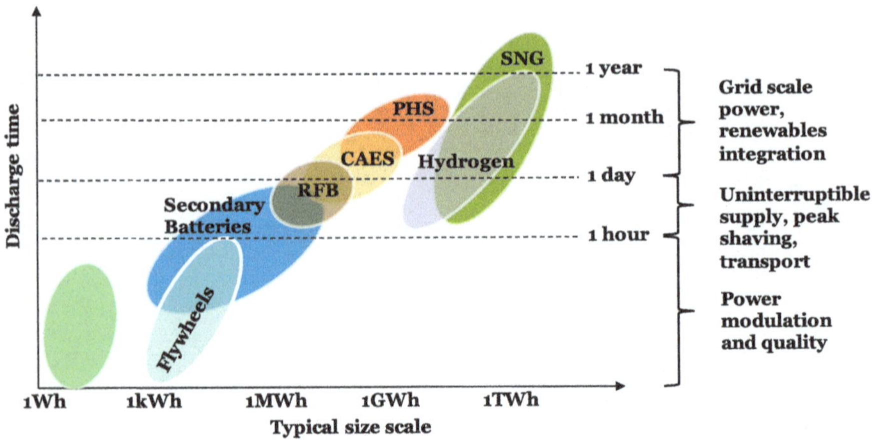

Fig. 1.4 **a** The range of capacities and discharge times for various energy storage technologies

cation areas for flow batteries are medium to large scale, especially for grid-scale backup power, standalone power and renewables integration.

The commercial status and adoption of RFBs has been variable over the past two decades, but with the increased urgency to integrate renewables into grids (in light of approaching deadlines for ambitious targets), there is expected to be renewed interest in the technology. Its primary advantages over other forms of storage for grid-scale power are its flexibility in terms of terrain and location, lack of emissions, the scalability of its capacity compared to conventional batteries, and potentially low cost with economies of scale.

In the next chapter we provide a detailed background on RFBs, including their operation, historical development, classification, components and materials, recent developments and open challenges.

References

1. A.Z.A. Shaqsi, K. Sopian, A. Al-Hinai, Review of energy storage services, applications, limitations, and benefits. Energy Rep. **6**, 288–306 (2020)
2. I. Dincer, M.A. Rosen, *Thermal Energy Storage: Systems and Applications* (John Wiley & Sons, 2021)
3. M.A. Miller, J. Petrasch, K. Randhir, N. Rahmatian, J. Klausner, Chemical energy storage, in *Thermal, Mechanical, and Hybrid Chemical Energy Storage Systems*, pp. 249–292 (2021)
4. P. Novak, K. Miller, K.S.V. Santhanam, O. Haas, Electrochemically active polymers for rechargeable batteries. Chem. Rev. **97**, 272 (1997)
5. M. Uddin, M.F. Romlie, M.F. Abdullah, S. Abd Halim, T.C. Kwang, A review on peak load shaving strategies. Renew. Sustain. Energy Rev. **82**, 3323–3332 (2018)
6. K.M. Tan, T.S. Babu, V.K. Ramachandaramurthy, P. Kasinathan, S.G. Solanki, S.K. Raveendran, Empowering smart grid: a comprehensive review of energy storage technology and application with renewable energy integration. J. Energy Storage. **39**, 102591 (2021)

7. S. Revankar, H. Bindra, *Storage and Hybridization of Nuclear Energy: Techno-economic Integration of Renewable and Nuclear Energy* (Academic Press, 2018)
8. M. Deymi-Dashtebayaz, A. Ebrahimi-Moghadam, S.I. Pishbin, M. Pourramezan, Investigating the effect of hydrogen injection on natural gas thermo-physical properties with various compositions. Energy. **167**, 235–245 (2019)
9. M. Thema, F. Bauer, M. Sterner, Power-to-gas: electrolysis and methanation status review. Renew. Sustain. Energy Rev. **112**, 775–787 (2019)
10. Ewelina Jankowska, Ashish K. Sahu, Piotr Oleskowicz-Popiel, Biogas from microalgae: review on microalgae's cultivation, harvesting and pretreatment for anaerobic digestion. Renew. Sustain. Energy Rev. **75**, 692–709 (2017)
11. S.S. Kumar, V. Himabindu, Hydrogen production by PEM water electrolysis: a review. Mater. Sci. Energy Technol. **2**(3), 442–454 (2019)
12. J. Davies, F. Dolci, D. Klassek-Bajorek, R. Ortiz Cebolla, E. Weidner Ronnefeld, Current status of chemical energy storage technologies, EUR 30159 EN, publications office of the European union. Technical report (Luxembourg, 2020)
13. T.M. Gur, Review of electrical energy storage technologies, materials and systems: challenges and prospects for large-scale grid storage. Energy & Environ. Sci. **11**(10), 2696–2767 (2018)
14. H. Chen, T.N. Cong, W. Yang, C. Tan, Y. Li, Y. Ding, Progress in electrical energy storage system: a critical review. Prog. Nat. Sci. **19**(3), 291–312 (2009)
15. Heng Tang, Junjun Yao, Yirong Zhu, Recent developments and future prospects for zinc-ion hybrid capacitors: a review. Adv. Energy Mater. **11**(14), 2003994 (2021)
16. P. Mukherjee, V.V. Rao, Design and development of high temperature superconducting magnetic energy storage for power applications-a review. Phys. C Supercond. Appl. **563**, 67–73 (2019)
17. E. Borri, A. Tafone, G. Comodi, A. Romagnoli, L.F. Cabeza, Compressed air energy storage: an overview of research trends and gaps through a bibliometric analysis. Energ. **15**, 7692 (2022)
18. A. Olympios, J. McTigue, P.F. Antunez, A. Tafone, A. Romagnoli, Y. Li, Y. Ding, W.-D. Steinmann, L. Wang, H. Chen et al., Progress and prospects of thermo-mechanical energy storage. a critical review. Prog. Energy. **3**, 022001 (2021)
19. M. Budt, D. Wolf, R. Span, J.A. Yan, Review on compressed air energy storage: basic principles, past milestones and recent developments. Appl. Energy. **170**, 250–268 (2016)
20. M. Aneke, M. Wang, Energy storage technologies and real life applications. state of the art review. Appl. Energy. **179**, 350–377 (2016)
21. grid-connected advanced compressed air energy storage plant comes online in Ontario. energy storage news (2019)
22. L. Blain, China turns on the world's largest compressed air energy storage plant. New Atlas, October 2022. Accessed 10 Jan 2023
23. Faramarz Faraji, Abbas Majazi, Kamal Al-Haddad, A comprehensive review of flywheel energy storage system technology. Renew. Sustain. Energy Rev. **67**, 477–490 (2017)
24. A.V. Filatov, E.H. Maslen, Passive magnetic bearing for flywheel energy storage systems. IEEE Trans. Magn. **37**(6), 3913–3924 (2001)
25. Stephentown, New York—beacon power
26. Canada's first grid storage system launches in Ontario—PV-Tech storage. PV-Tech storage
27. Guruprasad Alva, Yaxue Lin, Guiyin Fang, An overview of thermal energy storage systems. Energy. **144**, 341–378 (2018)
28. P. Denholm, J.C. King, C.F. Kutcher, P.P. Wilson, Decarbonizing the electric sector: combining renewable and nuclear energy using thermal storage. Energy Policy. **44**, 301–311 (2012)
29. Ugo Pelay et al., Thermal energy storage systems for concentrated solar power plants. Renew. Sustain. Energy Rev. **79**, 82–100 (2017)
30. A. Gil, M. Medrano, I. Martorell, A. Lázaro, P. Dolado, B. Zalba, L.F. Cabeza, State of the art on high temperature thermal energy storage for power generation. part 1: concepts, materials and modellization. Renew. Sustain. Energy Rev. **14**(1), 31–55 (2010)
31. J. Pacio, A. Fritsch, C. Singer, R. Uhlig, Liquid metals as efficient coolants for high-intensity point-focus receivers: implications to the design and performance of next-generation CSP systems. Energy Procedia. **49**, 647–655 (2014)

32. B. Cardenas, N. Leon, High temperature latent heat thermal energy storage: phase change materials, design considerations and performance enhancement techniques. Renew. Sustain. Energy Rev. **27**, 724–737 (2013)
33. Guruprasad Alva et al., Thermal energy storage materials and systems for solar energy applications. Renew. Sustain. Energy Rev. **68**, 693–706 (2017)
34. J.S. Prasad, P. Muthukumar, F. Desai, D.N. Basu, M.M. Rahman, A critical review of high-temperature reversible thermochemical energy storage systems. Appl. Energy. **254**(11373), 3 (2019)
35. K.E. N'tsoukpoe et al., A review on long-term sorption solar energy storage. Renew. Sustain. Energy Rev. **13**(9), 2385–2396 (2009)

Chapter 2
Electrochemical Theory and Overview of Redox Flow Batteries

2.1 Introduction

Due to the rapid growth in power generation from intermittent sources, the requirement for low cost and flexible energy storage systems has given rise to many opportunities [1, 2]. Electrochemical redox flow batteries (RFBs) have emerged as a promising and practical technology for storing energy at large scales [3, 4]. Their scales range from kW to multiples of MW, making them suitable for load levelling, power quality control, coupling with renewable energies and uninterrupted power supply [3]. This can be attributed to their design flexibility, allowing for them to be readily scaled up in power and energy output [5]. This chapter provides a concise overview of RFB systems, covering the fundamental theory behind their operation, their historical development, components and materials, applications and latest developments.

Redox flow batteries have been in development since the 1970s [6, 7], and enjoy advantages over conventional energy storage systems. Firstly, in contrast to pumped-hydro and compressed-air systems, they do not have any particular terrain requirements. Compared to conventional (static) lead-acid batteries, RFBs are less costly to maintain and have longer lifetimes, exceeding 10 years. The modular nature of redox flow batteries enhances their portability and renders their construction and maintenance costs the lowest among the energy storage systems available.

Redox flow batteries can be discharged completely without damaging the electrodes, allowing for more flexible charge/discharge cycles [8]. Typically, RFBs store energy in the electrolytes, so that their capacities can be increased by using greater volumes of electrolyte or higher concentrations of the electro-active species. Due to their fast response time, they are ideal for power quality control applications and often serve as uninterruptible power supply (UPS) units [9].

A. A. Shah et al., *New Paradigms in Flow Battery Modelling*,
Engineering Applications of Computational Methods 16,
https://doi.org/10.1007/978-981-99-2524-7_2

2.2 Properties of Redox Flow Batteries

The classical RFBs involve soluble redox couples that are oxidised or reduced during the charge and discharge processes. In most systems, the negative and positive electrodes are separated by an ion-exchange membrane/separator, in common with proton exchange membrane fuel cells (PEMFCs). The membrane serves to facilitate ion exchange and to prevent direct mixing of the two electrolytes, which are stored in separated reservoirs and recirculated by pumps through the cells/stacks during operation, as illustrated in Fig. 2.1. Although a membrane is necessary for most flow batteries, there exist a number of membrane-less systems currently at the early stages of development [10, 11].

In practice, a flow battery does not operate as a unit cell; multiple cells are combined in series or parallel to form a stack. In order to reduce weight, volume and cost, bipolar plates are often used in these stacks. The power output can be adjusted simply by increasing the numbers of cells in the stack and the size of the electrodes. An example of a RFB stack with four cells using bipolar electrodes is shown in [12] Fig. 2.2.

In contrast to other electrochemical energy storage systems, classical redox flow batteries store the energy in the form of reduced and oxidised electro-active species that are dissolved in the electrolyte. Conventional static batteries store energy within the electrode structure. The chemical reactions involved in a flow battery should be reversible to enable both charge and discharge. As a result, the active species in the electrolyte should always remain within the system in reduced or oxidised form, while in a fuel cell the reactants, which are stored externally to the cell, are consumed

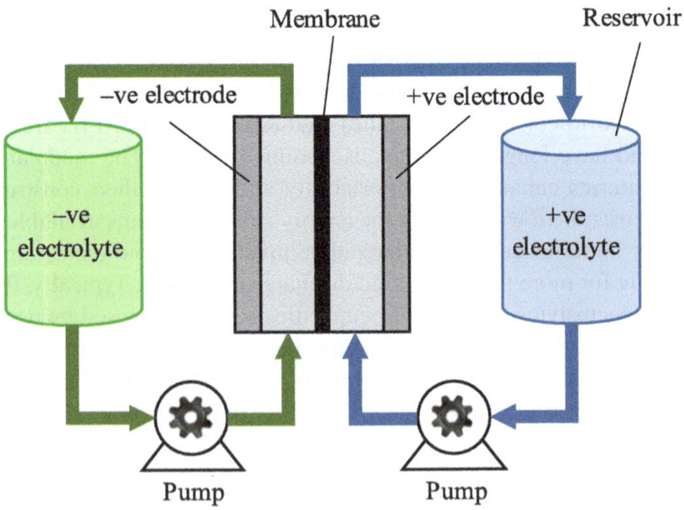

Fig. 2.1 Schematic of a typical redox flow battery

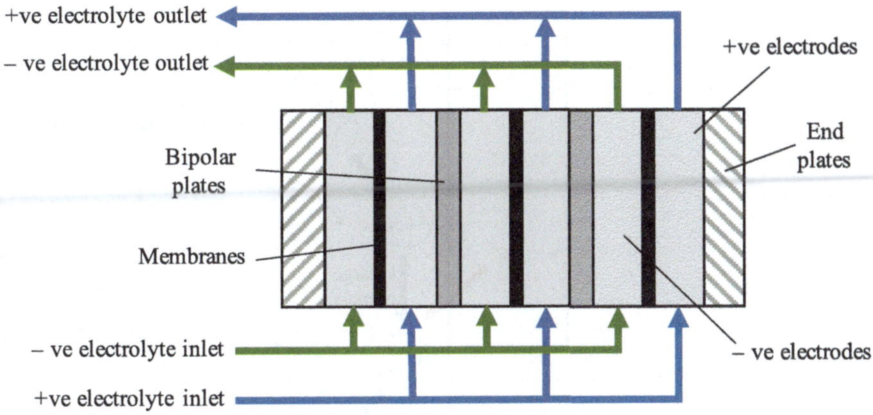

Fig. 2.2 A 3-cell stack design with a bipolar-plate series arrangement

Table 2.1 Comparisons between conventional static battery, redox flow battery and fuel cell

Electrochemical device	Reaction site	Electrolyte conditions	Separator
Static battery	Active electrode material	Static and held within cell	Microporous polymer separator
Redox flow battery	Aqueous electrolytes in reservoirs	Electrolyte recycles through the cell	Ion-exchange membrane (Cationic or anionic)
Fuel cell	Gaseous or liquid fuel plus air	Solid polymer or ceramic acts as solid electrolyte within cell	Ion-exchange membrane polymer or ceramic

during reactions to produce products that must be continually removed from the cell. Table 2.1 summarises the differences between static batteries, RFBs and fuel cells.

2.3 Fundamental Electrochemical Principles of Flow Batteries

2.3.1 Redox Reactions at the Electrodes

Electrochemical reactions occur *via* the transfer of charge across the interface between an electrode and an electrolyte. Electrochemical cells used to study such reactions usually consist of three electrodes: a working electrode, a counter electrode and a reference electrode. The working electrode is the electrode at which the reaction of interest takes place, the counter electrode is used to close the circuit with the

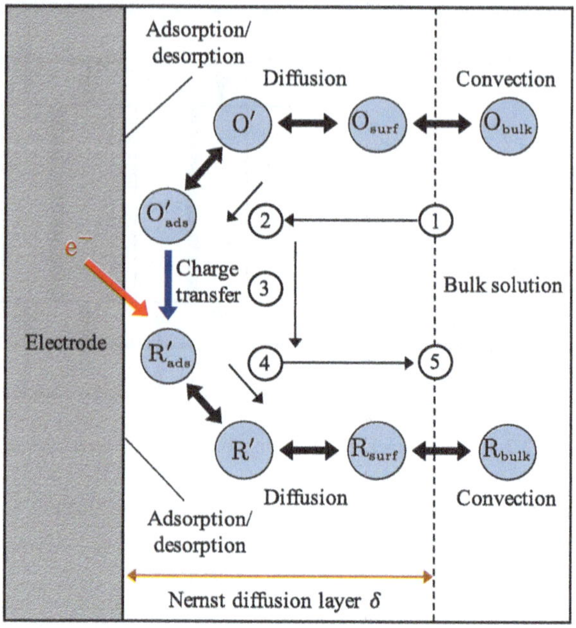

Fig. 2.3 Typical processes involved in an electrode reaction

electrolyte and the reference electrode is used to accurately measure the potential of the working electrode. The cell voltage of an electrochemical system is the potential across the working and counter electrodes.

A simple one-electron transfer reaction in an aqueous solution may be represented as

$$O + e^- \rightleftharpoons R \tag{2.1}$$

in which O and R are the oxidised and reduced species, respectively. Figure 2.3 shows the typical pathway of a general electrode reaction, which involves several stages: (1) transport of the reactant towards the electrode; (2) surface adsorption of the reactant; (3) charge (electron) transfer; (4) surface desorption and (5) removal of the product.

A typical electron-transfer reaction requires the reactant to be located within molecular scale distances of the electrode surface. For a cathodic process, the reactant O is reduced by gaining one or more electrons to form a product species R. In addition to the charge-transfer process, the rates of reactant supply to the electrode surface and removal of the product (i.e., their mass transport) can limit the rate of the overall reaction. Mass transport is primarily due to diffusion, convection and electro-migration.

Diffusion dominates within a thin *Nernst diffusion layer*, the effective thickness of which depends on the bulk solution concentration, the diffusion coefficient of the reactant and the solution viscosity. Forced convection, such as stirring or mechanical agitation, effectively enhances the transport of the material towards the electrode.

In the following, the theoretical influence of the potential and reactant concentrations on the current density (rate of charge transfer) are discussed and quantified.

2.3.2 *Faraday's Law*

Faraday's law states that the electrical charge transferred at the electrode surface is proportional to the mass of the species generated or consumed (at the surface)

$$c = \frac{Q}{nFV} = \frac{It}{nFV} \tag{2.2}$$

in which c is the concentration [mol dm^{-3}] of the species generated or consumed over a period of time t [s], I [A] is the current generated, V is the electrolyte volume [m^3], F is the Faraday constant, [9,6485.3 C mol^{-1}] and n is the number of electrons involved in the reaction.

2.3.3 *Thermodynamics and Nernst's Equation*

According to thermodynamics, the maximum (theoretical) amount of energy that can be extracted from a chemical reaction is the Gibbs free energy change ΔG. In a redox flow battery, this chemical energy is transformed into electrical energy. When the reaction is at equilibrium, no net current flows and the cell attains its maximum cell voltage. At this cell voltage E_{cell} [V], the (maximum) work done W_{max} [J mol^{-1}] should be equal to the Gibbs free energy [J mol^{-1}] according to

$$W_{\text{max}} = \Delta G = -nFE_{\text{cell}} \tag{2.3}$$

E_{cell} is also known as the open-circuit potential (OCP) or the equilibrium potential. When the activities of the product and reactant are unity, Eq. (2.3) can be written as

$$\Delta G^0 = -nFE^0_{\text{cell}} \tag{2.4}$$

E^0_{cell} is known as the standard electrode potential of the cell (also written E^0) and is related to the standard Gibbs free energy change ΔG^0 *via* Eq. (2.4). When $\Delta G^0 < 0$ the reaction is spontaneous and no energy is required for it to proceed, while when $\Delta G^0 > 0$ an external energy source is required to drive the reaction forward. The law of mass action states that the rate of a chemical reaction is proportional to the product of all species concentrations, each raised to the power of ν_i, where ν_i is the stoichiometric coefficient of species i.

At chemical equilibrium, the participating species concentrations are unchanged (the backward and forward rates are equal)

$$aA + bB \rightleftharpoons cC + dD + \ldots \tag{2.5}$$

in which $a, b, c, d, \ldots$ are the stoichiometric coefficients and A, B, C, D denote the species. The equilibrium constant K can be expressed as

$$K = \frac{a_C^c + a_D^d + \ldots}{a_A^a + a_B^b + \ldots} \tag{2.6}$$

in which a_i is the activity of species i. In the case of chemical equilibrium, the Gibbs free energy change ΔG can be expressed as the activity change in the mixture

$$\Delta G = \Delta G^0 + RT \log K \tag{2.7}$$

in which R is the gas constant (8.3145 J mol^{-1} K^{-1}) and T [K] is the temperature. The activity can be expressed in terms of the concentration as

$$a = \gamma c \tag{2.8}$$

in which c is the species concentration and γ is the activity coefficient. The activity coefficient approaches unity at low concentrations

$$\log_{c \to 0} \gamma = 1 \tag{2.9}$$

For a simple half-cell reaction, as in (2.1), the Gibbs free energy change can therefore be expressed as

$$\Delta G = \Delta G^0 + RT \log \left(\frac{[R]}{[O]} \right) \tag{2.10}$$

The Gibbs free energy change comprises the Gibbs free energy change in a standard state ΔG^0 at unit activities and a variable term that is a function of temperature and the species concentrations. By combining (2.4) and (2.10), the relationship between the cell potential and the concentrations of the electro-active species at chemical equilibrium (the Nernst equation) is obtained

$$E = E^0 + RT \left(\frac{RT}{nF} \right) \log \left(\frac{[R]}{[O]} \right) \tag{2.11}$$

During battery discharge, the redox reactions in an all-vanadium redox flow battery are as follows: the negative electrode becomes an anode due to the oxidation of V^{2+} to form V^{3+}

$$\text{Oxidation: } V^{2+} - e^- \rightarrow V^{3+}, \quad E^0 = -0.26 \text{ V } vs \text{ SHE} \tag{2.12}$$

and the positive electrode becomes a cathode due to the reduction of VO_2^+ to VO^{2+}

$$\text{Reduction: } VO_2^+ + 2H^+ + e^- \rightarrow VO^{2+} + H_2O, \quad E^0 = +1.0\ \text{V}\ vs\ \text{SHE} \tag{2.13}$$

The cell voltage of this system is the difference between the standard electrode potential values of the cathodic and the anodic half-cell reactions

$$E^0_{\text{cell}} = E^0_{\text{cathode}} = E^0_{\text{anode}} \tag{2.14}$$

Assuming that the battery is operating under standard conditions, the cell voltage during battery discharge can be expressed as

$$E^0_{\text{cell}} = E^0_{\text{cathode}} = E^0_{\text{anode}} = 1 - (-0.26) = 1.26\,\text{V} \tag{2.15}$$

During battery charge, the reactions occur in reverse and are driven by an applied current. Hence, the negative electrode becomes a cathode and the positive electrode becomes an anode

$$E^0_{\text{cell}} = E^0_{\text{cathode}} = E^0_{\text{anode}} = -0.261 - (1) = -1.26\,\text{V} \tag{2.16}$$

By using Eq. (2.4), the predicted standard free energy change to charge and discharge an all-vanadium battery can be estimated

$$\begin{aligned} \text{Battery charge: } \Delta G^0 &= -(1)F(-1.26) = +121.6\ \text{kJ} \\ \text{Battery discharge: } \Delta G^0 &= -(1)F(+1.26) = -121.6\ \text{kJ} \end{aligned} \tag{2.17}$$

The negative value indicates that the discharge reactions are spontaneous, while the positive value indicates that the charging reactions require an energy input.

2.3.4 *Charge-Transfer Reaction*

Considering the simple electron-transfer reaction (2.1), a flux of ions must occur in the electrolytic phase to balance the negative charge accumulation or deficit due to electron transfer at the electrode/electrolytic interface. The net current density due to electrochemical reaction j [A m^{-2}] can be viewed as the difference between the partial cathodic (reduction) current density $\overleftarrow{j}$ and the partial anodic (oxidation) current density $\overrightarrow{j}$, since both the forward and backward reactions occur, although generally at different rates

$$j = \overleftarrow{j} - \overrightarrow{j} \tag{2.18}$$

Each of these current densities is proportional to a corresponding heterogeneous rate constant k_f and k_b [cm sec^{-1}], for the forward reduction and backward oxidation reactions, respectively. The partial cathodic current density can be expressed as

$$\overleftarrow{j} = nFk_f C_O(0, t) \tag{2.19}$$

while the partial anode current density can be expressed as

$$\overrightarrow{j} = nFk_b C_R(0, t) \tag{2.20}$$

in which $C_i(x, t)$ [mol cm^{-3}] is the concentration of species i at a distance x [cm] from the electrode surface at time t [s] ($x = 0$ is the location of an assumed planar surface in 1D).

The forward and reverse rate constants depend on the electrode potential with respect to a reference electrode and both follow an Arrhenius law with a common heterogeneous rate constant k^0

$$\begin{aligned} k_f &= k^0 e^{-\alpha f(E-E_{\text{eq}})} \\ k_b &= k^0 e^{(1-\alpha) f(E-E_{\text{eq}})} \end{aligned} \tag{2.21}$$

in which $f = nF/(RT)$, E_{eq} is the equilibrium potential and α is a dimensionless transfer coefficient, with values between 0 and 1, often estimated to be 1/2. By dividing the second of Eq. (2.21) by the first, we obtain

$$\frac{k_b}{k_f} = e^{f(E-E_{\text{eq}})} \tag{2.22}$$

Combining Eqs. (2.18)–(2.21), the net current is given by

$$j = nFk^0 \left[C_O(0, t) e^{-\alpha f(E-E_{\text{eq}})} - C_R(0, t) e^{(1-\alpha) f(E-E_{\text{eq}})} \right] \tag{2.23}$$

2.3.5 An Electrode Surface Under Equilibrium Conditions

When an electrochemical reaction at an electrode/electrolyte surface is at equilibrium, the cathodic and anodic currents balance

$$j_0 = \overleftarrow{j} = \overrightarrow{j} \tag{2.24}$$

in which j_0 is the exchange current density, a kinetic parameter of a particular reaction on a particular electrode. From (2.21), we then obtain

$$j_0 = nFk^0 C_O^* e^{-\alpha f(E-E_{\rm eq})}$$
$$j_0 = nFk^0 C_R^* e^{(1-\alpha) f(E-E_{\rm eq})} \tag{2.25}$$

in which C_i^* is the bulk concentration of the reactant or oxidant. Dividing the second of (2.25) by the first yields

$$e^{f(E-E_{\rm eq})} = \frac{C_O^*}{C_R^*} \tag{2.26}$$

If both sides of this equation are raised to the power $-\alpha$ and substituted into the first of (2.25), the exchange current density can be written as

$$j_0 = nFk^0 \left(C_O^*\right)^{1-\alpha} \left(C_R^*\right)^{\alpha} \tag{2.27}$$

The exchange current density is the rate at which oxidised and reduced species engender electron transfer to or from the electrode at equilibrium. It is a measure of the electrocatalytic properties of the electrode material for a given reaction.

2.3.6 *An Electrode Surface Under Non-equilibrium Conditions*

For a non-spontaneous cell reaction to occur, an overpotential, η [V], must be applied over and above the equilibrium potential $E_{\rm eq}$ to drive the reaction. The overpotential can be written as

$$\eta = E - E_{\rm eq} \tag{2.28}$$

Figure 2.4 illustrates three cases: (1) net cathodic current, (2) no net current and (3) net anodic current. The anodic and cathodic currents are dependent on the sign and the magnitude of the overpotential. The resulting net current density is the algebraic sum of the partial cathodic and anodic currents.

In order to calculate the current at a given value of the overpotential, the *Butler-Volmer* relationship can be obtained by combining (2.23), (2.27) and (2.28)

$$j = j_0 \left[\frac{C_O(0,t)}{C_O^*} e^{-\alpha f \eta} - \frac{C_R(0,t)}{C_R^*} e^{(1-\alpha) f \eta} \right] \tag{2.29}$$

With efficient mass transport, the concentrations in the bulk solution and at the electrode surface ($x = 0$) are equal and Eq. (2.29) simplifies to

$$j = j_0 \left[e^{-\alpha f \eta} - e^{(1-\alpha) f \eta} \right] \tag{2.30}$$

This equation can be used to approximate the current when mass transport limitations are eliminated. This approximation is useful when j is less than 10 % of the limiting

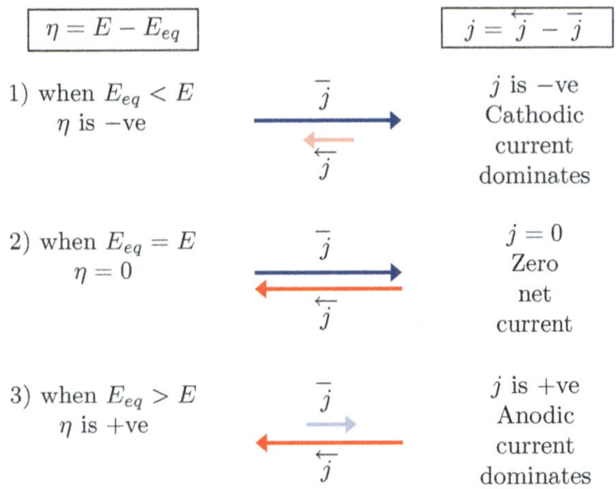

Fig. 2.4 Three cases of different magnitude and size of an overpotential

current density j_{lim}. It requires a well-stirred solution in which diffusion of the electroactive species to the electrode surface is not a limiting factor in the experiment.

The exchange current density can be estimated experimentally from the plot of $\log|j|$ *vs.* η, known as a *Tafel plot*. By using data corresponding to small overpotentials (<118 s mV), and therefore corresponding to charge-transfer control, a Tafel plot is useful for estimating kinetic parameters such as α and j_0. For large values of η (either positive or negative), $e^{(1-\alpha)f\eta}$ becomes negligible and Eq. (2.30) simplifies to

$$\eta = \frac{RT}{\alpha F}\left(\log\ j_0 - \log j\right) \tag{2.31}$$

By converting to the more commonly used decimal base, this result can be written as

$$\eta = \frac{2.3RT}{\alpha F}\left(\log\ j_0 - \log j\right) \tag{2.32}$$

As illustrated in Fig. 2.5, there is an anodic branch with a gradient of $\alpha F/(2.3RT)$ and a cathodic branch with a gradient of $(1-\alpha)F/(2.3RT)$. Both linear segments extrapolate to an intercept $\log\ j_0$. Therefore, the exchange current density (at zero overpotential) can be calculated. At large overpotentials (>118 mV), the reaction is under mixed or mass transport control and the gradient obtained experimentally will be smaller than the Tafel slope due to mass transport limitations.

2.3.7 *Mass Transport*

Mass transport involves the supply of reactants to the electrode surface and the removal of the products. The rate of mass transport is usually important when the

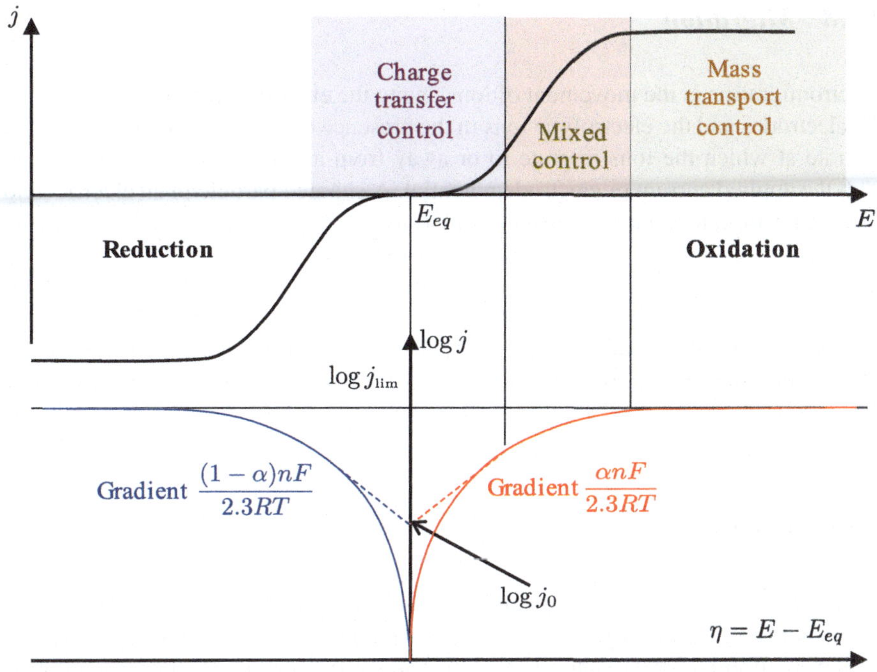

Fig. 2.5 Logarithm plot of current density *versus* overpotential under charge transfer, mixed and mass transport control

Table 2.2 The driving force behind and the nature of the three mass transport modes

Mechanism	Driving force	Comment
Migration	Potential gradient	Ions move due to the potential difference between electrodes. Migration is not specific to electro-active species
Diffusion	Concentration gradient	Always occurs close to electrode when current passes and chemical change occurs
Convection	External mechanical forces	Flow of solution, movement of electrode, gas sparking, natural convection due to thermal or density differences

reactant concentration is low or a high rate of reaction is required. In general, the contribution to mass transport includes: (1) electro-migration, (2) diffusion and (3) convection. The driving forces behind, and the nature of these three transport modes are summarised in Table 2.2.

2.3.8 *Migration*

Electromigration is the movement of ions due to the electrostatic attraction between the electrodes and the electrolytic ions in the presence of an electric field. In general, the rate at which the ions migrate to or away from an electrode surface increases with the applied current or electrode potential. A charged particle in an electric field accelerates until it reaches a constant drift velocity V_d [m s^{-1}]

$$V_d = uE \tag{2.33}$$

in which u is the electrical mobility [m^2 V^{-1} s^{-1}] and E is the electric field [V m^{-1}]. The current due to migration is negligible for oxidised and reduced species at low concentrations.

2.3.9 *Diffusion*

Diffusion is a spontaneous and random movement of a species from regions of higher concentration to regions of lower concentration. Fick's first law, which can be written in different forms, states that the diffusion is proportional to the concentration gradient (Fig. 2.6). Hence, the rate of movement by diffusion can be estimated as

$$N = \frac{j}{nF} = -D\frac{dc}{dx} \tag{2.34}$$

in which dc/dx is the concentration gradient and D is the diffusion coefficient of the species. In a diffusion-controlled process, as illustrated in Fig. 2.7, the flux of electrons at the electrode and the flux of soluble species in the electrolyte are related *via* mass conservation

$$N = \frac{j}{nF} = -D_O\frac{dc_O}{dx}\bigg|_{x=0} = \frac{j}{nF} = D_R\frac{dc_R}{dx}\bigg|_{x=0} \tag{2.35}$$

In the case of hybrid flow batteries, metal electrodeposition is involved. Electrodeposits accumulate on the substrate surface; hence no reaction product is removed from the electrode surface during the charging processes.

2.3.10 *Convection-Diffusion*

Reactants and products can be transferred to and from an electrode by mechanical mechanisms such as natural and forced convection. Mechanical agitation includes movement of an electrode or an electrolyte by reciprocation, vibration, rotation,

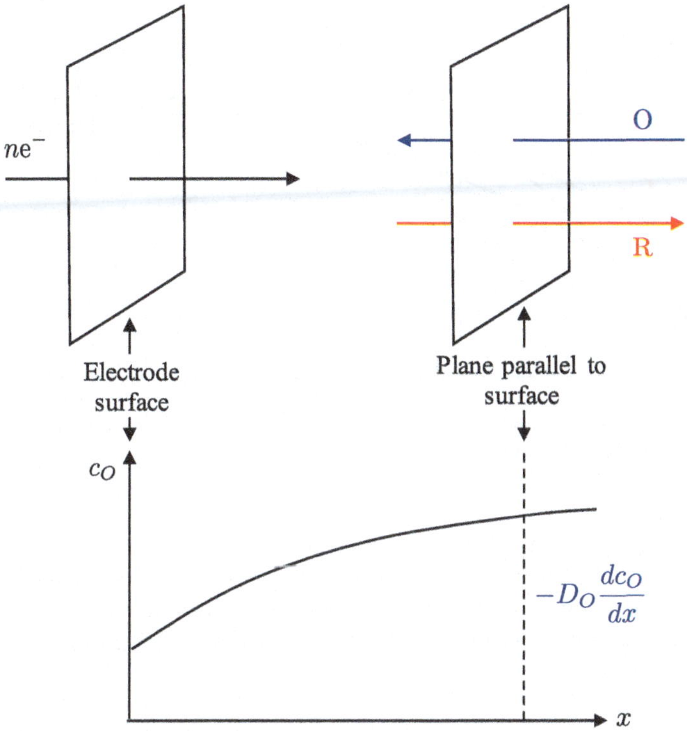

Fig. 2.6 Fick's law for linear diffusion

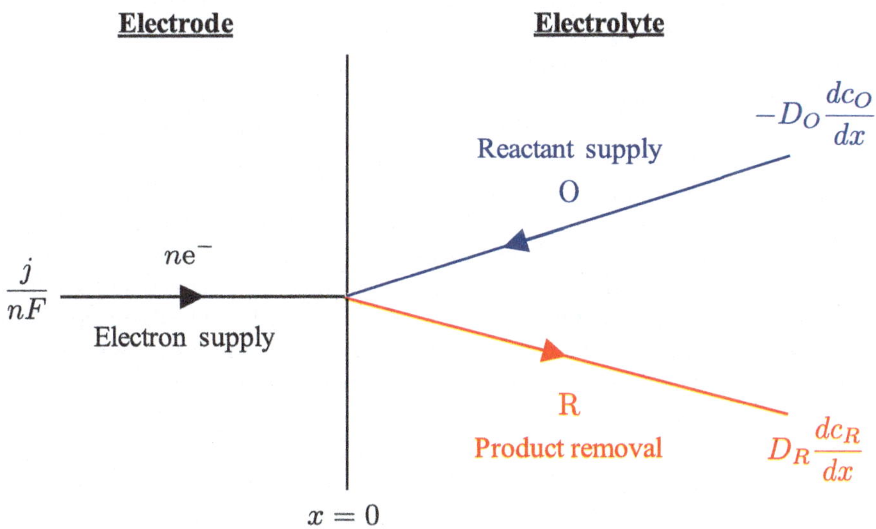

Fig. 2.7 Flux balance perpendicular to the surface

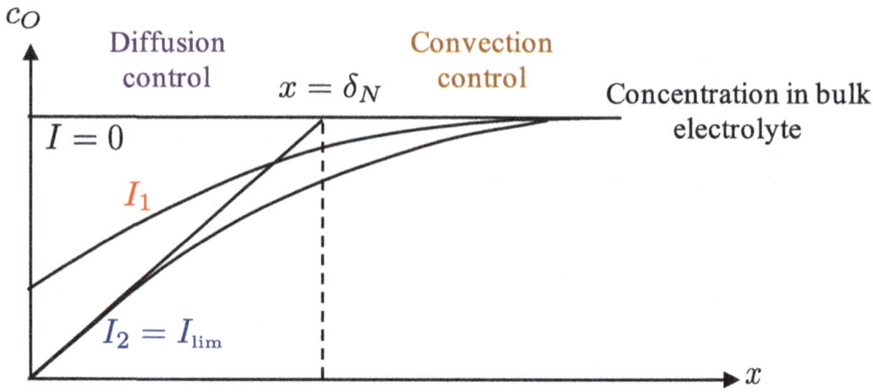

Fig. 2.8 The Nernst diffusion layer profile and the reactant concentration versus distance profiles at different applied current densities

pumping, stirring and gas/liquid flow. Natural convection is generated by small thermal or density differences in the electrolyte layer near the electrode, in which the movement is random and unpredictable. Forced convection can be introduced *via* rotating electrodes or controlled fluid flow, and can be described by a Nernst *diffusional layer* model.

As illustrated in Fig. 2.8, the electrolyte region close to the electrode surface can be divided into two zones. Close to the electrode surface, there is a stagnant layer with thickness δ_N [cm] (called the Nernst diffusion layer thickness), in which diffusion dominates by virtue of concentration differences. Outside this region (at $x \geq \delta_N$), mass transport is convection controlled.

In a real scenario, there is a gradual transition rather than two distinct regions since mass transport is neither pure diffusion nor pure convection controlled at $x = \delta_N$. At open circuit, when there is a zero net rate of reaction, the reactant concentrations are equal to the bulk electrolyte values C_O and C_R. However, if a current I_1 is applied, O is converted to R and the reactant concentration eventually decreases near the electrode surface. Increasing the applied current to I_2 will further decrease the reactant concentration and lead to more pronounced concentration gradients. In the limiting case, a zero concentration is reached at the electrode surface. The current in this limit is called the *limiting current* I_L, which is ideally independent of the electrode potential and can be expressed as a function of the bulk reactant concentration and the Nernst diffusion layer thickness as follows:

$$I_L = \frac{AnF}{\delta_N} D_O C_O \tag{2.36}$$

in which A is the active electrode surface area [cm^2] and D_O is the diffusion coefficient of O.

2.4 Brief Overview of Redox Flow Battery Developments

The redox flow battery was first introduced in the 1970s by The National Aeronautics and Space Administration (NASA, United States), Energy Development Associates (EDA) and Exxon Corp. One of the earliest systems was the iron-chromium system introduced by Thaller in 1974 [13]. According to NASA, the earliest studied redox couples include Fe^{2+}/Fe^{3+}, Ti/TiO^{2+}, Cr^{2+}/Cr^{3+}, V^{2+}/VO^{2+}, $Fe(O_3)^{3-}/Fe(O)_3{}^{4-}$, V^{4+}/V^{5+}, Br^-/Br^{3-} and $Cu(NH_3)^{2+}/Cu(NH_3)_4{}^{2+}$.

The main focus of NASA was energy storage systems for large-scale applications [6]. EDA and Exxon Inc. focused on a series of zinc-based hybrid systems, such as zinc-chlorine and zinc-bromine, to compete with conventional lead-acid batteries for vehicle power systems [7] and for large-scale storage of electricity [14].

Iron-chromium batteries have been scaled-up to kW level in the United States and Japan. However, these systems suffer from technical issues, such as cross contamination, poor reversibility of the chromium redox reactions and a high degree of hydrogen evolution during the charging process. The concept of an all-vanadium RFB was initially proposed by Pellegri and Spaziante as early as 1978 [15]. Significant technological advances were subsequently made by Skyllas-Kazacos et al. [16, 17] at the University of New South Wales, Australia.

All-vanadium RFBs have become the most studied systems and have been successfully commercialised for various applications, such as load levelling, power quality control and renewable coupling [18]. More than 100 large-scale plants have been installed globally by different manufacturers, at a scale of up to 100 MW [18]. Alongside the developments in all-vanadium RFBs, tremendous improvements in electrode and membrane materials for RFBs have been made. In order to enhance the cell voltages and energy densities, variations of the vanadium-based system, including vanadium-cerium [19] and vanadium-polyhalide [20], were introduced, in 2002 and 2008, respectively.

Hybrid flow batteries have also been investigated extensively. In these systems at least one electrode undergoes a solid/liquid transformation (solid electrode) or liquid/gas transformation (gas electrode). Zinc-based and metal-air flow batteries are typical examples of these systems. The advantages of conventional RFBs based on soluble active species are their scalability and flexibility due to the complete decoupling of energy and power. Hybrid flow batteries offer high energy density and reasonable discharge durations (>4 h), but their energy and power outputs are not fully decoupled.

Together with all-vanadium batteries, zinc-bromine and zinc-ferricyanide systems are among the few flow batteries that have been commercialised at over kW scales. Long-term development of the zinc-bromine RFB was carried out under the Moonlight Project in Japan during the 1980s. In the United States, zinc-bromine has been studied by ZBB Energy, the Department of Energy (D.O.E) and Sandia National Laboratories [21]. In the 1990s, a 1 MW/4MW h zinc-bromine RFB was installed in Fukuoka, Japan [21]. Since then, a number of RFB systems with all-vanadium and zinc-bromine chemistries have been installed [18].

In the early 2000s, two 15 MW/MW h bromine-polysulfide flow battery demonstration plants were constructed [18]. During this time, several single-flow 'membrane-less' batteries, such as the soluble lead-acid and zinc-nickel systems, were developed. These batteries have the advantages of simpler designs and lower costs compared to most other flow battery systems, particularly those involving expensive membranes and active materials [22].

Recent efforts have focused on the use of low-cost, abundant elements as the active materials rather than expensive metal species. For effective market penetration, the suggested cost targets of the European Union and United States are € 0.05 $(\mathrm{kW\,h})^{-1}$ cycle^{-1} and \$ 150 $(\mathrm{kW\,h})^{-1}$, respectively, in the current decade (2020–2030). Since the 2010s, various types of new redox active species based on organic, inorganic and polymer materials have been evaluated.

Organic molecules can be tuneable and are theoretically capable of higher solubilities and electrode potentials. Notable examples are the quinone-bromine [23] and quinone-ferricyanide systems [24] introduced by Aziz and co-workers at Harvard University. Currently, there are several start-ups in Europe using organic molecules as active materials, such as Kemiwatt, Jena Batteries, Green Energy Storage and CMBlu Energy. For instance, Jena Batteries has demonstrated a 20 kW/400 kWh pilot scale system in their testing facility in Germany.

The key cell parameters and the performance metrics of flow batteries are summarised in Table 2.3 [25]. In general, the cell voltages of redox flow batteries quoted are the open-circuit values $E_{\mathrm{cell}}^{\mathrm{OCV}}$ at a 50% state-of-charge (SOC). Calculations of the

Table 2.3 Key parameters and typical figure of metrics for redox flow batteries. [a] I is the current [mA] and A is active area [cm^{-2}]; [b] c is the concentration [$\mathrm{mol\,L^{-1}}$], n is the number of electrons transferred per mole, and υ is the electrolyte volume [mL]; [c] E_{cell} is the cell voltage under an applied current [V]

Key cell parameters	
$E_{\mathrm{cell}}^{\mathrm{OCV}}$ at 50 % SOC	$E_{\mathrm{cell}}^{\mathrm{OCV}} = E_{+} - E_{-}$ [V]
Current density[a]	$I_a = I/A$ [$\mathrm{mA\,cm^{-2}}$][a]
State-of-charge (SOC)	$\mathrm{SOC} = Q_{\mathrm{char}}/Q_t$ [%]
Solid area capacity	$Q_{\mathrm{a(solid)}} = Q_{\mathrm{rev}}/A$ [$\mathrm{mA\,h\,cm^{-2}}$]
Key electrolyte parameters	
Theoretical electrolyte capacity[b]	$c_{\mathrm{th}} = Q_t - nc\upsilon F/3600$ [C]
Electrolyte utilisation (E.U.)	$E.U. = Q/Q_t$ [%]
Performance metrics	
Volumetric capacity	$Q_{\mathrm{v,+/-}} = Q_{\mathrm{dis,+/-}}/\upsilon_{\mathrm{+/-}}$ [$\mathrm{A\,h\,L^{-1}}$]
Coulombic efficiency (C.E.)	$C.E. = Q_{\mathrm{dis}}/Q_{\mathrm{char}}$ [%]
Voltage efficiency (V.E.)	$V.E. = V_{\mathrm{dis}}/V_{\mathrm{char}}$ [%]
Energy efficiency (E.E.)	$E.E. = C.E. \times V.E.$ [%]
Power density	$P = I E_{\mathrm{cell}}/A$ [$\mathrm{mW\,cm^{-2}}$]

theoretical capacity c_{th} for either the negative or positive active electrolytes are based on Faraday's law, taking account of the number of electrons transferred involved n, the concentration of electrolyte c and the volume of electrolyte v. The current density is expressed per membrane area I_a. The volumetric capacity of the positive/negative electrolyte $Q_{v,+/-}$ is an important parameter since it represents the discharge capacity stored (reversible) per unit volume of the electrolyte.

In the case of hybrid batteries, the reactions involve liquid/solid transformations, and the areal capacity A_c is the main parameter controlling the reversible capacity Q_{rev} stored per unit membrane area A. A_c is a critical factor limiting this type of battery: if A_c is reached in practical applications, the battery cannot store more energy in spite of any further available capacity in the electrolyte.

2.5 Types of Flow Batteries

RFBs store energy entirely in the electrolyte, entirely on the electrodes or a combination of these two in the hybrid case, as illustrated in Fig. 2.9. Table 2.4 summarises the electrochemical reactions and performance metrics of various RFB systems.

The components and performance metrics of various liquid-phase systems are summarised in Table 2.5. Since energy is stored in the electrolyte, the energy storage capacity can be increased by using higher volumes of electrolytes and/or higher concentrations of species in the electrolyte. As early as the 1970s, NASA investigated various possible couples for liquid-phase systems, which include Fe, Cr

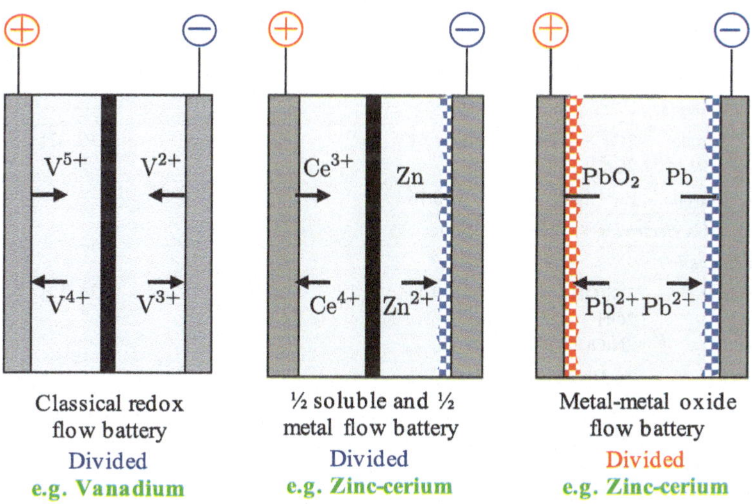

Fig. 2.9 Redox flow batteries store energy in the electrolyte, on the electrodes or a combination of the two (hybrid) [8]

Table 2.4 Typical redox couples used in redox and hybrid flow batteries

Electrochemical reactions		Electrode potential versus SHE/V	Cell voltage/V	Refs.
Energy stored in electrolytes				
Iron-chromium couple				
Positive electrode	$Fe^{2+} - 2e^- \leftrightarrow Fe^{3+}$	+0.77	1.18	[31, 32]
Negative electrode	$Cr^{3+} + e^- \leftrightarrow Cr^{2+}$	−0.41		
All-vanadium couple				
Positive electrode	$VO^{2+} + H_2O - e^- \leftrightarrow VO_2^+ + 2H^+$	+1.00	1.26	[33]
Negative electrode	$V^{3+} + e^- \leftrightarrow V^{2+}$	−0.26		
Vanadium-cerium couple				
Positive electrode	$2Ce^{3+} - 2e^- \leftrightarrow 2Cr^{4+}$	+1.61	1.87	[34]
Negative electrode	$V^{3+} + e^- \leftrightarrow V^{2+}$	−0.26		
Vanadium-bromine couple				
Positive electrode	$2Br_3 + Cl^- - 2e^- \leftrightarrow 2ClBr_2^-$	+0.8	1.3	[35]
Negative electrode	$VBr_3 + 2e^- \leftrightarrow VBr_2 + Br^-$	−0.5		
Polysulfide-bromine couple				
Positive electrode	$3Br^- - 2e^- \leftrightarrow Br_3^-$	+1.09	1.36	[35]
Negative electrode	$S_4^{2-} + 2e^- \leftrightarrow 2S_2^{2-}$	−0.27		
Hybrid flow batteries				
Zinc-bromine couple				
Positive electrode	$2Br_3 - 2e^- \leftrightarrow 2Br_3^-$	+1.09	1.85	[35]
Negative electrode	$Zn^{2+} + 2e^- \leftrightarrow Zn$	−0.76		
Zinc-cerium couple				
Positive electrode	$2Ce^{3+} - 2e^- \leftrightarrow 2Cr^{4+}$	+1.61	2.37	[36, 37]
Negative electrode	$Zn^{2+} + 2e^- \leftrightarrow Zn$	−0.76		
Energy stored in electrodes				
Zinc-nickel couple				
Positive electrode	$2Ni(OH)_2 + 2OH^- - 2e^- \leftrightarrow 2NiOOH + 2H_2O$	+0.49	1.71	[36, 37]
Negative electrode	$Zn(OH)_2^{2-} + 2e^- \leftrightarrow Zn + 4OH^-$	−1.22		
Soluble Lead-lead dioxide couple				
Positive electrode	$Pb^{2+} + 2H_2O - 2e^- \leftrightarrow PbO_2 + 4H^+$	+1.49	1.62	[10]
Negative electrode	$Pb^{2+} + 2e^- \leftrightarrow Pb$	−0.13		

Table 2.5 The cell components and operating parameters of various redox flow battery systems that store chemical energy in the electrolytes

System	Electrode material	−ve electrolyte	+ electrolyte	Background electrolyte	Membrane	Charge voltage	Discharge voltage	Current density (mA cm^{-2})	Temperature	Energy efficiency	Refs.
Fe/ Cr	Cr: Carbon felt + catalyst Fe: carbon fibre	1 M CrCl3	1 M FeCl2	2 M HCl	Cationic Nafion®117	1.05 V	1.03 V	21.5 mA cm^{-2}	25 °C	Nil	[31, 32]
Fe- V	Graphite foam	1 M V (II)/ V(III)	1 M Fe (III)/ Fe (II)	1 M H_2SO_4	Anion-exchange membrane	Nil	0.26 V	Nil	Nil	Nil	[6]
All-vanadium	Thick felt-heat bonded graphite impregnated polyethylene plate	2 M $VOSO_4$	2 M $VOSO_4$	2M H_2SO_4	Cationic polystyrene sulphonic acid membrane	1.47 V	1.30 V	30 mA cm^{-2}	35 °C	83%	[33]
Vanadium chloride-polyhalide	Graphite felts compressed	1M VCl3	1 M NaBr	1.5 M HCl	Cationic Nafion®112 membrane	1.2 V	0.98 V	20 mA cm^{-2}	Nil	66.4%	[38]
V-Ce	Carbon fibres	0.5 M V (III)/ V (II)	0.5 M V (III/IV)	1 M H_2SO_4	Vycor glass membrane (Asahi Glass Co. Ltd.)	1.83 V	1.51 V	22 mA cm^{-2}	40 °C	67.8 %	[19]
V- Mn	Spectral pure graphite	0.3 M V (III)/ V (II)	0.3 M V (II/III)	5 M H_2SO_4	Cationic Nafion®117	1.83 V	1.66 V	20 mA cm^{-2}	Ambient temperature	62.7%	[20]
Br-Polysulfide	Polysulfide: Nickel foam, Bromine: Poly-acrylonitrile (PAN)-based carbon felt	1.3 M Na_2S_4	4 M NaBr	1 M NaOH	Cationic Nafion®117	1.75 V	1.38 V	40 mA cm^{-2}	26 °C	77.2 %	[35]
AQ-BQ	Carbon papers	0.2 M AQ	0.2 M BQ	1 M H_2SO_4	Cationic Nafion®117	0.72 V	0.35 V	10 mA cm^{-2}	Ambient temperature	34 %	[39]
AQ-BQ											
AQ-Br	Carbon papers	0.1 - 1 MAQ	0.5 - 2.5 M HBr	1 M H_2SO_4	Cationic Nafion®212	0.85 V	0.65 V	200 mA cm^{-2}	Nil	76 %	[23]
AQ-Br	Carbon papers	0.5 M AQ	0.4 M$Fe(CN)_6^{4-}$	1 M H_2SO_4	Cationic Nafion®212	1.3 V	1.1 V	100 mA cm^{-2}	Nil	84 %	[24]

and Ti. The early liquid-phase systems, including iron-chromium and iron-titanium, were not particularly reversible with metallic electrodes. Porous, three-dimensional electrodes, such as graphite felt, graphite cloth, graphite foam and reticulated vitreous carbon, were therefore introduced in the 1980s [6].

The iron-titanium system has a relatively low OCP and suffers from passivation of the TiO_2. Iron-chromium RFBs were among the most studied systems in the early development phase. Although no catalyst was required for the iron reaction, the reduction of chromium was slow at most electrode surfaces. This system also suffers from cross contamination of the active species and hydrogen evolution [26]. Soluble electro-active species, such as Fe(III)/Fe(II) [27, 28], Ce(IV)/Ce(III) [29], Cr(III)/Cr(II) and Cr(V)/Cr(III) [30], were reported to have improved reaction rates when they are complexed with triethanolamine (TEA), diethylenetriaminepentaacetate (DPTA) and ethylenediaminetetra-acetate (EDTA) ligands, respectively. This enhanced the range of possibilities for redox couples, including all-chromium, vanadium-cerium and iron-bromine.

The all-chromium system was proposed as early as 1985 [40] but the relatively slow kinetics of the chromium reactions were a major challenge. In the same year, the all-vanadium RFB was developed by Skyllas-Kazacos at the University of New South Wales [41]. This system did not involve any significant gaseous evolutions. In addition, cross mixing of the electrolytes across the membrane does not lead to any electrolyte contamination. Due to its high energy efficiency (>80%) [42], this system has been successfully scaled-up in many industrial applications and has been studied extensively.

The major drawback of the all-vanadium RFB is its low specific energy of 25–35 Wh kg^{-1}, limiting its applications. In order to increase its energy density (up to 50 W h kg^{-1} [43]), a vanadium/polyhalide flow battery was proposed by Skyllas-Kazacos in 2003 [38]. This system uses VCl_2/VCl_3 and $Br^-/ClBr_2^-$ as the electro-active species in the negative and positive half-cells, respectively. In order to achieve higher potentials, Mn(II)/Mn(III) [20] and Ce(III)/Ce(IV) [34] have been used as the positive electro-active species in combination with vanadium. Vanadium-cerium flow batteries have the advantages of high Coulombic efficiency (87%), high cell potential (1.87 V) and low self-discharge rate, but low solubility remains the greatest obstacle [34].

Paulenova et al. [44] suggested that the slow redox kinetics of the Ce(III)/Ce(IV) reaction on carbon makes the species unsuitable for use in redox flow batteries. However, Kreh et al. reported that cerium is highly soluble in methanesulfonic acid and exhibits a highly positive electrode potential of >1.6 V *vs.* SHE, making it attractive if stable electrode materials for cerium reactions are available at reasonable costs [45].

In addition to the all-vanadium systems, the bromine/polysulfide battery is another well-known system that was initially developed by Innogy Technologies in the 1990s. Due to its low cost and high energy density, large-scale bromine/polysulfide batteries were demonstrated at MW scale in the early 2000s [46]. Shiokawa et al. also reported that the redox couples of uranium [47, 48] and neptunium [49–51] are remarkably similar to vanadium, and can potentially be used for RFBs.

Since the 2010s, the use of abundant and low-cost soluble organic active species has been explored extensively. These species include carbonyls, quinones, fluorenone, imide, nitroxide radical, heterocyclic molecules (e.g. viologen, phenazine, alloxaine, phenothiazine) and organo-metallic complexes [39]. Among these systems, the quinone-bromine [23] and quinone-ferricyanide systems [24] have the highest power densities due to impressive current and voltage outputs, respectively. The capital costs of these aqueous systems were estimated to be \$ 400 $(kW\ h)^{-1}$, mainly due to their high current densities ($\geq$100 mA cm^{-2}) and cell voltages ($\geq$1.3 V). Although organic active materials are reasonably low cost (around \$ 5 kg^{-1}), the electrolyte costs may not necessarily be lower than the cost of a conventional vanadium electrolyte (\$ 300 – 500 $(kW\ h)^{-1}$). Since these organic systems are mainly based on large molecules and yield lower cell voltages and/or current densities compared to non-organic systems, further improvements are necessary.

2.5.1 Systems with Energy Stored on the Electrodes

In some systems, the energy is stored entirely in the form electrodeposits at both the positive and negative electrodes, while dissolutions of the electrodeposits occur on discharge. All of these systems have a lower cost than conventional RFBs since only one electrolyte is needed and they have no membrane requirement. Furthermore, the chemicals involved in these membrane-less systems are generally low cost. They are, therefore, attractive for large-scale energy storage. On the other hand, incomplete discharge may lead to capacity degradation over prolonged periods of operation [52]. Table 2.6 summarises the cell components and performance metrics of these membrane-less systems.

Lead-acid batteries have been used extensively for over a century. A new type of 'soluble' lead-acid flow battery was introduced by Pletcher et al. in 2004 [53–57]. This soluble system differs from the conventional system in that it uses methanesulfonic acid rather than sulfuric acid, which enables lead and lead dioxide electrodeposition from soluble lead (II) species in the electrolyte. In contrast, both the negative and positive electrode reactions of conventional lead-acid batteries involve a solid-phase transformation rather than metal electrodeposition from soluble bulk species, since the solubility of Pb^{2+} ions is almost negligible in sulfuric acid. In the soluble lead-acid system, additives or surfactants are required to suppress dendritic growth of the electrodeposits during the charging process.

Similar to the soluble lead-acid batteries, a zinc-nickel, single-flow, membrane-less battery was introduced by Zhang et al. in 2004 [58, 59]. This technology was inspired by the conventional zinc-nickel secondary battery, but with flowing electrolytes to suppress the dendritic growth of zinc electrodeposits. It enjoys a number of advantages over the conventional system, including a high specific energy, good cycleability and low cost. A high-concentration alkaline zincate electrolyte is used. During charge, zinc is electrodeposited from the zincate ions and $Ni(OH)_2$ is oxidised to NiOOH at the negative and positive electrodes, respectively. The battery exhibits

Table 2.6 The cell components and operating parameters of various redox flow battery systems that store chemical energy on the electrodes

System	Electrode material	Electrolyte	Background electrolyte	Membrane	Charge voltage	Discharge voltage	Current density (mA cm^{-2})	Temperature	Energy efficiency	Refs.
Zn-Ni	Zinc: Cadmium-plated copper, Ni: Sintered nickel hydroxide electrode	1 M ZnO	10 M KOH	Membrane-less	1.85 V	1.72 V	10 mA cm^{-2}	Room temperature	86%	[58, 59]
Pb- PbO_2	Lead: Reticulated nickel foam, Lead dioxide: Scraped reticulated vitreous carbon	1.5 M $Pb(CH_3SO_3)2$	0.9 M CH_3SO_3H	Membrane-less	2.07 V	1.45 V	20 mA cm^{-2}	25 °C	65 %	[10]
Cu- PbO_2	Copper: High purity graphite, Lead dioxide: 98% lead dioxide, 1.2% graphite fibre	1 M $CuSO_4^-$ 1.9 M H_2SO_4	nil	Membrane-less	1.45 V	1.29 V	20.8 mA cm^{-2}	Room temperature	83%	[22]

a higher cell voltage (*ca.* 1.7 V), energy efficiency (*ca.* 88%) and a longer cycle life (≥1000 cycles) than the static counterpart.

2.5.2 Hybrid Flow Batteries

A hybrid flow battery stores energy in one of its negative electro-active components by mean of metal electrodeposition, while the positive electrode reactions are either liquid phase or gas phase. Table 2.7 summarises the cell components and performance metrics of various hybrid redox flow battery systems.

Zinc anodes are common in existing rechargeable batteries. Inspired by this, zinc-halogen hybrid flow batteries, including zinc-chlorine [60, 61] and zinc-bromine [14], were introduced in 1973 and 1975, respectively. These hybrid systems have high energy densities, high cell potentials and low costs. Zinc-chlorine flow batteries were initially intended as a competitor to conventional lead-acid batteries for vehicle power systems [7]. Zinc-bromine batteries have been commercialised and are available for various applications, from load levelling to electric vehicles [21, 62]. Since chlorine and bromide tend to evolve during charge, special designs for dissolving chlorine and bromide are necessary, which makes these systems bulky and complex.

In the 2000s, the zinc-cerium hybrid flow battery was developed by Plurion Inc. and AIC Inc. [36, 37, 66]. The operational current densities were claimed to be as high as 500 mA cm^{-2}, while the open circuit cell voltage was at least 2.4 V, which was nearly double of that of all-vanadium RFBs. A key feature of this system was the use of an organic acid (e.g., methanesulfonic acid), which is less corrosive than sulfuric acid and has a lower impact on human health and the environment. More importantly, cerium species are highly soluble in organic acids (up to 1 M), compared to conventional hydrochloric and sulfuric acids [45]. Although the cost of cerium is not particularly high (< $ 20 kg^{-1}), the use of platinised titanium or other dimensionally stable electrodes leads to a high overall cost [37].

It is important to note that the electronegativity of zinc anodes (<−0.76 V *vs.* SHE) tends to lead to hydrogen evolution, resulting in low current efficiencies. Zinc-based hybrid flow batteries often have faster self-discharge rates than their all-vanadium counterparts due to the corrosion of zinc electrodeposits in acidic or alkaline electrolytes. Other than zinc-based systems, all-iron hybrid flow batteries were also introduced, using Fe(II/0) and Fe(II/III) as the negative and positive electro-active species [67]. However, these systems received less attention, possibly due to their lower efficiencies and/or energy content compared to zinc-based systems [67].

Table 2.7 The cell components and operating parameters of various hybrid redox flow battery systems

System	Electrode material	Negative electrolyte	Positive electrolyte	Background electrolyte	Membrane	Charge voltage	Discharge voltage	Current density (mA cm^{-2})	Temperature	Energy efficiency	Refs.
Zn- Cl	Zinc: Dense graphite electrode Chlorine: Porous flow-through electrode	2 M $ZnCl_2$	2 M $ZnCL_2$	4 M KCl	Kynar separator	2.30 V	1.80 V	22 mA cm^{-2}	2 5 °C	66%	[63]
ZnBr	Carbon elec-trodes	$ZnBr_2$ in excess of Br2	$ZnBr_2$ in excess of Br2	KCl, NaCl	Cationic Nafion®125	1.92 V	1.64 V	15 mA cm^{-2}	25 °C	>80 %	[64]
Zn-Ce	Zinc-Carbon Cerium-Platinised Titanium	0.6 M $Ce(CH_3SO_3)_3$+ 1.3 M $Zn(CH_3SO_3)_2$	0.72 M $Ce(CH_3SO_3)_3$+ 0.9 M $Zn(CH_3SO_3)_2$	CH_3SO_3H	Cationic Nafion®mem-brane	2.75 V - 3.15 V	1.8 V (constant voltage dis-charge)	50 - 100 mA cm^{-2} (constant voltage charge) 5-50 mA cm^{-2} (constant voltage dis-charge)	60 °C	Nil	[36, 37]
Fe-Fe	Negative: Titanium, Positive: Porous carbon	2 M $FeCl_2$		NH_4Cl	microporous plastic separator	1.57 V	0.87 V	60 mA cm^{-2}	60 °C	50 %	[65]

2.6 Design Considerations and Components of Flow Batteries

2.6.1 Construction Materials

A typical flow battery unit cell is illustrated in Fig. 2.10, composed of negative and positive electrodes separated by an ion-exchange membrane. In practice, plastic gasket seals and steel tie bolts are used to compress cells in order to prevent leakage. Metallic end plates, such as aluminium and copper, may be used to provide better

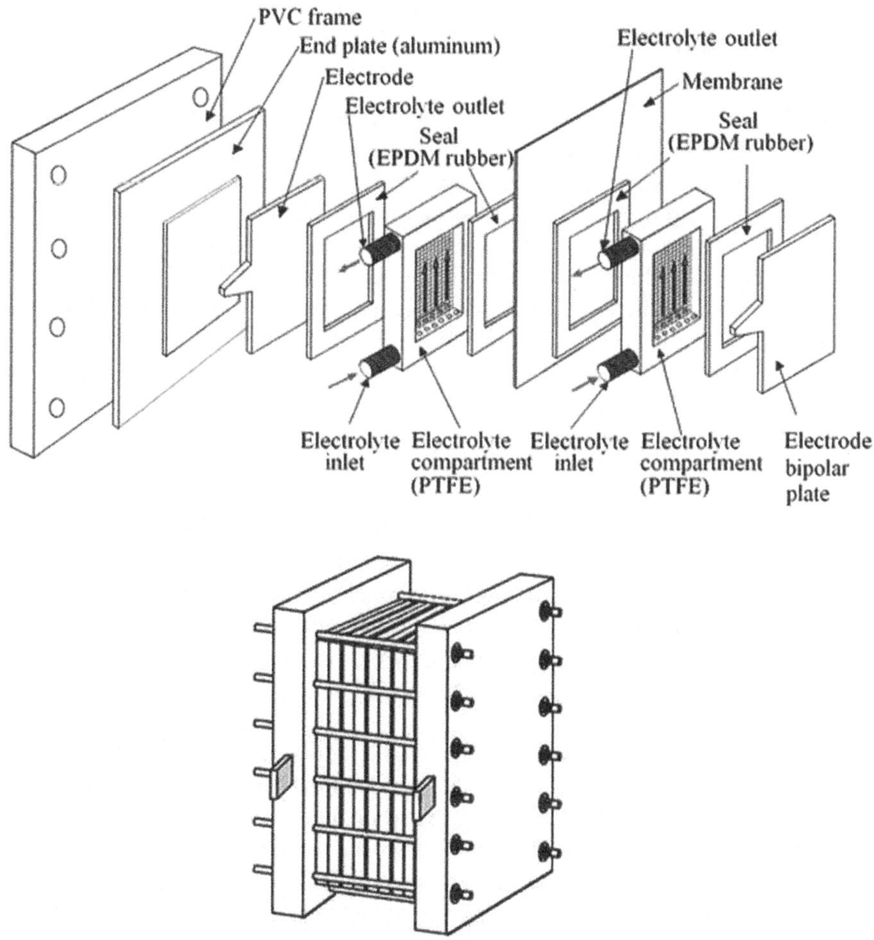

Fig. 2.10 Typical components of a redox flow battery [68]

electrical contacts. In order to enhance mass transport and promote active species exchange, turbulence promoters can be added to the compartments [68].

Since electro-active species are highly oxidative, no metallic component should be in contact with the electrolytes. Chemically resistant polymers, such as polytetrafluoroethylene (PTFE), ethylene-polypropylene-diene (EPDM) and polyvinylchloride (PVC), are the main materials for producing the cell components (excluding the metallic end plate and the electrode). By connecting a number of unit cells in series in a bipolar manner to form a battery stack, a large cell voltage can be obtained. Easily folded rubber reservoirs are sometimes used to effectively utilise spaces that are usually unoccupied or unused, such as underground cisterns in buildings [2].

2.6.2 Electrode Materials

An ideal electrode has a high electrical conductivity, good mechanical durability, high chemical resistance, a reasonable cost and a long cycle life in a highly oxidising medium. Electrodes for RFBs are carbon based, metal based or composites. Carbon-based electrodes are common, since they are chemically inert and have reasonable catalytic properties. Unlike metal-based electrodes, carbon-based electrodes do not undergo dissolution, formation of oxide layers and corrosion. Since pure carbon/graphite is brittle, composite and other materials have been introduced. The different electrode materials used in RFBs are listed in Table 2.8.

2.6.3 Carbon-Based Electrodes

Carbon-based electrodes are the most commonly used for RFBs. In highly oxidising environments, such as when V^{5+} and Ce^{4+} are present, carbon electrodes tend to degrade at elevated temperatures and sulfuric acid concentrations [44]. Although graphite has superior reversibility compared to pure carbon for vanadium reactions [69], it is still not stable enough for the highly oxidising species [70–72]. Developments in carbon-based materials have been carried out continuously to address these issues.

Two-dimensional carbon-based electrodes, including carbon black [73] and activated carbon [68], have been used in different redox flow batteries. In addition, three-dimensional carbon-based material, such as carbon felt, carbon cloths and reticulated vitreous carbon (RVC) [72, 74, 75], have also been used. Due to their high specific area, three-dimensional electrodes can reduce polarisation losses significantly [76].

Graphite felts are regarded as the most promising electrodes for vanadium RFBs [77], particularly polyacrylonitrile (PAN)-based carbon felt electrodes. PAN-based electrodes possess wide potential ranges, good electrochemical activities, high chemical stabilities and high surface areas, at relatively low costs [78]. Despite these advantages, modifications of graphite felt materials have been made to improve their

Table 2.8 Electrode materials that have been used in redox flow batteries

Electrode material	Manufacturer	Flow type	Thickness	Polarity	Flow battery system	Refs.
Carbon polymer	Nil	Flow-by	Nil	Negative	Zinc-cerium	[36, 37]
Graphite felt	Le Carbone	Flow-by	Nil	Positive and negative	All-vanadium	[84]
PAN-based Graphite felt	Shanghai XinXing Carbon Corp. China	Flow-by	5 mm	Positive	Polysulfide Bromine	[65]
Cobalt coated PAN-based Graphite felt	Shanghai XinXing Carbon Co. Ltd., China	Flow-by	5 mm	Negative	Polysulfide Bromine	[65]
70 ppi Reticulated Vitreous carbon	Nil	Flow-by	1.5 mm	Positive	Soluble lead acid	[22]
Graphite felt bonded bipolar electrode assembly with nonconducting plastic substrate	Graphite felt: sheet: FMI Graphite, USA	Flow-by	Nil	Bipolar	All-vanadium	[85]
Porous graphite	Union Carbide Porous Grade 60	Flow-through	2 mm	Positive and negative	Zinc-chlorine	[63]
Carbon felt type CH	Fibre Materials Incorporated	Flow-through	2.8 mm		Zinc-bromine	[86]
Nickel foam	Changsha LuRun Material Co. Ltd., China	Flow-by	2.5 mm	Negative	Polysulfide-bromine	[35]
Cadmium-plated Copper	Nil	Flow-by	Nil	Negative	Zinc-nickel	[58, 59]
Sintered nickel hydroxide	Nil	Flow-by	Nil	Positive	Zinc-nickel	[58, 59]
40 ppi Nickel foam	Nil	Flow-by	1.5 mm	Positive	Soluble lead acid	[22]
Platinised-titanium mesh	Nil	Flow-by	Nil	Positive	Zinc-cerium	[36, 37]
Iridium oxide-coated, dimensionally stable anode	Nil	Flow-by	Nil	Positive	All-vanadium	[87]

electrochemical performance [79–83]. Pristine carbon felts tend to have poor kinetics for both of the vanadium electrode reactions, requiring functionalisation treatments.

The vanadium half-cell mechanism was initially proposed by Skylla-Kasakos, emphasising the critical role of the C–OH functional groups as active sites for oxidation of the VO^{2+} species [88]. For the vanadium positive electrode reaction, the first step involves proton exchange. Following this, an electron and oxygen atom from the C–OH functional groups are transferred to the VO^{2+} species by forming VO_2^+ through C–O–V intermediates. Similarly, for the negative electrode reaction, C–OH functional groups are known to be crucial for catalysing the reduction of V^{3+} to V^{2+}, although suppressing hydrogen evolution and prevention of V^{2+} oxidation were also considered major challenges.

In order to further facilitate the vanadium reactions, the use of oxygen functionalities in a carbon network has also been explored in recent years. For instance, catalytic routes *via* O–catalyst sites, including C–OH and C–OOH groups, have been identified using carbon-based materials, such as graphite, graphite oxide and multi-walled carbon nanotubes. These materials tend to exhibit strong absorption of vanadium ions, accelerating electron and oxygen transfers during the positive electrode reaction. Furthermore, the proton within the C–COOH groups can be replaced much faster than that within the C–OH groups. On the other hand, C=C groups have also been proposed as alternative catalysts for the positive electrode reactions, which can be grown directly within the carbon felt electrodes [89].

Recently, a gradient-pore-oriented carbon felt electrode was developed by combining three types of porous structures, from micro- to nano-scale, on the carbon fibre surface, as illustrated in Fig. 2.11 [90]. Micropores of less than ~20 μm in the carbon felt enhance the surface area for electrolyte contact, while the nanopores (~20 nm) on the fibre surface increase the number of active sites for the redox reactions. The mesoscale pores (~0.5 μm) have connect the micropores and nanopores with an optimal distribution. Different sized pores can be created by thermal treatments and etching methods, with FeOOH, KOH, $FeCl_3$, ZnO, CuO and CH_3COOCo. Among these, KOH etching is the most common approach and has been used to generate nanopores for high current operations (>200 mA cm^{-2}).

The electrical conductivities of PAN-based graphite felt electrodes have also been improved by the deposition of metals (or metal oxides) on the fibre surface [76]. Skyllas-Kazacos and co-workers [91] have modified carbon felt electrodes by impregnation or ion exchange with solutions containing Pt^{4+}, Pd^{2+}, Te^{4+}, Mn^{2+}, In^{3+}, Au^{4+} and Ir^{3+}. Modified Ir^{3+} electrodes in particular exhibit superior performance as positive electrodes, in terms of electrocatalytic activity and stability in acidic media. Excessive hydrogen was observed with Pt-, Pd- and Au-modified electrodes. The incorporation of metal oxides (Ti-, W- as well as Nd-based oxides) as catalysts for vanadium half-cell reactions has received significant attention. These elements are useful in suppressing gas evolution but also enable adsorption of polar molecules to increase the wettability and hydrophilicity of the electrode, facilitating reactions at the electrode/electrolyte interface. For instance, RuO_2 can improve the reaction rate and eliminate the side reactions [92]. The cell resistance can be reduced by 25% with the use of Ir-modified electrodes [76].

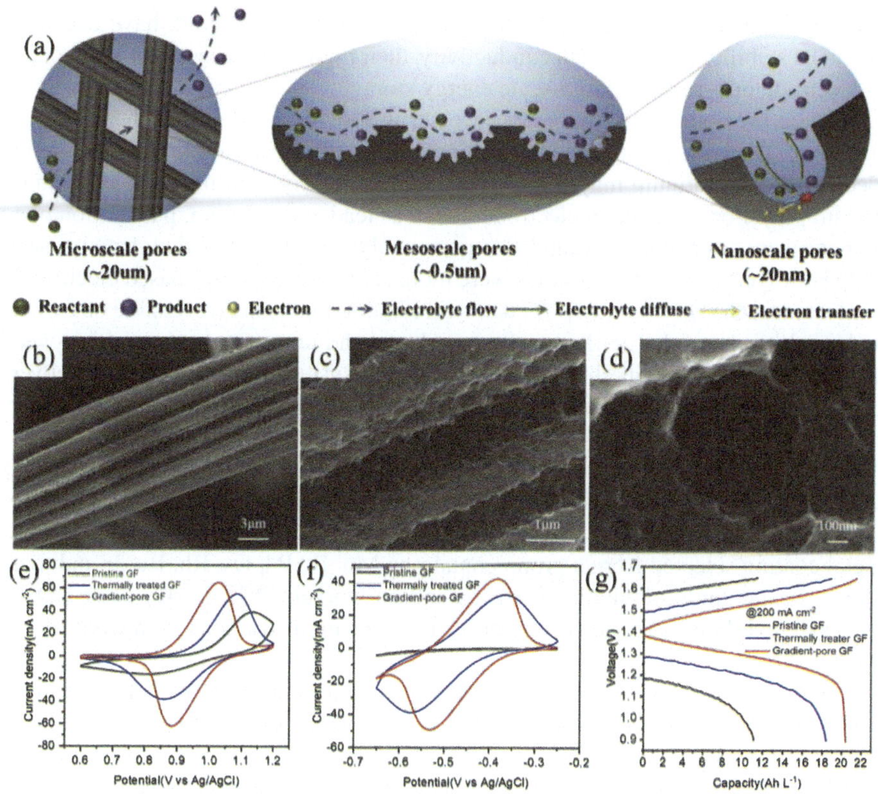

Fig. 2.11 Schematic of the gradient-pore graphite felt electrode. **a** Scanning Electron Microscope images of treated gradient-pore graphite felt; **b–d** cyclic voltammetry of pristine, thermally treated and gradient-pore graphite felts in the potential windows of **e** 0.6 to 1.2 V and **f** −0.7 to −0.2 V at a scan rate of 5 mV s^{-1}. **g** Cyclic voltammetry of gradient-pore graphite felt for VO^{2+}/VO_2^+ [90]

Reticulated vitreous carbon (RVC) electrodes have been used to provide a much larger surface area in zinc-bromine [93] and soluble lead-acid batteries [22]. In early studies the porosity of this material was exploited to retain the solid complex of bromide during charging of zinc-bromine batteries [94]. The surface of an electrode with a reticulated vitreous carbon layer is relatively rough, which is beneficial for establishing adherent layers. The compressed foam structure allows the electrodeposits to be formed within, and be supported by the three-dimensional structure [22].

2.6.4 Metal-Based Electrodes

Due to the issues associated with cell corrosion, metal-based electrodes are not commonly used in RFB applications unless coated with precious metals. Although it is known that some precious metals, such as platinum and gold, have excellent chemical stability and electrical conductivity, their high cost makes them impractical as electrode materials for large-scale energy storage.

Certain precious metals may not necessarily provide good electrochemical performance. For instance, the redox reactions of vanadium are not reversible at gold electrodes and platinum may form a non-conductive oxide during the oxidation of cerium [95]. For other metal electrodes, such as lead and titanium, passivation phenomena were often observed and led to poor catalytic properties and high resistances [96]. However, some redox couples were more reversible on metal-based electrodes. For instance, Cr^{2+}/Cr^{3+} and Ti^{3+}/TiO^{2+} were both found to be irreversible on carbon and graphite electrodes, while amalgamated lead and tungsten-rhenium alloys were used as the electrodes in chromium and titanium systems, respectively. Moreover, $Fe(O)_3^{31}/Fe(O)_3^{4-}$ exhibited a more reversible behaviour on platinum than it does on carbon electrodes [6].

Dimensionally stable anodes (DSA) have been used in highly oxidising media, such as the positive half-cells of the all-vanadium [87, 97, 98] and zinc-cerium RFBs [37]. A DSA electrode is cored by titanium (or an alloy of titanium) with a noble metal coating selected from a group containing Mn, Pt, Pd, Os, Rh, Ru, Ir and their alloys. In the study of Skyllas-Kazacos and co-workers [87], a IrO_2-coated DSA electrode exhibited superior reversibility for vanadium reactions compared to the others.

Another major development is the use of three-dimensional nickel foam electrodes. This material has been studied for use in sodium polysulfide [35] and soluble lead-acid flow batteries [22]. In common with other three-dimensional materials, nickel foam provides a large surface area and is electrocatalytically active for both electrode reactions in a typical sodium polysulfide system [35]. The small pores (*ca.* 150–250 μm) of nickel foam electrodes enable the electrolyte to flow smoothly, leading to relatively fast transport of the active species to the electrode reaction sites. Ni/C and NiS_x were also found to be suitable catalytic materials for polysulfide redox reactions [99].

2.6.5 Composite Electrodes

Carbon-polymer composite materials offer many benefits over solid carbon or graphite substrates, such as lower cost and reduced weight. These materials are typically composed of polymer binders and carbon particles. They have been reported to possess advantages compared to pure carbon in terms of their mechanical properties. Carbon is too brittle for scale-up purposes, in which large compressive forces are applied over large areas. Carbon-polymer composites have been widely used in redox flow batteries, particularly the all-vanadium, zinc-bromine, iron-chromium and soluble lead-acid systems [53, 56, 100–105].

A suitably conductive carbon-polymer composite was obtained by the addition of 20–30% conductive filler material. However, a high carbon content may lead to poor mechanical properties. The carbon black fillers used in conventional carbon-polymer composites may result in some side reactions during overcharge of the cells, such as gas evolution, water decomposition and irreversible cell deactivation. Oxidation of the carbon black filler also tends to increase the electrical resistance. The relatively low melt flow indices of carbon black materials lead to poor penetration of fibres into the substrate. Improved conductivity of composite substrates is required to overcome the long distances between the graphite felt fibres through the substrate material.

In order to reduce this distance and maintain a high conductivity, Skyllas-Kazacos and co-workers [85] proposed an alternative approach using a non-conductive polymer material with good melt flow properties, to replace the carbon black materials as substrates. High penetration and interconnection of the felt fibres were achieved, not only maintaining the high electrical conductivities but also improving the mechanical properties due to the reduced carbon loading. Carbon nanotubes (CNTs) have also been used for RFBs due to their high electrical conductivity and good mechanical properties [96]. However, the sole use of carbon nanotubes for the vanadium reactions exhibited a reduced reversibility and activity compared to graphite electrodes. In order to combine the advantages of both graphite and carbon nanotubes, a composite of graphite with 5 wt. % CNT has been investigated. The resulting composite material was found to be an effective material for both electrodes of the all-vanadium RFBs.

2.6.6 Membranes

Ideal membranes restrict diffusion of the electro-active species to and from the two electrolytes and permit transport of non-electro-active species with their associated water molecules in order to maintain electro-neutrality in both half-cells. The membrane, either anionic or cationic, should be appropriately selected to maintain a constant pH at both sides. In spite of being rather expensive, the most commonly used membrane is Nafion®, since it offers low electrical resistivity and has a long life span due to its excellent chemical stability. Lower-cost alternatives in the form of non-fluorinated membranes, such as poly ether ether ketone (PEEK) membranes, have been reported. For instance, sulfonated PEEK membranes were evaluated extensively. Of particular importance is the stabilities of these non-fluorinated membranes in highly oxidising or corrosive environments under prolonged operation [106, 107].

An ideal membrane offers not only a high ionic conductivity and low crossover of the active species, but is also economically viable. A high ion-exchange capacity (IEC) facilitates the passage of charge-carrying ions, enabling a low cell resistance [108]. In practice, the ionic conductivity of a membrane is determined by various factors, such as ionic group concentration, the degree of cross-linking and the size and valence of the counter ions [109]. As reported by Skyllas-Kazacos and co-workers, there is no direct correlation between the IEC, resistivity and diffusivity. Solution concentration and ion exchange groups within the membranes have a large influence on water uptake and hence the degree of swelling. A higher degree of cross-linking was found to lower the swelling effect and provide a higher selectivity.

Cation- and anion-exchange membranes should maintain the pH in both electrolytes. In general, cation-exchange membranes are more chemically stable and tend to be more efficient than anion counterparts [110]. Cation-exchange membranes also exhibit higher conductivities and electro-active species diffusivities. Cation-exchange membranes conduct H^+ ions, while anion exchange membranes in acidic solutions conduct both H^+ and SO_4^{2-} from the negative to the positive electrode in all-vanadium systems. The lower resistivity of cation-exchange membranes is due to the higher mobility of H^+ ions compared with that of SO_4^{2-}. The lower diffusivity of vanadium ions through the anion-exchange membranes can be attributed to the Donnan exclusion effect [111]. Anionic/non-ionic membranes tend to show net volumetric transfer towards the negative half-cell, while cation-exchange membrane volumetric transfer is towards the positive electrolyte compartment.

2.6.7 Commercially Available Membranes

Nafion® membranes remain the most commonly used membranes in fuel cell and flow battery applications. Most recent developments in membranes for RFBs were motivated by the performance of all-vanadium systems, and are summarised in Table 2.9. Perfluorinated sulfonated acid (PFSA) membranes are used as a consequence of their high ionic conductivities and chemical stability in the highly oxidising/corrosive electrolytes. Other cation-exchange membranes, such as Selemion® CMV and the DMV Asahi membrane, were found to degrade in vanadium electrolytes, which was not the case with Nafion® [113–115].

A drawback of Nafion® is the permeation of active species through the membrane, which causes mixing of the negative and positive electrolytes, resulting in lowered current efficiency [106]. In addition, Nafion® membranes are of high cost, up to \$ 500 m^{-2} [116]. Most commercial membranes exhibit good selectivity (with the exception of the Dow membrane) but apart from Nafion® 117 and Flemion, most do not possess good chemical stability. The majority have a high conductivity, but simultaneously a low volumetric transfer [111]. Examples include K142, Selemion® CMV, CMS, AMV, DMV, ASS, DSV, CMF (Flemion®), New Selemion® (polysulfone), Nafion® 112, 117 and 324, and RAI R1010 and R4010.

In addition to the aforementioned membranes, Hipore®, Selemion® HSV, HSF, Neosepta® CM-1, AM-1, ABT, HZ cation, HZ anion, Gore Select L01854 and M04494 among others have been evaluated for vanadium-bromine flow batteries. The best performance was seen with ABT3, L01854 and M04494, in terms of cell cycling and chemical stability. Due to their high membrane resistance, poor cell cycling was observed with CM-1, AM-1 and ABT-1. The cycle lives of these membranes were limited to less than 40 cycles due to rapid degradation.

Apart from ion-exchange membranes, polybenzimidazole (PBI) membranes have been commercialised by Fumatech, Germany and Celazole, United States, while a

Table 2.9 Commercially available membranes used in all-vanadium redox flow batteries

Membrane	Supplier	Type	Thickness (mm)	Area resistance (Ω cm^2)	V(IV) permeability (10^{-7} cm^2 min^{-1})	IEC (mmol g^{-1})	Refs.
Nafion® 117	Dupont, USA	Cation	0.165	2.5	8.63	Nil	[112]
Gore L01854	Gore & Associates, USA	Cation	0.03	0.38	0.36	0.69	[109]
Gore M04494	Gore & Associates, USA	Cation	0.04	0.41	0.96	1.00	[109]
ABT3	Australian Battery Tech. & Trading	Cation	0.02	3.24	0.11	6.01	[109]
ABT4	Australian Battery Tech. & Trading	Cation	0.04	9.97	1.44	3.77	[109]
ABT5	Australian Battery Tech. & Trading	Cation	0.06	5.39	1.44	3.92	[109]
SZ	Guangzhou Delong Technologies Pry Ltd. China	Cation	0.13	19.03	2.34	2.5	[109]
Hipore®	Asahi Kasei, Japan	Microporous separator	0.62	1.4	148	1.14	[109]

PVC-silica microporous separator is another low-cost alternative that has been commercialised by Amer-Sil S.A., Luxembourg. These membranes have been reported to have reasonable performances for several systems, with costs that are 20–40 % of the costs of Nafion® membranes.

2.6.8 *Modified and Composite Membranes*

Since Nafion® membranes are of high cost and suffer from permeation of the electroactive species, modified and composite membranes have been proposed. Table 2.10 summarises recent developments in the membranes used in redox flow batteries.

Lower-cost ion-exchange membranes, such as Daramic® [97, 117] and low-density polyethylene (LDPE) [117, 118], have been modified by employing grafting and sulfonating processes. Although these membranes exhibit high conductivi-

Table 2.10 Modified membranes used in redox flow battery systems

Membrane	Supplier	Type	Thickness (mm)	Area resistance (Ω cm2)	V(IV) permeability (10-7 cm2 min-1)	IEC (mmol g-1)	Refs.
PVDF-g-PSSA-111	Nil	Nil	0.151	Nil	2.20	0.82	[116]
PVDF-g-PSSA-222	Nil	Nil	0.115	Nil	2.53	1.2	[116]
PVDF-g-PSSA-co-PMAc3	Kureha Co. (Japan)	Cation	0.07	Nil	0.73	Nil	[119]
SPEEK	PEEK: Victrex, PEEK450PF	Nil	0.100	1.27	2.432	1.80	[121]
Nafion®/SPEEK	Nafion; PEEK; Victrex, PEEK450PF	Cation	0.100	1.6	1.928	1.67	[121]
ETFE-g-PDMAEMA	ETFE: Kureha Engineering Ltd, Japan DMAEMA, Acros Organic, USA	Anion	0.070	2.3	0.36	Nil	[112]
PSSS-composite5 (concentration 75 g l-1)	Daramic: W.R. Grace & Co. PSSS solution: Aldrich Chemical Company Inc., USA	Nil	Nil	1.09	4.48	Nil	[122]
PSSS-composite6 (concentration 75 g l-1)	Daramic: W.R. Grace & Co. PSSS solution: Aldrich Chemical Company Inc., USA	Nil	Nil	1.36	3.31	Nil	[122]

ties and IECs, ion permeability is still significant [119]. Poly(vinylidene difluoride) (PVDF) was therefore used as a matrix membrane with styrene and maleic anhydride grafting followed by a sulfonation processes. Maleic anhydride was added to reduce irradiation of the membrane. The resulting membranes exhibited reasonable chemical stability and low ion permeability. The degree of grafting tended to increase with water uptake, ion-exchange capacity and conductivity [116] (Table 2.10).

Due to the coulomb repulsion between the cation groups within the membranes and electro-active species in the electrolyte, anion exchange membranes were suggested to further reduce the permeability. Anion-exchange membranes consisting of polysulfone and poly(phenylenesulfidesulfone) have been introduced but the IEC was still less than that of Nafion® 117 [120]. Qiu et al. [101] suggested that an anion monomer can be grafted onto a copolymer membrane by UV-induced grafting. The area resistance of the membrane was reduced by increasing the grafting yield up to 20%. However, Hwang et al. [106] reported that a high degree of cross-linking of the anion New-Selemion® membrane by accelerated electron radiation may lead to failure due to a deterioration in the mechanical properties.

Multi-layered composite membranes have been prepared to increase the chemical resistance. This type of membrane is produced by hot-pressing or/and immersing the substrate membranes in the Nafion®-containing solution. The substrate membranes are usually low cost and have good thermal conductivity. Since SPEEK has a low ion permeability, it was used as substrate for making the Nafion-SPEEK composite membrane (N/S), while a thin Nafion layer was used to prevent oxidation by the highly oxidising species in the electrolytes. Additionally, diamine was used to cross-link the sulphonic acid groups of the Nafion® and SPEEK ionomers. The combination of Nafion and SPEEK can be enhanced by cross-linking the sulphonic acid groups of the Nafion® and SPEEK with diamine [121].

2.6.9 *Flow Distributor and Turbulence Promoter*

A filter-press configuration has frequently been used in redox flow batteries, especially in a stack design. Flow distributors and turbulence promoters are often employed within a filter-press cell to improve mass transport and promote the exchange of species between the bulk solution and electrode surfaces. Turbulence promoters are typically in the form of insulating nets and ribs, although mesh, foam or fibrous-bed electrodes can also be used [110, 114].

Frías-Ferrer et al. [68, 123] studied the effect of four types of polyvinylchloride (PVC) turbulence promoters on mass transport within the rectangular channel of a practical filter-press reactor. Turbulence promoters were particularly useful in large-scale cells, increasing the global mass transport coefficient significantly. The opposite effect was observed in small filter-press cells since the electrolyte flow was not fully developed due to the entrance/exit manifold effect. The decrease in mass transport coefficient had no correlation with either the projected area of the open spaces or the surface blocked by the promoter strands in contact with the electrodes. This could be due to the geometrical features of the fibres, such as their shape and size [68, 123].

The holes in the manifolds were aligned to form the cell inlet and outlet. The resulting change of geometry at the inlet and outlet can generate a high degree of turbulence [123]. Since the presence of the conductive electrolyte in the manifold can lead to a shunt current, electrolyte channels in the frames are sometimes designed to be long and narrow to increase the electrolyte resistance [124]. In the system of

Regenesys Technologies, the by-pass current was reduced by creating a labyrinth in the spiral-shaped paths [125]. Tsuda et al. [126] suggested that the electrical resistance of the electrolyte stream could be increased by flushing inert gas bubbles through the pipelines, without increasing fluid pressure.

The reaction environment in a filter-press polysulfide-bromine battery containing spiral-shaped paths in the manifolds has been investigated. A higher pressure drop across the bromine compartment than with conventional architectures was established, since the spiral in the manifold restricted the electrolyte flow [127]. Although a high pressure drop is often accompanied by enhanced mass transport, it is not favourable since more pumping power is required [125].

2.7 Current Developments in Flow Batteries

2.7.1 Electrolyte Formulation

The electrolyte composition is a crucial factor in battery performance with regard to the specific energy, chemical stability and electrochemistry. In the past decade, mixed acid electrolytes have been used to enhance the solubilities and stabilities of the species in several redox flow battery systems. A notable example is the mixed sulphuric/chloric acid all-vanadium flow battery developed by the Pacific Northwest National Laboratory (PNNL), US, in which 2.5 M vanadium concentrations were achieved for both the negative and positive electrolytes, with high stabilities in the temperature range 5 and 50 °C. KW class prototypes of the system have been commercialised, achieving stable performances with energy efficiencies of *ca.* 85%. More importantly, there was almost a twofold increase in the energy density compared to systems with conventional vanadium electrolytes [128].

Organic acids (e.g., methanesulfonic acid) were also used in several hybrid flow batteries, e.g., zinc-cerium and soluble lead-acid, to enhance the ionic conductivity and solubility of the active species. Methanesulfonic acid allows for higher solubilities of typical metal species, resulting in an increase in the energy density [129]. For instance, cerium (III/IV) methanesulfonates have solubilities higher than 2 M, compared to less than 1 M in sulfuric acid [45].

Complexing agents and additives are often added to the electrolytes to improve the cell performance. For instance, quaternary ammonium salts were added to the zinc-bromine battery to form a low-solubility second liquid phase by associating with polybromide ions. This prevents self-discharge and lowers the bromine vapour pressure [21, 130]. Ce (IV)/(III) [43], Cr (III)/Cr (II) [26] and Fe(III)/Fe(II) [27, 28] were found to have higher reaction rate constants when they were complexed with diethylenetriaminepentaacetate (DPTA), ethylenediaminetetra-acetate (EDTA) and triethanolamine (TEA) ligands, respectively.

Additives are primarily used in redox or hybrid flow batteries to enhance the electrochemical reactions or enable regular and flat electrodeposition layers. Dendritic growth has long been a major issue in redox flow batteries, e.g., the soluble lead-acid and zinc-bromine batteries [110]. When a dendrite on an electrode makes contact with the other electrode or punctures the membrane, an electrical short circuit can result. Additives have been employed to ameliorate dendrite formation, as well as facilitate electrochemical reactions and avoid side reactions. For instance, nickel(II) was reported to reduce the overpotential of lead dioxide deposition in soluble lead-acid batteries [56], while indium compounds were added to reduce the hydrogen overpotential in zinc-cerium flow batteries [37].

2.7.2 *Improvement in Battery Efficiencies*

Battery efficiencies can be improved with a number of recent approaches. Enhancing the reversibility of electrochemical reactions and reducing the overall cell resistance are effective ways to improve the Coulombic and voltage efficiencies, respectively. The introduction of organic materials and their acids tends to increase the solubility of the active species, resulting in higher energy densities [39, 129]. Reduction in the cell resistance can be achieved with the use of improved electrode and membrane materials.

It is often even more effective, however, to minimise the cell resistance using improved or optimised cell architectures. For instance, a zero-gap flow-field design can facilitate mass transport and minimise the Ohmic voltage drop (a major component) across the battery, effectively increasing the power density [131]. However, larger pump powers are often required to overcome the pressure drop. A few membrane-less systems [53–56] have also been introduced to reduce the Ohmic resistance afforded by elimination of membrane. Improved reaction reversibility was observed with the incorporation of some the complexing agents. Due to faster reaction rates and smaller peak separations, higher voltage efficiencies have been achieved [29].

In addition to improving voltage efficiencies, reducing self-discharge, electrolyte mixing, electrodeposit corrosion and gas evolution are common ways to improve Coulombic efficiencies of the RFB systems. Using cross-linked or grafted membranes [106, 116] was effective in preventing electrolyte mixing, while corrosion inhibitors [132] decreased the rate of corrosion. Current efficiency can be enhanced by promoting cell reactions and reducing side reactions with the use of suitable catalytic electrodes [53–56]. In order to avoid short circuit, organic additives have been employed so as to suppress the dendritic growth during charge [37, 56].

2.7.3 *Electrical Distribution System*

Electrical equipment for RFBs often consists of an alternating current (AC)/direct current (DC) converter, pump system, monitoring system, protective relay system and transformer. Economies of scale can be achieved by linking the cells together in series with a bipolar configuration. Modules are linked electrically in series to form a string with the required DC voltage, and linked hydraulically in parallel. The required storage capacities are dependent on the amount and concentration of the electrolytes stored in the reservoirs.

Large-scale flow batteries have been connected directed to the AC distribution system, as shown in Fig. 2.12. A key component is the power conversion system (PCS), which consists of two functional separate and autonomous converter systems: the chopper unit (DC/DC converter) providing the links with the variable voltages of the cell modules and the DC/AC inverter unit (three phase). The two converter units are interconnected by a DC link with a fixed DC voltage level. A control unit is used to maintain the required energy exchange between the energy storage system

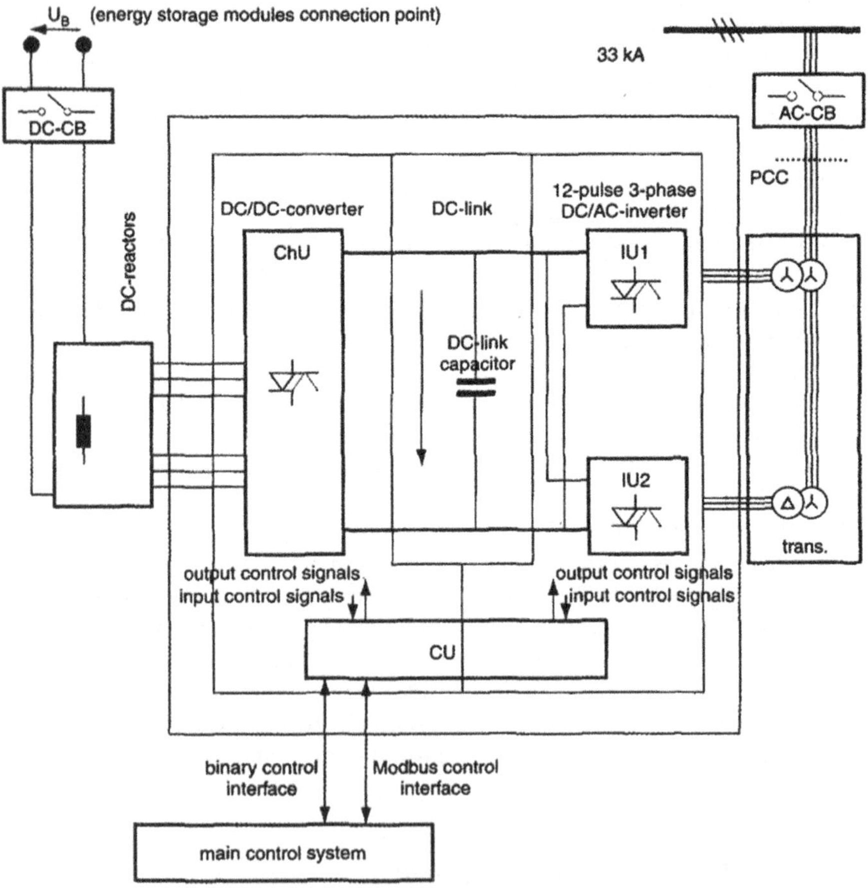

Fig. 2.12 Typical power conversion system for large-scale redox flow batteries [133]

and the grid. The power conversion system is also used for the operator(s) to control the flow battery operation [133].

2.8 Prototypes of Redox Flow Batteries

RFB prototypes are constructed with two key components: (i) the cell stack used to convert the chemical energy into electricity and (ii) the reservoirs that store energy within the electrolytes. Two hydraulic circuits continuously recirculate the electrolyte between the cell stack and the reservoirs. Each electrolyte is recirculated by a centrifugal pump, through which the flow rate is variable. A by-pass pipeline maintains a balance of the electrolyte levels in the reservoirs, off-setting vanadium ion and water crossover, avoiding any pressure differences and re-establishing the desired capacity. The reservoirs are well-sealed and filled with an inert gas (e.g., nitrogen) to avoid oxidation of the reduced species with air.

kW stacks of RFBs consist of a number of cells assembled in a filter-press configuration. These cells are hydraulically parallel-connected thorough internal manifolds, and electrically connected in series with bipolar plates. In order to distribute the electrolyte solutions within the stack, two-dimensional bipolar-plate designs are used with the following choices: 'flow-by' designs, in which the flow channels allow the penetration of the electrolyte solutions into the internal structures of the electrodes, and 'flow-through' designs without flow channels allowing the electrolyte solutions to penetrate by percolation. Commercial stacks adopt one of several flow-field patterns, such as serpentine, interdigitated, equal path length, aspect ratio, corrugated and tapered-interdigitated. These patterns are engraved in the graphite plate. Serpentine and interdigitated flow-field designs are often reported to lead to superior electrochemical performance compared to parallel flow-field designs.

2.9 Applications of Redox and Hybrid Flow Batteries

Despite the introduction of numerous RFBs, only a handful of chemistries (e.g., all-vanadium, zinc-bromine, zinc-iron, zinc-cerium, iron-chromium, bromine-polysulfide and recently all-organic) have been successfully commercialised and operated at kW scales. They are typically used for stationary applications, such as load levelling, power quality control, coupling with renewable energies and uninterrupted power supply [3].

Driven by the increasing adoption of renewable energy generation technologies, the demand for RFBs is expected to increase rapidly in the following decades. There are currently over 50 active flow battery companies, adopting different chemistries and developing systems at kW and MW scales. Recent major installations of these systems have been seen in North America, Europe and Asia.

To date, the largest installations of all-vanadium flow batteries are

1. A 5 MW/10 MW h system developed by Rongke Power in China in 2012.
2. A15 MW/60 MW h system from Hokkaido Electric Power Inc. in Minami Hayakita Substation, Japan in 2015.
3. A 2 MW/20 MW h energy station at Fraunhofer ICT in Germany in 2019.
4. A 100–200 MW/400–800 MW h system built by Rongke Power in Dalian in 2022.

For other chemistries, the notable installations are as follows:

1. Zinc-bromine batteries: 2 MW h built by Redflow Ltd. in California in 2021.
2. Zinc-iron batteries: 1 MW h under construction by ViZn Energy in Puducherry, India.
3. Zinc-cerium batteries: 2 kW built by Plurion Inc. in Glenrothes, Scotland in 2007.
4. Iron-chromium batteries: 250 kW/1.5 MW h built by State Power Investment Corp. in Zhangjiakou, China in 2020.

In addition to the aforementioned mature chemistries using metallic active materials, several start-ups have adopted organic materials. For instance, Jena Batteries has successfully demonstrated a 20 kW/400 kW h organic flow battery prototype in Germany in 2020.

2.10 Summary

RFBs have significant advantages in terms of flexibility, simplicity, low costs and safety over other energy storage systems, such as lithium, nickel-metal hydride and sodium-nickel-chloride (ZEBRA) batteries. Discharge times range from minutes to many hours. Most flow batteries are capable of overload and total discharge without any risk of damage.

Since the 1970s, various RFB systems have been developed. Recent advances in RFBs include the use of organic active species and advanced cell architectures. Tremendous improvements in membranes, electrode materials and electrolyte compositions have been made. Although various RFBs have been introduced, only the all-vanadium, zinc-bromine, zinc-iron, zinc-cerium, iron-chromium, bromine-polysulfide and, more recently, all-organic flow batteries have been successfully scaled up for industrial applications.

RFBs have been used in applications such as load levelling, power quality control and facilitating renewable energy deployment. Due to their capabilities of large power and discharge duration, flow batteries are attractive as enablers of variable renewable energy delivery. The rapidly growing penetration of renewables into grids is expected to give rise to massive market opportunities for RFBs.

References

1. Bottling electricity: Storage as a strategic tool for managing variability and capacity concerns in the modern grid. Technical report, The Electricity Advisory Committee (2008)
2. N. Tokuda, T. Kanno, T. Hara, T. Shigematsu, Y. Tsutsui, A. Ikeuchi, T. Itou, T. Kumamoto, Development of a redox flow battery system. SEI Tech. Rev. **50**, 88–94 (1998)
3. J. Abboud, J. Makansi, Energy storage-the missing link in the electricity value chain, energy storage council white paper (2002)
4. J. Kondoh, I. Ishii, H. Yamaguchi, A. Murata, K. Otani, K. Sakuta, N. Higuchi, S. Sekine, M. Kamimoto, Electrical energy storage systems for energy network - energy conversion and management. Energy Convers. Manag. **41**(17), 1863–1874 (2000)
5. E. McKeogh, A. Gonzalez, B. Gallachir, Study of electricity storage technologies and their potential to address wind energy intermittency in Ireland, sustainable energy Ireland, 2004. Sustainable energy research group, university college cork, 2004, final report (2004)
6. Redox flow cell development and demonstration project, redox flow cell development and demonstration project, calendar year 1977. u.s. dept. of energy, national aeronautics and space administration, nasa tm-79067 1-53. Technical report (1979)
7. T.R. Crompton, *Battery Reference Book Battery Reference Book*, Chap. 14 (Elsevier Science & Technology Books, Boston, Newnes, Oxford, England, 2000)
8. C. Ponce de Leon, A. Frias-Ferrer, J. Gonzalez-Garcia, D.A. Szanto, F.C. Walsh, Redox flow cells for energy conversion. J. Power Sources **160**, 716–732 (2006)
9. A. Joseph, Battery storage systems in electric power systems. ieee power engineering society general meeting (2006)
10. J.Q. Pan, Y.Z. Sun, J. Cheng, Y.H. Wen, Y.S. Yang, P.Y. Wan, Study on a new single flow acid cu-pbo2 battery. Electrochem. Commun. **10**(9), 1226–1229 (2008)
11. A. Hazza, D. Pletcher, R. Wills, A novel flow battery: a lead acid battery based on an electrolyte with soluble lead (ii) part i. preliminary studies. J. Phys. Chem. **6**, 1773–1778 (2004)
12. P.K. Leung, X. Li, C. Ponce de Leon, L. Berlouis, C.T.J. Low, F.C. Walsh, Progress in redox flow batteries, remaining challenges and their applications in energy storage. RSC Adv. **2**, 10125–10156 (2012)
13. L.H. Thaller, Electrically rechargeable redox flow cells, us patent 3996064 (1976)
14. P.C. Butler, D.W. Miller, A.E. Verardo, Flowing-electrolyte-battery testing and evolution, in *Energy Conversion Eng, editor, 17th Intersoc, Los Angeles, Conf* (1982)
15. P.M. Spaziante, A. Pelligri, To oronzio de nori impianti elettrochimici s.p.a., gb patent 2030349. (1978)
16. M. Skyllas-Kazacos, F. Grossmith, Efficient vanadium redox flow cell. J. Electrochem. Soc. **134**(12), 2950 (1987)
17. M. Skyllas-Kazacos, M. Rychcik, R.G. Robins, A. Fane, M. Green, New all-vanadium redox flow cell. J. Electrochem. Soc. **133**(5), 1057 (1986)
18. V-fuel pty ltd., "status of energy storage technologies as enabling systems for renewable energy from the sun, wind, waves and tides." House of representatives standing committee on industry and resources
19. B. Fang, S. Iwasa, Y. Wei, T. Arai, M. Kumagai, A study of the ce (iii)/ce (iv) redox couple for redox flow battery. Electrochim. Acta **47**, 3971–3976 (2002)
20. F.Q. Xue, Y.L. Wang, W.H. Wang, X.D. Wang, Investigation on the electrode process of the mn(ii)/mn(iii) couple in redox flow battery. Electrochim. Acta **53**, 6636–6642 (2008)
21. R.F. Koontz, R.D. Lucero, *Handbook of Batteries*, Chap. 39 (McGraw Hill, 1995)
22. D. Pletcher, R. Wills, A novel flow battery: a lead acid battery based on an electrolyte with soluble lead (ii) part ii. Flow cell studies. Phys. Chem. **6**, 1779–1785 (2004)
23. B. Huskinson, M.P. Marshak, C. Suh, S. Er, M.R. Gerhardt, C.J. Galvin, X. Chen, A. Aspuru-Guzik, R.G. Gordon, M.J. Aziz, A metal-free organic-inorganic aqueous flow battery. Nature **505**(7482), 195–198 (2014)

24. K.X. Lin, Q. Chen, M.R. Gerhardt, L.C. Tong, S.B. Kim, L. Eisenach, A.W. Valle, D. Hardee, R.G. Gordon, M.J. Aziz, M.P. Marshak, Alkaline quinone flow battery. Science **349**(6255), 1529–1532 (2015)
25. B. Yang, L. Hoober-Burkhardt, F. Wang, G.K. Surya Prakash, S.R. Narayanan, An inexpensive aqueous flow battery for large-scale electrical energy storage based on water-soluble organic redox couples. J. Electrochem. Soc. **161**(9), A1371–A1380 (2014)
26. M. Futamata, S. Higuchi, O. Nakamura, I. Ogino, Y. Takeda, S. Okazaki, S. Ashimura, S. Takahashi, J. Power Sources **24**, 137 (1988)
27. Y.H. Wen, H.M. Zhang et al., a study of the fe (iii)/ fe (ii) - triethanolamine complex redox couple flow battery application. Electrochim. Acta **51**(18), 3769–3775 (2006)
28. Y.H. Wen, H.M. Zhang et al., Studies on iron (fe3+/ fe2+)-complex/ bromine (br2/ br-) redox flow cell in sodium acetate solution. J. Electrochem. Soc **153**(5), A929–A934 (2006)
29. P. Modiba, A.M. Crouch, Electrochemical study of cerium(iv) in the presence of ethylenediaminetetraacetic acid (edta) and diethylenetriaminepentaacetate (dtpa) ligands. J. Appl. Electrochem. **38**(9), 1293–1299 (2008)
30. C.H. Bae, E.P.L. Roberts, R.A.W. Dryfe, Chromium redox couples for application to redox flow batteries. Electrochim. Acta **48**(3), 279–287 (2002)
31. G. Codina, J.R. Perez, M. Lopez-Atalaya, J.L. Vazquez, A. Aldaz, J. Power Sources **48**, 293 (1994)
32. P. Garces, M.A. Climent, A. Aldaz, An. Quim. Sistemas de almacenamiento de energıa **83**, 9 (1987)
33. M. Kazacos, M. Skyllas-Kazacos, Performance characteristics of carbon plastic electrodes in the all-vanadium redox cell. J. Electrochem. Soc **136**, 2759–2760 (1989)
34. B. Fang, S. Iwasa, Y. Wei, T. Arai, M. Kumagai, A study of the ce (iii)/ ce (iv) redox couple for redox flow battery application. Electrochim. Acta **47**(24), 3971–3976 (2002)
35. P. Zhao, H.M. Zhang, H.T. Zhou, B.L. Yi, Nickel foam and carbon felt applications for sodium polysulfide/bromine redox flow battery electrodes **51**(6), 1091–1098 (2005)
36. R.L. Clarke, B.J. Dougherty, S. Harrison, J.P. Millington, S. Mohanta, Battery with bifunctional electrolyte, us 2006/0063065 a1 (2005)
37. R.L. Clarke, B.J. Dougherty, S. Harrison, J.P. Millington, S. Mohanta, Cerium batteries, us 2004/ 0202925 a1 (2004)
38. M. Skyllas-Kazacos, Novel vanadium chloride/polyhalide redox flow battery. J. Power Sources **124**(1), 299–302 (2003)
39. P. Leung, A.A. Shah, L. Sanz, C. Flox, J.R. Morante, Q. Xu, M.R. Mohamed, C.P.d. Leon, F.C. Walsh, Recent developments in organic redox flow batteries: a critical review **360**, 243–283 (2017)
40. J. Doria, M.C.D. Andres, C. Armenta, Proc. 9th solar energy soc. **3**, 1500 (1985)
41. M. Skyllas-Kazacos, M. Rychcik, R. Robins, Au patent 575247 (1986)
42. M. Skyllas-Kazacos, C. Menictas, The vanadium redox battery for emergency back-up applications, in *19th International Telecommunications Energy Conference, INTELEC 97* (1997), pp. 463–471
43. M. Kazakos, M. Skyllas-Kazacos, A. Mousa, Metal bromide redox flow cell. pct application, 2003, pct/gb2003/001757 (2003)
44. A. Paulenova, S.E. Creager, J.D. Navratil, Y. Wei, Redox potentials and kinetics of the ce3+/ce4+ redox reaction and solubility of cerium sulfates in sulfuric acid solutions. J. Power Sources **109**(2), 431–438 (2002)
45. R.P. Kreh, R.M. Spotnitz, J.T. Lundquist, Mediated electrochemical synthesis of aromatic aldehydes, ketones, and quinones using ceric methanesulfonate. J. Org. Chem. **54**(7), 1526–1531 (1989)
46. F.C. Walsh, Electrochemical technology for environmental treatment and clean energy conversion. Pure Appl. Chem **73**(12), 1819–1837 (2001)
47. T. Yamamura, Y. Shiokawa, H. Yamana, H. Moriyama, Electrochemical investigation of uranium?-diketonates for all-uranium redox flow battery. Electrochim. Acta **48**(1), 43–50 (2002)

48. Y. Shiokawa, T. Yamamura, K. Shirasaki, Energy efficiency of an uranium redox-flow battery evaluated by the butler-volmer equation. J. Phys. Soc. Jpn. **75**, 137–142 (2006)
49. T. Yamamura, N. Watanabe, Y. Shiokawa, Energy efficiency of neptunium redox battery in comparison with vanadium battery. J. Alloys Compd. **408**, 1260–1266 (2006)
50. T. Yamamura, N. Watanabe, T. Yano, Y. Shiokawa, Electron-transfer kinetics of np [sup 3+] np [sup 4+], npo [sub 2][sup+]? npo [sub 2][sup 2+], v [sup 2+] v [sup 3+], and vo [sup 2+]? vo [sub 2][sup+] at carbon electrodes. J. Electrochem. Soc **152**(4), A830 (2005)
51. K. Hasegawa, A. Kimura, T. Yamamura, Y. Shiokawa, Estimation of energy efficiency in neptunium redox flow batteries by the standard rate constants. J. Phys. Chem. Solids **66**(2–4), 593–595 (2005)
52. C. Lotspeich, A comparative assessment of flow battery technologies. Proceedings of the electrical energy storage systems applications and technologies, in San Francisco, editor, *International Conference 2002 (EESAT2002)* (2002)
53. D. Pletcher, R. Wills, A novel flow battery: a lead acid battery based on an electrolyte with soluble lead (ii) part ii. flow cell studies. Phys. Chem. Chem. Phys. **6**(8), 1779–1785 (2004)
54. A. Hazza, D. Pletcher, R. Wills, A novel flow battery: a lead acid battery based on an electrolyte with soluble lead(ii) part i: preliminary studies. Phys. Chem. Chem. Phys **6**, 1773–1778 (2004)
55. D. Pletcher, R. Wills, A novel flow battery-a lead acid battery based on an electrolyte with soluble lead(ii): Iii. the influence of conditions on battery performance. J. Power Sources **149**, 96–102 (2005)
56. A. Hazza, D. Pletcher, R. Wills, A novel flow battery-a lead acid battery based on an electrolyte with soluble lead(ii): Iv. the influence of additives. J. Power Sources **149**, 103–111 (2005)
57. D. Pletcher, H.T. Zhou, G. Kear, C.T.J. Low, F.C. Walsh, R.G.A. Wills, A novel flow battery - a lead-acid battery based on an electrolyte with soluble lead(ii) part vi. studies of the lead dioxide positive electrode. J. Power Sources **180**(1), 630–634 (2008)
58. J. Cheng, L. Zhang, Y.S. Yang, Y.H. Wen, G.P. Cao, X.D. Wang, Preliminary study of single flow zinc-nickel battery. Electrochem. Commun. **9**(11), 2639–2642 (2007)
59. L. Zhang, J. Cheng, Y.S. Yang, Y.H. Wen, X.D. Wang, G.P. Cao, Study of zinc electrodes for single flow zinc/ nickel battery application. J. Power Sources **179**(1), 381–387 (2008)
60. P.C. Symons, Soc. electrochem, in *International Conference on electrolytes for power sources, Brighton*. Soc. Electrochem (1973)
61. P.C. Symons, Process for electrical energy using solid halogen hydrates, usp- 3713,888 (1970)
62. http://www.zbbenergy.com/
63. J. Jorn, J.T. Kim, D. Kralik, The zinc-chlorine battery: half-cell overpotential measurements. J. Appl. Electrochem. **9**, 573–579 (1979)
64. H.S. Lim, A.M. Lackner, R.C. Knechtli, Zinc-bromine secondary battery. J. Electrochem. Soc. **124**(8), 1154–1157 (1977)
65. H.T. Zhou, H.M. Zhang, P. Zhao, B.L. Yi, A comparative study of carbon felt and activated carbon based electrodes for sodium polysulfide/bromine redox flow battery. Electrochim. Acta **51**(28), 6304–6312 (2006)
66. V-fuel pty ltd., house of representatives standing committee on industry and resources
67. L.W. Hruska, R.F. Savinell, Investigation of factors affecting performance of the iron-redox battery. J. Electrochem. Soc. **128**(1), 18–25 (1981)
68. A. Frias-Ferrer, J. Gonzalez-Garcaa, V. Suez, C. Ponce de Leon, F.C. Walsh, The effects of manifold flow on mass transport in electrochemical filter-press reactors. AIChE J. **54**(3), 811–823 (2008)
69. Y.M. Zhang, Q.M. Huang, W.S. Li, H.Y. Peng, S.J. Hu, Graphite-acetylene black composite electrodes for all vanadium redox flow battery. J. Inorg. Mater **22**, 1051–1055 (2007)
70. M. Rychcik, M. Skyllas-Kazacos, Evaluation of electrode materials for vanadium redox cell. J. Power Sources **19**(1), 45–54 (1987)
71. B. Sun, M. Skyllas-Kazacos, Modification of graphite electrode materials for vanadium redox flow battery application i. thermal treatment. Electrochimica acta **37**(7), 1253–1260 (1992)
72. H. Kaneko, K. Nozaki, Y. Wada, T. Aoki, A. Negishi, M. Kamimoto, Vanadium redox reactions and carbon electrodes for vanadium redox flow battery. Electrochimica Acta **36**(7), 1191–1196 (1991)

73. J. Cathro, K. Cedzynska, D.C. Constable, Preparation and performance of plastic-bonded-carbon bromine electrodes. J. Power Sources **19**, 337 (1987)
74. H. Zhou, H. Zhang, P. Zhao, B. Yi, A comparative study of carbon felt and activated carbon based electrodes for sodium polysulfide/bromine redox flow battery. Electrochimica Acta **51**(28), 6304–6312 (2006)
75. V. Haddadi-Asl, M. KAZACos, M. Skyllas-Kazacos, Conductive carbon-polypropylene composite electrodes for vanadium redox battery. J. Appl. Electrochem. **25**(1), 29–33 (1995)
76. W.H. Wang, X.D. Wang, Investigation of ir-modified carbon felt as the positive electrode of an all-vanadium redox flow battery (ir-modified carbon felt). Electrochim. Acta **52**(24), 6755–6762 (2007)
77. L. Joerissen, J. Garche, C. Fabjan, G. Tomazic, Possible use of vanadium redox-flow batteries for energy storage in small grids and stand-alone photovoltaic systems. J. Power Sources **127**, 98–104 (2004)
78. H. Kaneko, K. Nozaki, A. Negishi, Y. Wada, T. Aoki, M. Kamimoto, Vanadium redox reactions and carbon electrodes for vanadium redox flow battery. Electrochimica Acta **36**(7), 1191–1196 (1991)
79. X. Li, K. Horita, Electrochemical characterization of carbon black subjected to rf oxygen plasma. Carbon **38**(1), 133–138 (2000)
80. M. Santiago, F. Stuber, A. Fortuny, A. Fabregat, J. Font, Modified activated carbons for catalytic wet air oxidation of phenol. Carbon **43**(10), 2134–2145 (2005)
81. K. Jurewicz, K. Babel, A. Ziolkowski, H. Wachowska, Ammoxidation of active carbons for improvement of supercapacitor characteristics. Electrochimica Acta **48**(11), 1491–1498 (2003)
82. N.S. Jacobson, D.M. Curry, Oxidation microstructure studies of reinforced carbon/carbon. Carbon **44**(7), 1142–1150 (2006)
83. X.G. Li, K.L. Huang, S.Q. Liu, L.Q. Chen, Electrochemical behavior of diverse vanadium ions at modified graphite felt electrode in sulphuric solution. J. Cent. South Univ. Technol. **14**(1), 51–56 (2007)
84. M. Skyllas-Kazacos, F. Grossmith, Efficient vanadium redox flow cell. J. Electrochem. Soc **134**(12), 2950–2953 (1987)
85. C.M. Hagg, M. Skyllas-Kazacos, Novel bipolar electrodes for battery applications. J. Appl. Electrochem **32**(10), 1063–1069 (2002)
86. K. Kinoshita, S.C. Leach, Mass transport of carbon-felt flow through electrode. Electrochem. Soc., J. **129**, 1993–1997 (1982)
87. M. Rychcik, M. Skyllas-Kazacos, Evaluation of electrode materials for vanadium redox cell. J. Power Sources **19**(1), 45–54 (1987)
88. K.J. Kim, M.S. Park, Y.J. Kim, J.H. Kim, S.X. Dou, M. Skyllas-Kazacos, A technology review of electrodes and reaction mechanisms in vanadium redox flow batteries. J. Mater. Chem. A **3**, 16913–16933 (2015)
89. Z. He, L. Liu, C. Gao, Z. Zhou, X. Liang, Y. Lei, Z. He, S. Liu, Carbon nanofibers grown on the surface of graphite felt by chemical vapour deposition for vanadium redox flow batteries. RSC Adv. **3**(43), 19774–19777 (2013)
90. R. Wang, Y.S. Li, Y.L. He, Achieving gradient-pore-oriented graphite felt for vanadium redox flow batteries: meeting improved electrochemical activity and enhanced mass transport from nano- to micro-scale. J. Mater. Chem. A **7**, 10962–10970 (2019)
91. B. Sun, M. Skyllas-Kazacos, Chemical modification and electrochemical behaviour of graphite fibre in acidic vanadium solution. Electrochim. Acta **36**, 513–517 (1991)
92. C. Fabjan, J. Garche, B. Harrer, L. Jorissen, C. Kolbeck, F. Philippi, G. Tomazic, F. Wagner, The vanadium redox-battery: an efficient storage unit for photovoltaic systems. Electrochim. Acta **47**(5), 825–831 (2001)
93. J.M. Friedrich, C. Ponce de Leon, G.W. Reade, F.C. Walsh, Reticulated vitreous carbon as an electrode material. J. Electroanal. Chem. **561**, 203–217 (2004)
94. M. Mastragostino, S. Valcher, Polymeric salt as bromine complexing agent in a zn-br 2 model battery. Electrochim. Acta **28**, 501–505 (1983)

95. Y. Liu, X. Xia, H. Liu, Studies on cerium (ce4+/ce3+) -vanadium (v2+/v3+) redox flow cell-cyclic voltammogram response of ce4+/ce3+ redox couple in h2so4 solution. J. Power Sources **130**(1–2), 299–305 (2004)
96. X.G. Li, K.L. Huang, S.Q. Liu, L.Q. Chen, Electrochemical behavior of diverse vanadium ions at modified graphite felt electrode in sulphuric solution. J. Cent. South Univ. Technol. (English Edition) **14**(1), 51–56 (2007)
97. B. Tian, F.H. Wang, C.W. Yan, Proton conducting composite membrane from daramic/nafion for vanadium redox flow battery. J. Membr. Sci. **234**(1-2), 51–54 (2004)
98. M. Skyllas-Kazacos, *wo/1989/005526, 47." PCT Int. Appl., 1989* (1989)
99. S.H. Ge, B.L. Yi, H.M. Zhang, Study of a high power density sodium polysulfide/bromine energy storage cell. J. Appl. Electrochem. **34**(2), 181–185 (2004)
100. C.M. Hagg, M. Skyllas-Kazacos, Novel bipolar electrodes for battery applications. J. Appl. Electrochem. **32**(10), 1063–1069 (2002)
101. K. Fushimi, H. Tsunakaw, K. Yonahara, Electrically conductive plastic complex material us pat, 4551267 (1985)
102. G. Tomazic, Process for the manufacture of bipolar electrodes and separators us pat, 4615108 (1986)
103. C. Herscovici, Porous and porous-nonporous composites for battery electrodes. *US Pat*, 4920017 (1990)
104. C. Herscovici, A. Leo, A. Charkey, Stable carbon-plastic electrodes and method of preparation thereof us pat. 4758473 (1988)
105. G. Iemmi, D. Macerata, Graphite-resin composite electrode structure, and a process for its manufacture, us pat. 4294893 (1981)
106. G.J. Hwang, H. Ohya, Crosslinking of anion exchange membrane by accelerated electron radiation as a separator for the all-vanadium redox flow battery. J. Membr. Sci. **132**(1), 55–61 (1997)
107. D.G. Oei, Permeation of vanadium cations through anionic and cationic membranes. J. Appl. Electrochem. **15**, 231–235 (1985)
108. S.C. Chieng, M. Kazacos, M. Skyllas-Kazacos, Preparation and evaluation of composite membrane for vanadium redox battery applications. J. Power Sources **39**, 11–19 (1992)
109. H. Vafiadis, M. Skyllas-Kazacos, Evaluation of membranes for the novel vanadium bromine redox flow cell. J. Membr. Sci. **279**(1–2), 394–402 (2006)
110. F.C. Walsh, *A First Course in Electrochemical Engineering* (Electrochemical Consultancy, UK, 1993)
111. S.C. Chieng, Ph.D. thesis, University of New South Wales, Sydney, Australia (1993)
112. J.Y. Qiu, M.Y. Li, J.F. Ni, M.L. Zhai, J. Peng, L. Xu, H.H. Zhou, J.Q. Li, G.S. Wei, Preparation of etfe-based anion exchange membrane to reduce permeability of vanadium ions in vanadium redox battery. J. Membr. Sci. **297**, 174–180 (2007)
113. H. Tasai, T. Horigome, N. Nozaki, H. Kaneko, A. Negishi, Y. Wada, Characteristics of vanadium redox flow cell, in *The 31th Denchi Touron Kouengai Yousisyu* (Japan, 1990), pp. 301–302
114. M. Skyllas-Kazacos, D. Kasherman, D.R. Hong, M. Kazacos, Characteristics and performance of 1 kw unsw vanadium redox battery. J. Power Sources **35**, 399–404 (1991)
115. T. Mohammadi, M. Skyllas-Kazacos, Evaluation of the chemical stability of some membranes in vanadium solution. J. Appl. Electrochem. **27**(2), 153–160 (1997)
116. X.L. Luo, Z.Z. Lu, J.Y. Xi, Z.H. Wu, W.T. Zhu, L.Q. Chen, X.P. Qiu, Influences of permeation of vanadium ions through pvdf-g-pssa membranes on performances of vanadium redox flow batteries. J. Phys. Chem. B **109**(43), 20310–20314 (2005)
117. T. Mohammadi, M. Skyllas-Kazacos, Preparation of sulfonated composite membrane for vanadium redox flow battery applications. J. Membr. Sci. **107**(1–2), 35–45 (1995)
118. G.J. Hwang, H. Ohya, Preparation of cation exchange membrane as a separator for the all-vanadium redox flow battery. J. Membr. Sci. **120**(1), 55–67 (1996)
119. J.Y. Qiu, L. Zhao, M.L. Zhai, J.F. Ni, H.H. Zhou, J. Peng, J.Q. Li, G.S. Wei, Pre-irradiation grafting of styrene and maleic anhydride onto pvdf membrane and subsequent sulfonation for application in vanadium redox batteries. J. Power Sources **177**(2), 617–623 (2008)

120. G.J. Hwang, H. Ohya, Preparation of anion-exchange membrane based on block copolymers. Part 1. amination of the chloromethylated copolymers. J. Membr. Sci. **140**, 195–203 (1998)
121. Q.T. Luo, H.M. Zhang, J. Chen, D.J. You, C.X. Sun, Y. Zhang, Preparation and characterization of nafion/speek layered composite membrane and its application in vanadium redox flow battery. J. Memb. Sci. **325**, 553–558 (2008)
122. T. Mohammadi, M. Skyllas-Kazacos, Use of polyelectrolyte for incorporation of ion-exchange groups in composite membranes for vanadium redox flow battery applications. J. Power Sources **56**(1), 91–96 (1995)
123. A. Fraas-Ferrer, J. Gonzalez-Garcia, V.S.E. Exposito, C.M. Sanchez-Sanchez, V. Montiel, A. Aldaz, F.C. Walsh, The entrance and exit effects in exit effects in small electrochemical filter-press reactors used in the laboratory. J. Chem. Edu. **82**, 1395–1398 (2005)
124. A. Leo, Status of zinc-bromine battery development, in Energy Conversion Engineering Conference, editor, *Proceedings of the 24th Intersociety , Energy Research Corporation, IECEC-89*, 3 (1989), pp. 1303–1309
125. A. Ponce de Leon, G.W. Reade, I. Whyte, S.E. Male, F.C. Walsh, Characterization of the reaction environment in a filter-press redox flow reactor. Electrochim. Acta **52**(19), 5815–5823 (2007)
126. I. Tsuda, K. Kurokawa, K. Nozaki, Development of intermittent redox flow battery for pv system, in *Photovoltaic Energy Conversion, Conference Record of the Twenty Fourth IEEE Photovoltaic Specialists Conference - 1994, 1994 IEEE First World Conference* 1, 1994 (1994), pp. 946–949
127. R.A. Scannell, F.C. Walsh, Comparative mass transfer and electrode area in electrochemical reactors. Inst. Chem. Engr. Symp. Ser. **112**, 59–71 (1989)
128. L.Y. Li, S.W. Kim, W. Wang, M. Vijaayakumar, Z.M. Nie, B.W. Chen, J.L. Zhang, G.G. Xia, J.Z. Hu, G. Graff, J. Liu, Z.G. Yang, A stable vanadium redox-flow battery with high energy density for large-scale energy storage. Adv. Energy Mater. **1**(3), 394–400 (2011)
129. M.D. Gernon, M. Wu, T. Buszta, P. Janney, Environmental benefits of methanesulfonic acid. Green Chem. **1**, 127–140 (1999)
130. K.V. Kordesch, C. Fabjan, J. Daniel-Ivad, J. Oliveira, Rechargeable zinc-carbon hybrid cells. J. Power Sources **65**, 77–80 (1997)
131. D.S. Aaron, Q. Liu, Z. Tang, G.M. Grim, A.B. Papandrew, A. Turhan, T.A. Zawodzinski, M.M. Mench, Dramatic performance gains in vanadium redox flow batteries through modified cell architecture. J. Power Sources **206**, 450–453 (2012)
132. P.R. Roberge, *Handbook of Corrosion Engineering*, Chap. 10 (McGraw-Hill,, 2000)
133. A. Price, S. Bartley, S. Male, G. Cooley, A novel approach to utility scale energy storage. Power Eng. J. **13**(3), 122–129 (1999)

Chapter 3
Modelling Methods for Flow Batteries

3.1 Introduction

Modelling and simulation can play a major role in the design, analysis, control, scale-up and optimisation of a host of technologies, including fuel cells and batteries [1, 2]. Experimental investigations and purely laboratory-based design can be both time consuming and costly. Modelling and simulation are tools that can help to reduce the timescales and financial costs, informing initial designs, rationalising observed phenomena to improve design concepts, and optimising designs or at least narrowing the window in which a search should be conducted in the laboratory.

There are, moreover, cases in which information is not available from experiments, such as the distributions of important quantities inside electrodes. Consider for example the problems of avoiding fuel starvation or local hotspots in temperature caused by elevated electric potentials in a flow battery. Solutions to such problems can be aided by modelling, since it is able to simulate (approximately) the evolution of the 3D reactant and potential distributions inside the electrodes and flow channels. Gaining such knowledge from experiments, on the other hand, is currently impossible.

Optimising flow battery designs with respect to performance, degradation and costs involves many variables and tradeoffs. The number of design parameters is vast, including those related to the component materials, redox species, geometrical configurations, electrolyte additives, flow field design, heat management strategies, and electrode designs and treatments. The costs and durability of the materials and components must also be taken into account at some stage. Moreover, the operating conditions under which the battery operates will affect the efficiency and longevity of the battery for a given design. Performing an optimisation over all or a subset of the design parameters and operating conditions is not feasible within reasonable timeframes and available budgets, necessitating computational approaches.

A. A. Shah et al., *New Paradigms in Flow Battery Modelling*,
Engineering Applications of Computational Methods 16,
https://doi.org/10.1007/978-981-99-2524-7_3

The latter are of course not a panacea, since computational models are only *approximations* of the actual physical processes inside the battery, limited by a number of factors. To quote the great George Cox

> '... *all models are approximations. Essentially, all models are wrong, but some are useful. However, the approximate nature of the model must always be borne in mind...*'

Another thing that has to be borne in mind is *Occam's razor*. Excessively elaborate models often fail to accurately depict the 'truth', or their predictive power fails to extend beyond the narrow windows of design and operating parameter spaces for which they were constructed and validated, i.e., they lack *transferability*. In all modelling there is a danger of reaching a point beyond which the model ceases to be useful or elucidating, as a result of its sheer complexity or over-parameterisation. This is especially true of models that depict complex devices such as batteries and fuel cells. Many of the phenomena are only partially understood, and in some cases hardly understood at all.

Examples of these phenomena include the detailed kinetics (rate limiting steps, side reactions, degradation phenomena), phase changes such as gas bubble formation, the transport of ions and water through membranes and separators, passive layer formation, the transport of species in multiphase flows, especially with concentrated solutions, the double-layer capacitance at electrode/electrolyte interfaces, fluid-structure interaction in porous electrodes, and the exact influence of surface treatments and functionalisation on charge transfer. Modelling should ideally be targeted towards specific problems so that whenever possible minimalist representations of the underlying physical processes are obtained, with a good understanding of the physics involved and access to parameter values or empirical methods to obtain them.

There is a vast array of available methods for solving scientific and engineering problems, some of which are applicable broadly and some of which are specialised to certain types of problems. The approach taken depends on several factors, including the ultimate goal of the modelling, the scales involved, the available computational budget and time, the required level of accuracy, the availability of software, and the numerical stability of a particular method for certain types of problems.

In Sect. 3.2 we provide an overview of physics-based models, in terms of the spatio-temporal scales involved. In Sect. 3.3, these methods are then discussed in detail, from the macroscopic level down to the individual-atom level. Further details on macroscopic models specific to flow batteries, which involve charge conservation in addition to mass, energy and momentum, and often flow in porous media, are provided in Chap. 4.

Physics-based models are the bedrock of theoretical analyses in science and engineering, but in the last decade there has been an enormous growth in the use of data-driven (or machine learning) methods to tackle a broad range of problems in these disciplines. This includes many problems that have traditionally been approached exclusively using physics-based methods, and some emerging problems such as classifying faults in cells and predicting their end-of-life. In fact, many of these data-driven methods have a long history in aerospace engineering, geospatial statistics and pharmaceutical science, but have recently found their way into other disciplines. In Sect. 3.7 we provide an outline of machine learning methods, which are at the heart of data science, analytics and the algorithmic component of artificial intelligence. This section is meant as a preparation for a more detailed presentation of machine learning methods in Chap. 6.

Some alternative approaches to machine learning for certain tasks, namely multi-fidelity and reduced-order models are covered in Chap. 7. These methods are a mixture of physics-based and data-driven approaches. Also covered in Chap. 7 is the specialised topic of time series methods, dealing with sequential data (usually but not always in time). Such methods can be approached using machine learning, or with more tailored (but related) approaches widely used in signal processing and financial forecasting. Finally, we refer to Appendices A–D for basic-level details on linear systems, time-stepping methods, methods for solving partial differential equations and gradient-based optimisation.

3.2 Overview of Available Physics-Based Modelling Approaches

Many different types of *physics-based* modelling approaches are available, from the nanoscale up to a systems scale. There is also a connection to the time scale involved but we will focus on the spatial scales. We will exclude the systems level in this book. With a vast hierarchy of methods and approaches available, as illustrated in Fig. 3.1, it would be an impossible task to faithfully review them all in depth, so we focus primarily on the device level down to the materials level.

What we mean by physics-based is that the models are built upon intuitive, axiomatic or empirically-determined physical laws. At the most fundamental level we have conservation principles such as those of mass, energy, momentum and charge, but there are also phenomenological laws established in areas such as heat and mass transfer, thermodynamics, solid mechanics, fluid mechanics, particle physics and reaction kinetics. Examples of these include Fick's law of diffusion, Arrhenius's law, Fourier's law of heat conduction, the ideal gas law, Hooke's law and countless others. These are experimentally verified or derived from fundamental principles, and while they may not hold in all cases, they have a firm basis in science.

For the device level, we would usually be interested in *macroscopic* models, almost always based on the principles of *continuum* mechanics, which can be subdivided into fluid and solid mechanics, as well as classical thermodynamics. Here, 'macroscopic' refers to what can be seen by the naked eye, so encompasses component level, device level and higher. At such levels, we require 'averaged' information about the matter contained in the system under consideration. These parsimonious (in some sense) descriptions serve two purposes. Firstly, a microscopic or smaller level description would involve an unwieldy number of variables. The number of molecules involved in a typical system at macroscopic level will be on the order of 10^{20} and

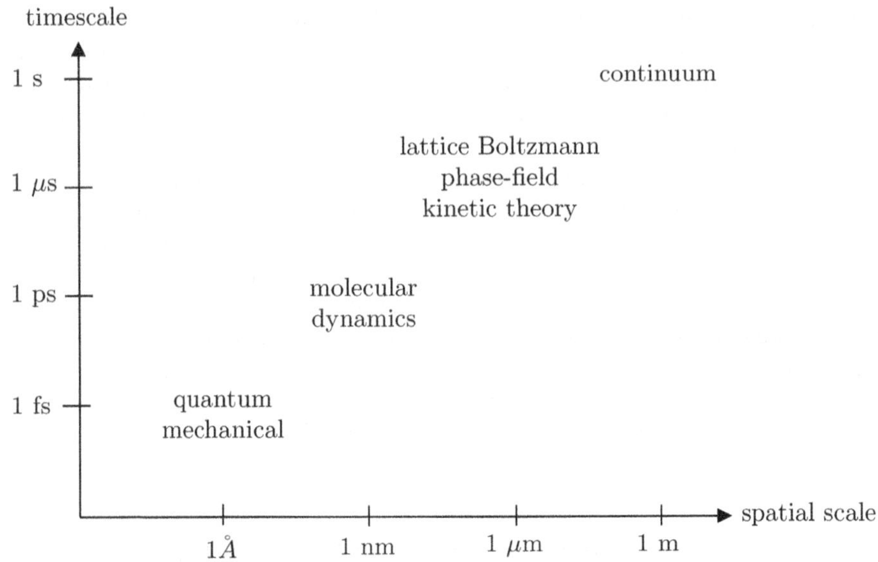

Fig. 3.1 Illustration of the scales involved in different physics-based modelling paradigms

higher, necessitating a microscopic description that includes 6×10^{20} positions and velocities, ignoring quantum effects.

Moreover, this level of information is generally not very useful in describing the system behaviour and response. What we are really interested in is observable quantities such as the temperature, velocity, current density and pressure. A continuum level model neglects the discrete-molecule nature of a material and adopts a smoothed-out approach of the conservation laws, representing microscopic information by averaging over a volume and introducing constitutive relations. These constitutive relations can be modelled empirically, but there may be cases in which the descriptions they provide are incomplete insofar capturing crucial information regarding the microstructure. Furthermore, they require accurate characterisation and parameterisation, which presents a challenge in itself.

Another point that needs to be made is that macroscopic models themselves are infeasible on very large 'systems-level' scales. What we mean here by systems level is the combination of various sub-technologies into a single 'system', as in an electrical system consisting of a network of electrical components to generate, distribute and use electric power. For example, it is impractical to study a system involving a stack of 100 flow batteries incorporating the hydraulic system, a heat management system, and the reservoirs using a detailed continuum model in 3D. For such problems, the macroscopic model requires simplification, primarily a simplification of the conservation laws, especially with regard to spatial distributions. As stated earlier, such models, including electrical systems, will not be covered.

Moving to a smaller scale, there are three main approaches, namely, kinetic theory [3], phase-field models [4] and the lattice-Boltzmann method (LBM) [5]. Such methods are often called *mesoscopic* or *pore scale*, in which mesoscopic refers to a lengthscale intermediate between the macroscopic and atomistic scales. Phase-field models are popular in problems involving moving interfaces or free boundaries, as may exist in a multiphase flow of a gas and liquid. The surfaces and interfaces are described by scalar fields (*order parameters*) that take on constant values in the bulk phases and have a continuous variation across diffuse fronts.

In a two-phase system, the microstructure is therefore represented using a single, continuous order parameter. The evolution of the phase-field is taken to be proportional to the functional derivative of the free energy, provided it can be expressed as a functional of the order parameter. Such methods would be well-suited to studying the evolution of deposition surfaces on metal-based electrodes, particularly the phenomenon of dendritic growth.

The LBM has become a popular approach for elucidating the behaviour of complex fluid flows, with extensions to describe mass, heat and charge transport phenomena. It is a bottom-up approach in which fluid particle interactions are simulated in a discrete phase space by defining a set of discrete velocities. Several choices exist for this set of velocities, which is dependent on the dimension of the domain and must satisfy certain requirements related to the recovery of the equivalent macroscopic equations. There is also a phase-field version of the LBM for multiphase flows, introducing an order parameter that is usually governed by the Cahn-Hilliard equation [6].

The LBM grew out of earlier lattice models, especially lattice gas cellular automata. It is able to handle complex boundaries more easily than conventional solutions to the Navier-Stokes equations, which is the primary reason for its popularity in the simulation of flow in porous media, including the gas diffusion layers in fuel cells and porous electrodes in batteries.

A third type of mesoscopic modelling is kinetic theory, based on the probability density of finding a given fluid 'particle' at a given position and time, moving with a given velocity. The most famous example is provided by the Boltzmann equation, which is derived under the assumption of a dilute mixture. The main challenge lies in modelling the particle-particle interactions through a collision operator, which is a nonlinear integral in velocity space, making numerical implementations difficult. It therefore remains a highly specialised topic. The LBM method is essentially a discrete version of the Boltzmann equation, in a discrete velocity space, with a simplified collision operator.

At the atomistic scale, the go-to-method is *molecular dynamics* (MD), which simulates the evolution of a system of particles based on Newton's laws of motion, relying on a description of the interatomic potential known as the force field [7]. MD simulations provide information at the microscopic level, namely the atomic positions and velocities, which can be converted to macroscopic variables such as a temperature or a stress tensor using statistical mechanics. MD has been used extensively to determine reaction rates, study defect formation, investigate the lithium insertion process in Li-ion batteries, characterise protein folding and much more. Although usually empirically derived, the force field can be determined from quantum level models, leading to *ab-initio MD*, which is capable of simulating chemical events. Such an approach, however, adds significant computational overheads to what is already a time-intensive technique.

A serious limitation of MD is the short times steps of $\mathcal{O}(10^{-15})$, which confines simulation timescales to sub-microsecond levels. Many of the processes of interest, such as diffusion, on the other hand, take place over much longer time (and associated length) scales. The use of statistical mechanics for such problems is well established using techniques that fall under the category of *Monte Carlo* (MC) methods [8]. These methods, which are a sub-class of the broader MC family, can sample from an intractable distribution *via* simple rules that are used to define a Markov Chain, constructed such that its stationary distribution is the distribution of interest. They form the basis of most simulations of the equilibrium properties of physical systems by generating states according to a Boltzmann distribution, rather than simulating the dynamics as in MD.

A dynamic version of the MC approach is *kinetic* MC or kMC [8], which exploits the long-time dynamics of a system, typically consisting of diffusive jumps from one state to another. It can therefore be used as an alternative to MD. By simulating atomic jumps between lattice sites, the timescales can be extended to seconds and beyond. It has been applied to such processes as surface diffusion, growth and adsorption, irradiation of materials, Li-ion transport in electrodes, ion transport in electrolytes, and solid-electrolyte interface (SEI) formation in Li-ion batteries.

It should also be mentioned that the time and spatial scales of MD can be increased by *coarse graining* [9]. Materials such as polymers and proteins often exhibit a hierarchy of length and time scales. In a coarse-grained approach the numbers of degrees of freedom and the frequency of motion are reduced by identifying groups of atoms (called pseudo-atoms or beads) that interact in an effective force field. Characterising this force field is the main challenge.

The most fundamental level of modelling focuses on the Schrödinger's equation for the many-body wavefunction of a quantum-mechanical system. From the ground-state wavefunction and energy, atomic configuration, highest occupied molecular orbital (HOMO), geometry, and other properties obtained from the calculations, properties such as a redox potential can be obtained. This allows for computational screening of molecules and even the design of new molecules with favourable properties. Moreover, these electronic-structure calculations are used in a variety of other methods, from MD, as already mentioned, to quantitative-structure-activity-relationship (QSAR) models, as descriptors of molecules that can act as predictors of vital properties.

Unfortunately, solving Schrödinger's equation for a many-body system is an intractable problem, which has led to a number of approximate techniques [10]. The archetypal method is density functional theory (DFT), which reformulates the problem as the minimisation of a functional of the electron density, after using the Born-Oppenheimer approximation to remove the nuclei from explicit consideration. The electron density is a function of space and spin, as opposed to the wavefunction, which is a function of the positions and spins of all of the electrons. Although it is not the most accurate of the methods available, most researchers in the area believe that DFT strikes an optimal balance between computational cost and accuracy.

Combining different scales into a single model leads to *multi-scale modelling* [11], which can be seen as an alternative to coarse graining. Ab-initio MD and quantum-mechanical/molecular-mechanical models are two of the earliest examples. However, it is generally not an easy matter to marry two or more approaches relating to different scales, which is achieved through a combination of matched asymptotics, homogenisation, scaling and renormalisation group methods.

3.3 Macroscopic Modelling

As mentioned in the previous section, macroscopic models are usually based on continuum mechanics [12], in which a body or material under investigation is treated as a continuous mass, rather than as a set of discrete particles (atoms or molecules). Provided that the length scales considered are much longer than the interatomic distances, this representation of the body can well describe its behaviour through a set of differential equations embodying conservation principles. The microscopic level of detail is expressed through certain averaged properties that ignore micro-structural inhomogeneity, with each material property defined *via* constitutive relationships specific to the material. These relationships are required in order to complete the model specification.

A *deformable* body $\mathcal{B}$ is considered to occupy a region in physical space, with each point in the region called a material point or *particle* (not to be confused with atomic particles). Collectively these particles form a *continuum*, meaning that the particles are infinitesimally small elements with local material properties. The region occupied by the body may change with time, and we say that the body is in a certain *configuration* or state at any time t. We can define the configuration by the mapping $\kappa_t(\mathcal{B}) : \mathcal{P} \mapsto \mathbf{r}(t)$ from particles $\mathcal{P} \in \mathcal{B}$ to a *position vector* $\mathbf{r}(t)$ written in some frame of reference with origin O

$$\mathbf{r}(t) = r_1(t)\mathbf{e}_1 + r_2(t)\mathbf{e}_1 + r_3\mathbf{e}_3 \tag{3.1}$$

The coordinates r_i are defined with respect to some basis $\{\mathbf{e}_i\}$ in the reference frame, usually the standard Euclidean basis.

The position vector $\mathbf{r}$ can be also be written as a function of the particle position $\mathbf{R}$ with respect to some *reference configuration*, such as the configuration $\kappa_0(\mathcal{B})$ at time $t = 0$. The function κ_t is required to be invertible, orientation-preserving and to lie in a subset of the space $C^2([0, \infty))$ of twice continuously differentiable functions (of time). This allows any position vector to be uniquely associated to a particle, and for velocities and accelerations of particles to be defined.

Forces acting on the body will engender motion, and they can be categorised as either body or surface (contact) forces, as well as either external or internal forces. Surface forces act on the external surface of the body *via* contact with other bodies, or they act on internal surfaces as a consequence of interactions between different parts of the body. External contact forces engender internal contact forces as a consequence of Newton's third law. In contrast to internal forces, external forces change the total energy of the body. Internal contact forces can be related to the deformation of the body *via* constitutive equations, and they are assumed to be continuously distributed throughout the body.

Body forces are external and act on the whole volume of the body. They can be inertial forces arising from the motion of the body, or they can be due to force fields, such as a gravitational or electromagnetic field. These forces are again assumed to be continuously distributed throughout the body. A change in the configuration $\kappa_t(\mathcal{B})$ leads to a displacement, which is comprised of a deformation of the body and a rigid-body displacement, in which the body is translated and rotated without changing its shape. A deformation, on the other hand, refers to a change in the shape. A particle, therefore, traces out a path in space as the body undergoes a displacement through time.

3.3.1 Eulerian and Lagrangian Descriptions

In fluid and solid mechanics, it is necessary to characterise the paths that particles follow as a consequence of displacement, for which we need to define a suitable reference frame with respect to a chosen reference configuration. All subsequent configurations are then written in terms of this configuration. There are broadly speaking two approaches [12]. In the first, called the *Lagrangian* description, we define a set of *material* or referential coordinates as those corresponding to $\kappa_0(\mathcal{B})$, and describe the motion in terms these coordinates. In the second, called the *Eulerian* description, the reference configuration is taken to be $\kappa_t(\mathcal{B})$, i.e., the current configuration.

In the Lagrangian description, the motion is given in terms of some mapping $\mathbf{r} = \chi(\mathbf{R}, t)$ of coordinates $\mathbf{R}$ in the original configuration to those in the current configuration $\mathbf{r}$. This means that the particle $\mathcal{P}$ with position vector $\mathbf{R}$ at $t = 0$ occupies the position $\mathbf{r}$ in the current configuration. In terms of the configuration mapping

$$\mathbf{r} = \kappa_t(\mathcal{P}) = (\kappa_t \circ \kappa_0^{-1})(\mathbf{R}) := \chi(\mathbf{R}, t) \tag{3.2}$$

in which $\circ$ denotes a composition. The components of $\mathbf{r}$ are referred to as *spatial coordinates*. Properties ϕ that define the body, such as the rate at which particles move, are expressed as functions of $\mathbf{R}$ and t, that is $\phi = \phi(\mathbf{R}, t)$.

The *material* or substantial (also called total) derivative of $\phi(\mathbf{R}, t)$, denoted $d\phi/dt$, is its rate of change in terms of a specific group of particles within the body, i.e., the rate of change as measured by an observer travelling with these particles. The position $\mathbf{r}$ is instantaneous, meaning that it refers only to a specific time and is a property of a particle in the continuum. Since $\mathbf{R}$ does not change with time, from a

Lagrangian point of view the material derivative is equal to the partial derivative of ϕ with respect to time

$$\frac{d}{dt}\phi(\mathbf{R}, t) = \frac{\partial}{\partial t}\phi(\mathbf{R}, t) \tag{3.3}$$

The material derivative of the instantaneous position $\mathbf{r}$ of a particular particle is referred to as the *instantaneous flow velocity* (or just flow velocity) $\mathbf{u}$ and is given by

$$\mathbf{u}(\mathbf{R}, t) = \left.\frac{d\mathbf{r}}{dt}\right|_{\mathbf{R}} = \frac{\partial \chi(\mathbf{R}, t)}{\partial t} \tag{3.4}$$

in which $|_{\mathbf{R}}$ makes clear that $\mathbf{R}$ is fixed.

In the Eulerian perspective, we use the transformation $\mathbf{R} = \chi^{-1}(\mathbf{r}, t)$ to relate the spatial coordinates of a particle to its coordinates in the reference configuration, so that

$$\phi(\mathbf{R}, t) = \phi(\chi^{-1}(\mathbf{r}, t), t) = \psi(\mathbf{r}, t) \tag{3.5}$$

Note that usually the symbol ϕ is used for ψ, i.e., we write $\phi(\mathbf{r}, t)$ since it refers to the same physical quantity. The material derivative in this case is given by

$$\frac{D}{Dt}\psi(\mathbf{r}, t) = \frac{d}{dt}\psi(\mathbf{r}, t) = \frac{\partial}{\partial t}\psi(\mathbf{r}, t) + \nabla\psi(\mathbf{r}, t) \cdot \frac{d\mathbf{r}}{dt}(\mathbf{r}, t) = \frac{\partial}{\partial t}\psi(\mathbf{r}, t) + \mathbf{u} \cdot \psi(\mathbf{r}, t) \tag{3.6}$$

with the first term equal to the local rate of change in ϕ at $\mathbf{r}$. The second term, called the *convective* rate of change, is the contribution made by the particle moving in space. The notations D/Dt and d/dt are interchangeable. The *deformation gradient* is defined by

$$\mathbf{D} = \nabla_{\mathbf{X}}\mathbf{r} \tag{3.7}$$

in which $\nabla_{\mathbf{X}}$ is the gradient with respect to $\mathbf{X}$.

Each of the Eulerian and Lagrangian descriptions has its advantages and shortcomings. The Lagrangian description is mainly employed in structural mechanics, since it allows for easy tracking of free surfaces and any interfaces between materials. However, algorithms based on a Lagrangian description are unable to easily follow large distortions of the body and require frequent remeshing, The Eulerian description, on the other hand, is the standard in fluid dynamics. Although it can handle large deformations, this can come at the expense of a precise resolution of the flow details. Strategies that combine these descriptions will be discussed in Sect. 3.3.9.

3.3.2 Conservation Laws

The equations of continuum mechanics express the (balance) laws of mass, momentum and energy conservation, with accompanying kinematic and constitutive relationships. The fundamental laws of thermodynamics must also be satisfied under all conditions. The balance laws embody the fact that the rates of change of mass, energy and momentum in the volume V that bounds the body are a consequence of three effects: the quantity $\phi(\mathbf{r}, t)$ being conserved flows through the surfaces ∂V of V, and there may be sources of the quantity at ∂V or within the interior of V.

The laws can be expressed in the following general form [12]

$$\frac{d}{dt}\int_V \phi(\mathbf{r},t)d\mathbf{r} = \int_{\partial V} \phi(\mathbf{r},t)\left[u_n(\mathbf{r},t) - \mathbf{u}(\mathbf{r},t)\cdot\mathbf{n}(\mathbf{r},t)\right]d\mathbf{r} + \int_{\partial V} s_s(\mathbf{r},t)d\mathbf{r} + \int_V s_b(\mathbf{r},t)d\mathbf{r} \tag{3.8}$$

in which $\mathbf{n}(\mathbf{r}, t)$ is the outwardly pointing unit normal to the surface, $s_b(\mathbf{r}, t)$ and $s_s(\mathbf{r}, t)$ are interior and surface sources of ϕ, respectively, and $u_n(\mathbf{r}, t)$ is the speed at which ∂V is moving in the direction $\mathbf{n}$.

In an Eulerian description, the balance laws for mass, linear momentum, angular momentum and energy are, respectively [12]

$$\begin{gathered}\frac{\partial \rho}{\partial t} + \nabla\cdot(\rho\mathbf{u}) = 0\\ \rho\left(\frac{\partial \mathbf{u}}{\partial t} + \mathbf{u}\cdot\nabla\mathbf{u}\right) - \nabla\cdot\boldsymbol{\sigma} - \rho\mathbf{b} = 0\\ \boldsymbol{\sigma} = \boldsymbol{\sigma}^T\\ \rho\left(\frac{\partial e}{\partial t} + \mathbf{u}\cdot\nabla e\right) - \boldsymbol{\sigma} : \nabla\mathbf{u} + \nabla\cdot\mathbf{q} - \rho\, s = 0\end{gathered} \tag{3.9}$$

in which $\rho(\mathbf{r}, t)$ is the mass density, $\boldsymbol{\sigma}(\mathbf{r}, t)$ is the Cauchy stress tensor (a matrix or second-order tensor), $\mathbf{b}(\mathbf{r}, t)$ is the density of the body force, $e(\mathbf{r}, t)$ is the specific internal energy, $\mathbf{q}(\mathbf{r}, t)$ is the heat flux and $s(\mathbf{r}, t)$ represents a source of specific energy. The term $\boldsymbol{\sigma} : \nabla\mathbf{u}$ is a double dot product, meaning the sum of the entries of the element-wise (Hadamard) product $\boldsymbol{\sigma} \odot \nabla\mathbf{u}$ of $\boldsymbol{\sigma}$ and $\nabla\mathbf{u}$.

In this notation in (3.9), $\nabla\mathbf{u}$ is treated like an outer product (∇ is an operator so it is not an outer product in a strict sense). In component form, in terms of a basis $\{\mathbf{e}_i\}$ for the current reference frame, it can be written as

$$\nabla\mathbf{u} = \sum_{i,j}\frac{\partial u_i}{\partial x_j}\mathbf{e}_i \otimes \mathbf{e}_j^T = \sum_{i,j}\frac{\partial u_i}{\partial x_j}\mathbf{e}_i \circ \mathbf{e}_j = \sum_{i,j}\frac{\partial v_i}{\partial x_j}\mathbf{e}_i\mathbf{e}_j^T \tag{3.10}$$

in the basis $\{\mathbf{e}_i \otimes \mathbf{e}_j^T\}_{i,j} \subset \mathbb{R}^{3\times 3}$, where u_i are the components of $\mathbf{u}$. We return to these notations and the meaning of the tensor product $\otimes$ in Sect. 6.11.1 of Chap. 6. We point out here that all of the above notations, in which ^{T} denotes a transpose and $\circ$ an outer product, refer to the same object. Moreover, $\mathbf{e}_i \otimes \mathbf{e}_j^T$, is sometimes written $\mathbf{e}_i \otimes \mathbf{e}_j$, which is formally incorrect since the latter is a column vector in $\mathbb{R}^9$, while the first is a 3×3 matrix, or second-order tensor, though the spaces in which both live are isometrically isomorphic.

Note that the first equation in (3.9) is usually called *mass continuity* or simply continuity. For an incompressible fluid the density is constant yielding the simpler continuity equation

$$\mathrm{div}(\mathbf{u}) = \nabla \cdot (\mathbf{u}) = 0 \tag{3.11}$$

We present in (3.11) an alternative notation for the divergence, often found in the physics literature and in some engineering texts.

We now need constitutive relationships for quantities such as the stress tensor $\boldsymbol{\sigma}$ in order to fully specify any model. These will depend on the material in question. We can express $\boldsymbol{\sigma}$ as

$$\boldsymbol{\sigma} = \boldsymbol{\tau} - p\mathbf{I} \tag{3.12}$$

in which $\boldsymbol{\tau}$ is the deviatoric stress (viscous term) and $-p\mathbf{I}$ is the volumetric stress, in which p is the fluid pressure. If we assume that the Cauchy stress is symmetric and Galilean invariant, with a linear dependence on $\nabla\mathbf{u}$, and that the fluid is isotropic, the deviatoric stress takes the form

$$\boldsymbol{\tau} = \zeta \nabla \cdot \mathbf{u}\mathbf{I} + \eta \left[\nabla\mathbf{u} + (\nabla\mathbf{u})^T - \frac{2}{3}(\nabla \cdot \mathbf{u})\mathbf{I} \right] \tag{3.13}$$

in which η is the dynamic viscosity and ζ is the bulk viscosity. If the dynamic viscosity is assumed to be constant and the bulk velocity is set to zero (Stoke's assumption), we arrive at the most general form of the *Navier-Stokes* equations [13] governing a fluid that satisfies the assumptions above

$$\rho \frac{\mathrm{D}\mathbf{u}}{\mathrm{D}t} = \rho \left(\frac{\partial \mathbf{u}}{\partial t} + \mathbf{u} \cdot \nabla\mathbf{u} \right) = -\nabla p + \eta \nabla^2 \mathbf{u} + \frac{1}{3} \eta \nabla(\nabla \cdot \mathbf{u}) + \rho \mathbf{F} \tag{3.14}$$

in which $\mathbf{F}$ includes external forces such as gravity. Note that $(\nabla\mathbf{u})^T = \nabla(\nabla \cdot \mathbf{u})$. For so-called *non-Newtonian* fluids, (3.13) is not valid and other formulations are required.

For incompressible fluids, Stoke's stress equation $\boldsymbol{\tau} = \eta \left(\nabla\mathbf{u} + \nabla\mathbf{u}^T\right)$ is often used, and a dimensionless form of the equations is

$$\frac{\partial \mathbf{u}}{\partial t} + \mathbf{u} \cdot \nabla\mathbf{u} - \frac{1}{Re} \nabla^2 \mathbf{u} = -\nabla p + \mathbf{f} \tag{3.15}$$

in which $\mathbf{f}$ is a dimensionless force density and the *Reynolds number* Re is defined by

$$\mathrm{Re} = \frac{\rho L U}{\eta} \tag{3.16}$$

L and U are a characteristic length of the system and the mean speed of the fluid, respectively. Re is a measure of the ratio between inertial and viscous forces. When Re is small, the flow is said to be *laminar*, flowing in parallel layers. For high values of Re, the flow becomes *turbulent* [13], meaning chaotic with random fluctuations in velocity and pressure. Solving the Navier-Stokes equations in such cases, incorporating all of the turbulent scales, is particularly challenging and there are various approximate approaches, such as the time *Reynolds-averaged Navier–Stokes* (RANS) and *large eddy simulation* (LES) methods [14]. In *direct numerical simulation*, the equations are solved exactly, down to the smallest so-called Kolmogorov scale of turbulence, which seems a simple enough task but requires extremely fine meshes and very small time steps due to the explicit time stepping used to reduce memory requirements.

For the energy balance, it is normally assumed that the heat flux is proportional to the gradient in temperature $-k\nabla T(\mathbf{r}, t)$ (Fourier's law) [15], in which k is called the *thermal conductivity*, not necessarily a constant. Once again we can use the relationship $\boldsymbol{\sigma} = \boldsymbol{\tau} - p\mathbf{I}$ to define the stress tensor, for now leaving open the question of the form of $\boldsymbol{\tau}$. The specific internal energy can be related to temperature *via* [15]

$$dh = C_p dT + \frac{1}{\rho}(1 - \beta T)dp, \quad e = h - \frac{p}{\rho} \tag{3.17}$$

in which h is the specific enthalpy, C_p is the specific capacity of the body at constant pressure and β is its bulk expansion coefficient. This leads us from the energy balance in (3.9) to the following temperature equation

$$\rho C_p \left(\frac{\partial T}{\partial t} + \mathbf{u} \cdot \nabla T \right) - \nabla \cdot (k \nabla T) = \beta \left(\frac{\partial p}{\partial t} + \mathbf{u} \cdot \nabla p \right) + \boldsymbol{\tau} : \nabla \mathbf{u} + \rho s \tag{3.18}$$

There are some special cases

1. For ideal gases, $\beta T = 1$
2. For incompressible fluids, $Dp/Dt = 0$
3. Except in cases involving high shear rates (e.g., bearing and hydraulics), the viscous heating term $\boldsymbol{\tau} : \nabla \mathbf{u}$ makes a negligible contribution and can be set to 0.

3.3.3 Conservation of Multiple Charged and Neutral Species

For the mass balance, we are usually interested not only in the overall continuity, but also mass balances for individual species contained in the body, e.g., the constituents of an electrolyte. We can denote the concentrations (moles per unit volume) of the species by c_i, $i = 1, \ldots, S$, and their molecular weights by W_i. The mole fraction of each species is defined as

$$X_i = \frac{c_i}{c}, \quad c = \sum_i c_i, \quad \sum_i X_i = 1 \tag{3.19}$$

in which c is the total concentration. The mass fraction is defined as

$$Y_i = \frac{n_i}{n} = \frac{\rho_i}{\rho}, \quad \sum_i Y_i = 1, \quad n = \sum_i n_i, \quad \rho = \sum_i \rho_i \tag{3.20}$$

in which n_i and ρ_i are the number of moles and partial density of species i, respectively, with n and ρ being the total number of moles and the density of the mixture. It must hold that $Y_i = X_i W_i / W$, in which $W = \sum_i X_i W_i$ is the mixture molecular weight.

The balance for a species i can be written in terms of its mass fraction as follows

$$\frac{\partial(\rho Y_i)}{\partial t} + \nabla \cdot (\rho Y_i \mathbf{u}) = \omega_i \tag{3.21}$$

in which ω_i is the source or sink for the species, e.g., through a chemical or electrochemical reaction, satisfying $\sum_i \omega_i = 0$ in order to conserve mass. Each species has an instantaneous velocity $\mathbf{u}_i$, from which we can recover the mass average velocity of the mixture $\mathbf{u}$ and define *diffusion velocities* $\mathbf{U}_i$

$$\mathbf{u} = \sum_i Y_i \mathbf{u}_i, \quad \mathbf{U}_i = \mathbf{u}_i - \mathbf{u} \tag{3.22}$$

which must satisfy $\sum_i Y_i \mathbf{U}_i = 0$. The diffusion velocities quantify the rate of dispersion of the species relative to the mean velocity and relate to drift *via* molecular collisions. Thus, we can write Eq. (3.21) as

$$\frac{\partial(\rho Y_i)}{\partial t} + \nabla \cdot (\rho Y_i \, (\mathbf{u} + \mathbf{U}_i)) = \omega_i \tag{3.23}$$

Summing over all equations (3.23) therefore returns us to the (mass) continuity equation in (3.9), noting also that $\sum_i \rho Y_i = \rho$. Expanding the gradient and time derivative terms and using the continuity equation (3.9) then leads to

$$\rho \frac{\partial Y_i}{\partial t} + \rho \mathbf{u} \cdot \nabla Y_i + \nabla \cdot (\rho Y_i \mathbf{U}_i) = \omega_i \tag{3.24}$$

The most common approximation for the diffusion velocities is *Fick's law* [16]

$$\mathbf{U}_i = -D_i \nabla \ln Y_i = -\frac{D_i}{Y_i} \nabla Y_i \tag{3.25}$$

in which D_i is called a *diffusion coefficient*. Fick's law is a mass analogue of Fourier's law, stating that the flux of a species is proportional to the negative gradient in its mass fraction. Placing (3.25) in (3.24) yields the final form

$$\rho \frac{\partial Y_i}{\partial t} + \rho \mathbf{u} \cdot \nabla Y_i - \nabla \cdot (\rho D_i \nabla Y_i) = \omega_i \tag{3.26}$$

Fick's law is valid for dilute mixtures, in which there is one dominating component. It is, nevertheless, used even in other cases, with remarkable success. There are cases in which it will not provide sufficient accuracy, for which more sophisticated theories such as the Maxwell-Stefan model [16] can be used.

When the species is charged (an ion), it also moves under the influence of a potential field, so that an additional *electro-osmotic* term is required. The flux $\mathbf{N}_i$ or diffusion velocity of the species in this case can be approximated for a dilute solution using the *Nernst-Planck* equation

$$\mathbf{N}_i = \rho Y_i \left(\mathbf{u} + \mathbf{U}_i\right) = -\rho D_i \nabla Y_i - \rho Y_i \frac{F z_i}{RT} D_i \nabla \phi + \rho Y_i \mathbf{u} \tag{3.27}$$

which is placed directly in (3.23). Here, z_i is the valence of the species, F is Faraday's constant and $\phi(\mathbf{r}, t)$ is the ionic potential. Note that we can convert between mass fraction, partial density, mole fraction and concentration using the relationships above, with concentration often being the preferred form.

3.3.4 Flow in Porous Media

Flow in porous media is highly relevant to flow batteries since they often contain porous electrodes, through which flows the electrolytes. The most often used momentum balance for porous media is a simplified version of the Navier-Stokes equations called *Darcy's law*. Actually, the law was originally derived empirically by Darcy [17], but was later found to be recoverable from the Navier-Stokes equations. It is assumed that the flow is viscous, incompressible and steady. Darcy's equations are as follows [18]

$$\begin{aligned} \mathbf{u} &= -\frac{K}{\eta} \nabla p \\ \nabla \cdot \mathbf{u} &= 0 \end{aligned} \tag{3.28}$$

in which K is called the intrinsic permeability of the porous medium. Many forms can be adopted for K, with a popular choice being the Kozeny-Carman law [19, 20]

$$K = \frac{d_f^2 \varepsilon^3}{K_{KC}(1-\varepsilon)^2} \tag{3.29}$$

in which ε is the electrode porosity, d_f is an average fibre diameter of the porous material and K_{KC} is a structural parameter dependent on the type of porous structure, called the Kozeny-Carman constant.

Another often used model for porous media flow, again a simplified version of the Navier-Stokes equations, is the Brinkman equations [21]

$$\frac{\eta}{K}\mathbf{u} + \nabla p - \frac{\eta}{\varepsilon}\nabla^2\mathbf{u} = 0$$
$$\nabla \cdot \mathbf{u} = 0 \tag{3.30}$$

which modifies Darcy's law by adding a standard viscosity term.

3.3.5 *Transport of Water and Ions in Membranes*

In most flow batteries, an ion-exchange membrane (almost always Nafion®) is employed to separate the electrodes, and the transport of ions such as protons across the membrane is vital to maintaining electroneutrality, as well as completing complementary (half cell) reactions in some systems. Moreover, water is simultaneously transported since the protons move as protonated water complexes.

A phenomenological model for water and proton transport across the anion-exchange membrane Nafion was developed by Springer et al., based on species transport driven by concentration gradients [22]. This model, however, was for PEM fuel cells, in which membranes are not fully humidified for much of the time. In flow batteries, on the other hand, the membrane is always in contact with the liquid electrolyte and is thus saturated.

Bernadi and Verbrugge [23, 24] developed an alternative model that is more appropriate in this case. The model enforces electroneutrality, with the negative charges in the Nafion membrane taken to be the fixed sulfonic acid groups, and the positive charges taken to be the protons. The dissolved water concentration $c_{\mathrm{H_2O}}$ satisfies the following mass conservation law

$$\frac{\partial c_{\mathrm{H_2O}}}{\partial t} - \nabla \cdot \left(D_{\mathrm{H_2O}}\nabla c_{\mathrm{H_2O}}\right) + \nabla \cdot \left(\mathbf{u}c_{\mathrm{H_2O}}\right) = 0 \tag{3.31}$$

incorporating diffusion, with an effective diffusion coefficient $D_{\mathrm{H_2O}}$, and convection, with a flow velocity $\mathbf{u}$.

Bulk flow is engendered by both potential and pressure gradients and the liquid velocity is governed by Schloegl's equation [25]

$$\mathbf{u} = -\frac{k_\phi}{\eta}Fc_{\mathrm{H^+}}\nabla\phi - \frac{k_p}{\eta}\nabla p \tag{3.32}$$

for electrokinetic permeability k_ϕ, hydraulic permeability k_p, proton concentration $c_{\mathrm{H^+}}$ and ionic potential ϕ. The latter must satisfy Laplace's equation

$$\nabla^2\phi = 0 \tag{3.33}$$

since there are no proton sources in the membrane. The liquid water is assumed to be incompressible, $\nabla \cdot \mathbf{u} = 0$.

With the electroneutrality assumption, the proton concentration is fixed as

$$c_{\mathrm{H}^+} = -z_f c_f \tag{3.34}$$

in which c_f is concentration of fixed (negative) charge sites in the membrane, with charge z_f. Therefore, we can derive the following equation for the pressure using (3.32), (3.33) and the incompressibility assumption

$$\nabla^2 p = 0. \tag{3.35}$$

3.3.6 Charge Balances

To derive charge balances in electrochemical systems, we must consider both the electronic and ionic phases, as well as their interaction. It is normally assumed that electroneutrality holds. While this may be violated locally, we would not expect large deviations from this condition, otherwise any charge-transfer process will not proceed. Electroneutrality demands that for N charged species with concentrations c_i and valences z_i

$$\sum_i z_i c_i = 0 \tag{3.36}$$

At any electrode/electrolyte interface, the principle of charge conservation implies that the charge entering the electrolyte $\mathbf{j}_e$ is balanced by the charge that leaves the electrode $\mathbf{j}_s$.

The total current (density) $\nabla \cdot \mathbf{j}$ transferred between the two phases therefore satisfies

$$\nabla \cdot \mathbf{j} = \nabla \cdot \mathbf{j}_e = -\nabla \cdot \mathbf{j}_s \tag{3.37}$$

and is equal to Faraday's constant F multiplied by the volumetric rate of electrochemical reaction (usually modelled by a Butler-Volmer law (2.23)) and the number of electrons transferred during the reaction. Using the Nernst-Planck equation (3.38), we can write the flux of ionic charge as

$$\mathbf{j}_e = \sum_i \mathbf{N}_i z_i = \sum_i \left(-z_i D_i \nabla c_i - c_i \frac{F z_i^2}{RT} D_i \nabla \phi + c_i z_i \mathbf{u} \right) \tag{3.38}$$

Making use of the electroneutrality condition, this yields

$$-\nabla \cdot \left(\sigma_e \nabla \phi + F \sum_i z_i D_i \nabla c_i \right) = \nabla \cdot \mathbf{j} \tag{3.39}$$

in which σ_e is an effective ionic conductivity defined by

$$\sigma_e = \frac{F^2}{RT} \sum_i z_i^2 D_i c_i \tag{3.40}$$

For a solid-phase electron conducting medium, Ohm's law is usually employed for charge conservation, neglecting double-layer capacitance and assuming a steady state

$$-\sigma_s \nabla^2 \phi_s = -\nabla \cdot \mathbf{j} \tag{3.41}$$

for an electronic conductivity σ_s.

3.3.7 The Volume-of-Fluid Method

For problems involving immiscible fluids and phase changes, a very powerful tool is the *volume-of-fluid* (VOF) method [26], which grew out of the earlier marker and cell (MAC) model. The MAC model uses mass-less *marker* particles distributed throughout the cells of the solution domain to distinguish between the fluids, with cells containing markers indicating the presence of one fluid and those without markers indicating the presence of the other fluid, in the case of a binary mixture [27]. Those cells containing markers and adjacent to empty cells are considered to contain the interface, and surface tension (pressure) terms are applied at the centres of these cells.

A high number of markers (more than the number of cells) is required to accurately identify the interface, and for interfaces that undergo significant deformation, markers must be added or removed continually. Moreover, numerical errors incurred when advancing the markers can lead to a loss of resolution of the interface. In general, the computational costs are high for MAC methods. The VOF method requires fine spatial resolutions, as well as small time steps, and is also therefore computationally expensive and limited to relatively small spatial scales.

Consider a liquid-vapour flow in which the goal is to track the liquid-vapour interface. This problem has, for over two decades, received a great deal of attention in the PEM fuel cell community, since the formation of liquid water (either directly or by condensation) in the cathode gas diffusion layer and subsequent blocking of the pores is a major issue. An implementation of VOF for PEM fuel cells can be found in [28].

The VOF method uses a *fraction function* (also called an indicator, colour or characteristic function) $0 \leq \phi(\mathbf{r}, t) \leq 1$ to denote the volume-fraction of one of the phases at any point in space and at time t. For example, we can set $\phi = 1$ for the pure liquid and $\phi = 0$ for the pure vapour, with intermediate values denoting a mixture of the two. Additional fraction functions ϕ_i are introduced when there are N phases or fluids, with $\sum_i \phi_i = 1$ to conserve volume, so that only $N - 1$ equations are required.

The computational domain is divided into cells (or elements) as in a finite-volume approach and the quantity ϕ at the centre of each cell is tracked in time. If $\phi = 1$ ($\phi = 0$) in the cell centre, the cell is occupied by the liquid (vapour) and the liquid (vapour) conservation laws apply, otherwise the cell contains both components, including a portion of the separating interface $\Gamma(t)$. In the latter case, we can use a volume averaging of the density and of parameters such as the specific heat capacity, viscosity and thermal conductivity in the conservation equations (3.14), (3.18) and either (3.26) when there are more than 2 fluids or the continuity equation in (3.9) otherwise. The volume averaging is given by

$$\chi(\mathbf{r},t) = \phi(\mathbf{r},t)\chi_l + (1-\phi(\mathbf{r},t))\chi_v, \quad \chi \in \{\rho, \rho c_p, \nu, k\} \tag{3.42}$$

in which the subscripts l and v denote bulk liquid and bulk vapour values.

The force density $\mathbf{F}$ in the momentum equation (3.14) includes a surface tension component $\mathbf{F}_s$; surface tension engenders a surface pressure, i.e., a force per unit surface area. $\mathbf{F}_s$ can be approximated using the continuous-surface-force (CSF) model of Brackbill et al. [29]

$$\mathbf{F}_s(\mathbf{r},t) = \sigma\kappa(\mathbf{r},t)\mathbf{n}_{\Gamma(t)}(\mathbf{r},t)\delta_{\Gamma(t)} \tag{3.43}$$

in which σ is the surface tension of the fluid, $\kappa(\mathbf{r},t)$ is the *curvature* of the interface $\Gamma(t)$, $\mathbf{n}_{\Gamma(t)}(\mathbf{r},t)$ is a unit normal to the surface $\Gamma(t)$ and $\delta_{\Gamma(t)}$ is the delta function, such that $\delta_{\Gamma(t)} = 1$ if $\mathbf{r} \in \Gamma(t)$ and is otherwise set to 0. Other models can be used and other forces incorporated according to the system under consideration. Note that $\mathbf{F}_s$ is volumetric by virtue of the delta function.

In [26], the surface force is incorporated by first defining a mollified (as opposed to discontinuous) fraction function for a finite-thickness interface

$$\widetilde{\phi}(\mathbf{r},t) = \frac{1}{h^3}\int_{\mathbb{R}^3} \phi(\mathbf{r}',t)k(\mathbf{r}-\mathbf{r}')d\mathbf{r}' \tag{3.44}$$

in which h is the interface thickness and the (specified) kernel function $k(\cdot)$ has bounded support, $k(\mathbf{r}) = 0$ for $\|\mathbf{r}\| \geq h/2$, and satisfies the normalisation constraint $\int_{\mathbb{R}^3} k(\mathbf{r})d\mathbf{r} = h^3$. In terms of this mollified fraction function, the surface tension force can be written as

$$\mathbf{F}_s(\mathbf{r},t) = \sigma\kappa(\mathbf{r},t)\nabla\widetilde{\phi}(\mathbf{r},t) \tag{3.45}$$

and for incompressible flows it takes the form

$$\mathbf{F}_s(\mathbf{r},t) = 2\sigma\kappa(\mathbf{r},t)\left(\frac{\rho(\mathbf{r},t)}{\rho_l+\rho_v}\right)\left(\frac{\nabla\rho(\mathbf{r},t)}{\rho_l-\rho_g}\right) \tag{3.46}$$

In the general case, the term $\kappa(\mathbf{r},t)\nabla\widetilde{\phi}(\mathbf{r},t)$ in (3.45) is given entirely in terms of $\widetilde{\phi}(\mathbf{r},t)$ as

$$\kappa(\mathbf{r},t)\nabla\widetilde{\phi}(\mathbf{r},t) = -\nabla\widetilde{\phi}(\mathbf{r},t)\left(\nabla\cdot\frac{\nabla\widetilde{\phi}(\mathbf{r},t)}{\|\nabla\widetilde{\phi}(\mathbf{r},t)\|}\right) \tag{3.47}$$

in which $\widetilde{\phi}/\|\nabla\widetilde{\phi}\|$ defines the unit normal $\mathbf{n}_{\Gamma(t)}$.

For two fluids, the evolution of the fraction function is given by the continuity equation (3.9)

$$\frac{\partial \phi}{\partial t} + \nabla \cdot (\phi \mathbf{u}) = \omega_l \tag{3.48}$$

in which ω_l is a source term for the liquid. This equation is essentially that for the liquid phase, with the gas phase equation eliminated by the fact that the volume fractions of the two phases sum to unity. The source ω_l is automatically balanced by a term $-\omega_l$ for the vapour phase in order to ensure mass conservation. In the case of N phases, it is simpler to work with Eq. (3.26), noting that $\rho Y_i = \rho_i \phi_i$, leading to

$$\frac{\partial \phi_i}{\partial t} + \nabla \cdot (\phi_i \mathbf{u}) = \omega_i, \quad i = 1, \ldots, N-1, \quad \phi_N = 1 - \sum_{i=1}^{N-1} \phi_i \tag{3.49}$$

This provides a full specification of the model, which can be solved by standard finite-difference, -volume or -element schemes, with a Lagrangian, Eulerian or ALE description (described later).

3.3.8 The Level-Set Method

The VOF method shares similarities with another interface tracking method called the *level set* method [30, 31], in which the interface $\Gamma(t)$ is represented by the level sets $\psi(\mathbf{r}, t) = 0$ of a function $\psi(\mathbf{r}, t)$. The level-set function evolves according to an equation that resembles (3.48)

$$\frac{\partial \psi}{\partial t} + \mathbf{u} \cdot \nabla \psi = 0 \tag{3.50}$$

for a velocity field $\mathbf{u}$. If $\mathbf{u}$ is consistently oriented along the normal to the level set, the equation can be written as

$$\frac{\partial \psi}{\partial t} + u(\mathbf{r}, t) \|\nabla \psi\| = 0 \tag{3.51}$$

for some speed u. This is the more familiar form, from the original work [30]. The velocity $\mathbf{u}$ may be relevant only to the interface, in which case it has to be extended to the whole domain. The form (3.50) is relevant to fluid flows, with $\mathbf{u}$ defined by the flow velocity. Note that Γ is a manifold, embedded in the physical domain $V \subset \mathbb{R}^3$.

The level-set function is defined in the whole domain V, usually as a *signed distance function*

$$t > 0, \ \mathbf{r} \in V, \quad \begin{cases} \psi(\mathbf{r}, t) = d(\mathbf{r}, \gamma) & \mathbf{r} \in W(t)^c = V \backslash W(t) \\ \psi(\mathbf{r}, t) = 0 & \mathbf{r} \in \Gamma(t) \\ \psi(\mathbf{r}, t) = -d(\mathbf{r}, \gamma) & \mathbf{r} \in W(t) \end{cases} \tag{3.52}$$

in which $W(t)$ is that part of the domain in which phase l resides (e.g., a liquid), with phase v in the complement $W(t)^c$. $d(\cdot)$ is a distance function (metric) defined to be the shortest Euclidean distance between the point $\mathbf{r}$ and the interface

$$d(\mathbf{r}, \gamma) = \inf_{\mathbf{r}' \in \Gamma(t)} \|\mathbf{r} - \mathbf{r}'\| \tag{3.53}$$

In practical implementations, the interface is not considered sharp and is instead defined by the region $\{\mathbf{r} \mid |\psi(\mathbf{r}, t)| \leq \epsilon\} \subset V$, in which ϵ is a small user-chosen constant. We can then define mollified Heaviside and delta functions as follows

$$H_\epsilon(\psi) = \begin{cases} 0 & \psi < -\epsilon \\ \frac{\psi+\epsilon}{2\epsilon} + \frac{1}{2\pi} \sin\left(\frac{\pi\psi}{\epsilon}\right) & |\psi| \leq \epsilon \\ 1 & \psi > \epsilon \end{cases} \tag{3.54}$$

$$\delta_\epsilon(\psi) = \frac{d}{d\psi} H_\epsilon(\psi) = \begin{cases} \frac{1}{2\epsilon} + \frac{1}{2\epsilon} \cos\left(\frac{\pi\psi}{\epsilon}\right) & |\psi| < \epsilon \\ 0 & |\psi| \geq \epsilon \end{cases} \tag{3.55}$$

Averaged quantities are defined using the Heaviside function (analogues to (3.56))

$$\chi(\mathbf{r}, t) = H_\epsilon(\psi)\chi_l + (1 - H_\epsilon(\psi))\chi_v, \quad \chi \in \{\rho, \rho c_p, \nu, k\} \tag{3.56}$$

The analogue of (3.43) for the surface tension in the CSF model is given by

$$\mathbf{F}_s(\mathbf{r}, t) = \sigma\kappa(\mathbf{r}, t)\mathbf{n}_{\Gamma(t)}(\mathbf{r}, t)\delta_\epsilon(\psi) \tag{3.57}$$

in which $\kappa(\mathbf{r}, t)$ and $\mathbf{n}_{\Gamma(t)}$ are calculated more easily than in the VOF method from the following

$$\mathbf{n}_{\Gamma(t)} = \frac{\nabla\psi(\mathbf{r}, t)}{\|\nabla\psi(\mathbf{r}, t)\|}, \quad \kappa(\mathbf{r}, t) = -\nabla \cdot \mathbf{n}_{\Gamma(t)} \tag{3.58}$$

closely resembling (3.47).

The easy calculation of the curvature and normal is one of the advantages of the level-set method over VOF. Since the volume-fraction function in VOF is a step function, obtaining accurate values of the curvature and smoothing discontinuous quantities close to the interface is challenging. In general, however, the level-set method is not as accurate as VOF when modelling interfaces that undergo extreme stretching and tearing. This leads to a violation of mass conservation, and several attempts have been made to remedy this problem without success [32, 33].

A combined level-set-VOF method, on the other hand, shows good performance by leveraging the advantages of both, while managing to conserve mass [34]. Both a level-set and a volume-fraction function are included in these models, advanced by the velocity field. The level-set function is used in the calculations of normals and curvatures, while the volume-fraction function is used to ensure mass conservation.

3.3.9 Arbitrary Lagrangian Eulerian Methods

Arbitrary Lagrangian Eulerian (ALE) methods [35] are based on combining the Eulerian and Lagrangian descriptions, with a reference frame that is neither the initial frame nor the current frame and is set by the user (hence 'arbitrary'). They are again used primarily for interface tracking and fluid-solid interaction problems. The mesh can be evolved to keep track of mobile boundaries or interfaces, e.g., between a solid and a fluid. In the usual framework, a Lagrangian description is used for interfaces, while an Eulerian description is used in regions far from the interface. Nodes in the transitional region between these extremes are interpolated to ensure smoothness of the mesh displacement and velocity field.

Any dependent variable such as a fluid density can be expressed in the Lagrangian (**R**), Eulerian (**r**) or ALE (**x**) coordinates, with well-defined mappings between the coordinate systems

$$\chi : \mathbf{R} \to \mathbf{r}, \quad \Phi : \mathbf{R} \to \mathbf{x}, \quad \Psi : \mathbf{x} \to \mathbf{r} \tag{3.59}$$

satisfying $\chi = \Phi \circ \Psi$. The *convective* velocity $\widetilde{\mathbf{u}}(\mathbf{x}, t)$ written in the ALE coordinates is defined as the difference between the material velocity $\mathbf{u}(\mathbf{R}, t)$ and the mesh velocity $\widehat{\mathbf{u}}(\mathbf{x}, t)$

$$\widehat{\mathbf{u}}(\mathbf{x}, t) = \left.\frac{d\mathbf{r}}{dt}\right|_{\mathbf{x}} = \frac{\partial \Psi(\mathbf{x}, t)}{\partial t} \tag{3.60}$$

namely

$$\widetilde{\mathbf{u}}(\mathbf{x}, t) = \mathbf{u}(\mathbf{R}, t) - \widehat{\mathbf{u}}(\mathbf{x}, t) = \mathbf{u}(\Phi^{-1}(\mathbf{x}, t), t) - \widehat{\mathbf{u}}(\mathbf{x}, t) \tag{3.61}$$

It can be shown using the definitions of the mappings above and the chain rule that [35]

$$\widetilde{\mathbf{u}}(\mathbf{x}, t) = \frac{\partial \Psi}{\partial \mathbf{x}}(\mathbf{x}, t)\frac{\partial \Phi}{\partial t}(\Phi^{-1}(\mathbf{x}, t), t) \tag{3.62}$$

in which $\partial \Psi / \partial \mathbf{x}$ is a Jacobian matrix. This yields the following expression for the material derivative of a quantity written $\phi(\mathbf{r}, t)$ in spatial coordinates or $\widehat{\phi}(\mathbf{x}, t)$ in ALE coordinates

$$\frac{d\phi}{dt}(\mathbf{x}, t) = \frac{\partial \widehat{\phi}}{\partial t}(\mathbf{x}, t) + \widetilde{\mathbf{u}}(\mathbf{x}, t) \cdot \nabla \phi(\Psi(\mathbf{x}, t), t) \tag{3.63}$$

in which the gradient is taken with respect to spatial coordinates. This is the fundamental equation used to derive the ALE formulation for the conservation equations (3.14), (3.18) and (3.9).

3.3.10 Immersed Boundary Methods

Closely related to the ALE method is the *immersed boundary method* [36], which does not require any changes of the grid for the fluid when fluid-structure interactions are modelled. The domain contains a set of boundary points that are connected by an elastic law. These boundary points define an immersed boundary or interface that interacts with the fluid by virtue of localised body forces applied at the points. The fluid flow uses an Eulerian description while the solid structure deformation is represented using a Lagrangian description.

The manner in which boundary conditions are imposed on the immersed boundary distinguishes different approaches. A forcing function $\mathbf{F}_s$ is introduced into the momentum equation to account for the effect of the boundary, with two main approaches, the CSF method of Sect. 3.3.8 or a *discrete* (direct) forcing method (there is also a third approach called the cut-cell method) [37].

Let $\Gamma(t)$ be the boundary at time t, which is parameterised by some $\mathbf{s} \in U \subset \mathbb{R}^2$ in a Lagrangian description. The Lagrangian to Eulerian map is given by an invertible function $\mathbf{r} = \mathbf{X}(\mathbf{s}, t)$, so that the velocity of a particle at $\mathbf{r}$ is given by

$$\frac{\partial \mathbf{X}(\mathbf{s}, t)}{\partial t} = u(\mathbf{X}(\mathbf{s}, t), t) = \int_V \mathbf{u}(\mathbf{r}, t)\, \delta\big(\mathbf{r} - \mathbf{X}(\mathbf{s}, t)\big)\, d\mathbf{r} \tag{3.64}$$

for which an initial condition $\mathbf{X}(\mathbf{s}, 0)$ is required. The volumetric surface force exerted by the solid on the liquid is given by

$$\mathbf{F}_s(\mathbf{r}, t) = \int_U \mathbf{f}(\mathbf{r}, t)\, \delta\big(\mathbf{r} - \mathbf{X}(\mathbf{s}, t)\big)\, d\mathbf{s} \tag{3.65}$$

in which $\mathbf{f}(\mathbf{r}, t)$ is the force density, which has to be specified. A common choice is one that corresponds to a boundary prevented from deviating from an initial configuration by a restoring force

$$\mathbf{f}(\mathbf{r}, t) = \beta\,(\mathbf{X}(\mathbf{s}, 0) - \mathbf{X}(\mathbf{s}, t)) \tag{3.66}$$

in which $\beta \gg 1$ is a constant. The delta function is usually mollified in order to spread the force imparted by the boundary from the Lagrangian points onto the fluid, which can lead to a violation of mass conservation.

In discrete forcing methods, the force term is instead applied after discretisation of the governing equations. This method can be further subdivided, but we shall not go into the details. The advection and viscous terms in the momentum balance are directly altered to incorporate the effect of the boundary. In this way the boundary effect is localised and only applied to a small number of cells, which is an advantage for high Re number flows, for which the force spreading described above can lead to artificially large boundary layers.

A popular and simple approach to direct forcing uses *ghost cells*. Cells that contain the boundary and neighbour a fluid cell are designated ghost cells. Values of quantities ϕ defined in these ghost cells are replaced by interpolated values such that the immersed boundary condition is enforced. For example, the velocities in the ghost cells are computed such that the interpolated values satisfy a no-slip condition. Various linear, quadratic and higher order interpolations have been used [38].

3.4 Mesoscopic Models

Mesoscopic models (on the order of a few μm to a few hundreds of μm) are used in cases where detailed micro-structural information cannot be ignored, rendering macroscopic models unsuitable. They can be used in flow batteries for detailed studies of transport phenomena in porous electrodes, the deposition of metals, including dendrite formation, and the flow of gas bubbles in an electrolyte. Below we introduce three popular mesoscopic modeling approaches.

3.4.1 Phase-Field Models

Phase-field models [39, 40] are used for interfacial problems, as a means of describing the evolution of microstructures and quantifying physical properties. These models are used extensively in materials science, e.g., for droplet formation, solidification, viscous fingering and fracture dynamics. They have much in common with the VOF method discussed in Sect. 3.3.8. Sharp interfaces are transformed to diffuse interfaces so that no explicit interface tracking is required. Phase-field modelling can solve problems with more than two phases and is readily extended to 3D. On the other hand, it is computationally intensive, so that its applications are limited.

Phase-field models introduce a variable $0 \leq \phi \leq 1$ that represents the state of the system, e.g., 1 for the pure liquid phase of a substance and 0 for the pure gas phase. Across an interface, this *field variable* or *order parameter* varies continuously, so that there is no need to track the interface [4]. The boundary between phases is therefore *diffuse*. A model for the phase-field can be constructed provided that an expression for the free energy of the system is available. The latter can be written in the general form as a functional of the order parameter and conserved quantities

$$F[c, \phi] = \int_{\Omega} d\mathbf{r} \left[f(\phi, c, T) + \frac{\epsilon_c}{2} \|\nabla c\|^2 + \frac{\epsilon_\phi}{2} \|\nabla \phi\|^2 + F(\phi, c, T) \right] \quad (3.67)$$

in which Ω is the volume occupied by the system, c is a conserved quantity such as a concentration (when there is more than one component, otherwise no c is required), T is temperature and ϵ_c, ϵ_ϕ are constants. The first term in the integrand is the bulk-phase free-energy density, which varies from one system to the next, while the second and third terms are the interface contributions. The last term represents any additional effects, such as a deformation or electrostatic energy. A functional is a mapping from some space $\mathcal{X}$ to the real or complex numbers. In this case, $\mathcal{X}$ is the joint space of allowable functions (of space and time) c and ϕ. Loosely speaking, we may think of a functional as a 'function of a function'.

The model can include more conserved quantities c_i, and more phases, with one order parameter ϕ_j for each phase, in which case the functional takes the form

$$F[\{c_i\}, \{\phi_j\}] = \int_\Omega d\mathbf{r} \Big[f(\{c_i\}, \{\phi_j\}, T) + \sum\nolimits_i \frac{\epsilon_c^i}{2} \|\nabla c_i\|^2 + \sum\nolimits_j \frac{\epsilon_\phi^j}{2} \|\nabla \phi_j\|^2 + F(\{c_i\}, \{\phi_j\}, T) \Big] \quad (3.68)$$

with constants ϵ_c^i and ϵ_ϕ^j for each of the variables.

The dynamics of conserved variables are governed by a generalised diffusion equation (Cahn-Hilliard equation [6]), while the dynamics of the non-conserved order parameters are governed by an equation that implies a linear response of the rate of evolution to a driving force (Allen-Cahn equation [41])

$$\begin{aligned} \frac{\partial c_i}{\partial t} &= \nabla \cdot \left(M_i \nabla \frac{\delta F[\{c_i\}, \{\phi_j\}]}{\delta c_i} \right) \\ \frac{\partial \phi_j}{\partial t} &= -L_j \frac{\delta F[\{c_i\}, \{\phi_j\}]}{\delta \phi_j} \end{aligned} \quad (3.69)$$

in which L and M are called mobility parameters and $\delta(\cdot)/\delta(\cdot)$ is a functional derivative. Additional stochastic terms may appear on the right-hand sides to incorporate thermal fluctuations. This leads to the following final form of the equations

$$\begin{aligned} \frac{\partial c_i}{\partial t} &= M_i \nabla \cdot \left(\nabla \left[\frac{\partial f}{\partial c_i} + \frac{\partial F}{\partial c_i} - \epsilon_c^i \nabla^2 c_i \right] \right) \\ \frac{\partial \phi_j}{\partial t} &= -L_j \left(\frac{\partial f}{\partial \phi_j} + \frac{\partial F}{\partial \phi_j} - \epsilon_\phi^j \nabla^2 \phi_j \right) \end{aligned} \quad (3.70)$$

The functional $f(\phi, c, T)$ or $f(\{c_i\}, \{\phi_j\}, T)$ can be constructed in different ways, depending on how much information is already available. For example, it can be built from a double-well function $g(\phi) = \phi^2(1-\phi)^2$ and some interpolating function $p(\phi)$, together with the free energies of the pure components in different phases, which are functions of temperature. Equation (3.70) can be solved by any number of standard methods, including the finite-volume, -element and -difference methods, mesh-free methods, and fast Fourier transforms, with various time-stepping schemes. We refer to [42] for a review of numerical implementations.

3.4.2 Kinetic Theory Models

Kinetic theory provides a bridge between the macroscopic and atomistic scales, i.e., the mesoscopic range. Ideally, we would like to track the evolution of a distribution function $f(\{\mathbf{r}_i\}, \{\mathbf{v}_i\}, t)$ describing the probability that a system of N particles of mass m will be found in the neighbourhood of some point in the $6N$-dimensional position-velocity phase space $\{\mathbf{r}_i\}$, $\{\mathbf{v}_i\}$ at some time t (ignoring relativistic effects). This, however, is too ambitious, so instead the fundamental quantity of interest is a one-particle distribution function $f(\mathbf{r}, \mathbf{v}, t)$ capturing the expected number of particles lying at some point $\mathbf{r}$, $\mathbf{v}$

$$f(\mathbf{r}, \mathbf{v}, t) = N \int_{\mathbb{R}^{3(N-1)} \times \mathbb{R}^{3(N-1)}} f(\{\mathbf{r}_i\}, \{\mathbf{v}_i\}, t) \prod_{i=2}^{N} d\mathbf{r}_i d\mathbf{v}_i \tag{3.71}$$

renaming $\mathbf{r}_1$ and $\mathbf{v}_1$ as $\mathbf{r}$ and $\mathbf{v}$, with $\int_{\mathbb{R}^3 \times \mathbb{R}^3} f(\mathbf{r}, \mathbf{v}, t) d\mathbf{r} d\mathbf{v} = M$, where M is the total mass. The first two moments yield the density and average velocity of the particles

$$\rho(\mathbf{r}, t) = \int_{\mathbb{R}^3} f(\mathbf{r}, \mathbf{v}, t) d\mathbf{v}, \quad \rho(\mathbf{r}, t)\mathbf{u}(\mathbf{r}, t) = \int_{\mathbb{R}^3} \mathbf{v} f(\mathbf{r}, \mathbf{v}, t) d\mathbf{v} \tag{3.72}$$

The forces on the particles are assumed to consist of an external force $\mathbf{F} = -\nabla V(\mathbf{r})$ for some potential V, and two-body collisions, the latter of which is sometimes referred to as the molecular chaos assumption. An analysis of the Hamiltonian dynamics of the system then leads to the following *Boltzmann transport equation*, which governs the evolution of f

$$\frac{\partial f}{\partial t} + \mathbf{v} \cdot \nabla f + \frac{\mathbf{F}}{m} \cdot \nabla_{\mathbf{v}} f = C(f, f) \tag{3.73}$$

In fact, derivation of this equation also assumes that there are two time scales in the problem, one being the time τ between collisions (scattering or relaxation time) and the other being the time τ' it takes for a collision to occur. The condition $\tau \gg \tau'$ is necessary. Therefore, we would expect that f follows its Hamiltonian evolution but with infrequent perturbations caused by the collisions, which is what happens in a dilute mixture.

The nonlinear integral in the velocity space on the right-hand side is called the *collision* term, representing inter-particle collisions. It is defined as

$$C(f,f) = \int_{\mathbb{R}^3\times\mathbb{S}^3} \sigma(\omega)\|\mathbf{w}-\mathbf{v}\|[f(\mathbf{r},\mathbf{v}',t)f(\mathbf{r},\mathbf{w}',t) - f(\mathbf{r},\mathbf{v},t)f(\mathbf{r},\mathbf{w},t)]\,d\omega d\mathbf{w} \tag{3.74}$$

in which $\mathbb{S}^3$ is the unit sphere, and ω is a scattering angle, treated as a random variable with density proportional to $\sigma(\omega)$, called a scattering cross section. $\mathbf{w}'$, $\mathbf{v}'$ are the velocities after collisions, with $\mathbf{w}$, $\mathbf{v}$ being the velocities pre-collision. For N species with particle masses m_i, we require N equations

$$\frac{\partial f_i}{\partial t} + \mathbf{v}_i\cdot\nabla f_i + \frac{\mathbf{F}}{m_i}\cdot\nabla_{\mathbf{v}_i} f_i = \sum\nolimits_j C_{ij}(f_i, f_j) \tag{3.75}$$

in which the collision operator is generalised to

$$C_{ij}(f_i,f_j) = \int_{\mathbb{R}^3\times\mathbb{S}^3} \sigma_{ij}(\omega)\|\mathbf{v}_i-\mathbf{w}\|[f_i(\mathbf{r},\mathbf{u}',t)f_j(\mathbf{r},\mathbf{w}',t) - f_i(\mathbf{r},\mathbf{v},t)f_j(\mathbf{r},\mathbf{w},t)]\,d\omega d\mathbf{w} \tag{3.76}$$

The collision operator is highly complex, so much work has been carried out to find simpler, manageable forms. The Bhatnagar-Gross-Krook (BGK) method [5] uses a linear approximation

$$\frac{\partial f}{\partial t} + \mathbf{v}\cdot\nabla f + \frac{\mathbf{F}}{m}\cdot\nabla_{\mathbf{v}} f = \frac{1}{\tau}(f-g) \tag{3.77}$$

in which τ is the collision time and $g(\mathbf{r},\mathbf{u},t)$ is a local Maxwell-Boltzmann distribution (also called a Maxwellian)

$$g(\mathbf{r},\mathbf{u},t) = \rho(\mathbf{r},t)\left(\frac{m}{2\pi k_B T(\mathbf{r},t)}\right)^{3/2}\exp\left(-\frac{m\|\mathbf{v}-\mathbf{u}(\mathbf{r},t)\|^2}{2k_B T(\mathbf{r},t)}\right) \tag{3.78}$$

where k_B is Boltzmann's constant. ρ and $\mathbf{u}$ are obtained from (3.72), while T is obtained from the second moment about the mean velocity

$$\frac{3}{2}k_B T(\mathbf{r},t) = \int_{\mathbb{R}^3}\frac{m}{2}\|\mathbf{v}-\mathbf{u}(\mathbf{r},t)\|^2 f(\mathbf{r},\mathbf{v},t)d\mathbf{v} \tag{3.79}$$

Close to equilibrium, an asymptotic analysis on the Boltzmann equation (Chapman-Enskog analysis [5]) yields the Euler equations and the Navier-Stokes equations at the first two terms in an expansion in terms of an effective Knudsen number. This demonstrates that the Boltzmann equation reproduces the correct macroscopic behaviour of fluids, at least to a large degree (higher order terms contain singularities). For dense gases, the Boltzmann equation does not apply and modifications are required.

3.4.3 The Lattice-Boltzmann Model

Initially see as a means to solve the Boltzmann equation, the *lattice-Boltzmann* method (LBM) now defines an area in its own right. It is a mesoscopic approach largely used for elucidating the behaviour of complex fluid flows, as well studying mass, heat and charge transport phenomena. In the LBM, the collision operator is simplified and the degrees of freedom are restricted by defining discrete velocities $\mathbf{e}_i$ on a lattice, meaning that the fluid particles can only move along directions defined by the $\mathbf{e}_i$. Provided sufficient symmetry is built into this discrete formulation, it is able (perhaps remarkably) to recover the corresponding macroscopic equations governing the fluid, which can again be demonstrated by a Chapman-Enskog analysis.

Several choices exist for the discrete set of velocities, or lattices, with a dependence on the dimension of the domain. The lattices used are denoted DLQM, in which M denotes the number of discrete velocities and L denotes the dimension of the physical space, with typical examples being D3Q19 and D3Q15. One-particle (density) distribution functions are advanced in time using the following *lattice-Boltzmann equation* (LBE)

$$f_i(\mathbf{r} + \mathbf{e}_i \Delta t, t + \Delta t) = f_i(\mathbf{r}; t) + \sum\nolimits_j \Omega_{ij}(f_j^{eq}(\mathbf{r}; t) - f_j(\mathbf{r}; t)) \tag{3.80}$$

in which $f_i(\mathbf{r}; t)$ is the probability distribution at the point $(\mathbf{e}_i, \mathbf{r})$ in phase space at time t. The term on the left-hand side of (3.80) approximates the operator $\partial_t f + \mathbf{v} \cdot \nabla f$ in the Boltzmann equation (3.73), while $C(f, f')$ is approximated by $\sum_j \Omega_{ij}(f_j^{eq}(\mathbf{r}; t) - f_j(\mathbf{r}; t))$, a simplified description of particle collisions with relaxation to a local equilibrium distribution $f_j^{eq}(\mathbf{r}; t)$. Controlling this relaxation is a *scattering matrix* Ω with elements Ω_{ij}. The eigenvalues of Ω determine the timescale in which the kinetic moments equilibrate locally.

The equilibrium distribution $f_j^{eq}(\mathbf{r}; t)$ is a discrete equivalent of a local Maxwellian in which the velocity and density are given by

$$\rho = \sum\nolimits_i f_i(\mathbf{r}, t), \quad \rho\mathbf{u} = \sum\nolimits_i \mathbf{e}_i f_i(\mathbf{r}, t) \tag{3.81}$$

respectively, i.e., moments of the currently computed discrete distribution. The scattering matrix is often assumed to be diagonal, which results in the Lattice BGK (LBGK) method, since it is discrete version of the BGK approach (3.77) [5]. Although this approximation will lead to artefacts in the viscosity, the aberrant term that appears can be absorbed into the true viscosity.

In the analytical characteristic integral LB, the velocity space discretisation is performed directly on the LBGK equation. The velocity space discretisation chosen is intimately linked to the equilibrium distribution used. The derivation starts with a small Mach-number approximation and must ensure that the macroscopic equations (at least mass continuity and momentum conservation) can be recovered by the discretised equation.

The approach is to match the relative velocity moment integrals arising from the discretised equilibrium distribution with those of the full Maxwellian. Using only a second-order Taylor expansion of the Maxwellian as the equilibrium distribution guarantees that moments up to second order match. This matching involves a quadrature that can be carried out with Hermite polynomials to yield the discrete velocity vectors, along with the weights used in the expansion defining $f_j^{eq}(\mathbf{r}; t)$ (in Chap. 4 we present the D3Q19 and D3Q7 velocities and weights). We do not provide the details of the quadrature since it is straightforward but lengthy.

In practice, the computations are performed in two parts: a *collision* step, in which the f_i are updated according to the collision operator

$$f_i^*(\mathbf{r}, t) = f_i(\mathbf{r}, t) - \frac{1}{\tau}[f_i(\mathbf{r}, t) - f_i^{eq}(\mathbf{r}, t)] \tag{3.82}$$

and a *streaming* step, in which the f_i are moved, with special attention paid to boundary lattice nodes

$$f_i(\mathbf{r} + \mathbf{e}_i\delta_t, t + \delta_t) = f_i^*(\mathbf{r}, t) \tag{3.83}$$

The grid is typically discretised in $\mathbf{r}$ such that the distance between neighbouring grid points is $\mathbf{e}_i\delta_t$, so that between t and $t + \delta_t$, $f_i(\mathbf{r}, t)$ will propagate along the direction $\mathbf{e}_i$ to arrive at the neighbouring point.

The BGK collision operator can be generalised to $\Omega(\mathbf{f} - \mathbf{f}^{eq})$, in which $\mathbf{f}$ is a vector containing the f_i (and likewise $\mathbf{f}^{eq}$), by introducing a *collision matrix* Ω rather than using a single relaxation time, the latter of which is equivalent to $\Omega = \tau^{-1}\mathbf{I}$. This formulation affords a greater number of degrees of freedom for a linearised model of the collision operator and is the basis of the so-called *multiple-relaxation-time* (MRT) model [43].

In the MRT model, in addition to the velocity space, a moment space is introduced, based upon the number of discrete velocities. These moments of the distribution function are physical quantities such as density (0th order), momentum (1st order), energy (2nd order), heat flow (3rd order), and so on. The advantage of this formulation is that it incorporates much of the physics directly into the model, facilitating the identification of different relaxation times with physical parameters.

Defining a vector of the moments $\mathbf{m}$, with corresponding equilibrium $\mathbf{m}^{eq}$, a transformation matrix $\mathbf{M}$ is used to relate the distribution function vector to the moment vector *via* $\mathbf{m} = \mathbf{M}\mathbf{f}$, transforming $\mathbf{f}$ from velocity to moment space. We can now write the discrete collision operator as

$$\mathbf{M}^{-1}\mathbf{S}(\mathbf{m} - \mathbf{m}^{eq}) \tag{3.84}$$

in which we set $\Sigma = \mathbf{M}^{-1}\mathbf{S}\mathbf{M}$, where the collision matrix $\mathbf{S}$ in moment space is diagonal. Essentially, the algorithm transforms the distribution to moment space, performs the collision step to obtain $\Delta\mathbf{m} = \mathbf{m} - \mathbf{m}^{eq}$, transforms back to velocity space to obtain $\Delta\mathbf{f} = \mathbf{f} - \mathbf{f}^{eq}$, and then performs the streaming step.

Application of the LB method to multiphase flows is more challenging, especially when the fluids (e.g., liquid and gas) exhibit large density ratios. There are roughly speaking 4 approaches [44]: the colour-gradient method, the pseudopotential method, the free-energy method, and the phase-field method. The phase-field approach is based on the phase-field theory introduced in Sect. 3.4.1, which describes the interface dynamics using an order parameter or phase-field ϕ obeying the Cahn-Hilliard equation (3.69)

$$\frac{\partial \phi(\mathbf{r},t)}{\partial t} + \nabla \cdot (\phi(\mathbf{r},t)\,\mathbf{u}) = \nabla \cdot (M \nabla \mu_{\phi}) \tag{3.85}$$

in which μ_{φ} is the chemical potential and M is the mobility parameter. In Chap. 4 we implement the MRT framework, while in Chap. 5, we implement the phase-field approach.

3.5 Molecular Dynamics Simulations

In classical molecular dynamics (MD) simulations, the motion of individual particles is tracked using the laws of classical mechanics, namely Newtons' second law [45]. This allows us to study the structural properties of molecular assemblies with the inclusion of microscopic interactions. The connection between this microscopic description and the macroscopic description of the system is obtained through statistical mechanics.

We appeal to the Born-Oppenheimer approximation, separating the molecular from the electronic motion (see Sect. 3.6). The nuclei are assumed to move slowly enough that classical laws of motion apply. We can quantify this notion in terms of the de Broglie wavelength of a single particle of mass m with kinetic energy proportional to $k_B T$

$$\lambda = \frac{h}{\sqrt{2\pi m k_B T}} \tag{3.86}$$

in which h is Planck's constant. If the average interatomic distance d_i is much greater than λ, we may treat the particles classically, whereas if $d_i = \mathcal{O}(\lambda)$, the wavelike behaviour of the particles is non-negligible.

For a system of N particles with positions $\mathbf{r}_i$ and masses m_i, we can write down the classical laws of motion in terms of a conservative force with associated potential energy function $V(\mathbf{r}_1, \ldots, \mathbf{r}_N)$

$$m_i \frac{d^2\mathbf{r}_i}{dt^2} = \mathbf{F}_i = -\nabla_i V(\mathbf{r}_1, \ldots, \mathbf{r}_N) \tag{3.87}$$

in which ∇_i denotes the gradient operator with respect to $\mathbf{r}_i$. In terms of the Hamiltonian, representing the energy of the system, we can write

$$\begin{aligned}\frac{d\mathbf{r}_i}{dt} &= \nabla_{\mathbf{p}_i} H \\ \frac{d\mathbf{p}_i}{dt} &= -\nabla_i H \\ H(\mathbf{r}_1, \ldots, \mathbf{r}_N, \mathbf{p}_1, \ldots, \mathbf{p}_N) &= \sum\nolimits_i \frac{\|\mathbf{p}_i\|^2}{2m_i} + V(\mathbf{r}_1, \ldots, \mathbf{r}_N)\end{aligned} \tag{3.88}$$

in which $\mathbf{p}_i$ are the momenta and H is the Hamiltonian.

There are two important properties of these equations. The first is that they do not change under a reversal of time $t \mapsto -t$. Therefore, MD is a deterministic technique—given an initial condition, the subsequent evolution is completely determined by the equations. A second important property is the conservation of energy or the Hamiltonian

$$\begin{aligned}\frac{dH}{dt} &= \sum\nolimits_i \left[\frac{d\mathbf{r}_i}{dt} \cdot \nabla_i H + \frac{d\mathbf{p}_i}{dt} \cdot \nabla_{\mathbf{p}_i} H\right] \\ &= \sum\nolimits_i \left[\nabla_{\mathbf{p}_i} H \cdot \nabla_i H - \nabla_i H \cdot \nabla_{\mathbf{p}_i} H\right] = 0\end{aligned} \tag{3.89}$$

This second property (conservation of total system energy) provides an important link between MD simulations and statistical mechanics, explained later when we discuss ensembles.

3.5.1 *Interatomic Potentials*

Characterising the interatomic potential function V is the major challenge in MD simulations, alongside finding efficient solution techniques. It summarises all of the information about electron interactions in the system and is written in a condensed form that depends only on the positions of the nuclei. It thus represents an effective *binding energy* for an atom contained in an N-atom configuration, in which individual atoms have positions $\mathbf{r}_i$. Although in principle V can be quantified using costly electronic-structure methods, leading to *ab-initio* MD, in most cases the interatomic potential is empirically derived for a particular system. The procedure usually involves writing down some functional form for V and then fitting undetermined parameters to empirical data, which leads to what is termed a *force field* [46].

To simplify the notation we set

$$\mathbf{r} := \{\mathbf{r}_i\}, \quad \mathbf{p} := \{\mathbf{p}_i\} \tag{3.90}$$

We can in principle decompose the potential into a sum of terms representing n−body, $n = 2, 3, \ldots,$ interactions between the particles [46]

$$V(\mathbf{r}) = \frac{1}{2}\sum\nolimits_{i\neq j} V_{\text{2-body}}(\mathbf{r}_i, \mathbf{r}_j) + \frac{1}{3!}\sum\nolimits_{i\neq j\neq k} V_{\text{3-body}}(\mathbf{r}_i, \mathbf{r}_j, \mathbf{r}_k) + \ldots \tag{3.91}$$

in which the factors of $1/2$ and $1/6$ correct for double counting etc.. We can also include a term representing an external force, e.g., for ions in an electric field. Retaining only the 2-body term leads to *pair potentials*, retaining the first two terms leads to 3-body potentials and so on. Pair potentials take the form

$$V(\mathbf{r}) = \frac{1}{2}\sum\nolimits_{i\neq j} V_{2\text{-body}}(r_{ij}), \quad r_{ij} := \|\mathbf{r}_i - \mathbf{r}_j\| \tag{3.92}$$

due to the invariance under rigid-body motions, so that the potential depends only on the relative distances between pairs of atoms. For example, for ions with charges Z_i, the Coulomb potential is

$$V(\mathbf{r}) = \frac{1}{2}\sum\nolimits_{i\neq j} \frac{1}{4\pi\epsilon_0}\frac{Z_i Z_j}{r_{ij}} \tag{3.93}$$

in which ϵ_0 is the permittivity of a vacuum.

The *Lennard-Jones* potential models soft repulsive and attractive, namely van der Waals, interactions

$$V(\mathbf{r}) = \frac{1}{2}\sum\nolimits_{i\neq j} 4\varepsilon_{ij}\left[\left(\frac{\sigma_{ij}}{r_{ij}}\right)^{12} - \left(\frac{\sigma_{ij}}{r_{ij}}\right)^{6}\right] \tag{3.94}$$

in which ε_{ij} and σ_{ij} are pairwise Lennard-Jones parameters. For a single substance ε_{ii} is the depth of the local minimum of a potential well, called the dispersion energy, while σ_{ii} is the distance at which the potential energy reaches a zero value. Pairwise values are obtained from these values by using one of several empirically derived relationships. The Lennard-Jones potential is short-range, since it decays rapidly by virtue of the exponent 6 in Eq. (3.94). Conversely, Coulomb interactions are considered long-range. Short range interactions can be simplified by defining a *cut-off* distance r_c, such that only those particles j with $\|\mathbf{r}_i - \mathbf{r}_j\| < r_c$ are included in the summation over i.

Pair potentials lead to efficient calculations, but are usually not sufficient for most materials, particularly liquids and solids. In that case, a 3-body or higher potential is required. For metals, the most common potential is derived from the *embedded atom method* (EAM), in the form

$$V(\mathbf{r}) = \sum\nolimits_i U(\rho_i) + \frac{1}{2}\sum\nolimits_{i\neq j} \phi(r_{ij}) \tag{3.95}$$

in which $U(\cdot)$ is called an embedding function, of the electron density ρ_i at $\mathbf{r}_i$ (see Eq. (3.155) in Sect. 3.6.3). $\phi(\cdot)$ is another pair potential representing interactions between nuclei. The first term can be interpreted as the energy required to embed an atom i into the electron density ρ, and ρ_i is normally approximated by some fitting function f such that

$$\rho_i = \sum\nolimits_j f(r_{ij}) \tag{3.96}$$

The functions U and ϕ are also treated as fitting functions, e.g., the Finnis-Sinclair potential $U(\rho_i) = K\sqrt{\rho}$ for some constant K.

The *Stillinger-Weber* potential is another example, used for silicon. It takes the general form

$$V(\mathbf{r}) = \frac{1}{2}\sum_{i\neq j}\psi(r_{ij}) + \frac{1}{6}\sum_{i\neq j\neq k}\chi(r_{ij}, r_{ik}, \theta_{ijk}) \tag{3.97}$$

by including, in addition to a pair potential $\psi(\cdot)$, a 3-body term with a function $\chi(r_{ij}, r_{ik}, \theta_{ijk})$ of pairwise distances r_{ij} and r_{ik}, as well as the *bond angle* θ_{ijk} between r_{ij} and r_{ik}.

3.5.2 Force Fields and Molecular Mechanics

Molecular mechanics is another way to study molecular structures, but in contrast to MD it does not perform simulations of the atomic motion. It is based on a description of the molecular system that relies on certain assumptions. Each particle is assigned a radius, a polarisability, and a constant net charge, and bonded interactions between particles in the system are modelled as spring-like forces. It calculates the potential energy as a function of the particle coordinates based on a force field, which when fitted can be used for MD simulations. A general 3-body model for a force field decomposes the forces as follows [46]

$$V(\mathbf{r}) = V_{\text{bonded}} + V_{\text{non-bonded}} \tag{3.98}$$

in which V_{bonded} is a term that accounts for covalent-bond interactions, while $V_{\text{non-bonded}}$ is a contribution arising from non-bonded interactions, such as the electrostatic and van der Waals interactions. The bonded interactions can be subdivided into contributions from changes in the bond lengths, the bond angles and the torsion (dihedral) angles. Thus, we may write

$$V(\mathbf{r}) = \frac{1}{2}\sum_{i\neq j}\frac{1}{4\pi\epsilon_0}\frac{Z_iZ_j}{r_{ij}} + \frac{1}{2}\sum_{i\neq j}4\varepsilon_{ij}\left[\left(\frac{\sigma_{ij}}{r_{ij}}\right)^{12} - \left(\frac{\sigma_{ij}}{r_{ij}}\right)^{6}\right]$$
$$+ \sum_{\text{bonds}} k^{\text{bond}}\left(r_{ij} - r_{ij}^{\text{eq}}\right)^2 + \sum_{\angle} k^{\angle}\left(\theta_k - \theta_k^{\text{eq}}\right)^2 + \sum_{\measuredangle} k^{\measuredangle}\left(\cos(\omega_n) - \cos(\omega_n^{\text{eq}})\right)^2 \tag{3.99}$$

which sums over all bonds, of length r_{ij}, angles ($\angle$) θ_k and torsion angles ($\measuredangle$) ω_n. The superscript *eq* represents equilibrium properties and the constants k^j are fitted through experimental measurements or electronic-structure calculations.

Aside from using force fields in MD simulations, they can also be used for *energy minimisation*, which discovers the molecular structures at local and global energy minima. At finite temperatures, the molecule adopts the structures corresponding to these minima, which therefore dominate the properties of the molecule. Methods for global optimisation include stochastic methods such as simulated annealing and Markov Chain Monte Carlo, as well as gradient-based methods such as steepest and conjugate gradient descent. Evolutionary algorithms are perhaps the best for finding global maxima and are frequently employed for energy minimisation.

3.5.3 Ensembles and Statistical Averages

As mentioned above, the link between the microscopic simulations in MD and the macroscopic observables, such as thermodynamic properties at equilibrium and transport coefficients, is statistical mechanics. Statistical mechanics is based on the concept of an *ensemble*, which is a family of microscopic configurations (microstates) of a system that all yield the same macroscopic properties (macrostates) [47]; the same macroscopic property can be consistent with multiple microstates. The implication of this hypothesis is that macroscopic observables of a given system can be determined by *ensemble averages*. Let us make the ansatz that the observables O are functions of the microstate, which is uniquely defined by $(\mathbf{r}, \mathbf{p})$, that is $O = O(\mathbf{r}, \mathbf{p})$.

Different ensembles can be characterised by fixed values of different thermodynamic variables [47].

1. The micro-canonical (NVE) ensemble pertains to an isolated system with a constant number of particles, a constant energy and a constant volume
2. The canonical (NVT) ensemble pertains to a system in contact with a heat bath, with a constant temperature, number of particles and volume
3. The grand canonical (μVT) ensemble pertains to a system in contact with heat and particle baths, with a constant temperature, volume and chemical potential
4. The isothermal-isobaric (NpT) ensemble pertains to a system in which the temperature, pressure and number of particles are kept constant

In an MD simulation, the particles move along some trajectory given some initial condition. To compute ensemble averages would therefore require many different MD simulations with different initial conditions. To avoid this issue, which would be too costly, we need to make use of an hypothesis that is central to statistical mechanics. As we have already seen from Eq. (3.89), a system containing a constant number of particles in a fixed volume and evolving according to the MD equations will conserve energy, i.e., the microstates of the system are samples from the microcanonical ensemble. To work with other ensembles required modification of the Hamiltonian system (discussed later).

The system evolves on some manifold called a *constant energy surface*, defined by $H(\mathbf{r}, \mathbf{p}) = E$ for a fixed energy E. At each time t, the microstate $(\mathbf{r}, \mathbf{p})$ lies on the surface. Under the *ergodic hypothesis*, the system will visit every possible microstate on this constant energy surface given infinite time. This implies that averages of some observable O over a single trajectory are equivalent to averages over the NVE ensemble

$$\langle O \rangle = \int_{\mathbb{R}^{3N} \times \mathbb{R}^{3N}} p(\mathbf{r}, \mathbf{p}) O(\mathbf{r}, \mathbf{p}) d\mathbf{r} d\mathbf{p} = \lim_{T \to \infty} \int_0^T O(\mathbf{r}(t), \mathbf{p}(t)) dt \tag{3.100}$$

Here $\langle O \rangle$ denotes the ensemble or phase average, which is determined by the probability density of (all) microstates $p(\mathbf{r}, \mathbf{p})$ (roughly, speaking, $p(\mathbf{r}, \mathbf{p}) d\mathbf{r} d\mathbf{p}$ is the probability of finding the system in the vicinity of the state $(\mathbf{r}, \mathbf{p})$), while the last integral is the time average of a single trajectory.

3.5.4 The Micro-canonical Ensemble and Macroscopic Observables

The volume enclosed by the hypersurface $H(\mathbf{r}, \mathbf{p}) = E$ is given by [48]

$$\Omega(E) = \int_{\mathbb{R}^{3N} \times \mathbb{R}^{3N}} H(E - H(\mathbf{r}, \mathbf{p})) d\mathbf{r} d\mathbf{p} \tag{3.101}$$

in which $H(\cdot)$ is the Heaviside function. Ω defines a *density of states* in the micro-canonical ensemble as follows

$$D(E) = \frac{d\Omega(E)}{dE} \tag{3.102}$$

In essence, D defines the density of microstates $(\mathbf{r}, \mathbf{p})$ that are enclosed by the manifolds or hypersurfaces defined by $H = E$ and $H = E + \delta E$. These hypersurfaces enclose a *hypershell*

$$\Sigma(E, \delta E) = \{\{\mathbf{r}, \mathbf{p}\} \mid E \leq H(\mathbf{r}, \mathbf{p}) \leq E + \delta E\} \tag{3.103}$$

We can define the probability distribution function $p(\mathbf{r}, \mathbf{p}; E, \delta E)$ for the micro-canonical as being constant within $\Sigma(E, \delta E)$ and equal to zero elsewhere

$$p(\mathbf{r}, \mathbf{p}; E, \delta E) = \begin{cases} \dfrac{1}{\Omega(E + \delta E) - \Omega(E)} & \{\mathbf{r}, \mathbf{p}\} \in \Sigma(E, \delta E) \\ 0 & \text{otherwise} \end{cases} \tag{3.104}$$

taking into account that $\int p \, d\mathbf{r} d\mathbf{p} = 1$.

For an observable $O(\mathbf{r}, \mathbf{p})$, we can write down the phase average as

$$\begin{aligned} \langle O \rangle &= \lim_{dE \to 0} \int_{\mathbb{R}^{3N} \times \mathbb{R}^{3N}} O(\mathbf{r}, \mathbf{p}) p(\mathbf{r}, \mathbf{p}; E, \delta E) d\mathbf{r} d\mathbf{p} \\ &= \lim_{dE \to 0} \frac{1}{\Omega(E + \delta E) - \Omega(E)} \int_{\mathbb{R}^{3N} \times \mathbb{R}^{3N}} O(\mathbf{r}, \mathbf{p}) d\mathbf{r} d\mathbf{p} \end{aligned} \tag{3.105}$$

from which, after noting that the derivative of a Heaviside function is a delta function, we obtain

$$\langle O\rangle = \frac{1}{D(E)}\int_{\mathbb{R}^{3N}\times\mathbb{R}^{3N}} O(\mathbf{r},\mathbf{p})\delta(E-H(\mathbf{r},\mathbf{p}))d\mathbf{r}d\mathbf{p} \tag{3.106}$$

which yields the micro-canonical distribution

$$p(\mathbf{r},\mathbf{p}) = \frac{\delta(E-H(\mathbf{r},\mathbf{p}))}{D(E)} \tag{3.107}$$

Macroscopic variables can now be obtained using this density. The internal energy U is given by $U = \langle H\rangle$ and for an isolated system $U = E$. A well-know thermodynamic relationship is

$$\frac{\partial s(E)}{\partial E} = \frac{1}{T} \tag{3.108}$$

in which $s(E)$ is the *entropy*. Boltzmann's postulate that $s = k_B \ln\Theta$, in which Θ is the number of microstates that are consistent with a given set of macroscopic observables, can be restated as (without proof)

$$s(E) = \lim_{N\to\infty} k_B \ln D(E) = k_B \ln \Omega(E) \tag{3.109}$$

Let us now pick a component x_i of any of the positions or momenta in $\{\mathbf{r},\mathbf{p}\}$ in phase space. Then

$$\begin{aligned}
\left\langle x_i\frac{\partial H}{\partial x_j}\right\rangle &= \frac{1}{D(E)}\frac{\partial}{\partial E}\int_{\Omega(E)} x_i\frac{\partial H}{\partial x_j}d\mathbf{r}d\mathbf{p}\\
&= \frac{1}{D(E)}\frac{\partial}{\partial E}\left[\int_{\Omega(E)}\frac{\partial}{\partial x_j}(x_i(H-E))\,d\mathbf{r}d\mathbf{p} - \int_{\Omega(E)}\delta_{ij}(H-E)d\mathbf{r}d\mathbf{p}\right]\\
&= \frac{1}{D(E)}\frac{\partial}{\partial E}\left[\int_{s(E)} x_i(H-E)n_j d\mathbf{r}d\mathbf{p} - \int_{\Omega(E)}\delta_{ij}(H-E)d\mathbf{r}d\mathbf{p}\right]\\
&= \frac{\delta_{ij}}{D(E)}\frac{\partial}{\partial E}\int_{\Omega(E)}(E-H)d\mathbf{r}d\mathbf{p}\\
&= \frac{\delta_{ij}}{D(E)}\left[\int_{\mathbb{R}^{3N}\times\mathbb{R}^{3N}} H(E-H)d\mathbf{r}d\mathbf{p} + \int_{\mathbb{R}^{3N}\times\mathbb{R}^{3N}}(E-H)\delta(E-H)d\mathbf{r}d\mathbf{p}\right]\\
&= \frac{\delta_{ij}}{D(E)}\int_{\mathbb{R}^{3N}\times\mathbb{R}^{3N}} H(E-H)d\mathbf{r}d\mathbf{p}
\end{aligned} \tag{3.110}$$

in which δ_{ij} is the Kronecker delta and n_j is the j-th component of the unit normal to the hypersurface $s(E)$, on which $H = E$. The integral on the last line of (3.110) is the volume $\Omega(E)$, so that (using (3.109))

$$\left\langle x_i\frac{\partial H}{\partial x_j}\right\rangle = \delta_{ij}\frac{\Omega(E)}{d\Omega/dE} = \frac{\delta_{ij}}{d\ln\Omega(E)/dE} = \delta_{ij}\frac{k_B}{ds(E)/dE} = \delta_{ij}k_B T \tag{3.111}$$

The result in (3.111) is called the *equipartition theorem*. Setting $i = j$ and $x_i = p_i$, which is the i-th component of one of the $\mathbf{p}$, in Eq. (3.111) yields

$$\left\langle p_i \frac{\partial H}{\partial p_i} \right\rangle = \left\langle \frac{p_i^2}{2m_i} \right\rangle = \frac{1}{2} k_B T \tag{3.112}$$

Summing over all $3N$ degrees of freedom (over all particles and all spatial directions) yields the following expression for the temperature

$$T = \frac{2}{3} \frac{\mathcal{K}_{\mathrm{vibr}}}{N k_B}, \quad \mathcal{K}_{\mathrm{vibr}} = \sum_{i=1}^{3N} \left\langle \frac{p_i^2}{2m_i} \right\rangle \tag{3.113}$$

in which $\mathcal{K}_{\mathrm{vibr}}$ is the average energy of vibration. Choosing instead $x_i = r_i$ (the i-th component of one of the $\mathbf{r}$), setting $i = j$ and summing over the 3 degrees of freedom for a particle n yields

$$3k_B T = \sum_{i=1}^{3} \left\langle r_i \frac{\partial H}{\partial r_i} \right\rangle = \sum_{i=1}^{3} \left\langle r_i \frac{\partial V}{\partial r_i} \right\rangle = -\left\langle \mathbf{r}_n \cdot \mathbf{f}_n \right\rangle, \quad \mathbf{F}_n = -\nabla_n V \tag{3.114}$$

in which $\mathbf{F}_n$ is the force experienced by particle n (3.87). Summing over all particles then leads to

$$-\sum_{n=1}^{N} \left\langle \mathbf{r}_n \cdot \mathbf{f}_n \right\rangle = -\mathcal{W} := 3N k_B T = 2\mathcal{K}_{\mathrm{vibr}} \tag{3.115}$$

which is called the *virial theorem*. Other macroscopic variables can be determined in a similar manner, e.g., the pressure

$$p = \frac{1}{3V} \sum_{n=1}^{N} \left\langle \frac{\mathbf{p}_n \cdot \mathbf{p}_n}{2m_n} + \mathbf{r}_n \cdot \mathbf{f}_n \right\rangle \tag{3.116}$$

3.5.5 *Solving the Hamiltonian System*

The goal of MD is to solve the system (3.87) together with suitable initial positions and velocities. A solution to this ODE problem is often referred to as 'integrating' the equations of motion since we essentially integrate the right-hand side twice with respect to t to obtain $\mathbf{r}_i$. Solutions are very challenging given the size of a typical system, leading to high memory requirements, and the existence of local energy minima.

The most popular method is called the *velocity-Verlet* algorithm. Like most other time-stepping methods in use for MD, it is an explicit method, with a low memory requirement compared to implicit schemes. We can choose to solve the problem in a *zero* temperature or *finite* temperature case. The former corresponds to molecular statics

$$\mathbf{F}_i = -\nabla_i V = \mathbf{0}, \quad i = 1, \ldots, N \tag{3.117}$$

With a constant time step Δt and times $t_n = n\Delta t$, we can use a Taylor expansion on Eq. (3.87) to obtain

$$\begin{aligned}\mathbf{r}_i(t_n \pm \Delta t) &= \mathbf{r}_i(t_n) \pm \Delta t \frac{d\mathbf{r}_i}{dt}(t) + \frac{1}{2}(\Delta t)^2 \frac{d^2\mathbf{r}_i}{dt^2}(t) + \mathcal{O}\left((\Delta t)^3\right) \\ &= \mathbf{r}_i(t_n) \pm \Delta t \frac{d\mathbf{r}_i}{dt}(t) + (\Delta t)^2 \frac{\mathbf{f}_i}{2m_i} + \mathcal{O}\left((\Delta t)^3\right)\end{aligned} \tag{3.118}$$

which when added together yields the Verlet algorithm [49]

$$\mathbf{r}(t + \Delta t) = 2\mathbf{r}(t) - \mathbf{r}(t - \Delta t) + (\Delta t)^2 \frac{\mathbf{f}_i}{m_i} + \mathcal{O}\left((\Delta t)^4\right) \tag{3.119}$$

noting that the third-order terms cancel. This presents two problems, the first being that at $n = 1$ (start of the algorithm to calculate positions are $n = 2$) we have the positions and velocities at $n = 0$ *via* the initial conditions, but still require the positions at $n = 1$. These can be approximated to within $\mathcal{O}\left((\Delta t)^3\right)$ using the forces and velocities at $n = 0$. The second problem is that in MD simulations we require the velocities (or momenta), whereas the Verlet algorithm furnishes only the positions. The velocities can be computed from

$$\mathbf{v}_i(t) = \frac{\mathbf{r}_i(t + \Delta t) - \mathbf{r}_i(t - \Delta t)}{2\Delta t} + \mathcal{O}\left((\Delta t)^2\right). \tag{3.120}$$

which introduces more error, albeit small.

An alternative method that avoids these issues leads to the velocity-Verlet algorithm [50], of the same order of accuracy and based on the equations

$$\begin{aligned}\mathbf{r}_i(t + \Delta t) &= \mathbf{r}_i(t) + \mathbf{v}_i(t)\,\Delta t + \frac{1}{2}\mathbf{f}_i(t)(\Delta t)^2, \\ \mathbf{v}_i(t + \Delta t) &= \mathbf{v}_i(t) + \frac{\mathbf{f}_i(t) + \mathbf{f}_i(t + \Delta t)}{2}\Delta t\end{aligned} \tag{3.121}$$

It is implemented using fractional steps, as follows

1. First calculate

$$\mathbf{v}_i\left(t + \frac{1}{2}\Delta t\right) = \mathbf{v}_i(t) + \frac{1}{2}\mathbf{f}_i(t)\,\Delta t$$

2. Next calculate the position at the next time step

$$\mathbf{r}_i(t + \Delta t) = \mathbf{r}_i(t) + \mathbf{v}_i\left(t + \frac{1}{2}\Delta t\right)\Delta t$$

3. Lastly calculate the velocity at the next time step

$$\mathbf{v}_i(t + \Delta t) = \mathbf{v}_i\left(t + \tfrac{1}{2}\,\Delta t\right) + \frac{1}{2}\,\mathbf{f}(t + \Delta t)\Delta t$$

Another popular method is the *leapfrog* algorithm, which again calculates velocities at $t + (1/2)\Delta t$ to update the positions

$$\begin{aligned}\mathbf{r}_i(t + \Delta t) &= \mathbf{r}_i(t) + \mathbf{v}_i\,(t + 12\Delta t)\,\Delta t \\ \mathbf{v}_i\,(t + 12\Delta t) &= \mathbf{v}_i\,(t - 12\Delta t) + \mathbf{f}_i(t)\Delta t\end{aligned} \tag{3.122}$$

Since the velocities are calculated at half steps, a further approximation is required

$$\mathbf{v}_i\,(t + \Delta t) = \frac{1}{2}\,(\mathbf{v}_i\,(t + 12\Delta t) + \mathbf{v}_i\,(t - 12\Delta t)) \tag{3.123}$$

3.5.6 Thermostats and Other Ensembles

Finite-temperature simulations are more involved, requiring that the temperature is controlled or monitored. One method involves rescaling the velocities, but it introduces discontinuities. The most common way to control temperature is to use a *thermostat*. In experiments, temperature is often used as a control *via* a heat bath, which is not consistent with the micro-canonical ensemble, but rather with the NVT ensemble. For simulations that pertain to this ensemble the Hamiltonian system needs to be modified. Note that in the canonical (NVT) ensemble the probability density $p(\mathbf{r}, \mathbf{p})$ is a Boltzmann distribution.

There are a number of ways to modify the Hamiltonian system, each method referred to as a thermostat. In the case of the Langevin thermostat, the following *Langevin equation* [51] is solved (with desired temperature T)

$$m_i\,\frac{d^2\mathbf{r}_i}{dt^2} = -\nabla_i V(\mathbf{r}) - \gamma_i\,m_i\,\frac{d\mathbf{r}_i}{dt} + \sqrt{2\,m_i\,\gamma_i\,k_B T}\,\mathbf{R}(t) \tag{3.124}$$

in which γ_i is a damping constant and $\mathbf{R}(t)$ is a Gaussian (random) process with mean 0 and covariance function given by $\delta(t - t')$ (Sect. 6.7 explains Gaussian processes). The additional two terms represent frictional forces (proportional to velocity) and particle collisions. The random term is introduced in order to force the temperature towards the desired value, while the damping term removes excess kinetic energy introduced into the system as a result of the forcing.

The most often used thermostat is the *Nosé-Hoover* thermostat [52], which incorporates a fictitious variable ψ that represents friction, decelerating or accelerating the motion of the particles until the desired temperature value T is attained. The equations in this case are

$$m_i \frac{d^2\mathbf{r}_i}{dt^2} = -\nabla_i V(\mathbf{r}) - m_i \psi(t) \frac{d\mathbf{r}_i}{dt}$$
$$\frac{d\psi(t)}{dt} = \frac{1}{Q}\left(\sum_i \frac{\|\mathbf{p}_i\|^2}{2m_i} - \frac{1}{2}(2N+1)k_B T\right) \tag{3.125}$$

Here Q is a relaxation constant that determines the relaxation dynamics.

The above equations for both Langevin dynamics and Nosé-Hoover dynamics can be solved by modifying the velocity-Verlet algorithm. There are a number of other thermostats, e.g., the Berendsen [53] and Andersen [54] thermostats. We can also implement barostats, i.e., simulate from the NpT ensemble.

3.6 Quantum Mechanical Calculations

3.6.1 Background in Many-Body Quantum Theory

Electronic structure calculations are based on solving the fundamental non-relativistic Schrödinger's equation, namely obtaining the many-body wavefunction $\widetilde{\Psi}(\{\mathbf{r}_i\}, \{\mathbf{r}_\alpha\}, t)$ for a system of N electrons (positions $\mathbf{r}_i$) and M nuclei (positions $\mathbf{r}_\alpha$), from which a variety of fundamental properties can be estimated [10]. In fact, as we shall see, in contrast to MD such calculations ignore the structure of the nuclei by assuming that they are fixed in space, and hence the name 'electronic structure'.

We work with the non-spin form for the purposes of presentation. In its most basic form the Schrödinger equation is

$$i\frac{\partial}{\partial t}|\Psi(\{\mathbf{r}_i\}, \{\mathbf{r}_\alpha\}, t)\rangle = \hat{H}|\Psi(\{\mathbf{r}_i\}, \{\mathbf{r}_\alpha\}, t)\rangle \tag{3.126}$$

If we assume that the wavefunction takes a separable form

$$\widetilde{\Psi}(\{\mathbf{r}_i\}, \{\mathbf{r}_\alpha\}, t) = \Psi(\{\mathbf{r}_i\}, \{\mathbf{r}_\alpha\}) \exp(-iEt) \tag{3.127}$$

for some E, the problem reduces to the following eigenvalue problem for the time-independent part of the state vector

$$\widehat{H}\,|\Psi\rangle = E\,|\Psi\rangle \tag{3.128}$$

with Hamiltonian

$$H(\{\mathbf{r}_i\}, \{\mathbf{r}_\alpha\}) = -\frac{1}{2}\sum_i \nabla_i^2 - \sum_\alpha \frac{1}{2m_\alpha}\nabla_\alpha^2 - \sum_{i,\alpha} \frac{Z_\alpha}{\|\mathbf{r}_i - \mathbf{r}_\alpha\|}$$
$$+ \frac{1}{2}\sum_{j\neq i} \frac{1}{\|\mathbf{r}_i - \mathbf{r}_j\|} + \frac{1}{2}\sum_{\beta\neq\alpha} \frac{Z_\alpha Z_\beta}{\|\mathbf{r}_\alpha - \mathbf{r}_\beta\|} \tag{3.129}$$

in which m_α are the masses of the nuclei, with atomic numbers Z_α. Those states that are eigenfunctions of the Hamiltonian are called stationary states. The first two terms of the Hamiltonian are the kinetic energies of the electrons and nuclei, in which ∇_i and ∇_α denote the gradient operators with respect to the corresponding spatial coordinate $\mathbf{r}_i$ and $\mathbf{r}_\alpha$. The remaining terms represent the electron-nuclear, electron-electron and inter-nuclear Coulomb interactions, respectively.

We have used the Dirac notation by the identification of the wavefunction $\Psi(\{\mathbf{r}_i\}, \{\mathbf{r}_\alpha\})$ with the components of the state vector $|\Psi\rangle$ in the position basis, defined as

$$|\mathbf{r}_1, \ldots, \mathbf{r}_N, \mathbf{r}_1, \ldots, \mathbf{r}_M\rangle = \bigotimes_i |\mathbf{r}_i\rangle \otimes \bigotimes_\alpha |\mathbf{r}_\alpha\rangle \tag{3.130}$$

in which $\otimes$ is a tensor product operation, with $\bigotimes_i \mathbf{r}_i = \mathbf{r}_1 \otimes \ldots \otimes \mathbf{r}_N$. With this notation

$$\Psi(\{\mathbf{r}_i\}, \{\mathbf{r}_\alpha\}) = \langle \mathbf{r}_1, \ldots, \mathbf{r}_N, \mathbf{r}_1, \ldots, \mathbf{r}_M|\Psi\rangle \tag{3.131}$$

and

$$H(\{\mathbf{r}_i\}, \{\mathbf{r}_\alpha\}) = \left\langle \mathbf{r}_1, \ldots, \mathbf{r}_N, \mathbf{r}_1, \ldots, \mathbf{r}_M|\widehat{H}|\mathbf{r}_1, \ldots, \mathbf{r}_N, \mathbf{r}_1, \ldots, \mathbf{r}_M\right\rangle \tag{3.132}$$

The state vector lies in a Hilbert space $\mathcal{H}$ formed from the tensor product $\mathcal{H}_1 \otimes \ldots \otimes \mathcal{H}_1 \otimes \mathcal{H}_2 \ldots \otimes \mathcal{H}_2$ of one-particle Hilbert spaces $\mathcal{H}_1$ for the electrons and $\mathcal{H}_2$ for the nuclei, with a basis given by the tensor products of the one-particle orthonormal orbital basis vectors.

We note that a position basis $|\mathbf{r}\rangle$ is not a basis in the usual mathematical sense since the position operator $\widehat{\mathbf{r}}$ is unbounded and so its eigenstates or eigenvectors do not form a basis for $\mathcal{H}_1$. Position basis vectors form what is termed a 'generalised continuous orthonormal' basis in the sense that

$$\left\langle \mathbf{r}|\mathbf{r}'\right\rangle = \delta(\mathbf{r} - \mathbf{r}') \tag{3.133}$$

where δ is the delta function or 'distribution'.

The square magnitude of the wavefunction can be interpreted as a probability density at the positions $\{\mathbf{r}_i\}$, $\{\mathbf{r}_\alpha\}$

$$p(\{\mathbf{r}_i\}, \{\mathbf{r}_\alpha\}) = |\Psi(\{\mathbf{r}_i\}, \{\mathbf{r}_\alpha\})|^2 \tag{3.134}$$

with

$$\langle\Psi|\Psi\rangle = \int_{\mathbb{R}^{3N} \times \mathbb{R}^{3M}} |\Psi(\{\mathbf{r}_i\}, \{\mathbf{r}_\alpha\})|^2 \prod_i d\mathbf{r}_i \prod_\alpha d\mathbf{r}_\alpha = 1 \tag{3.135}$$

Additional constraints on $\Psi(\{\mathbf{r}_i\}, \{\mathbf{r}_\alpha\})$ are required in the form of symmetry under exchange of particle labels for bosons and anti-symmetry for fermions (e.g., electrons and protons).

Since the nuclei are much bigger than the electrons, their velocities will be much smaller, so that the electrons will relax to their ground-state configuration on a much shorter timescale than that of nuclear motion. In the *Born-Oppenheimer approximation*, it is assumed that the nuclei are stationary, and therefore that the solution to the electronic ground state can be sought first. With $\Psi(\{\mathbf{r}_i\}, \{\mathbf{r}_\alpha\}) = \Phi(\{\mathbf{r}_i\})\Psi_\alpha(\{\mathbf{r}_\alpha\})$, the problem becomes

$$H\left|\Phi\right\rangle = \mathcal{E}\left|\Phi\right\rangle \tag{3.136}$$

in which $\mathcal{E}$ is called the adiabatic contribution to the system energy of the electrons, with the remaining part being much smaller. Both this energy and $\Phi(\{\mathbf{r}_i\})$ will depend on the $\mathbf{r}_\alpha$ but this dependence is usually suppressed in the notation. The new Hamiltonian can be written in a compact form as follows

$$H(\mathbf{r}^{(N)}) = \sum\nolimits_i h(\mathbf{r}_i) + \frac{1}{2}\sum\nolimits_{j \neq i} \frac{1}{\|\mathbf{r}_i - \mathbf{r}_j\|}, \quad \mathbf{r}^{(N)} = \{\mathbf{r}_i\} \tag{3.137}$$

in which

$$h(\mathbf{r}) = -\frac{1}{2}\nabla^2 + v_{ne}(\mathbf{r}), \quad v_{ne}(\mathbf{r}) = -\sum\nolimits_\alpha \frac{Z_\alpha}{\|\mathbf{r} - \mathbf{r}_\alpha\|} \tag{3.138}$$

is a one-electron Hamiltonian and $v_{ne}(\mathbf{r})$ is the nuclei-electron potential.

Even with the simplifications above, solving (3.136) is a formidable task, which has led to a number of approximate techniques. These techniques form a hierarchy, in terms of the associated accuracy and computational burden. The high time cost even for the simpler models has led to efforts to replace all or part of their formulation with machine learning equivalents, which we cover in Sect. 6.19 of Chap. 6.

3.6.2 Hartree-Fock, Semi-empirical and Post-Hartree-Fock Methods

The early *Hartree-Fock* (HF) approach was based on approximating the many-electron wavefunction with a Slater determinant of N orthonormal orbitals $\chi_i(\mathbf{r}_i)$

$$\Phi(\{\mathbf{r}_i\}) = \frac{1}{\sqrt{N!}} \begin{vmatrix} \chi_1(\mathbf{r}_1) & \chi_2(\mathbf{r}_1) & \cdots & \chi_N(\mathbf{r}_1) \\ \chi_1(\mathbf{r}_2) & \chi_2(\mathbf{r}_2) & \cdots & \chi_N(\mathbf{r}_2) \\ \vdots & \vdots & \ddots & \vdots \\ \chi_1(\mathbf{r}_N) & \chi_2(\mathbf{r}_N) & \cdots & \chi_N(\mathbf{r}_N) \end{vmatrix} \tag{3.139}$$

which satisfies the anti-symmetry property embodying the Pauli exclusion principle. It is important to note that the state vector has a basis expansion of the form described earlier

$$\left|\Phi\right\rangle = \sum\nolimits_{\mathrm{i}=1}^{\infty} c_{\mathrm{i}} \left|\chi_{i_1}\right\rangle \wedge \ldots \wedge \left|\chi_{i_N}\right\rangle \tag{3.140}$$

in which $\wedge$ denotes an *antisymmetric* tensor product since electrons are fermions, and $\mathbf{i} = (i_1, \ldots, i_N)$ is a multi-index. Thus, (3.139) represents only a single basis function in (3.140), which will obviously lead to large errors in many cases.

The orbitals are found by minimisation of the system energy given by the expected value of the Hamiltonian over the wavefunction (3.139)

$$\arg\min_{\chi^{(N)}} \langle \Phi | \widehat{H} | \Phi \rangle \quad \text{such that } \langle \chi_i(\mathbf{r}_i) | \chi_i(\mathbf{r}_i) \rangle = \delta_{ij}, \quad \chi^{(N)} := \{\chi_i(\mathbf{r}_i)\} \tag{3.141}$$

in which δ_{ij} is the Kronecker delta. $E[\chi^{(N)}] = \langle \Phi | \widehat{H} | \Phi \rangle$ is a functional of the orbitals and $E_{HF} = \min_{\chi^{(N)}} E[\chi^{(N)}]$ is called the HF total energy. Using the definition (3.137)–(3.138) of the Hamiltonian, we obtain

$$E[\chi^{(N)}] = \langle \Psi | \widehat{H} | \Psi \rangle = \sum_i \langle \chi_i | \widehat{h} | \chi_i \rangle + \sum_{i \neq j} \langle \chi_i \chi_j | \chi_i \chi_j \rangle - \langle \chi_i \chi_j | \chi_j \chi_i \rangle \tag{3.142}$$

in which

$$\langle \chi_i | \widehat{h} | \chi_i \rangle = \int_{\mathbb{R}^3} \chi_i^*(\mathbf{r}) h(\mathbf{r}) \chi_i^*(\mathbf{r}) d\mathbf{r} \tag{3.143}$$

is a one-electron integral and

$$\begin{aligned} \langle \chi_i \chi_j | \chi_i \chi_j \rangle &= \int_{\mathbb{R}^3 \times \mathbb{R}^3} \frac{\chi_i^*(\mathbf{r}) \chi_j^*(\mathbf{r}') \chi_i(\mathbf{r}) \chi_j(\mathbf{r}')}{\|\mathbf{r} - \mathbf{r}'\|} d\mathbf{r} d\mathbf{r}' \\ \langle \chi_i \chi_j | \chi_j \chi_i \rangle &= \int_{\mathbb{R}^3 \times \mathbb{R}^3} \frac{\chi_i^*(\mathbf{r}) \chi_j^*(\mathbf{r}') \chi_j(\mathbf{r}) \chi_i(\mathbf{r}')}{\|\mathbf{r} - \mathbf{r}'\|} d\mathbf{r} d\mathbf{r}' \end{aligned} \tag{3.144}$$

are two-electron integrals, where $*$ denotes the complex conjugate. The first of (3.144) corresponds to a Coulomb interaction between the charge distributions $\|\chi_i\|^2$ and $\|\chi_j\|^2$, while the second is an exchange interaction resulting from quantum interferences.

The minimisation (3.141) can be performed by transforming the problem into one that is unconstrained, with Lagrangian

$$L[\chi^{(N)}] = E[\chi^{(N)}] - \sum_i \epsilon_i \left(\langle \chi_i | \chi_i \rangle - 1 \right) \tag{3.145}$$

in which ϵ_i are Lagrange multipliers. Setting the functional derivatives $\delta L[\chi^{(N)}]/\delta \chi_i^*$ to 0 using the expression (3.142) for $E[\chi^{(N)}]$ then yields the *Hartree-Fock equations*

$$h(\mathbf{r}) \chi_i(\mathbf{r}) + v_H(\mathbf{r}) \chi_i(\mathbf{r}) + \int_{\mathbb{R}^3} v_x(\mathbf{r}, \mathbf{r}') \chi_i(\mathbf{r}') d\mathbf{r}' = \epsilon_i \chi_i(\mathbf{r}), \quad i = 1, \ldots, N \tag{3.146}$$

in which

$$v_H(\mathbf{r}) = \sum_j \int_{\mathbb{R}^3} \frac{\chi_j^*(\mathbf{r}') \chi_j(\mathbf{r})}{\|\mathbf{r} - \mathbf{r}'\|} d\mathbf{r}' \tag{3.147}$$

is the local Hartree potential and

$$v_x(\mathbf{r}, \mathbf{r}') = -\sum\nolimits_j \frac{\chi_j^*(\mathbf{r}')\chi_j(\mathbf{r})}{\|\mathbf{r} - \mathbf{r}'\|} \tag{3.148}$$

is called the non-local exchange or Fock potential. Setting

$$H_{HF}(\mathbf{r}, \mathbf{r}') = \delta(\mathbf{r} - \mathbf{r}')\,(h(\mathbf{r}) + v_H(\mathbf{r})) + v_x(\mathbf{r}, \mathbf{r}') \tag{3.149}$$

we can write

$$\int_{\mathbb{R}^3} H_{HF}(\mathbf{r}, \mathbf{r}')\chi_i(\mathbf{r}')d\mathbf{r} = \epsilon_i \chi_i(\mathbf{r}), \quad \text{or} \quad \widehat{H}_{HF}\,|\chi_i\rangle = \epsilon_i\,|\chi_i\rangle \tag{3.150}$$

in which $\langle \mathbf{r}|\widehat{H}_{HF}|\mathbf{r}'\rangle = H_{HF}(\mathbf{r}, \mathbf{r}')$.

The HF equations are an eigenvalue problem for a one-electron effective Hamiltonian $\widehat{H}_{HF}$ (the HF or Fock operator), with solutions χ_i and associated eigenvalues ϵ_i. The HF operator is composed of a noninteracting part $\widehat{h}(\mathbf{r})$ with a remainder $\widehat{v}_{HF} = \widehat{v}_H + \widehat{v}_x$ called the HF potential. $\widehat{H}_{HF} : \mathcal{H}_1 \to \mathcal{H}_1$ turns out to be a self-adjoint operator with a discrete spectrum, so its eigenfunctions are orthogonal and form a basis for $\mathcal{H}_1$. The equations have to be solved iteratively using a procedure called the *self-consistent field* (SCF) method since the HF potential depends on the orbitals. This involves a basis function expansion of the one-electron orbital in terms of K *atomic orbitals* ϕ_k

$$\chi_i = \sum\nolimits_{k=1}^{K} c_{ki}\phi_k \tag{3.151}$$

called a linear combination of atomic orbitals (LCAO). The atomic orbitals are usually based on hydrogen-like atoms, for which analytical forms are available, and are called *Slater-type* orbitals. However, Gaussian functions are also frequently employed. The N eigenfunctions $|\chi_i\rangle$ corresponding to the lowest eigenvalues are the occupied HF spin-orbitals. The remaining eigenfunctions are virtual or unoccupied.

The main weakness of HF is the assumption of a single Slater determinant, which neglects correlations between electrons. As a result it overestimates energies but has an attractive computational cost of $O(N^3)$. A number of improvements have been developed over the years, termed *post-HF* methods. Notable examples are *configuration interaction* (CI), Møller-Plesset (MP) perturbation theory and *coupled cluster* (CC) theory, which combines CI and MP. Common to all post-HF wavefunction methods is that they massively increase the computational costs to well beyond $\mathcal{O}(N^3)$. The state-of-the-art coupled cluster theory [55] can scale-up to $\mathcal{O}(N^7)$, while Møller-Plesset [56] can be up to $\mathcal{O}(N^5)$.

Semi-empirical methods reduce the computational costs by ignoring or approximating the two-electron integrals in the original framework. These integrals can be approximated in a mechanistic fashion with parameters that are fitted to either to experimental data or to *ab-initio* calculations. Due to their low accuracy, the use of such methods is primarily for large-scale molecular systems for which more advanced methods are not feasible. Prominent examples are the complete/intermediate neglect of differential overlap (CNDO/INDO) [57], the method of Austin [58], and Zerner's variant of INDO (ZINDO).

The most obvious extension to HF that can incorporate the electron interactions is to write the wavefunction as a linear combination of multiple Slater determinants Ψ_i

$$|\Phi\rangle = \sum_{i=1}^{I} c_i |\Psi_i\rangle \tag{3.152}$$

with $\langle\Phi|\Phi\rangle = 1$, which is called the *full configuration interaction* (FCI) method. The most common way of choosing the determinants is to use L of the HF occupied and virtual orbitals, leading to $I = L!/(N!(L-N)!)$ determinants in total. Notice that the Lagrangian

$$L(\{c_i\}) = \sum_{i,j=1}^{I} c_i^* c_j \langle\Psi_i|\widehat{H}|\Psi_j\rangle - E\left(\sum_{i=1}^{I} |c_i|^2 - 1\right) \tag{3.153}$$

in which E is the multiplier, is now a function of the coefficients and the variational principle takes the partial derivatives of $L(\{c_i\})$ with respect to the c_i. The solution corresponding to the smallest eigenvalue gives the ground-state wavefunction, while those corresponding to the larger eigenvalues provide excited-state energies and wavefunctions. FCI is not feasible on systems with more than a few electrons since $I = \mathcal{O}(L^N)$, so for larger systems a truncated version is used. The determinants can be generated from the HF state vector $|\Phi\rangle_{HF}$ by swapping a single one-electron HF orbital χ_i for a virtual HF orbital.

3.6.3 Hohenberg-Kohn and Levy-Leib Formulations and Functionals

By far the most common method in use is DFT, which scales much better than the post-HF methods while being more accurate than HF, although it has the same $\mathcal{O}(N^3)$ cost. At the heart of the method is an exchange-correlation energy functional, approximated using various models of differing complexity [59]. DFT was enabled by the two highly-celebrated Hohenberg-Kohn (HK) theorems [60], with a particular form of (3.137)–(3.138) that made the HK theory more tractable. When the electronic correlations are strong and there are non-local van der Waals interactions, DFT can fail [61]. Nevertheless, its optimal balance between accuracy and computational burden for a large class of problems explains its enduring appeal.

Due to the Born-Oppenheimer approximation, the Coulomb potential engendered by the nuclei is treated as a static external potential v_{ne}, defined in (3.138). The Hamiltonian in (3.137) can be written as

$$
\begin{aligned}
H(\mathbf{r}^{(N)}) &= -\frac{1}{2}\sum\nolimits_i \nabla_i^2 + \frac{1}{2}\sum\nolimits_{j\neq i} \frac{1}{\|\mathbf{r}_i - \mathbf{r}_j\|} + \sum\nolimits_i v_{ne}(\mathbf{r}_i) \\
&:= T + W_{ee} + V_{ne}
\end{aligned} \tag{3.154}
$$

in which T is the kinetic energy operator and W_{ee} is an electron-electron interaction operator. Neither of these operators change when we change the N-electron system, so that the ground-state must be completely determined by the external potential $V_{ne}(\mathbf{r}^{(N)})$, for a fixed N. The ground-state wavefunction $\Phi_0(\mathbf{r}^{(N)})$ defines a corresponding ground-state *electron density* $\rho_0(\mathbf{r})$

$$
\rho_0(\mathbf{r}) = N \int_{\mathbb{R}^{3(N-1)}} |\Psi_0\,(\mathbf{r}, \mathbf{r}_2 \ldots \mathbf{r}_N)|^2 \prod\nolimits_{i=2}^{N} d\mathbf{r}_i \tag{3.155}
$$

To make progress, we can appeal to the two Hohenberg-Kohn theorems [60], which state that the ground-state energy E_0 is a functional of the ground-state density, $E_0 = E_{HK}[\rho_0]$, and is obtained by minimising this functional

$$
E_0 = \min_{\rho} E_{HK}[\rho] \quad \text{such that } \rho \text{ is } v\text{-representable} \tag{3.156}
$$

The set of n v-representable densities is defined as those corresponding to a ground-state wavefunction of the Hamiltonian (3.154) with the potential v_{ne} and satisfying $\int_{\mathbb{R}^3} \rho(\mathbf{r})d\mathbf{r} = N$. This is a rather restrictive assumption, which we will touch upon later.

The ground-state energy is given by the expected value of the Hamiltonian corresponding to the ground-state wavefunction

$$
\begin{aligned}
E_{HK}[\rho_0] &= \langle \Psi_0|\widehat{T}|\Psi_0\rangle + \langle \Psi_0|\widehat{W}_{ee}|\Psi_0\rangle + \langle \Psi_0|\widehat{V}_{ne}|\Psi_0\rangle \\
&:= T[\rho_0] + W[\rho_0] + \int_{\mathbb{R}^3} \rho_0(\mathbf{r}) v_{ne}(\mathbf{r}) d\mathbf{r}
\end{aligned} \tag{3.157}
$$

in which the second step follows from the symmetry properties of $|\Psi_0|^2$ so that each $\langle \Psi_0|\widehat{v}_{ne}|\Psi_0\rangle$ is identical and of the form

$$
\begin{aligned}
\langle \Psi_0|\widehat{v}_{ne}|\Psi_0\rangle &= \int_{\mathbb{R}^3} v_{ne}(\mathbf{r})d\mathbf{r} \int_{\mathbb{R}^{3(N-1)}} |\Psi_0\,(\mathbf{r}, \mathbf{r}_2 \ldots \mathbf{r}_N)|^2 \prod\nolimits_{i=2}^{N} d\mathbf{r}_i \\
&= \frac{1}{N} \int_{\mathbb{R}^3} v_{ne}(\mathbf{r})\rho(\mathbf{r})d\mathbf{r}
\end{aligned} \tag{3.158}
$$

The sum of first two terms, which are not given explicitly as functionals, $F_{HK}[\rho_0] = T[\rho_0] + W[\rho_0]$, is called the *universal* Hohenberg-Kohn functional, since it is independent of the external potential.

Conceptually, therefore, we are faced will a simple problem, which is to perform the minimisation in (3.156). Unfortunately, we have no systematic way of writing down the Hohenberg-Kohn functional exactly for a given system. In the

Thomas-Fermi model, a simplified version was obtained by assuming that the electrons are uniformly distributed inside each small element of volume, but that the electron density is allowed to vary across the volume elements

$$\begin{aligned} T[\rho] &= \frac{3}{10}\left(3\pi^2\right)^{\frac{2}{3}} \int_{\mathbb{R}^3} [\rho(\mathbf{r})]^{5/3}\, d\mathbf{r} \\ W[\rho] &= \frac{1}{2} \int \frac{\rho(\mathbf{r})\rho(\mathbf{r}')}{\|\mathbf{r}-\mathbf{r}'\|}\, d\mathbf{r}d\mathbf{r}' \end{aligned} \tag{3.159}$$

for a general density ρ. Such a model, however, is highly numerically unstable, as well as being a severe approximation. Various attempts to correct the model essentially failed.

The major breakthrough came with the *Kohn-Sham* DFT (KS-DFT) formulation, which is based on a fictitious system of noninteracting electrons, the electron density of which is equal to the true ground-state electron density $\rho_0(\mathbf{r})$. The kinetic energy functional $T[\rho]$ accounts for the majority of the total electronic energy, and the KS approach allows for the kinetic energy part to be treated with good accuracy using only one slater determinant. KS-DFT was originally motivated by the HK theorems, which say little about the allowable spaces for the external potential and electron density. The modern derivation uses the constrained-search procedure of Levy [62] and Lieb [63], which is more precise in this regard and provides some theoretical guarantees.

Rather than directly minimising (3.157), the Levy-Lieb formulation splits the minimisation into two steps: minimisation over the wave function under the constraint of reproducing the density, followed minimisation over the density

$$\begin{aligned} E_0 &= \inf_{\rho} \left\{ \inf_{\Phi\to\rho} \left\langle \Psi \middle| \widehat{T} + \widehat{W}_{ee} + \widehat{V}_{ne} \middle| \Psi \right\rangle \right\} \\ &:= \inf_{\rho} \left\{ \inf_{\Phi\to\rho} \left\langle \Psi \middle| \widehat{T} + \widehat{W}_{ee} \middle| \Psi \right\rangle + \int_{\mathbb{R}^3} \rho(\mathbf{r}) v_{ne}(\mathbf{r}) d\mathbf{r} \right\} \end{aligned} \tag{3.160}$$

in which the notation $\Phi \to \rho$ denotes a minimisation over normalised wave functions Φ that yield some fixed density ρ. Here, the space for the wavefunctions is taken to be

$$\Phi \in \mathcal{D} = \bigwedge^{N} H^1(\mathbb{R}^3), \quad H^1(\mathbb{R}^3) = \{\Phi \mid \|\Phi\|_{L_2(\mathbb{R}^3)} < \infty,\ \|\nabla\Phi\|_{L_2(\mathbb{R}^3)} < \infty\} \tag{3.161}$$

in which $L^p(\mathbb{R}^3)$ is the Lebesgue space of equivalent classes of functions ϕ on $\mathbb{R}^3$ satisfying

$$\|\phi(\mathbf{r})\|_p = \left(\int_{\mathbb{R}^3} |\phi(\mathbf{r})|^p d\mathbf{r} \right)^{\frac{1}{p}} < \infty \tag{3.162}$$

with equivalence defined as differing only on sets of measure zero. The densities corresponding to these wavefunctions $\Phi \in \mathcal{D}$ lie in the space

$$\rho \in \mathcal{S} = \{\rho \mid \rho \geq 0, \int_{\mathbb{R}^3} \rho(\mathbf{r})d\mathbf{r} = N \ \sqrt{\rho} \in H^1(\mathbb{R}^3)\} \subset L^1(\mathbb{R}^3) \cap L^3(\mathbb{R}^3) \tag{3.163}$$

which ensures integrability and a finite kinetic energy. Such densities are said to the *N-representable*. This broadens the definition of the HK functional to densities in $\mathcal{S}$ and replaces the energy functional in the HK formulation with the following

$$\begin{aligned} E_L[\rho] &= \inf_{\Phi \to \rho} \langle \Psi | \widehat{T} + \widehat{W}_{ee} | \Psi \rangle + \int_{\mathbb{R}^3} \rho(\mathbf{r}) v_{ne}(\mathbf{r}) d\mathbf{r} \\ &:= F_L[\rho] + \int_{\mathbb{R}^3} \rho(\mathbf{r}) v_{ne}(\mathbf{r}) d\mathbf{r} \end{aligned} \tag{3.164}$$

in which $F_L[\rho]$ is the (universal) Levy-Lieb functional and $E_0 = \inf_\rho E_L[\rho]$. Lieb [63] provides a proof that $\rho \in \mathcal{S}$ implies the existence of at least one $\Phi \in \mathcal{D}$ such that $\Phi \to \rho$, ensuring that the constrained minimisation is well-defined. The potential is constrained to ensure finite variation

$$\int_{\mathbb{R}^3} \rho(\mathbf{r}) v_{ne}(\mathbf{r}) d\mathbf{r} < \infty \tag{3.165}$$

and is thus chosen to lie in the dual space $(L^1(\mathbb{R}^3) \cap L^3(\mathbb{R}^3))^*$.

3.6.4 Kohn-Sham Density Functional Theory

We can now use the Levy-Lieb formulation formalism to derive the KS approach. There are a number of ways to explain this approach, and we use that found in Lindgren and Salomonson [64]. The HK or Levy-Lieb functional F_L is the major obstacle in the respective formulations. KS simplifies the task by considering a (fictitious) system of *noninteracting* electrons that move in a some *effective local* potential $v_{KS}(\mathbf{r})$, with Hamiltonian [64]

$$H_{KS}(\mathbf{r}^{(N)}) = -\frac{1}{2}\sum_i \nabla_i^2 + \sum_i v_{KS}(\mathbf{r}_i) := T + V_{KS} \tag{3.166}$$

and the goal is to determine the form of $v_{KS}(\mathbf{r}_i)$ so that the ground-state density of this noninteracting system is equal to that of the real system. This requires noninteracting v-representability of the density, meaning that it is reproduced by the ground-state density of a noninteracting system with some local potential, which it turns out is always achievable with arbitrary accuracy [65].

Within the Levy-Lieb formalism, the kinetic energy part alone defines F_L

$$E_{KS}[\rho] = \inf_{\Phi \to \rho} \langle \Psi | \widehat{T} | \Psi \rangle + \int_{\mathbb{R}^3} \rho(\mathbf{r}) v_{KS}(\mathbf{r}) d\mathbf{r} := T_{KS}[\rho] + \int_{\mathbb{R}^3} \rho(\mathbf{r}) v_{KS}(\mathbf{r}) d\mathbf{r} \tag{3.167}$$

in which $T_{KS}[\rho]$ corresponds to a noninteracting kinetic energy functional defined over a single Slater determinant (3.139). The ground-state energy is found by minimising E_{KS} over $\rho \in \mathcal{S}$

$$\frac{\delta E_{KS}[\rho]}{\delta \rho} = \frac{\delta T_{KS}[\rho]}{\delta \rho} + v_{KS}(\mathbf{r}) = 0 \tag{3.168}$$

in which we ignore constants. This can be compared with the minimisation of the Levy-Lieb energy functional (3.164) [64], which shows that

$$v_{KS}(\mathbf{r}) = \frac{\delta F_L[\rho]}{\delta \rho} - \frac{\delta T_{KS}[\rho]}{\delta \rho} + v_{ne}(\mathbf{r}) \tag{3.169}$$

We can go further and express F_L as

$$F_L[\rho] = T_{KS}[\rho] + E_H[\rho] + E_{xc}[\rho] \tag{3.170}$$

in which

$$E_H[\rho] = \int d\mathbf{r}' d\mathbf{r} \frac{\rho(\mathbf{r}')\rho(\mathbf{r})}{\|\mathbf{r} - \mathbf{r}'\|} \tag{3.171}$$

is the Hartree energy functional, representing the classical Coulomb repulsion energy for a charge distribution ρ. The remaining term $E_{xc}[\rho]$ is called the *exchange-correlation* energy functional, and it essentially represents all of the non-classical interactions after removing the Hartree part and a large portion of the kinetic part.

The motivation for this particular choice is that the kinetic and Hartree terms can account for over 95% of the total energy. Thus, although $E_{xc}[\rho]$ is unknown, and has to be approximated, this task is far simpler than attempting to approximate F_L directly. Moreover, it permits an exact treatment of a major portion of the kinetic energy using only a single-determinant wavefunction Ψ. The exchange-correlation energy functional is often further decomposed into an exchange and a correlation part, $E_x[\rho]$ and $E_c[\rho]$.

With the KS potential (3.169) placed in the KS model (3.166), we expect to obtain the same ground-state density as that of the original problem, and inserting the resulting density into the HK functional (3.157) should, in theory, furnish the ground-state energy. The *KS equations* representing the KS approach read as

$$H_{KS}\chi_i(\mathbf{r}) = \left(-\frac{1}{2}\nabla_n^2 + v_{ext}(\mathbf{r}) + v_H(\mathbf{r}) + v_{xc}(\mathbf{r}) \right) \chi_i(\mathbf{r}) = \epsilon_i \chi_i(\mathbf{r}), \quad i = 1, \ldots, N \tag{3.172}$$

in which

$$v_H(\mathbf{r}) = \int_{\mathbb{R}^3} \frac{d\mathbf{r}'\rho(\mathbf{r}')}{\|\mathbf{r}-\mathbf{r}'\|}, \quad v_{xc}(\mathbf{r}) = \frac{\delta E_{xc}[\rho]}{\delta \rho} \tag{3.173}$$

is the Hartree potential in terms of the density and the exchange-correlation potential, respectively.

As in the HF method, we seek orbitals χ_i forming a single slater determinant, with $\rho(\mathbf{r}) = \sum_i |\chi_i(\mathbf{r})|^2$ by virtue of the orthonormality of the orbitals. Again, the solution uses an SCF procedure, which can be summarised as follows

1. initialise: $j = 0$, density ρ^j
2. solve the KS equations with ρ^j to obtain $\{\chi_i^j\}_i$
3. calculate a new Kohn–Sham electron density: $\rho^j \leftarrow \sum_i |\chi_i^j|^2$
4. $j \leftarrow j + 1$
5. if $\|\rho^j - \rho^{j-1}\| > \varepsilon$ and $j < j_{max}$ return to step 2

The ground-state energy E_0 is calculated from the converged orbitals

$$E_0 = \sum_i \langle \chi_i | \widehat{T} | \chi_i \rangle + E_H[\rho] + E_{ne}[\rho] + E_{xc}[\rho] \tag{3.174}$$

with $E_{ne}[\rho] = \int_{\mathbb{R}^3} \rho(\mathbf{r}) v_{ne}(\mathbf{r}) d\mathbf{r}$.

3.6.5 Exchange-Correlation Functional Hierarchy

The main difficulty consists of finding appropriate forms of $E_{xc}[\rho]$. There is a vast choice of functional available, distinguished by the level of theory incorporated. As the functional become more complex, the computational costs increase. The simplest possible form of $E_{xc}[\rho]$ is known as the *local density approximations* (LDA), or local spin-density approximation (LSDA) if spin is included. This approximation treats $E_{xc}[\rho]$ as that of a homogeneous electron gas of the same density $\rho(\mathbf{r})$ [66, 67]. It takes the form

$$E_{xc}^{\mathrm{LDA}}[\rho] = \int_{\mathbb{R}^3} \epsilon_{xc}^{\mathrm{LDA}}[\rho(\mathbf{r})] \, d\mathbf{r} \tag{3.175}$$

with a kernel $\epsilon_{xc}^{\mathrm{LDA}}$, normally decomposed as

$$\epsilon_{xc}^{\mathrm{LDA}}[\rho] = \epsilon_c^{\mathrm{LDA}}[\rho] + \epsilon_x^{\mathrm{LDA}}[\rho] \tag{3.176}$$

$\epsilon_x^{\mathrm{LDA}}[\rho]$ has an exact form, derived by Dirac [68]

$$\epsilon_x^{\mathrm{LDA}}[\rho] = -\frac{3}{4}\left(\frac{3}{\pi}\right)^{1/3} \rho(\mathbf{r})^{4/3} \tag{3.177}$$

while $\epsilon_c^{\mathrm{LDA}}[\rho]$ does not usually have an explicit form and can be approximated using quantum Monte Carlo models [69].

A refined version called the *generalised gradient approximation* (GGA) is dependent locally on both $\rho(\mathbf{r})$ and its gradient $\nabla\rho(\mathbf{r})$ [70, 71]

$$E_{xc}^{\mathrm{GGA}}[\rho] = \int_{\mathbb{R}^3} \epsilon_{xc}^{\mathrm{GGA}}(\rho(\mathbf{r}), \nabla\rho(\mathbf{r}))\rho(\mathbf{r})\,\mathrm{d}\mathbf{r} \tag{3.178}$$

with some kernel $\epsilon_{xc}^{\mathrm{GGA}}$, normally constructed as a correction to LDA

$$\epsilon_{xc}^{\mathrm{GGA}}[\rho] = \epsilon_{xc}^{\mathrm{LDA}}[\rho] + \widehat{\epsilon}_{xc}\left[\frac{|\nabla\rho(\mathbf{r})|}{\rho(\mathbf{r})^{4/3}}\right] \tag{3.179}$$

Many different GGA functionals have been proposed, including functionals containing fitted parameters and functionals without such parameters, with the most common being the Perdew-Burke-Ernzerhof (PBE) functional [70], which falls into the latter category. Including the Laplacian $\nabla^2\rho(\mathbf{r})$, leads to *meta-GGA* functionals such as TPSS [72], which is superior to the PBE functional in terms of approximating atomisation and surface energies [73].

Hybrid functionals are formed by combining the Hartree-Fock exchange energy functional in terms of KS orbitals

$$E_{\mathrm{X}}^{\mathrm{HF}} = -\frac{1}{2}\sum\nolimits_{i,j} \int_{\mathbb{R}^3\times\mathbb{R}^3} \chi_i^*(\mathbf{r})\chi_j^*(\mathbf{r}')\frac{1}{\|\mathbf{r}-\mathbf{r}'\|}\chi_j(\mathbf{r})\chi_i(\mathbf{r}')\,d\mathbf{r}\,d\mathbf{r}' \tag{3.180}$$

with local GGA and meta-GGA correlations in various mixtures, involving weights for each component that are determined through fitting. The two most prominent examples are the B3LYP functional [74–77] and the PBE0 functional [78]. New functionals for specific problems continue to be developed.

3.7 Data Driven or Machine Learning Approaches

There are many different methods in machine learning, data science and analytics, falling into different categories. Some topics will not, therefore, be covered. For example, we will not discuss generative vs. discriminative models, and we will not provide any details of semi-supervised and reinforcement learning.

Supervised machine learning involves examples of inputs and outputs from an experiment or a computer model, with the outputs often referred to as *labels* or *targets*, while the inputs are often called *patterns* in the case of classification. We will use these terms interchangeably. The given examples of inputs and outputs form the *data set*, which is typically split into a part for *training* and a part for *testing*. Training refers to the task of learning certain machine learning model parameters, whereas testing refers to independent assessment of the accuracy of the learned model. In *regression*, the outputs lie on a continuous interval, e.g., a temperature, whereas in *classification* the outputs are categorical, e.g., 'membrane rupture' or 'leakage' in the case of failure modes of a redox flow battery.

One of the main applications of machine learning in science and engineering is in the development of *surrogate models* (also called metamodels), which are computationally cheap replacements for time-intensive, complex computer codes of scientific and engineering phenomena, systems or devices. The primary motivation for surrogate models is parametric problems involving repeated executions of the complex codes for different parameters of the system under investigation. In certain applications, especially optimisation, sensitivity analysis and uncertainty analysis, use of the complex code is rendered infeasible. We outline surrogate modelling in the next section, including the procedure for generating data (design-of-experiment) and its normalisation. Alternative surrogate-modelling approaches, namely, reduced-order and multi-fidelity models, will be covered in Chap. 7.

Semi-supervised learning is somewhat intermediate between supervised and unsupervised learning, involving a small portion of inputs that have corresponding outputs or are *labelled*, while having a larger volume of unlabelled data that is too time consuming to label. An initial model can be trained using the labelled data, and subsequently used to make predictions on the unlabelled data. For example, in the *self-learning* version, confident predictions are added to the initial training data and training is repeated, with this process conducted iteratively. Semi-supervised learning can be used for a variety of tasks, including classification and regression, *clustering* and association.

Clustering involves the grouping of data points that are in some (well-defined) sense similar, using a measure called a similarity score. It is an unsupervised method, meaning that there are no outputs corresponding to the inputs (in fact, whether we call them inputs or outputs is immaterial). The other main unsupervised method is *dimension reduction*, in which the goal is either to reduce the size of a data set or (using the same techniques) to find relationships among different attributes (components) of the data points. It can also be used for a change of basis to one that is more convenient, as in principal component analysis and the Karhunen-Loeve expansion, which we will meet in Chaps. 6 and 7, respectively. Both of these methods involve a change of basis and are also used to approximate the data points by projecting them onto a low-dimensional subspace of the original space.

The other major category of machine learning is *reinforcement learning*, which in common with supervised learning involves inputs and outputs. However, there is a feedback loop in reinforcement learning, with the goal of an *agent* (essentially the algorithm) to learn in an interactive environment through trial and error, based on the feedback from chosen actions. The feedback comes in the form of rewards or punishments, which are used by the algorithm to learn a 'correct' behaviour by maximising a *cumulative reward*. Reinforcement learning is appropriate in situations where sequential decision-making is required and the goal is long-term, e.g., robotics and general optimal control problems. It could, therefore, find application in the control of flow battery systems, especially when integrated into a grid with intermittent sources of power.

In Chap. 6 we cover in detail a number of regression, classification, dimension reduction and clustering methods, including both linear and nonlinear methods, Bayesian approaches, kernel methods and methods for multivariate (including tensor-variate) data. Deep learning, which has become highly popular in recent years, is also covered. Sequential data, including time series, are covered in Chap. 7.

After covering surrogate models, we introduce the reader to the basic terminology and the basic principles of machine learning, together with an outline of the framework followed in most machine learning problems.

3.7.1 Surrogate Models

The data set used in machine learning applications for flow batteries and other technologies can be from experimental measurements or from a model. When originating from a model, the goal is to build a replacement for the originating model, which is deemed too costly to run repeatedly for different parameter/input values. The replacement model based on machine learning is termed a *surrogate model* or *metamodel*. It typically provides results for a given parameter value in sub-second time, which makes tasks such as optimisation, sensitivity analysis and uncertainty quantification feasible within a typical computational budget.

As an example, we may consider a numerical model requiring a moderate 20 min to provide one result, relating to a single model parameter. Hundreds to tens of thousands of such results may be required, especially for sensitivity and uncertainty analyses (e.g., Monte Carlo estimates). For 1000 parameter values, around 14 days are required (without parallelisation) to complete all of the simulations, making some form of approximation desirable, if not necessary. There are actually a number of ways in which surrogate models can be developed, although machine learning approaches are by far the most common. The two other approaches, termed *multi-fidelity models* and *reduced-order models* will be detailed in Chap. 7.

The numerical model is usually one that solves a system of differential equations that embody one or more conservation laws. Such a system can be steady-state or transient, with usually transient models more appropriate for flow batteries since their operation involves charge and discharge occurring over time. The system of equations, together with accompanying constitutive relations for phase changes, reactions and transport/transfer phenomena, is typically solved using finite-difference, -volume or -element formulations alongside a time-stepping scheme for dynamic problems.

We can denote any vector of model parameters by $\boldsymbol{\xi} \in \mathbb{R}^l$; that is, there are l parameters ξ_i, which form the components of $\boldsymbol{\xi}$. These parameters can be any variable quantity of interest related to the design or operating conditions, such as inlet concentrations, porosities and channel dimensions. In the statistics literature, they are frequently called *predictor variables*. The outputs or *quantities of interest* (QoIs) from the numerical model are usually decided *a-priori*. These are the quantities that require optimising or controlling to meet a particular objective, or simply the quantities that are the subject of any exploratory analysis. Given values of the QoIs are called targets or outputs in machine learning terminology, while the corresponding inputs can also be called *design points*.

In almost all applications of surrogate models the QoIs are scalars, such as the cell voltage or Coulombic efficiency. More challenging are spatial or spatio-temporal field outputs such as temperature, current density and electric or ionic potential. The latter are typically recorded from the numerical model results at a specified set of spatial locations $\mathbf{x}_j$, $j = 1, \ldots, d$, in the numerical grid, at some specified times t_k, $k = 1, \ldots, T$. These values can be vectorised to form vector-valued outputs. In detail, for a steady-state problem at a given parameter value $\boldsymbol{\xi}$, the values of an output field $u(\mathbf{x}; \boldsymbol{\xi})$ at each of the d grid points $\mathbf{x}_j$ is placed in a vector

$$\mathbf{y} = (u(\mathbf{x}_1; \boldsymbol{\xi}), \ldots, u(\mathbf{x}_d; \boldsymbol{\xi}))^T \tag{3.181}$$

The notation $u(\mathbf{x}; \boldsymbol{\xi})$ makes it explicit that the field is a function of the spatial coordinate $\mathbf{x}$, while it is parameterised by $\boldsymbol{\xi}$.

In spatio-temporal problems. the vectorized spatial profiles at each discrete time t_k can be concatenated, e.g.,

$$\begin{aligned}\mathbf{y} = (u(\mathbf{x}_1, t_1; \boldsymbol{\xi}), \ldots, u(\mathbf{x}_d, t_1; \boldsymbol{\xi}), u(\mathbf{x}_1, t_2; \boldsymbol{\xi}), \ldots, u(\mathbf{x}_d, t_2; \boldsymbol{\xi}),\\ \ldots, u(\mathbf{x}_1, t_T; \boldsymbol{\xi}), \ldots, u(\mathbf{x}_d, t_T; \boldsymbol{\xi}))^T\end{aligned} \tag{3.182}$$

The ordering of the components with respect to the spatio-temporal grid is arbitrary, but each time the parameter $\boldsymbol{\xi}$ is changed the same ordering must be used. There are alternative choices, which are tied to the assumptions made with regard to the covariances or correlations between the field values at different spatio-temporal locations and different $\boldsymbol{\xi}$. It may be more convenient, under certain assumptions to be discussed in Chap. 6, to use tensor or multi-dimensional array formulations, rather than the vector-based formulations above.

There can be a single target or multiple targets. The targets can be learned independently (i.e., a separate model for each) or learned together, a subset of machine learning termed *multi-task* learning. In the latter case, an attempt is made to capture the correlations between the different targets. For the time being, we focus on single QoIs, either scalar or vector-valued. There is no practical difference between a vector-valued output, multiple vector-valued outputs and multiple scalar quantities placed inside a single vector. The methods for treating each of these cases are the same and boil down to choosing a correlation structure (Chap. 6 presents the full details).

An added complication arises in the case of time-dependent data. Such problems are not necessarily suited to a purely supervised machine learning approach. A host of *time series* methods have been developed for such problems, and these methods often rely on machine learning but require additional assumptions and a model of the dynamic process, which defines the way in which the data is shaped. This issue is discussed briefly in Sect. 6.12, while a detailed discussion is provided in Chap. 7. We next describe how to generate the data set for developing a surrogate model based on machine learning.

3.7.2 Design-of-Experiment and Data Generation

To apply supervised machine learning in the case of surrogate models (and more generally) a *design* is first required, meaning a set of values $\Xi = \{\boldsymbol{\xi}_n\} \subset \mathcal{X}$, $n = 1, \ldots, N$, from the so-called *design space*. The design space $\mathcal{X} \subset \mathbb{R}^l$ is the set of feasible values of $\boldsymbol{\xi}$; the parameter values will normally be constrained by physical considerations, such as a minimum and maximum temperature. The parameter values (or inputs) $\boldsymbol{\xi}_n$ are called *design points*, and they are used to generate corresponding outputs from the numerical model in order to form the data set. In most cases, Ξ is user-defined, based on a *design-of-experiment* (DOE).

The basic goal of a DOE is to sufficiently capture variations in the QoI such that the surrogate model is able to accurately predict the QoI inside the region of input space that is of interest. This is non-trivial since the effect of the input on the QoI is usually nonlinear, and the different components of $\boldsymbol{\xi}$ will impact the QoI to different degrees. More importantly, the input-output map is at best only partially understood at this stage. In the absence of any prior information, the points in Ξ are usually selected in a random, quasi-uniform manner [79]. Options for this include Latin hypercube sampling (LHS) and low-discrepancy sequences such as a Sobol sequence. A Sobol sequence [80] generates inputs over a unit hypercube as uniformly as possible.

In LHS, the range of any assumed cumulative distribution over a scalar parameter ξ can be divided into a given number M of equal length segments. Numbers can be drawn at random from each segment and the cumulative distribution inverted in each case to yield M samples of ξ, one from each segment. For multiple parameters ξ_i, the same procedure is performed individually for each, and subsequently these values are combined at random to form vector inputs. Such *stratified* sampling has been shown to be more efficient than random (Monte Carlo) sampling, in which no intervals are defined and numbers are generated at random without restrictions.

With a design $\Xi \subset \mathcal{X}$, the numerical model is executed at each design point $\boldsymbol{\xi}_n$ to yield the outputs (QoIs) of interest. The same procedure can be applied to laboratory experiments—generating the data using a DOE is not unique to surrogate models. A single output can be labelled $y_n \in \mathbb{R}$ for the scalar case or $\mathbf{y}_n \in \mathbb{R}^d$, $n = 1, \ldots, N$, for the vector case. This completes the data generation procedure for applying machine learning; namely, we now have input-output pairs $\boldsymbol{\xi}_n, y_n$ or $\boldsymbol{\xi}_n, \mathbf{y}_n$, $n = 1, \ldots, N$.

3.7.3 Data Normalisation

In many instances, the data will need to be scaled or normalised, especially when it takes on very large or small values, or when the components of the input and/or output have different scales. The difference between the performance of a machine learning algorithm with normalised and unnormalised data can be dramatic. The data can be normalised in a number of ways, e.g., division by a constant, transformation *via* a function such as a log, or by using one of a standard set of methods based on empirical estimates of the means and variances.

The *z-score* normalisation of the inputs and outputs is performed as follows

$$\xi_{jn} \mapsto \frac{\xi_{jn} - \widehat{\mathbb{E}}[\xi_j]}{\sqrt{\widehat{\text{var}}(\xi_j)}}, \quad y_{kn} \mapsto \frac{y_{kn} - \widehat{\mathbb{E}}[y_k]}{\sqrt{\widehat{\text{var}}(y_k)}} \tag{3.183}$$

for each $j = 1, \ldots, l, k = 1, \ldots, d$, in which ξ_{jn} and y_{kn} are, respectively, the j–th coefficient of $\boldsymbol{\xi}_n$ and the k–th coefficient of $\mathbf{y}_n$. $\widehat{\mathbb{E}}[\cdot]$ and $\widehat{\text{var}}(\cdot)$ represent empirical mean and variance estimates. This normalisation scales each component such that it has a mean of 0 and a standard deviation of 1. Another method is the *min-max* normalisation, e.g., for the inputs

$$\xi_{jn} \mapsto \frac{\xi_{jn} - \min(\xi_j)}{\max(\xi_j) - \min(\xi_j)} \tag{3.184}$$

which constrains each component to lie in the range [0, 1].

3.7.4 Basic Framework for Supervised Machine Learning

Machine learning involves a number of choices and assumptions, which we now briefly outline so that the reader is familiarised with the basic framework before we delve into the details in Chap. 6. The first assumption is that underlying the data there exists a function describing the relationship between the inputs/design points and outputs/targets. That is

$$y = \eta(\boldsymbol{\xi}) \quad \text{or} \quad \mathbf{y} = \boldsymbol{\eta}(\boldsymbol{\xi}) \tag{3.185}$$

for the scalar and vector case, respectively. The data represents examples of this function, e.g., $y_n = \eta(\boldsymbol{\xi}_n)$ and once an approximation of this function is found, predictions can be made at any value of $\boldsymbol{\xi}$. The scalar or vector function $\eta(\boldsymbol{\xi})$ or $\boldsymbol{\eta}(\boldsymbol{\xi})$ is a *deterministic* function of the inputs $\boldsymbol{\xi}$, i.e., it is not random. Since the data is generated by a physical process (whether through experiments or a model) the existence of such a function is assured—we assume that the model does not contain any stochastic terms. This function is often termed a *latent* function. There may be some random components within the data, which we shall gather together into a single generalised error term.

The error is almost always assumed to be additive. A 'statistical model' for the data then takes the form

$$\text{target} = \widehat{\eta}(\boldsymbol{\xi}) + \text{error} \tag{3.186}$$

$\widehat{\eta}(\boldsymbol{\xi})$ (or $\widehat{\boldsymbol{\eta}}(\boldsymbol{\xi})$) is an approximation of the latent, true or underlying function $\eta(\boldsymbol{\xi})$ (or $\boldsymbol{\eta}(\boldsymbol{\xi})$). For simplicity, we shall consider the scalar case. We shall also abuse notation by using the same symbol η or $\boldsymbol{\eta}$ for both the latent function and the approximating function, unless the distinction is necessary to make in the discussion (as below on the bias vs. variance tradeoff). The error or *noise* accounts for a number of possible sources of data error, which can include sensor measurement errors for laboratory data or numerical errors arising from a numerical formulation and solver settings. These errors can be bounded by functions of the mesh parameters, time step and dimension of the approximating subspace.

In terms of the learning problem, there are broadly speaking two types of error, which we discuss next. *Model bias* is defined as the difference between the expected (average) prediction of our model $\widehat{\eta}(\boldsymbol{\xi})$ and the correct value that we are trying to predict $\eta(\boldsymbol{\xi})$. $\widehat{\eta}(\boldsymbol{\xi})$ depends on some unknown weights or parameters $\mathbf{w}$, the estimation of which depends on the data. If we were able to construct $\widehat{\eta}(\boldsymbol{\xi})$ using different training data or different initial guesses for $\mathbf{w}$, the resulting models would yield a range of predictions. Bias measures how far on average these model predictions are from the correct value.

Model variance, on the other hand, is the error due to the variability of a model prediction for a given point $\boldsymbol{\xi}$. Again, if we were able to construct $\widehat{\eta}(\boldsymbol{\xi})$ more than once, the model variance would measure how much the model predictions vary at $\boldsymbol{\xi}$. Suppose we have estimated $\widehat{\eta}(\boldsymbol{\xi})$ using training data $\boldsymbol{\xi}_n, y_n, n = 1, \ldots, N$. The expected squared prediction error at a new point $\boldsymbol{\xi}$ is

$$E_P = \mathbb{E}[(y - \widehat{\eta}(\boldsymbol{\xi}))^2] \tag{3.187}$$

in which $y = \eta(\boldsymbol{\xi}) + \epsilon$, where ϵ is zero-mean error with variance σ^2. We can derive the so-called *bias-variance decomposition* as follows

$$\begin{aligned}
E_P &= \mathbb{E}\left[(y - \widehat{\eta}(\boldsymbol{\xi}))^2\right] \\
&= \mathbb{E}\left[(\eta(\boldsymbol{\xi}) + \epsilon)^2 - 2(\eta(\boldsymbol{\xi}) + \epsilon)\widehat{\eta}(\boldsymbol{\xi}) + \widehat{\eta}(\boldsymbol{\xi})^2\right] \\
&= \mathbb{E}\left[(\eta(\boldsymbol{\xi})^2 + 2\eta(\boldsymbol{\xi})\epsilon + \epsilon^2\right] - 2\mathbb{E}[(\eta(\boldsymbol{\xi}) + \epsilon)\widehat{\eta}(\boldsymbol{\xi})] + \mathbb{E}\left[\widehat{\eta}(\boldsymbol{\xi})^2\right] \\
&= \eta(\boldsymbol{\xi})^2 + 2\eta(\boldsymbol{\xi})\mathbb{E}[\epsilon] + \mathbb{E}[\epsilon^2] - (2\eta(\boldsymbol{\xi}) + 2\mathbb{E}[\epsilon])\mathbb{E}[\widehat{\eta}(\boldsymbol{\xi})] + \mathbb{E}\left[\widehat{\eta}(\boldsymbol{\xi})^2\right]
\end{aligned} \tag{3.188}$$

using the linearity of expectation and the fact that $\eta(\boldsymbol{\xi})$ is not random, together with the fact that any two independent random variables X and Y satisfy $\mathbb{E}[XY] = \mathbb{E}[X]\mathbb{E}[Y]$.

From the assumptions on ϵ, we have $\mathbb{E}[\epsilon] = 0$ and $\mathbb{E}[\epsilon^2] - \mathbb{E}[\epsilon]^2 = \text{var}(\epsilon) = \sigma^2$. Therefore

$$\begin{aligned}
E_P &= \eta(\boldsymbol{\xi})^2 + \sigma^2 - 2\eta(\boldsymbol{\xi})\mathbb{E}[\widehat{\eta}(\boldsymbol{\xi})] + \mathbb{E}[\widehat{\eta}(\boldsymbol{\xi})^2] \\
&= \eta(\boldsymbol{\xi})^2 + \sigma^2 - 2\eta(\boldsymbol{\xi})\mathbb{E}[\widehat{\eta}(\boldsymbol{\xi})] + \left(\mathbb{E}[\widehat{\eta}(\boldsymbol{\xi})^2] - \mathbb{E}[\widehat{\eta}(\boldsymbol{\xi})]^2\right) + \mathbb{E}[\widehat{\eta}(\boldsymbol{\xi})]^2 \\
&= \eta(\boldsymbol{\xi})^2 - 2\eta(\boldsymbol{\xi})\mathbb{E}[\widehat{\eta}(\boldsymbol{\xi})] + \mathbb{E}[\widehat{\eta}(\boldsymbol{\xi})]^2 + \text{var}(\widehat{\eta}(\boldsymbol{\xi})) + \sigma^2 \\
&= (\eta(\boldsymbol{\xi}) - \mathbb{E}[\widehat{\eta}(\boldsymbol{\xi})])^2 + \text{var}(\widehat{\eta}(\boldsymbol{\xi})) + \sigma^2
\end{aligned} \tag{3.189}$$

in which $(\eta(\boldsymbol{\xi}) - \mathbb{E}[\widehat{\eta}(\boldsymbol{\xi})])^2$ is the square of the model bias and $\text{var}(\widehat{\eta}(\boldsymbol{\xi}))$ is the model variance.

Of course we cannot compute these terms since we do not know the true function, but this bias vs. variance decomposition illustrates that there are three components to the error. For particular models we can derive the asymptotic bias and variance, meaning their dependence on the sample size N as $N \to \infty$, and on other parameters in $\widehat{\eta}(\boldsymbol{\xi})$ as these parameters approach zero or ∞. The noise component σ^2 is called *irreducible* error since it is beyond our control. The first two parts are due to modelling assumptions and therefore under our control. We call these sources of error *reducible*.

In machine learning we seek to minimise the bias and the variance simultaneously, which in practise means (for a fixed data set) selecting different functional forms of the model and tuning their complexity. The complexity is, as the name suggests, the complexity of the approximating function used, a proxy for which is the number of parameters it contains. Models that are too simple tend to have a high bias because they are not flexible enough to capture the underlying function well. On the other hand, simple models do not vary much for different data sets, i.e., they possess low variance. Models that are too complex have a high variance because they have more parameters that need to be learned, the estimates of which can change dramatically with different data sets or initial guesses. Conversely, they tend to have a low bias since they are more expressive.

This is the essence of what is called the *bias-variance tradeoff*, a fundamental issue that we face in choosing models in machine learning. If the estimate $\widehat{\eta}(\boldsymbol{\xi})$ approaches $\eta(\boldsymbol{\xi})$ as $N \to \infty$, we say that the estimate is *consistent*. If $\mathbb{E}[\widehat{\eta}(\boldsymbol{\xi})] = \eta(\boldsymbol{\xi})$, we say that the estimate is *unbiased*, otherwise it is said to be *biased*.

It is possible to neglect the error entirely, and in fact many methods, including basic implementations of neural and deep neural networks do precisely this. There are, however, cases in which the error must be treated with care, either because it represents a large component of the targets or because the data is scarce. In Chap. 6, we shall discuss Bayesian and regularisation methods, which are able to incorporate error terms either directly or indirectly, leading (in theory) to better approximations of the latent function.

The main set of choices and assumptions concerns the approximating function and the error. As mentioned above, the choice of approximating function introduces unknown parameters (or weights) $\mathbf{w}$ associated with the chosen form, and assumptions on the error introduce further parameters, which are also usually unknown. The data is then used to *train* the model, which consists of estimating the parameters in order to fully specify the model. This is achieved by minimising a function of the parameters called the *loss function*, which is a measure of the error between the approximating function predictions at the design points and the given targets. Equivalently, the parameters are estimated by maximising a *likelihood function*, which expresses the probability of the data given the assumed model, and is also a function of the parameters. This can be viewed as the basic framework in all types of machine learning.

In general, the latent function lies in an infinite dimensional space, characterised by a countably infinite set of (non-unique) basis functions. It is not possible to find an infinite set of coefficients in such a basis, thus we must resort to finite-dimensional approximations (the approximating function), with coefficients determined using only a finite set of data points. The approximating function is defined as the 'best' approximation available within a class of possible candidates, which first necessitates a definition of 'best', leading to further choices.

Returning to our scalar valued targets y_n, the simplest possible method ignores the error and assumes a polynomial approximating function for the latent function $\eta(\boldsymbol{\xi})$. A linear polynomial, i.e., a straight line in 1D, is the most familiar example. The polynomial coefficients (or weights) can be estimated by minimising the total square error between the values given by the polynomial at each $\boldsymbol{\xi}_n$ and the targets y_n, which forms the loss function.

Assuming instead the model (3.186) and a normally (Gaussian) distributed error, the problem can be posed in an alternative way. With the same polynomial approximating function, the probability distribution over each target is Gaussian and the joint probability of the targets conditioned on the weights is given by the product of these probability distributions, assuming that the error is independent and identically distributed (i.i.d.); this product is the likelihood of the data. The *maximum likelihood* solution consists of finding the set of weights that maximises the likelihood. This optimisation problem is in fact equivalent to the least-squares solution, provided the error is Gaussian and i.i.d..

The aforementioned approximation is termed *linear*, meaning that the approximating function is *linear in the weights* (generally it is nonlinear in $\boldsymbol{\xi}$). Quite often, more powerful nonlinear and *non-parametric* methods are required. Here, nonlinear means that the approximating function is not linear in the weights. As an example, consider the following approximating functions for an input $\xi \in \mathbb{R}$

$$
\begin{aligned}
\eta_1(\xi; \mathbf{w}) &= w_0 + w_1 e^{w_2 \xi} \\
\eta_2(\xi; \mathbf{w}) &= w_0 + w_1 \xi + w_2 \xi^2
\end{aligned}
\tag{3.190}
$$

with weights w_0, w_1 and w_2, collected inside a vector $\mathbf{w} = (w_0, w_1, w_2)^T$. The notation $\eta(\xi; \mathbf{w})$ explicitly indicates the dependence of the approximating functions on the weights. $\eta_2(\xi; \mathbf{w})$ can be written as $\eta_2(\xi; \mathbf{w}) = \mathbf{w}^T(1, \xi, \xi^2)$, which, considered as a function of $\mathbf{w}$, is linear. The approximating function $\eta_1(\xi)$, however, has no such representation in terms of $\mathbf{w}$, and would therefore constitute a nonlinear regression. Non-parametric methods do not assume an explicit form, such as a polynomial, but instead use an implicit form that introduces associated parameters, called *hyperparameters*. A prominent example of this type of method is Gaussian process models, discussed in Chap. 6.

3.8 Summary

As we have seen in this chapter, there are many available modelling methods for studying and developing flow battery systems—indeed, we did not cover many prominent methods such as Monte Carlo simulation, kinetic Monte Carlo, and coarse graining. Aside from physics-based models, which range from the nanoscale up to the macroscale, including multi-scale approaches, there is a huge variety of data-driven approaches that were briefly described in this chapter and which we will discuss in detail in Chap. 6.

The choice of method depends on the particular aspect of the system under consideration, including its spatio-temporal scale, as well as the goal of the study. In general, there is more than one appropriate method (whether physics based or data-driven), and it is recommended that the different methods are compared. In the next chapter we present a coupled macroscopic-mesoscopic approach to the modelling of a vanadium-iron flow battery.

References

1. A.A. Shah, K. Luo, T.R. Ralph, F.C. Walsh, Recent trends and developments in polymer electrolyte membrane fuel cell modelling. Electrochim. Acta **56**, 3731–3757 (2011)
2. V. Ramadesigan, P.W.C. Northrop, S. De, S. Santhanagopalan, R.D. Braatz, V.R. Subramanian, Modeling and simulation of lithium-ion batteries from a systems engineering perspective. J. Electrochem. Soc. **159**, R31–R45 (2012)
3. N. Bellomo, M. Pulvirenti, *Modeling in Applied Sciences: A Kinetic Theory Approach* (Springer Science & Business Media, 2013)
4. I. Steinbach, Phase-field models in materials science. Model. Simul. Mater. Sci. Eng. **17**(7), 073001 (2009)
5. R. Benzi, S. Succi, M. Vergassola, The lattice boltzmann equation: theory and applications. Phys. Rep. **222**(3), 145–197 (1992)
6. J.W. Cahn, J.E. Hilliard, Free energy of a nonuniform system. i. interfacial free energy. J. Chem. Phys. **28**(2), 258–267 (1958)
7. K. Zhou, B. Liu, *Molecular Dynamics Simulation: Fundamentals and Applications* (Academic Press, 2022)
8. K. Binder, D. Heermann, L. Roelofs, A.J. Mallinckrodt, S. McKay, Monte carlo simulation in statistical physics. Comput. Phys. **7**(2), 156–157 (1993)
9. S.Y. Joshi, S.A. Deshmukh, A review of advancements in coarse-grained molecular dynamics simulations. Mol. Simul. **47**(10–11), 786–803 (2021)
10. R.M. Martin, *Electronic Structure: Basic Theory and Practical Methods* (Cambridge university press, 2020)
11. E Weinan, *Principles of multiscale modeling* (Cambridge University Press, 2011)
12. J.N. Reddy, *An Introduction to Continuum Mechanics* (Cambridge university press, 2013)
13. R. Temam, *Navier-Stokes Equations: Theory and Numerical Analysis*, vol. 343 (American Mathematical Soc., 2001)
14. H.K. Versteeg, W. Malalasekera, *An Introduction to Computational Fluid Dynamics: The Finite Volume Method* (Pearson education, 2007)
15. F.P. Incropera, D.P. DeWitt, T.L. Bergman, A.S. Lavine et al., *Fundamentals of Heat and Mass Transfer*, vol. 6 (Wiley, New York, 1996)
16. R. Taylor, R. Krishna, *Multicomponent Mass Transfer*, vol. 2 (Wiley, 1993)

17. H. Darcy, *Les fontaines publiques de la ville de Dijon: exposition et application des principes à suivre et des formules à employer dans les questions de distribution d'eau... un appendice relatif aux fournitures d'eau de plusieurs villes au filtrage des eaux*, vol. 1 (Victor Dalmont, éditeur, 1856)
18. S. Whitaker, Flow in porous media i: a theoretical derivation of darcy's law. Transp. Porous Media **1**(1), 3–25 (1986)
19. J. Kozeny, Uber kapillare leitung des wassers im boden-aufstieg, versickerung und anwendung auf die bewasserung, sitzungsberichte der akademie der wissenschaften wien. Mathematisch Naturwissenschaftliche Abteilung **136**, 271–306 (1927)
20. P.C. Carman, Permeability of saturated sands, soils and clays. J. Agric. Sci. **29**(2), 262–273 (1939)
21. H.C. Brinkman, A calculation of the viscous force exerted by a flowing fluid on a dense swarm of particles. Flow Turbul. Combust. **1**, 27–34 (1949)
22. T.E. Springer, T.A. Zawodinski, S. Gottesfeld, Polymer electrolyte fuel cell model. J. Electrochem. Soc. **138**(8), A2334–A2341 (1991)
23. D.M. Bernadi, M.W. Verbrugge, AIChE J. **37**, 1151 (1991)
24. D.M. Bernadi, M.W. Verbrugge, A mathematical model of the solid-polymer-electrolyte fuel cell. J. Elecrochem. Soc. **139**(9), A2477–A2490 (1992)
25. R. Schlögl, Zur theorie der anomalen osmose. Z. Phys. Chem (Munich) **3**, 73 (1955)
26. C.W. Hirt, B.D. Nichols, Volume of fluid (vof) method for the dynamics of free boundaries. J. Comput. Phys. **39**(1), 201–225 (1981)
27. F.H. Harlow, J.E. Welch, Numerical calculation of time-dependent viscous incompressible flow of fluid with free surface. Phys. Fluids **8**(12), 2182–2189 (1965)
28. X. Zhu, P.C. Sui, N. Djilali, Dynamic behaviour of liquid water emerging from a gdl pore into a pemfc gas flow channel. J. Power Sources **172**(1), 287–295 (2007)
29. J.U. Brackbill, D.B. Kothe, C. Zemach, A continuum method for modeling surface tension. J. Comput. Phys. **100**(2), 335–354 (1992)
30. J.A. Sethian, *Level Set Methods and Fast Marching Methods: Evolving Interfaces in Computational Geometry, Fluid Mechanics, Computer Vision, and Materials Science*, vol. 3 (Cambridge university press, 1999)
31. S. Osher, R. Fedkiw, K. Piechor, Level set methods and dynamic implicit surfaces. Appl. Mech. Rev. **57**(3), B15–B15 (2004)
32. M. Sussman, E. Fatemi, P. Smereka, S. Osher, An improved level set method for incompressible two-phase flows. Comput. Fluids **27**(5–6), 663–680 (1998)
33. R.R. Nourgaliev, S. Wiri, N.T. Dinh, T.G. Theofanous, On improving mass conservation of level set by reducing spatial discretization errors. Int. J. Multiph. Flow **31**(12), 1329–1336 (2005)
34. D.L. Sun, W.Q. Tao, A coupled volume-of-fluid and level set (voset) method for computing incompressible two-phase flows. Int. J. Heat Mass Transf. **53**(4), 645–655 (2010)
35. J. Donea, A. Huerta, J.-P. Ponthot, A. Rodriguez-Ferran, Arbitrary Lagrangian-Eulerian methods, in *Encyclopedia of Computational Mechanics*, vol. Part 1 Fundamentals, Chap. 10, 2nd edn., ed. by E. Stein, R. de Borst, T.J.R. Hughes (Wiley, Chichester, 2017)
36. C.S. Peskin, Flow patterns around heart valves: a numerical method. J. Comput. Phys. **10**(2), 252–271 (1972)
37. R. Mittal, G. Iaccarino, Immersed boundary methods. Annu. Rev. Fluid Mech. **37**, 239–261 (2005)
38. G. Iaccarino, R. Verzicco, Immersed boundary technique for turbulent flow simulations. Appl. Mech. Rev. **56**(3), 331–347 (2003)
39. C. Miehe, F. Welschinger, M. Hofacker, Thermodynamically consistent phase-field models of fracture: variational principles and multi-field fe implementations. Int. J. Numer. Methods Eng. **83**(10), 1273–1311 (2010)
40. M.J. Borden, C.V. Verhoosel, M.A. Scott, T.J.R. Hughes, C.M. Landis, A phase-field description of dynamic brittle fracture. Comput Methods Appl Mech Eng **217**, 77–95 (2012)

41. S.M. Allen, J.W. Cahn, Coherent and incoherent equilibria in iron-rich iron-aluminum alloys. Acta Metallurgica **23**(9), 1017–1026 (1975)
42. X. Zhuang, S. Zhou, G.D. Huynh, P. Areias, T. Rabczuk, Phase field modeling and computer implementation: a review. Eng. Fract. Mech. **262**, 108234 (2022)
43. D. d'Humières, Multiple-relaxation-time lattice boltzmann models in three dimensions. Philos. Trans. R. Soc. Lond. Ser. A: Math. Phys. Eng. Sci. **360**(1792), 437–451 (2002)
44. Q. Li, K.H. Luo, Q.J. Kang, Y.L. He, Q. Chen, Q. Liu, Lattice boltzmann methods for multiphase flow and phase-change heat transfer. Prog. Energy Combust. Sci. **52**, 62–105 (2016)
45. B. Leimkuhler, C. Matthews, Molecular dynamics. Interdiscip. Appl. Math. **39**, 443 (2015)
46. I. Torrens, *Interatomic Potentials* (Elsevier, 2012)
47. R.K. Pathria, Statistical Mechanics (Elsevier, 2016)
48. D.M. Kochmann, *Computational Multiscale Modeling* (ETH Zurich, Zurich, Switzerland, 2018)
49. L. Verlet, Computer "experiments" on classical fluids. i. thermodynamical properties of lennard-jones molecules. Phys. Rev. **159**(1), 98 (1967)
50. W.C. Swope, H.C. Andersen, P.H. Berens, K.R. Wilson, A computer simulation method for the calculation of equilibrium constants for the formation of physical clusters of molecules: application to small water clusters. J. Chem. Phys. **76**(1), 637–649 (1982)
51. T. Schlick, *Molecular Modeling and Simulation: An Interdisciplinary Guide*, vol. 2 (Springer, 2010)
52. W.G. Hoover, B.L. Holian, Kinetic moments method for the canonical ensemble distribution. Phys. Lett. A **211**(5), 253–257 (1996)
53. H.J.C. Berendsen, J.P.M. Postma, W.F. van Gunsteren, A. DiNola, J.R. Haak, Molecular dynamics with coupling to an external bath. J. Chem. Phys. **81**(8), 3684–3690 (1984)
54. H.C. Andersen, Molecular dynamics simulations at constant pressure and/or temperature. J. Chem. Phys. **72**(4), 2384–2393 (1980)
55. P. Cársky, J. Paldus, J. Pittner. Recent Progress in Coupled Cluster Methods: Theory and Applications (2010)
56. D. Cremer, Møller-plesset perturbation theory: from small molecule methods to methods for thousands of atoms. Wiley Interdiscip. Rev. Comput. Mol. Sci. **1**(4), 509–530 (2011)
57. J.A. Pople, D.P. Santry, G.A. Segal, Approximate self-consistent molecular orbital theory. i. invariant procedures. J. Chem. Phys. **43**(10), S129–S135 (1965)
58. M.J.S. Dewar, E.G. Zoebisch, E.F. Healy, J.J.P. Stewart, Development and use of quantum mechanical molecular models. 76. am1: a new general purpose quantum mechanical molecular model. J. Am. Chem. Soc. **107**(13), 3902–3909 (1985)
59. A. Jain, Y. Shin, K.A. Persson, Computational predictions of energy materials using density functional theory. Nat. Rev. Mater. **1**(1), 1–13 (2016)
60. P. Hohenberg, W. Kohn, Inhomogeneous electron gas. Phys. Rev. **136**(3B), 864–871 (1964). (November)
61. A.J. Cohen, P. Mori-Sánchez, W. Yang, Challenges for density functional theory. Chem. Rev. **112**(1), 289–320 (2012)
62. M. Levy, Universal variational functionals of electron densities, first-order density matrices, and natural spin-orbitals and solution of the v-representability problem. Proc. Natl. Acad. Sci. **76**(12), 6062–6065 (1979)
63. E.H. Lieb, Density functionals for coulomb systems. Int. J. Quantum Chem. **24**(3), 243–277 (1983)
64. I. Lindgren, S. Salomonson, Differentiability in density-functional theory. Adv. Quantum Chem. **43**, 95–119 (2003)
65. R. Van Leeuwen, Density functional approach to the many-body problem: key concepts and exact functionals. Adv. Quantum Chem. **43**(25.10), 1016 (2003)
66. R.G. Parr, Density functional theory of atoms and molecules, in *Horizons of Quantum Chemistry*. (Springer, 1980), pp.5–15
67. J.P. Perdew, Y. Wang, Accurate and simple analytic representation of the electron-gas correlation energy. Phys. Rev. B **45**(23), 13244 (1992)

68. P.A.M. Dirac, Note on exchange phenomena in the thomas atom. Math. Proc. Camb. Philos. Soc. **26**(3), 376–385 (1930)
69. D.M. Ceperley, B.J. Alder, Ground state of the electron gas by a stochastic method. Phys. Rev. Lett. **45**, 566–569 (1980). (Aug)
70. J.P. Perdew, K. Burke, M. Ernzerhof, Generalized gradient approximation made simple. Phys. Rev. Lett. **77**(18), 3865 (1996)
71. J.P. Perdew, K. Burke, Y. Wang, Generalized gradient approximation for the exchange-correlation hole of a many-electron system. Phys. Rev. B **54**(23), 16533 (1996)
72. J. Tao, J.P. Perdew, V.N. Staroverov, G.E. Scuseria, Climbing the density functional ladder: nonempirical meta–generalized gradient approximation designed for molecules and solids. Phys. Rev. Lett. **91**(14), 146401 (2003)
73. J.P. Perdew, J. Tao, V.N. Staroverov, G.E. Scuseria, Meta-generalized gradient approximation: explanation of a realistic nonempirical density functional. J. Chem. Phys. **120**(15), 6898–6911 (2004)
74. P.J. Stephens, F.J. Devlin, C.F. Chabalowski, M.J. Frisch, Ab initio calculation of vibrational absorption and circular dichroism spectra using density functional force fields. J. Phys. Chem. **98**(45), 11623–11627 (1994)
75. A.D. Becke, Density-functional exchange-energy approximation with correct asymptotic behavior. Phys. Rev. A **38**(6), 3098 (1988)
76. C. Lee, W. Yang, R.G. Parr, Development of the colle-salvetti correlation-energy formula into a functional of the electron density. Phys. Rev. B **37**(2), 785 (1988)
77. S.H. Vosko, L. Wilk, M. Nusair, Accurate spin-dependent electron liquid correlation energies for local spin density calculations: a critical analysis. Can. J. Phys. **58**(8), 1200–1211 (1980)
78. C. Adamo, V. Barone, Toward reliable density functional methods without adjustable parameters: the pbe0 model. J. Chem. Phys. **110**(13), 6158–6170 (1999)
79. A. Saltelli, P. Annoni, I. Azzini, F. Campolongo, M. Ratto, S. Tarantola, Variance based sensitivity analysis of model output. Design and estimator for the total sensitivity index. Comput. Phys. Commun. **181**(2), 259–270 (2010)
80. I.M. Sobol, Uniformly distributed sequences with an additional uniform property. USSR Comput. Math. Math. Phys. **16**(5), 236–242 (1976)

Chapter 4
Numerical Simulation of Flow Batteries Using a Multi-scale Macroscopic-Mesoscopic Approach

4.1 Introduction

The development of redox flow batteries presents challenges in terms of scale-up, optimization, improvements in electrolyte stability, and the development of new materials [1]. These challenges can be tackled using a combination of laboratory analysis and modelling, which lowers the financial costs and timescales. Modelling of the cells can provide insights into the coupled physico-chemical processes, including heat and mass transfer, the redox reactions, charge transfer and any phase changes, all of which are difficult to quantify experimentally. For design and optimization, an accurate model can be invaluable.

The different modelling tasks (cell and component design, optimization, sensitivity analysis, control, materials design and screening) rely on different approaches. Macroscopic models are relevant to the device (and possibly small stack) level. They can be employed for designing and optimising components and cells with respect to the geometries, averaged component characteristics such as conductivity and porosity, as well the operating conditions. The aim of this chapter is to present a general macroscopic modelling approach for RFBs and to provide a detailed case study of its application.

The all-vanadium RFB (VRFB) with an aqueous electrolyte is the most mature RFB technology. In recent years, nonaqueous solvents were used as electrolytes for RFBs to overcome the lower operating voltage of aqueous electrolyte-based RFBs. As one of the promising nonaqueous solvents, deep eutectic solvents (DESs), which consist of an organic halide salt and a hydrogen bond donor, have shown promise as solvents for electrochemical systems due to their low cost, high electrochemical stability, non-toxicity and wider potential window [2].

For the majority of RFB systems, whether with aqueous or nonaqueous electrolytes, the porous electrode is a key component that not only provides active sites for electrochemical reaction, but also affects the electrolyte flow and ion transport within its pore structure. Due to its simple and convenient implementation method

A. A. Shah et al., *New Paradigms in Flow Battery Modelling*,
Engineering Applications of Computational Methods 16,
https://doi.org/10.1007/978-981-99-2524-7_4

for treating the irregular boundaries in porous electrodes, the lattice-Boltzmann (LB) method has become a powerful numerical tool to elucidate the pore-scale transfer mechanisms and electrochemical reactions that occur during the operation of batteries, including RFBs. In recent years, a series of novel LB models [3–5] have been proposed to reveal the effects of pore morphologies and operating conditions on the performance of RFBs with aqueous or nonaqueous electrolytes, contributing to the design of RFB electrodes with optimal pore structures.

Taking the DES-electrolyte nonaqueous vanadium-iron RFB as an example, a coupled macroscopic and LB model will be developed to study the reactive transfer process within a double-layer gradient electrode. This heterogeneous electrode is comprised of graphite felt (GF) and carbon paper (CP) electrodes with different micro-structural parameters. A three-dimensional multiple-relaxation-time (MRT) LB model is established to solve the pore-scale transport problem, incorporating electrochemical reactions, in a composite electrode reconstructed by X-ray micro-computed tomography. By comparison with commercial GF and CP electrodes, the specific transport and reactive characteristics of the double-layer electrode during the galvanostatic discharging of the negative half cell of a vanadium-iron RFB are elucidated. Moreover, the simulations reveal the effects of electrolyte feeding modes and flow rates on the power losses.

4.2 Macroscopic Modelling Approaches

In Sect. 3.3 of Chap. 3 we provided a general introduction to macroscopic models. In this section we provide an in-depth introduction to such models for flow batteries. Macroscopic modelling for flow batteries includes one or more of the conservation laws for mass, charge, fluid momentum and thermal energy. These laws and their mathematical embodiment are described below, before we detail the coupling with a pore-scale Lattice-Boltzmann model.

4.2.1 Conservation of Momentum and Fluid Flow

The macroscopic superficial velocity of the electrolyte $\mathbf{u}$ within the porous electrode is assumed to follow Darcy's law (3.28), which represents the conservation of fluid momentum

$$\frac{\mu}{K}\mathbf{u} = -\nabla p \tag{4.1}$$

where μ is the kinematic viscosity, K is the permeability and p is the liquid pressure. The permeability of a porous medium is a function of the material properties, as detailed in the next section. The liquid pressure is determined from an overall mass balance $\nabla \cdot \mathbf{u} = 0$ and Eq. (4.1), which result in the following

$$\frac{K}{\mu}\nabla^2 p = 0 \tag{4.2}$$

The permeability is usually assumed to follow some law that relates it to the micro-structural characteristics of the electrode, e.g., the Kozeny-Carman law (3.29)

$$K = \frac{d_f^2 \varepsilon^3}{K_{KC}(1-\varepsilon)^2} \tag{4.3}$$

in which ε is the electrode porosity, d_f is the average fibre diameter and K_{KC} is the Kozeny-Carman constant.

During practical operation, when the SOC is higher than 0.85, side reactions are prone to occur. If the side reactions, notably the evolution of oxygen in the positive electrode and the evolution of hydrogen in the negative electrode on charge are considered, bubble evolution leads to a two phases (liquid-gas) solution and the flow equations should be rewritten as

$$\mathbf{u}_l = -K\frac{k_{rl}}{\mu_l}\nabla p_l \tag{4.4}$$

$$\mathbf{u}_g = -K\frac{k_{rg}}{\mu_g}\nabla p_g \tag{4.5}$$

for the liquid and gas phases, respectively, in which the subscripts l and g refer to the properties of the liquid and gas phases, while k_r is the relative permeability of a phase and is a function of phase saturation alone.

The difference between the pressures of the gas phase and liquid phases is defined as the capillary pressure, a result of the interfacial tension existing at the interface separating the two fluids

$$p_c = p_g - p_l = \sigma \cos\theta_c \sqrt{\frac{\varepsilon}{K}} J(s) \tag{4.6}$$

in which σ is the interfacial tension, θ_c is the contact angle, $J(s)$ represents the Leverette function and s is the liquid saturation in the porous electrode. Taking into account bubble evolution, the mass balance for the gas phase is

$$\varepsilon\rho_g \frac{\partial}{\partial t}(1-s) + \varepsilon\rho_g \nabla\cdot\big((1-s)\mathbf{u}_g\big) = S_g \tag{4.7}$$

in which ρ_g is the gas density and S_g is the rate of gas evolution. The mixture density and velocity of the mass centre (mixture velocity) are defined by

$$\rho_m = s\rho_l + (1-s)\rho_g, \quad \rho_m\mathbf{u}_m = s\mathbf{u}_l + (1-s)\mathbf{u}_g \tag{4.8}$$

respectively, in which ρ_l is the liquid density. The liquid phase and gas phase velocities are related by the slip velocity $\mathbf{u}_{slip}$ in the following manner

$$\mathbf{u}_{slip} = \mathbf{u}_g - \mathbf{u}_l \tag{4.9}$$

An equation for the slip velocity can be derived by performing a balance of forces on a bubble and the final form can be expressed as

$$\mathbf{u}_{slip} = \frac{d_b^3}{18\mu_l} \nabla p \tag{4.10}$$

in which d_b is the averaged bubble diameter.

In the case of an open flow channel, as in many systems, including membrane-less systems such as the soluble lead-acid flow battery in which the reactions take place at planar electrodes rather than porous electrodes, Darcy's law is replaced by the Navier-Stokes Eq. (3.14) (assuming a single phase)

$$\rho \left(\frac{\partial \mathbf{u}}{\partial t} + \mathbf{u} \cdot \nabla \mathbf{u} \right) = -\nabla p + \eta \nabla^2 \mathbf{u} + \frac{1}{3} \eta \nabla(\nabla \cdot \mathbf{u}) + F \tag{4.11}$$

in which ρ is the liquid density, η is the dynamic viscosity, and F is an external force such as gravity. These equations are usually accompanied by continuity $\nabla \cdot \mathbf{u} = 0$ under the assumption of incompressibility.

4.2.2 Conservation of Mass

The mass balance for a charged species i in the electrolyte is

$$\begin{aligned} \frac{\partial}{\partial t}(\epsilon s c_i) &= -\nabla \cdot \mathbf{N}_i + S_i \\ &= \nabla \cdot \left(D_i^{eff} \nabla c_i + \frac{F z_i}{RT} D_i^{eff} c_i \nabla \phi_l - \mathbf{u}_l c_i \right) + S_i \end{aligned} \tag{4.12}$$

in which c_i is the concentration, D_i^{eff} is the effective diffusion coefficient, ϕ_l is the electrolyte potential, z_i is the valence and S_i is the source term due to chemical/electrochemical reactions for species i in the porous electrode. R is the universal gas constant and T is the temperature, while F is Faraday's constant. $\mathbf{N}_i$ is the flux of species i, which is expressed in (4.12) using the Nernst-Planck equation (3.27) the latter includes terms for convection, diffusion and electromigration if the species is charged ($z_i \neq 0$).

The effective diffusivity can be expressed using the Bruggemann equation

$$D_i^{eff} = (\epsilon s)^{\frac{3}{2}} D_i \tag{4.13}$$

in which D_i is the bulk diffusivity of species i.

4.2.3 Conservation of Charge

Conservation of charge usually assumes a pseudo-steady state for charge transport and neglects double-layer effects.

The electrolyte can be assumed to obey electroneutrality, i.e., $\sum_i z_i c_i = 0$. Following the derivation in Sect. 3.3.6, the ionic conductivity satisfies (3.39)

$$\nabla \cdot \mathbf{j} = \nabla \cdot \left(-\sigma_l \nabla \phi - F \sum_i z_i D_i^{eff} \nabla c_i \right) \tag{4.14}$$

in which

$$\sigma_l = \frac{F^2 \sum_i z_i^2 D_i^{eff} \nabla c_i}{RT} \tag{4.15}$$

is an effective electrolyte conductivity and $\nabla \cdot \mathbf{j}$ is the total current density.

Ohm's law (3.41) can be used for the charge balance in the solid-phase for the solid-state potential ϕ_s

$$-\sigma_s^{eff} \nabla^2 \phi_s = -\nabla \cdot \mathbf{j}_s \tag{4.16}$$

in which the effective conductivity can be approximated using a Bruggeman correction

$$\sigma_s^{eff} = (1 - \epsilon)^{\frac{3}{2}} \sigma_s \tag{4.17}$$

for a bulk conductivity σ_s.

4.2.4 Equations Specific to the Membrane

The model of Bernadi and Verbrugge [6, 7] is used for proton and water transport in a Nafion membrane (Sect. 3.3.5). The concentration of dissolved water c_{H_2O} is governed by (3.31)

$$\frac{\partial c_{H_2O}}{\partial t} - \nabla \cdot \left(D_{H_2O} \nabla c_{H_2O} \right) + \nabla \cdot \left(\mathbf{u}_l c_{H_2O} \right) = 0 \tag{4.18}$$

in which D_{H_2O} is the dissolved water diffusion coefficient [8]

$$D_{H_2O} = 4.17 \times 10^{-8}\lambda(1 + 161e^{-\lambda}) \exp\left(-\frac{2436}{T}\right) \quad \text{in m}^2\ \text{s}^{-1} \tag{4.19}$$

λ is the water content, which for a liquid saturated Nafion membrane has a value of 22. The velocity $\mathbf{u}_l$ satisfies incompressibility $\nabla \cdot \mathbf{u}_l = 0$ and is modelled using Schlogl's Eq. (3.32)

$$\mathbf{u}_l = -\frac{\kappa_\phi}{\mu} F c_{H^+} \nabla \phi_e - \frac{\kappa_p}{\mu_{H_2O}} \nabla p \tag{4.20}$$

in which κ_ϕ is the electrokinetic permeability, while κ_p is the hydraulic permeability. The equation for current conservation is given by (3.33)

$$\nabla^2 \phi_e = 0 \tag{4.21}$$

and the pressure satisfies (3.35)

$$\nabla^2 p = 0. \tag{4.22}$$

4.2.5 Conservation of Thermal Energy

The energy balance accounts for heat conduction, heat convection and heat generation by electrochemical reaction and Joule heating. As an approximation, it can be assumed that the two phases in the electrodes attain the same temperature instantaneously (infinitely fast heat exchange). The energy balance can then be expressed as (3.18)

$$\frac{\partial}{\partial t}\left(\overline{\rho C_p} T\right) + \nabla \cdot \left(\mathbf{u}_m \overline{\rho C_p} T\right) = \overline{\lambda} \nabla^2 T + \sum_k Q_k \tag{4.23}$$

in which $\overline{\rho C_p}$ is the volume-averaged thermal capacity, $\overline{\lambda}$ is the volume-averaged thermal conductivity and the Q_k are the source terms (Ohmic heating and electrochemical reaction). The volume averaging is over all phases, including the solid.

It is worth pointing out that isothermal conditions cannot be maintained and thermal effects during operation should be taken into account if the flow rate of electrolyte is not high.

4.2.6 Electrochemical Kinetics

The Butler-Volmer Eq. (2.23) is the most frequently used model for electrochemical reactions. The nonaqueous redox flow battery under consideration uses a DES electrolyte, which is prepared by mixing choline chloride and urea at a molar ratio of 1:2.

Adding 0.1 Mol $FeCl_2$ and 0.1 Mol VCl_3 to the prepared DES leads to the positive and negative electrolytes, respectively. During charging and discharging, the redox reaction occurring in the positive and negative electrodes can be expressed as

$$\begin{aligned}&\text{positive: } Fe^{3+} + e^- \leftrightarrow Fe^{2+}\\&\text{negative: } V(III) \leftrightarrow V(II) + e^-\end{aligned} \tag{4.24}$$

Although it could be a severe approximation, for the reasons given above, we that the redox flow battery is operating in isothermal state. We also assume that the ion-exchange membrane only allows chloride ions to pass.

The exchange current densities engendered by the redox reactions in the positive and negative electrodes can be written as follows

$$\nabla \cdot \mathbf{j} = a\varepsilon s F k_{pos} \left(\widehat{c}_{Fe(II)}\right)^{\alpha_{1,c}} \left(\widehat{c}_{Fe(III)}\right)^{1-\alpha_{1,c}} \left(e^{F(1-\alpha_{1,c})\eta/(RT)} - e^{-\alpha_{1,c}F\eta/(RT)}\right) \tag{4.25}$$

$$\nabla \cdot \mathbf{j} = a\varepsilon s F k_{neg} \left(\widehat{c}_{V(II)}\right)^{\alpha_{2,c}} \left(\widehat{c}_{V(III)}\right)^{1-\alpha_{2,c}} \left(e^{F(1-\alpha_{2,c})\eta/(RT)} - e^{-\alpha_{2,c}F\eta/(RT)}\right) \tag{4.26}$$

in which a is the specific surface area, η is the overpotential, k_{neg} and k_{pos} are rate constants and $\alpha_{i,c}$ are the cathodic transfer coefficients. Note that the bulk concentrations c_i are different from the surface concentration $\widehat{c}_i$.

The overpotential caused by redox reactions in the positive and negative electrodes can be defined as

$$\eta = \phi_s - \phi_l - E_{eq,pos} \tag{4.27}$$

$$\eta = \phi_s - \phi_l - E_{eq,neg} \tag{4.28}$$

in which $E_{eq,pos}$ and $E_{eq,\ neg}$ are the equilibrium potentials of the reactions, which can be estimated from the Nernst equation

$$E_{eq,neg} = E^0_{neg} + \frac{RT}{nF} \ln\left(\frac{c_{V(III)}}{c_{V(II)}}\right) \tag{4.29}$$

$$E_{eq,pos} = E^0_{pos} + \frac{RT}{nF} \ln\left(\frac{c_{Fe(III)}}{c_{Fe(II)}}\right) \tag{4.30}$$

in which E^0_{neg} and E^0_{pos} represent the standard electrode potentials of the negative and positive electrode reactions, respectively.

The relationship between c_i and $\widehat{c}_i$ can be derived through a balance between the diffusion rate and the reaction rate on the electrode surface [9]

$$\widehat{c}_{V(II)} = \frac{c_{V(II)} + \varepsilon s k_{neg} e^{-F\alpha_{2,c}(\phi_s-\phi_l-E_{eq,neg})/RT}\left(c_{V(II)}/\gamma_{V(III)} + c_{V(III)}/\gamma_{V(II)}\right)}{1 + \varepsilon s k_{neg}\left\{\frac{e^{-F\alpha_{2,c}(\phi_s-\phi_l-E_{eq,neg})/RT}}{\gamma_{V(III)}} + \frac{e^{-F(1-\alpha_{2,c})(\phi_s-\phi_l-E_{eq,neg})/RT}}{\gamma_{V(II)}}\right\}} \tag{4.31}$$

in which $\gamma_{V(III)} = D_{V(III)}/d_f$ and $\gamma_{V(III)} = D_{V(III)}/d_f$. Similar expressions can be derived for the positive electrode surface concentrations.

4.2.7 *Reservoirs and Inlet Conditions*

The active species in the circulating electrolyte undergo reaction in the electrode, thus lowering or increasing their concentration over time. If the simulations are conducted at steady state or over a short time period, or the reservoir volume is sufficiently large, this may not be important. In other cases, a model for the concentrations of species at the inlet will be required.

From the conservation of volume, at an outlet of cross-sectional area A_{out}, the volumetric flow rate will be given by

$$\omega = \upsilon_{\text{in}} \epsilon A_{\text{out}} \tag{4.32}$$

if the inlet velocity is unidirectional with speed υ_{in}. The average concentration of a species i at the outlet can be calculated from

$$c_i^{\text{out}} = \int_O c_i dA \tag{4.33}$$

in which O is the outlet surface and dA is an area differential. We can then approximate the inlet concentrations from a mass balance, which assumes instantaneous mixing

$$\frac{dc_i^{\text{in}}}{dt} = \frac{\omega}{V}\left(c_i^{\text{out}} - c_i^{\text{in}}\right), \quad c_i^{\text{in}}(0) = c_i^0 \tag{4.34}$$

in which c_i^0 is the inlet concentration of a species i at $t = 0$ and V is the reservoir volume.

4.3 Lattice-Boltzmann Models

A comprehensive pore-scale lattice-Boltzmann method (LBM) is employed to model the reactive transfer processes in porous electrode. The LBM is able to handle complex boundaries more easily than conventional solutions to the Navier-Stokes equations, which is the primary reason for its popularity in the simulation of flow in porous media. In Sect. 3.4.3 of Chap. 3 we introduced the LBM.

A three-dimensional multiple-relaxation-time (MRT) LBM is used to model the flow of the DES electrolyte in the porous electrode. The evolution equation for $N = 19$ discrete velocities in the D3Q19 model can be expressed as [10]

$$|\mathbf{f}(\mathbf{x}+\mathbf{e}_i\delta t, t+\delta t)\rangle - |\mathbf{f}(\mathbf{x}, t)\rangle = -\mathbf{M}^{-1}\mathbf{S}\mathbf{M}\left(|\mathbf{f}(\mathbf{x}, t)\rangle - |\mathbf{f}^{eq}(\mathbf{x}, t)\rangle\right) \tag{4.35}$$

in which δt is the time step and

$$|\mathbf{f}(\mathbf{x}, t)\rangle = \left(f_0(\mathbf{x}, t), \ldots, f_{N-1}(\mathbf{x}, t)\right)^{\mathrm{T}} \tag{4.36}$$

$$|\mathbf{f}^{eq}(\mathbf{x}, t)\rangle = \left(f_0^{eq}(\mathbf{x}, t), \ldots, f_{N-1}^{eq}(\mathbf{x}, t)\right)^{\mathrm{T}} \tag{4.37}$$

$\mathbf{M}$ is a transformation matrix projecting $f_i(\mathbf{x}, t)$ and $f_i^{eq}(\mathbf{x}, t)$ onto the velocity moment space, given for D3Q19 by

$$\mathbf{M} = \begin{pmatrix}
1 & 1 & 1 & 1 & 1 & 1 & 1 & 1 & 1 & 1 & 1 & 1 & 1 & 1 & 1 & 1 & 1 & 1 & 1 \\
-30 & -11 & -11 & -11 & -11 & -11 & -11 & 8 & 8 & 8 & 8 & 8 & 8 & 8 & 8 & 8 & 8 & 8 & 8 \\
12 & -4 & -4 & -4 & -4 & -4 & -4 & 1 & 1 & 1 & 1 & 1 & 1 & 1 & 1 & 1 & 1 & 1 & 1 \\
0 & 1 & -1 & 0 & 0 & 0 & 0 & 1 & -1 & 1 & -1 & 1 & -1 & 1 & -1 & 0 & 0 & 0 & 0 \\
0 & -4 & 4 & 0 & 0 & 0 & 0 & 1 & -1 & 1 & -1 & 1 & -1 & 1 & -1 & 0 & 0 & 0 & 0 \\
0 & 0 & 0 & 1 & -1 & 0 & 0 & 1 & 1 & -1 & -1 & 0 & 0 & 0 & 0 & 1 & -1 & 1 & -1 \\
0 & 0 & 0 & -4 & 4 & 0 & 0 & 1 & 1 & -1 & -1 & 0 & 0 & 0 & 0 & 1 & -1 & 1 & -1 \\
0 & 0 & 0 & 0 & 0 & 1 & -1 & 0 & 0 & 0 & 0 & 1 & 1 & -1 & -1 & 1 & 1 & -1 & -1 \\
0 & 0 & 0 & 0 & 0 & -4 & 4 & 0 & 0 & 0 & 0 & 1 & 1 & -1 & -1 & 1 & 1 & -1 & -1 \\
0 & 2 & 2 & -1 & -1 & -1 & -1 & 1 & 1 & 1 & 1 & 1 & 1 & 1 & 1 & -2 & -2 & -2 & -2 \\
0 & -4 & -4 & 2 & 2 & 2 & 2 & 1 & 1 & 1 & 1 & 1 & 1 & 1 & 1 & -2 & -2 & -2 & -2 \\
0 & 0 & 0 & 1 & 1 & -1 & -1 & 1 & 1 & 1 & 1 & -1 & -1 & -1 & -1 & 0 & 0 & 0 & 0 \\
0 & 0 & 0 & -2 & -2 & 2 & 2 & 1 & 1 & 1 & 1 & -1 & -1 & -1 & -1 & 0 & 0 & 0 & 0 \\
0 & 0 & 0 & 0 & 0 & 0 & 0 & 1 & -1 & 1 & -1 & 0 & 0 & 0 & 0 & 0 & 0 & 0 & 0 \\
0 & 0 & 0 & 0 & 0 & 0 & 0 & 0 & 0 & 0 & 0 & 0 & 0 & 0 & 0 & 1 & -1 & -1 & 1 \\
0 & 0 & 0 & 0 & 0 & 0 & 0 & 0 & 0 & 0 & 0 & 1 & -1 & -1 & 1 & 0 & 0 & 0 & 0 \\
0 & 0 & 0 & 0 & 0 & 0 & 0 & 1 & -1 & 1 & -1 & -1 & 1 & -1 & 1 & 0 & 0 & 0 & 0 \\
0 & 0 & 0 & 0 & 0 & 0 & 0 & -1 & -1 & 1 & 1 & 0 & 0 & 0 & 0 & 1 & -1 & 1 & -1 \\
0 & 0 & 0 & 0 & 0 & 0 & 0 & 0 & 0 & 0 & 0 & 1 & 1 & -1 & -1 & -1 & -1 & 1 & 1
\end{pmatrix} \tag{4.38}$$

$\mathbf{S}$ is a diagonal relaxation matrix

$$\mathbf{S} = diag\left(s_0, \ldots, s_{N-1}\right) \tag{4.39}$$

in which s_i are the relaxation factors, given by

$$s_0 = s_3 = s_5 = s_7 = 0 \tag{4.40}$$

$$s_1 = s_2 = s_{9:15} = \tau^{-1} \tag{4.41}$$

$$s_4 = s_6 = s_8 = s_{16:18} = 8\left(\frac{2\tau - 1}{8\tau - 1}\right) \tag{4.42}$$

in which τ is the relaxation time. The relationship between the relaxation time and fluid kinematic viscosity ν is given by [11]

$$\tau = \frac{\nu}{c_s^2 \delta t} + \frac{1}{2} \tag{4.43}$$

in which $c_s = 1/\sqrt{3}$ denotes the sound speed in the LBM. The equilibrium distribution equation is given by

$$f_i^{eq} = \rho\omega_i\left[1 + \frac{\mathbf{e}_i \cdot \mathbf{u}}{c_s^2} + \frac{(\mathbf{e}_i \cdot \mathbf{u})^2}{2c_s^4} - \frac{\mathbf{u} \cdot \mathbf{u}}{2c_s^2}\right] \tag{4.44}$$

in which ω_i denote the weights corresponding to the discrete velocities $\mathbf{e}_i$ in D3Q19. These velocities and weight coefficients are specified as follows [10]

$$\omega_i = \begin{cases} 1/3, & i = 0 \\ 1/18, & 1 \le i \le 6 \\ 1/36, & 7 \le i \le 18 \end{cases} \tag{4.45}$$

$$\mathbf{e}_i = \begin{cases} (0,0,0), & i = 0 \\ (\pm 1,0,0), (0,\pm 1,0), (0,0,\pm 1), & 1 \le i \le 6 \\ (\pm 1,\pm 1,0), (\pm 1,0,\pm 1), (0,\pm 1,\pm 1), & 7 \le i \le 18 \end{cases} \tag{4.46}$$

As explained in Chap. 3, the macroscopic variables can be recovered using [12] $\rho = \sum_i f_i(\mathbf{x}, t)$ and $\rho\mathbf{u} = \sum_i \mathbf{e}_i f_i(\mathbf{x}, t)$.

This MRT-LB model of electrolyte flow employs the no-slip bounce-back scheme to treat the irregular momentum boundary at the fluid-solid interface between the electrolyte and electrode. Moreover, the nonequilibrium extrapolation method is adopted for treating the macroscopic momentum boundary of the computational domain.

To simulate ion species transport in the pores, a D3Q7 MRT-LBM coupled to the electrochemical reaction is employed. The evolution equation of species transfer for $N = 7$ discrete velocities is written as [13]

$$|\mathbf{g}^\alpha(\mathbf{x} + \mathbf{e}_i\delta t, t + \delta t)\rangle - |\mathbf{g}^\alpha(\mathbf{x}, t)\rangle = -\Gamma^{-1}\Lambda\Gamma\left(|\mathbf{g}^\alpha(\mathbf{x}, t)\rangle - |\mathbf{g}^{\alpha,eq}(\mathbf{x}, t)\rangle\right) \tag{4.47}$$

with

$$|\mathbf{g}^\alpha(\mathbf{x}, t)\rangle = \left(g_0^\alpha(\mathbf{x}, t), \ldots, g_{N-1}^\alpha(\mathbf{x}, t)\right)^{\mathrm{T}} \tag{4.48}$$

$$|\mathbf{g}^{\alpha,eq}(\mathbf{x}, t)\rangle = \left(g_0^{\alpha,eq}(\mathbf{x}, t), \ldots, g_{N-1}^{\alpha,eq}(\mathbf{x}, t)\right)^{\mathrm{T}} \tag{4.49}$$

in which $g_i^\alpha(\mathbf{x}, t)$denotes the mass distribution function for species α. The corresponding equilibrium distribution function can be expressed as [14]

$$g_0^{\alpha,eq}(\mathbf{x}, t) = c_\alpha\left[J_i + \frac{\mathbf{e}_i \cdot \mathbf{u}}{2}\right] \tag{4.50}$$

with

$$J_i = \begin{cases} J_0, & i = 0 \\ (1 - J_0)/6, & 1 \le i \le 6 \end{cases} \tag{4.51}$$

in which J_0 is the rest fraction, taking a value between 0 and 1 depending on the diffusivity. Γ denotes the orthogonal transformation matrix and Λ denotes the diagonal relaxation matrix of the D3Q7 model, which are given by

$$\Gamma = \begin{pmatrix} 1 & 1 & 1 & 1 & 1 & 1 & 1 \\ 0 & 1 & -1 & 0 & 0 & 0 & 0 \\ 0 & 0 & 0 & 1 & -1 & 0 & 0 \\ 0 & 0 & 0 & 0 & 0 & 1 & -1 \\ 6 & -1 & -1 & -1 & -1 & -1 & -1 \\ 0 & 2 & 2 & -1 & -1 & -1 & -1 \\ 0 & 0 & 0 & 1 & 1 & -1 & -1 \end{pmatrix} \tag{4.52}$$

$$\Lambda = diag\,(\lambda_0, \ldots, \lambda_{N-1}) \tag{4.53}$$

with relaxation rates $\lambda_i = 1/\tau_{g,\alpha}$. The diffusivity of species α is obtained from

$$D_\alpha = (1 - J_0)\left(\tau_{q,\alpha} - \frac{1}{2}\right)\frac{(\delta x)^2}{3\delta t} \tag{4.54}$$

while the mass concentration of species α can be calculated from

$$C_\alpha = \sum_i g_i^\alpha \tag{4.55}$$

To describe the electrochemical reactions involving the V(II) and V(III) species, a Neumann boundary condition is employed at the fluid-solid interface of the carbon fibres *via* the Butler-Volmer equation

$$\begin{aligned} N &= -D_{V(II)}\frac{\partial c_{V(II)}}{\partial n} = D_{V(III)}\frac{\partial c_{V(III)}}{\partial n} \\ &= a\varepsilon s F k_{neg}\left(\widehat{c}_{V(II)}\right)^{\alpha_{2,c}}\left(\widehat{c}_{V(III)}\right)^{1-\alpha_{2,c}}\left(e^{F(1-\alpha_{2,c})\eta/(RT)} - e^{-\alpha_{2,c}F\eta/(RT)}\right) \end{aligned} \tag{4.56}$$

N denotes the mass flux caused by the electrochemical reaction and $\partial(\cdot)/\partial n \equiv \nabla(\cdot)\cdot\mathbf{n}$ denotes a derivative normal to the boundary. If the deviation between $\widehat{c}_\alpha$ and c_α is ignored, the activation overpotential can be calculated using Eq. (4.56).

4.4 Pore Structure of Electrode

The actual topological structure of the porous electrode (including the GF and CP components) can be characterised using X-ray micro-computed tomography (Micro-CT). The specimens of commercial graphite felt (GFA6 EA, SGL carbon) and carbon paper (TGP-H-090, Toray) were illuminated by a high-resolution

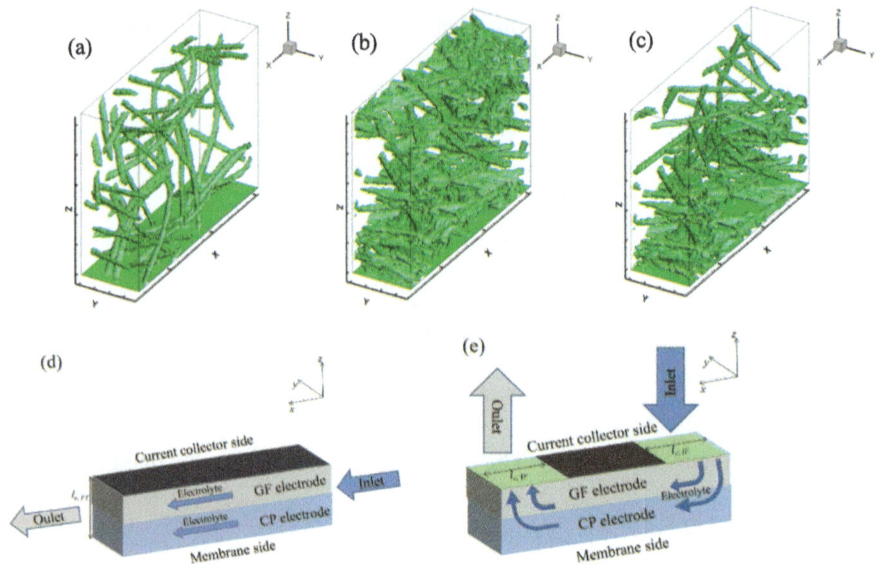

Fig. 4.1 Electrode pore structures and feeding modes: **a** GF electrode, **b** CP electrode, **c** DL electrode, **d** FT mode, **e** IF mode

micro-tomograph (SkyScan 2211X-ray, Bruker). Based on the scanning results, a series of acquired 2D segmented slices were assembled into a 3D pore structure of the electrode.

A portion of the 3D reconstructed structure, which represents the typical pore topological features of the electrode, was chosen as the computational domain for the numerical study. The pore structures of the GF and CP electrodes are displayed in Figs. 4.4a and b, respectively. In this chapter we investigate the pore-scale reactive transfer mechanisms of the vanadium-iron RFB. The composite double-layer (DL) electrode is composed of a GF electrode near the current collector side and a CP electrode near the membrane side, as shown in Fig. 4.1c.

To exclude the effect of electrode size on the operational performance, all electrodes have the same dimensions. The length (L), width (W) and height (H) are 280 μm, 80 μm and 220 μm, respectively. For the composite gradient electrode, the thicknesses of the CP and GF portions are 120 μm and 100 μm, respectively.

Moreover, the numerical study also considers the role of electrolyte feeding modes on the pore-scale performance. Figure 4.1d illustrates an electrolyte flow-through porous electrode, without flow field channels, which is termed the full flow-through (FT) mode [15]. On the other hand, Fig. 4.1e illustrates an electrolyte fed into the porous electrode with interdigitated flow field channels, which is termed the interdigitated flow (IF) mode. It should be noted that lc in Figs. 4.1d and e refers to the characteristic lengths of the inlet and outlet in the two feeding modes, which are not equal. In addition, the compressibility of these electrodes is not considered in the

Table 4.1 Structural characteristics of porous electrode

Quantity	Value
Length L	280 μm
Width W	80 μm
Height H	220 μm
Porosity of GF ϵ_g	~0.95
Porosity of CP ϵ_c	~0.78
Fibre diameter d_f	~10 μm
Grid resolution	2 μm voxel

simulations (see Chap. 5 for a study of these effects). The structural characteristics of the porous electrodes are listed in Table 4.1.

4.5 Boundary Conditions

We assume that the battery discharges under galvanostatic conditions; hence a constant-current density is applied to the current collector on the positive electrode side. In addition, a fixed species concentration, according to the state of charge (SOC), is set at the inlet boundary. The full set of boundary conditions is given in Table 4.2. The numerical simulation was carried out at steady state in COMSOL Multiphysics® and the LB model was implemented in Fortran 95.

Table 4.2 Boundary conditions applied in the model

Types	Boundary	Expression
Charge conservation	Applied current density +ve side	$-\sigma_s^{eff} \nabla\phi_s \cdot \mathbf{n} = -I$
	Ground at −ve negative side	$\phi_s = 0$
	Insulation on other sides	$\mathbf{j}_l \cdot \mathbf{n} = \mathbf{j}_s \cdot \mathbf{n} = 0$
Momentum conservation	Flow velocity at inlet in terms of flow rate Q and geometric area A	$u_x = Q/(\varepsilon A)$
	Pressure condition at outlet	$p_{out} = 0$
	Non-slip at other boundaries	$\nabla p \cdot \mathbf{n} = 0$
	Concentration from SOC at inlet and total concentration c_0	$c_{V(II),in} = c_{0,neg} SOC$ $c_{V(III),in} = c_{0,neg}(1 - SOC)$ $c_{Fe(II),in} = c_{0,pos}(1 - SOC)$ $c_{Fe(III),in} = c_{0,pos} SOC$ $c_{Cl,neg} = 2c_{0,neg} SOC$ $+3c_{0,neg}(1 - SOC)$ $c_{Cl,neg} = 2c_{0,pos}(1 - SOC)SOC$ $+3c_{0,pos} SOC$
	All diffusive fluxes 0 at outlet	$D_i^{eff} \nabla c_i \cdot \mathbf{n} = 0$
Mass conservation	The fluxes are zero at other boundaries	$\left(-D_i^{eff} \nabla - \frac{F z_i}{RT} c_i \nabla\phi_l - \mathbf{u}_l c_i\right) \cdot \mathbf{n} = 0$

4.6 Validation and Numerical Details

All transport and physico-chemical parameters are listed in Table 4.3. In order to validate the model, a validation case was considered based on a carbon felt electrode without a flow channel, as illustrated in Fig. 4.2.

Figure 4.3 compares the simulated polarisation curve with the experimental data reported in reference [16]. It is found that the average relative error between the simulated and experimental values is 5.7% under the same flow rate, temperature and SOC conditions, which can be taken as validation of the proposed model. We now use the model to analyse the convective mass transfer and electrochemical performance of the DES flow battery.

In order to explore the influence of different flow field configurations on the mass transfer process, parallel, serpentine and interdigital flow channels were considered based on the above carbon felt electrodes, as shown in Fig. 4.4. In addition, for investigating the effect of electrodes with different porosity distributions, three electrodes with: (i) a linearly decreasing porosity, (ii) a linearly increasing porosity and (iii) a uniform porosity are considered. Using the positive electrode with a parallel flow channel as an example, the distribution of the average porosity is shown in Fig. 4.4d.

The porosity distributions along the x axis for the decreasing porosity (PD), increasing porosity (PI) and uniform porosity (AP) cases are shown in Fig. 4.5. In the PD case the porosity decreases linearly from the flow channel ($x = 0$) to the membrane ($x = 0.004$). The porosity of the PI electrode increases linearly from the flow channel to the membrane. It should be pointed out that the three electrodes have the same average porosity ($\varepsilon = 0.9$). Moreover, the permeability of the electrode will change accordingly. The same porous electrode is employed for the negative half cell due to the structural symmetry of the flow battery.

4.7 Analysis of the Performance of a Vanadium-Iron Flow Battery

In this section, the numerical model is used to investigate the coupling effects of flow channel and electrode structure on the performance of the DES-based RFB. Unless otherwise specified, all simulation results are based on a temperature of 298K, a discharge current density of 3.0 mA cm^{-2} and a flow rate of 25 mL min^{-1}.

4.7.1 *Influence on Flow Field*

During operation of the flow battery, the ion transport characteristics of the active material will be affected by the velocity distribution of the electrolyte, which will ultimately influence the charge and discharge performance. In order to investigate

Table 4.3 Parameter values used for the simulations

Symbol	Parameter	Value	Units
$c^0_{V2}/c^0_{V3}/c^0_{Fe2}/c^0_{Fe3}$	Initial concentrations	100	mol m^{-3}
c^0_{Cl}	Initial concentration of Cl^-	300	mol m^{-3}
D_{V2}/D_{V3}	Diffusion coefficient of V^{2+}/V^{3+}	1.16×10^{-13}	m^2 s^{-1}
D_{Fe2}/D_{Fe3}	Diffusion coefficient of Fe^{2+}/Fe^{3+}	1.83×10^{-13}	m^2 s^{-1}
D_{Cl}	Diffusion coefficient of Cl^-	1.55×10^{-12}	m^2 s^{-1}
E^0_{neg}	Equilibrium potential: negative	–0.26	V
E^0_{pos}	Equilibrium potential: negative	0.769	V
σ_s	Electrode conductivity	220	S m^{-1}
σ_m	Membrane conductivity	0.3	S m^{-1}
σ_{pos}	Positive electrolyte conductivity	0.22	S m^{-1}
σ_{neg}	Negative electrolyte conductivity	0.1332	S m^{-1}
T	Temperature	298	K
i_{app}	Applied current density	30	A m^{-2}
d_f	Carbon fibre diameter	1.76×10^{-5}	m
k_{neg}	Reaction rate constant of negative	1.02×10^{-7}	m s^{-1}
k_{pos}	Reaction rate constant of positive	6.35×10^{-8}	m s^{-1}
K_{CK}	Carman-Kozeny constant	5.55	–
μ_{neg}	Viscosity of negative electrolyte	0.5224	Pa s
μ_{pos}	Viscosity of positive electrolyte	0.312	Pa s
ρ	Density of electrolyte	1246.74	kg m^{-3}
α_2	transfer coefficient of negative	0.445	–
α_1	transfer coefficient of positive	0.419	–
ϵ	Electrode porosity	0.9	–
a	Specific surface area	3.5×10^4	m^{-1}
Q	Volumetric flow rate	25	mL min^{-1}
η_P	Pump efficiency	0.99	–

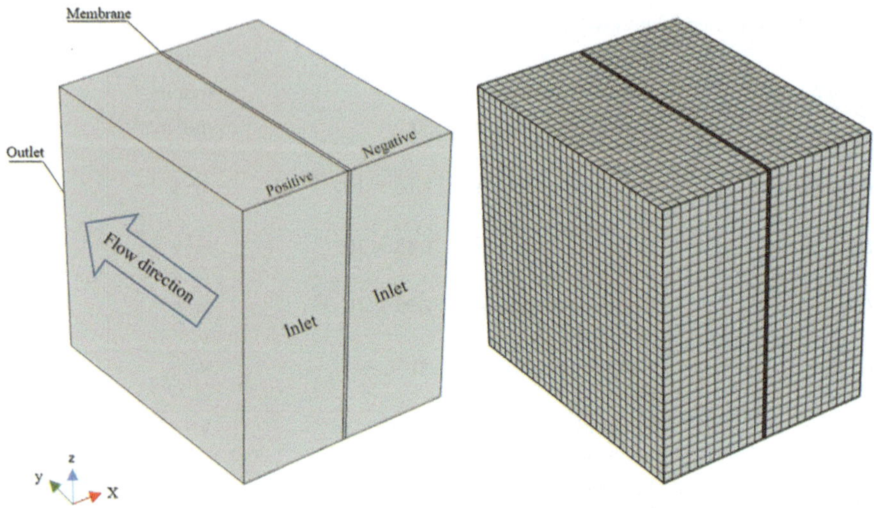

Fig. 4.2 Carbon felt electrode without flow channel and the computational mesh

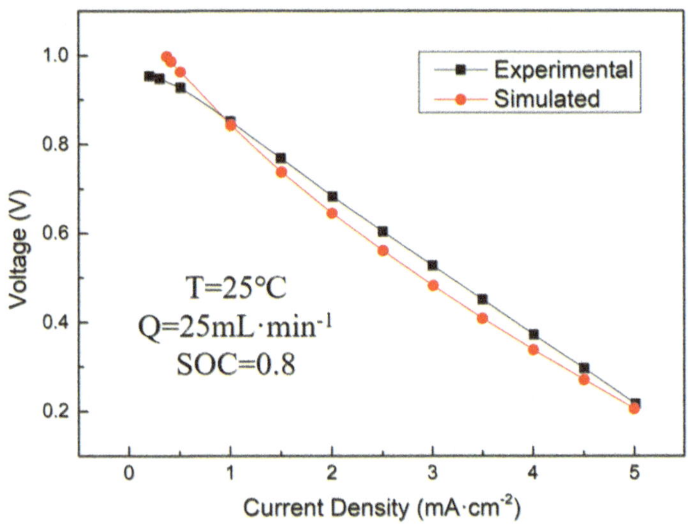

Fig. 4.3 Comparison of simulated and experimental values

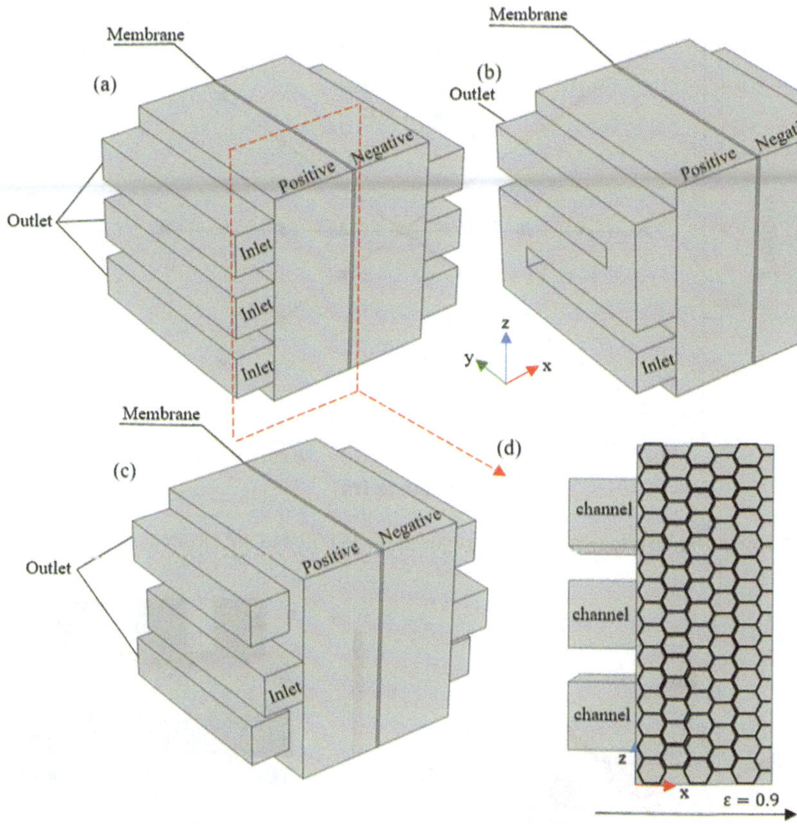

Fig. 4.4 Structures of flow battery with different flow channels, **a** parallel flow channel, **b** serpentine flow channel, **c** interdigital flow channel and schematic of porosity in the AP electrodes (**d**)

the effect of different flow channels on mass transfer, this section compares the velocity distributions of the positive electrolyte in porous electrodes with different flow channels with the same inlet flow rate and porosity distribution.

Figures 4.6a, 4.7a and 4.8a show the three-dimensional velocity distributions of the electrolyte in the AP porous electrode with a parallel flow channel, serpentine flow channel and interdigital flow channel, respectively. The right-hand sides show the flow velocity contours of the electrolyte at two cross sections from the flow field side to membrane side (cross-plane) and from the inlet to the outlet (in-plane), in which black arrows indicate the direction of the velocity vector.

The three-dimensional velocity contours indicate the non-uniformity of the velocity in the internal electrolyte flow. All flow channels exhibit a tendency in the velocity to decrease from the flow field side to membrane side, as observed from the cross-plane contour plots (upper right sides of Figs. 4.6, 4.7 and 4.8). This result can be explained

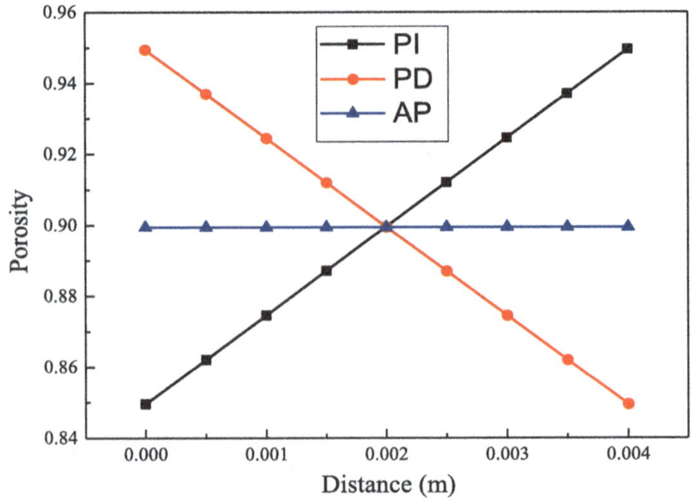

Fig. 4.5 The porosity distribution along the x axis in the PD, PI and AP electrodes

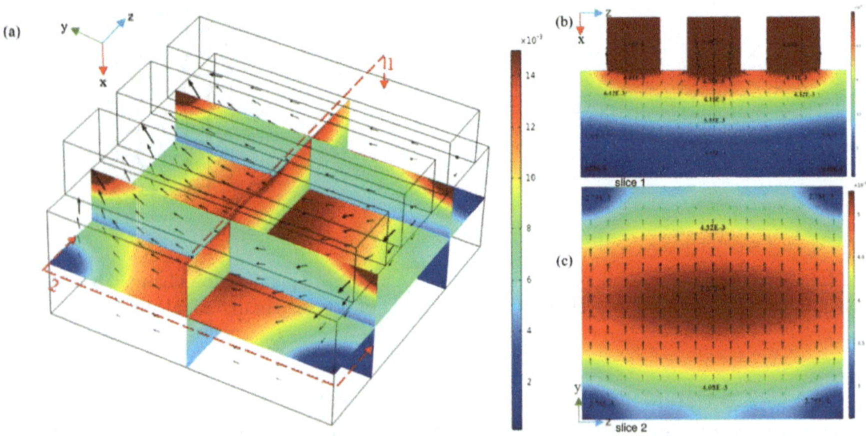

Fig. 4.6 Flow velocity distribution of the electrolyte in the porous electrode with a parallel channel structure in units of $m\ s^{-1}$, **a** three-dimensional contours, **b** flow velocity distribution of cross section 1 and **c** cross section 2

by the fact that the electrolyte flow through the electrode tends to follow the path of least resistance. Since the percolating resistance increases when the electrolyte flows from the flow field side to the membrane side, a high flow velocity is observed near the channel, while the flow performance near the membrane side is poor.

On the other hand, these figures clearly show a difference in the flow velocity under different channel structures with a fixed inlet flow rate. The flow direction of electrolyte in different channel structures is also different (as seen in right sides of these figures), which will lead to differences in the concentration distributions of the

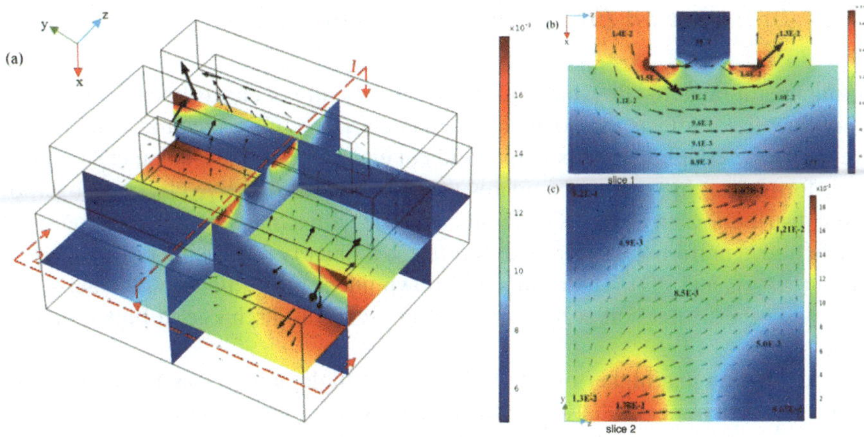

Fig. 4.7 Flow velocity distribution of the electrolyte in the porous electrode with a serpentine channel structure, in units of $m\ s^{-1}$, **a** three-dimensional contours, **b** flow velocity distribution of cross section 1 and **c** cross section 2

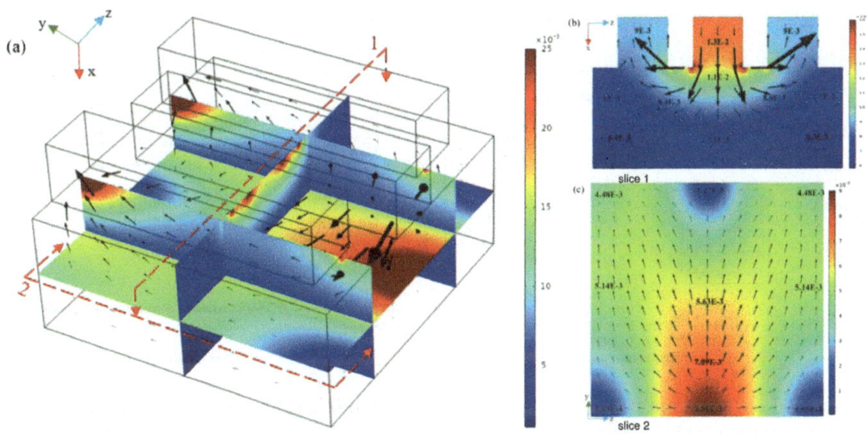

Fig. 4.8 Flow velocity distribution of the electrolyte in the porous electrode with an interdigital channel structure in unit of $m\ s^{-1}$, **a** three-dimensional contours, **b** flow velocity distribution of cross section 1 and **c** cross section 2

species, and eventually influence the electrochemical performance. Compared with parallel and serpentine channel structures, the interdigitated structure has a higher electrolyte flow rate at the membrane side because it has a higher flow velocity towards the membrane, which can enhance the mass transfer process and eventually reduce the concentration overpotential in this region.

4.7.2 Influence on Electrochemical Performance

During operation of the flow battery, the non-uniformity of electrolyte flow with different flow fields affects the polarisation characteristics of the electrode. In order to investigate the coupling effects of the gradient electrode and flow channel structures on performance, this section compares the positive and negative electrode overpotential distributions in a porous gradient electrode with a linearly increasing porosity distribution, a uniform porosity distribution and a linearly decreasing porosity distribution (PI, AP, and PD). The results are shown under three different channel structures along the x direction at the same constant-current discharge in Figs. 4.9, 4.10 and 4.11.

It can be seen that higher overpotentials (magnitude) appear mainly at the membrane side, which means that the redox reactions occur more rapidly in this region. The lower concentration of V(II) in this region leads to higher concentration losses and therefore higher overpotentials (according to the Butler-Volmer equation). Furthermore, the numerical results reveal differences in the overpotential distribution for different gradient electrode structures. As seen in the overpotential distribution

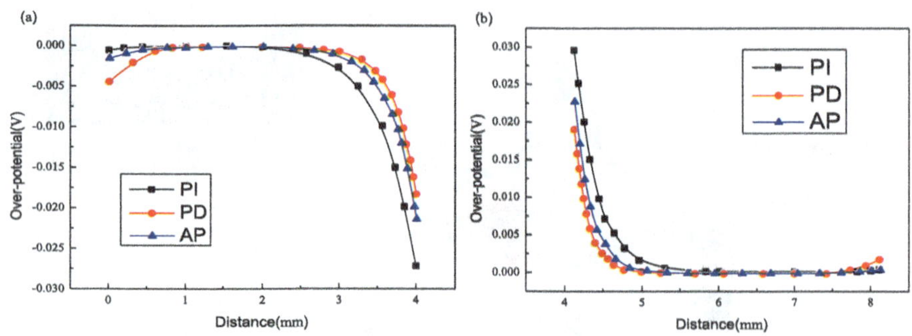

Fig. 4.9 **a** Positive electrode overpotential distribution and **b** negative electrode overpotential distribution (x direction) in different gradient electrodes with a parallel flow channel

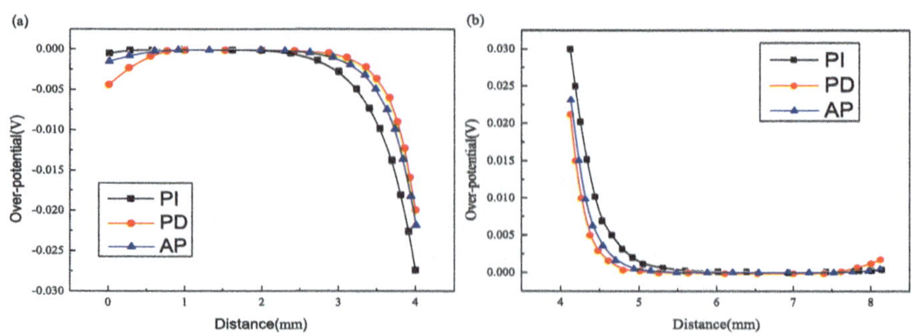

Fig. 4.10 **a** Positive overpotential distribution and **b** negative overpotential distribution (x direction) in different gradient electrodes with a serpentine flow channel

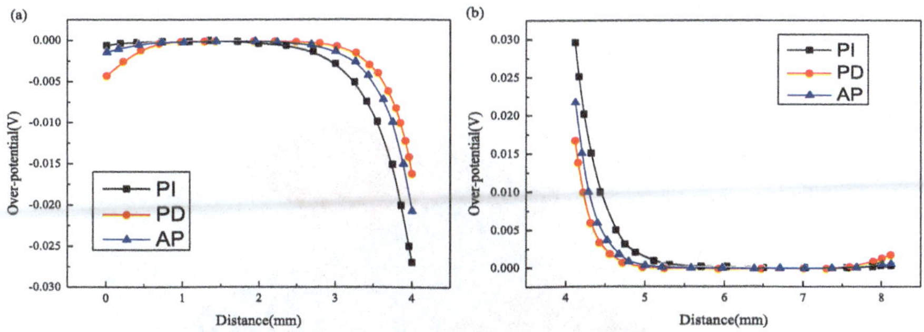

Fig. 4.11 **a** Positive overpotential distribution and **b** negative overpotential distribution (x direction) in different gradient electrodes with an interdigital flow channel

for the PI case, a higher overpotential magnitude is established, despite the higher porosity, and therefore higher volume for the active species. This suggests that the flow effects for this electrode are dominant, restricting the ingress of species by virtue of the low porosity at the flow field side. The lower transport resistance from the flow field to membrane side inside the PI electrode reduces the overpotential near the flow field side.

Figure 4.12 shows the output voltage of flow batteries with different flow channels and porosity distributions. As can be observed from this figure, the flow battery with an interdigital flow channel and linearly decreasing porosity has the highest voltage, which is consistent with the findings above. In Fig. 4.12, compared with the flow channel effects, it can be observed that the porosity distribution has a greater influence on battery performance, which may be explained by the fact that the porosity distribution not only affects mass transfer, but also affects the specific surface area and ion transport.

4.7.3 *Effect of Electrode Structures and Feeding Modes*

Using the LB model, numerical computations were carried out to elucidate the reactive mass transfer processes during galvanostatic discharging at the negative side of the RFB with different electrode structures and electrolyte feeding modes. When the applied external current density is fixed at 100 Am^{-2} and the Reynolds number of the inlet DES electrolyte flow is fixed at $Re = 0.5 \times 10^{-4}$, the iso-surfaces of the dimensionless flow rate (i.e, U/U_0, $U = \sqrt{u_x^2 + u_y^2 + u_z^2}$) in different electrodes and with different flow patterns are depicted in Fig. 4.13. These results indicate the zones of higher electrolyte flow rate in the porous structures. In the GF electrode with a FT feeding mode, owing to the uniform distribution and isotropy of pores, the electrolyte flows fully into the pores between carbon fibres, thus leading to a uniform flow rate

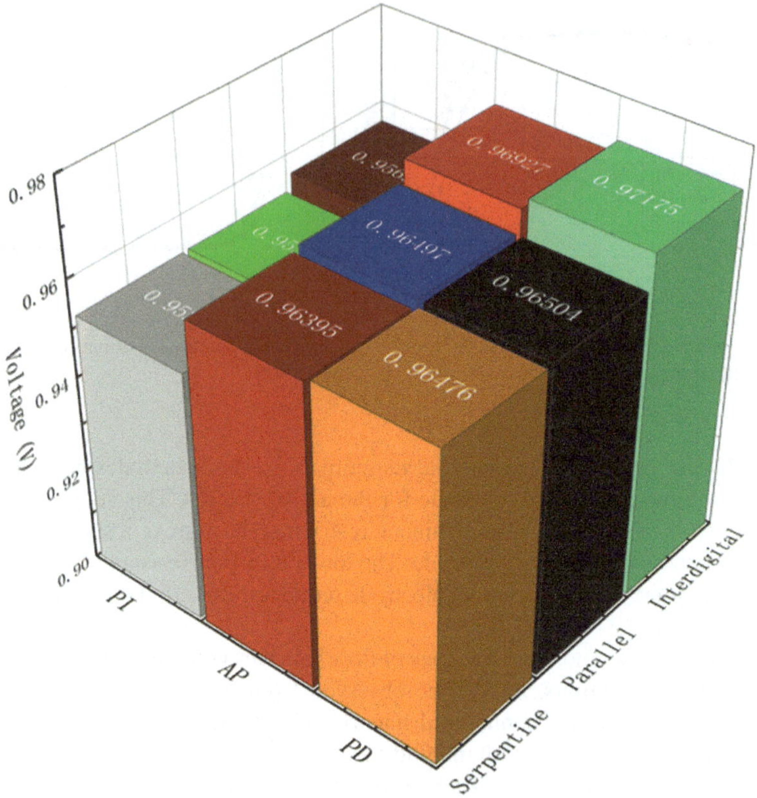

Fig. 4.12 Comparison of the output voltage of RFBs with different flow channels and porosity distributions

distribution. The low flow rate zones are mainly close to the current collector and the membrane, as shown in Fig. 4.11a.i.

Figure 4.13b.i depicts the flow rate distribution in the CP electrode with a FT feeding mode, which is different from that in the CP electrode. Due to the lower through-plane permeability for the anisotropic CP material, a low resistance flow path forms between the woven carbon-fibre layers along the flow direction (i.e., in-plane direction), leading to preferential electrolyte flow in this direction. For the double-layered gradient electrode (DL) with a FT feeding mode, the electrolyte flow exhibits an obvious difference between the CP and GF sections, as shown in Fig. 4.13c.i. The higher flow rate zone primarily exists in the pores of the GF section, which is attributable to a higher porosity, leading to a lower flow resistance.

Figure 4.14a plots the distribution of the electrolyte flow rate from the membrane to the current collector along z axis. As shown by the solid lines in this figure, the CP electrode with a FT feeding mode has the highest flow rate in the pores due to the lower porosity. A relatively homogeneous flow rate is observed for the GF electrode,

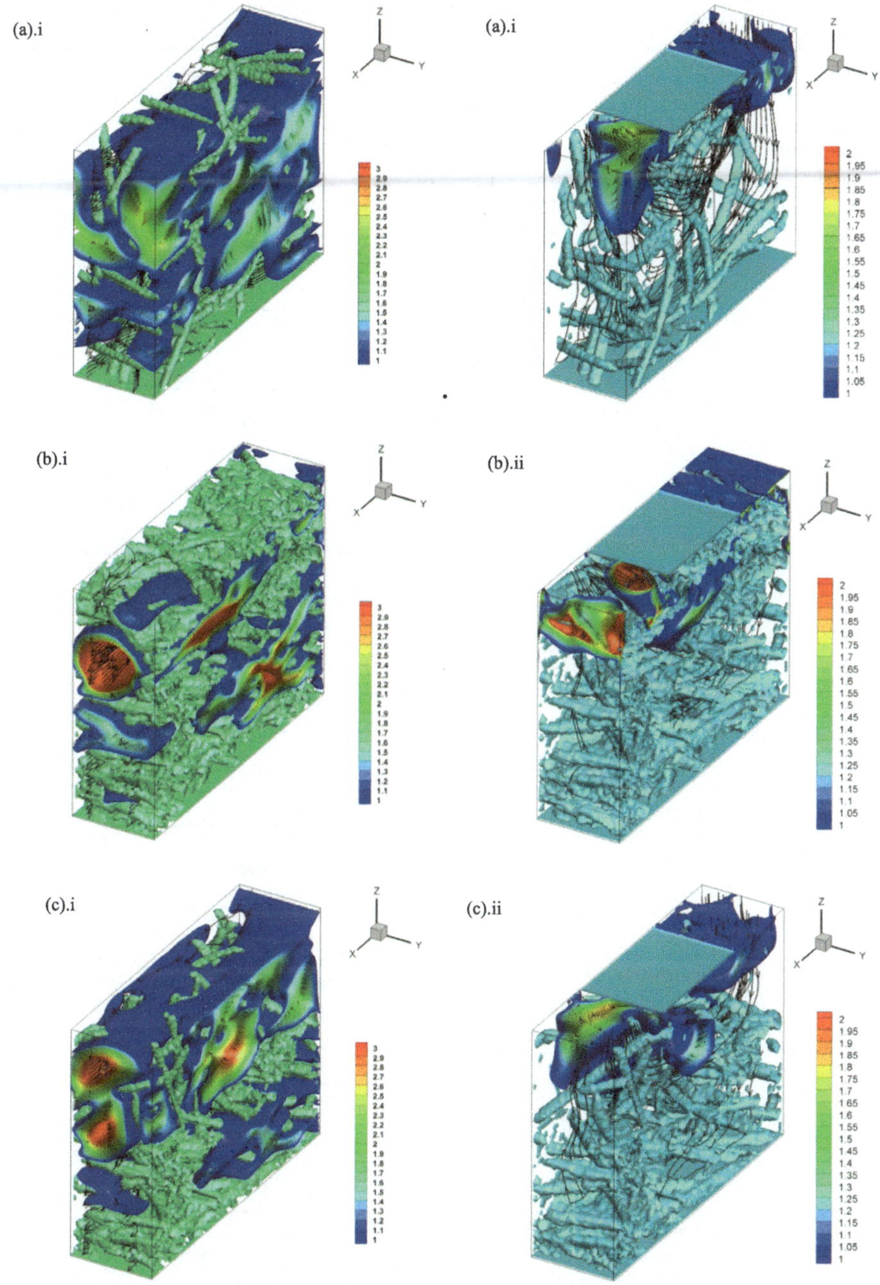

Fig. 4.13 Iso-surfaces of dimensionless flow rate in different electrodes: **a** GF electrode, **b** CP electrode, **c** DL electrode, (i) FT mode, (ii) IF mode

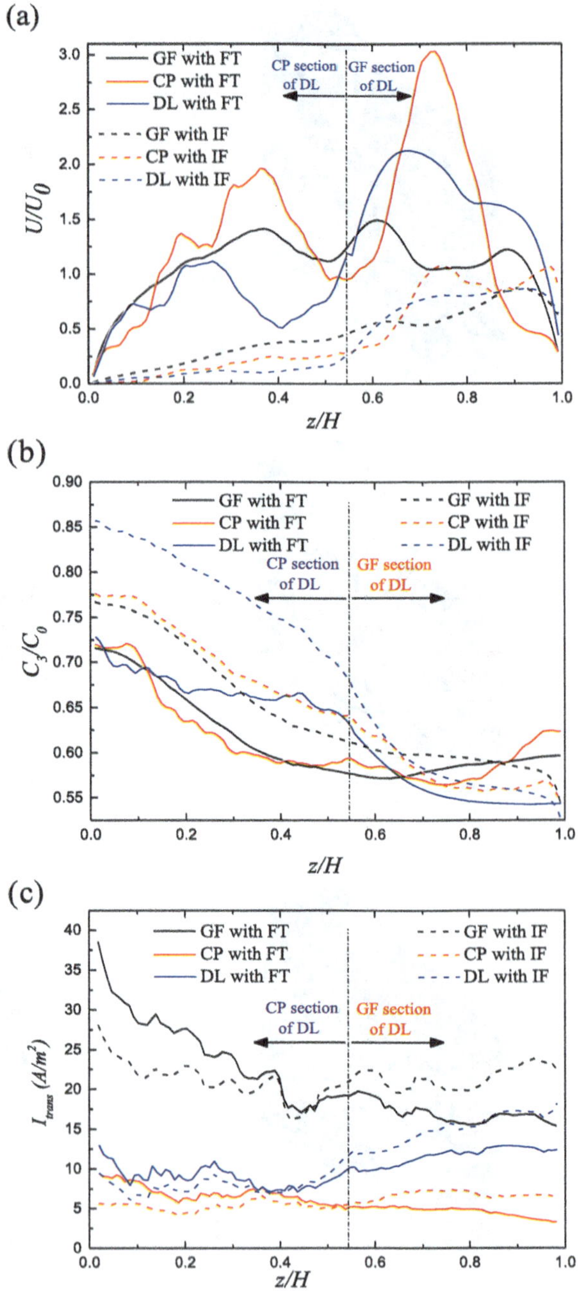

Fig. 4.14 **a** electrolyte flow rate, **b** V(III) concentration and **c** transfer current density distributions from the membrane side to the current collector side with different electrodes

which is consistent with the previous results. With regards to the DL electrode in FT mode, the low porosity of the CP section provides a larger specific surface area for electrochemical reaction, but also results in a high flow resistance near the membrane. As a result, compared with the GF section, the CP section of the DL electrode has the lowest flow rate into the pores, which may lower the ingress of redox species near the membrane side.

In contrast to the full FT mode, the higher electrolyte flow rate occurs primarily within the pores near the inlet and outlet of the different electrode structures with the interdigitated flow (IF) mode, as shown in Figs. 4.13a-c.ii. It should be noted that the inlet flow rate U_0 of the IF mode is higher than that of the FT mode, since the IF mode has a lower inlet characteristic length when the same inlet Reynolds number (i.e., mass flux) is selected for the two feeding modes. Since a higher electrolyte flux is located near the current collector for the different pore structures in the IF mode, the electrolyte flow rates decrease gradually from the current collector to membrane, as shown in Fig. 4.14a.

It is noticeable that the sluggish electrolyte flow slows the convective mass transfer of redox species near the membrane, which may cause a higher concentration overpotential during operation. In comparison to the homogeneous CP and GF electrodes, the numerical results suggest that the DL electrode attains the lowest electrolyte flow rate in the pores of the CP section, with the main reaction region near the membrane. The DL pore structure leads to a heterogeneous flow distribution from the current collector to the membrane, with a greater momentum transfer in the pores of the GF section. Consequently, the DL electrode in IF mode exhibits an inhomogeneous flow distribution, and has the most sluggish convective mass transfer intensity for a fixed inlet flow rate.

The momentum transfer behaviour of the electrolyte has a great influence on the concentration of redox species in the pores of the electrode. Figure 4.15 shows the concentration contours of V(III) in the electrode with different porous structures and feeding modes. For porous electrodes in FT mode, the V(III) concentration (the product) increases gradually along the in-plane direction (i.e., x axis) from inlet to outlet. On the other hand, the DL electrode in FT mode exhibits a distinct layered concentration distribution between the GF and CP sections, which is caused by the difference in convective mass transfer in the different pore structures. Due to the sluggish electrolyte flow in the pores of the CP section, the reaction-generated V(III) species collects in the pores of the electrode.

Furthermore, the effect of feeding modes on the V(III) concentration distribution is depicted in Figs. 4.15 and 4.14b. Due to the decreasing electrolyte flux from the current collector the membrane, a higher V(III) concentration appears near the membrane in IF mode. As shown in Fig. 4.14b, the V(III) concentration in this case (dash line) is higher than that with the FT mode (solid line) when $z/H < 0.5$. Therefore, owing to the more sluggish convective mass transfer, the DL electrode with IF mode has the highest reactant product content in the pores near the membrane. In particular, at this low electrolyte flow rate condition ($Re = 0.5 \times 10^{-4}$), the dimensionless V(III) concentration is close to 1 in the local pores of the DL electrode near the mem-

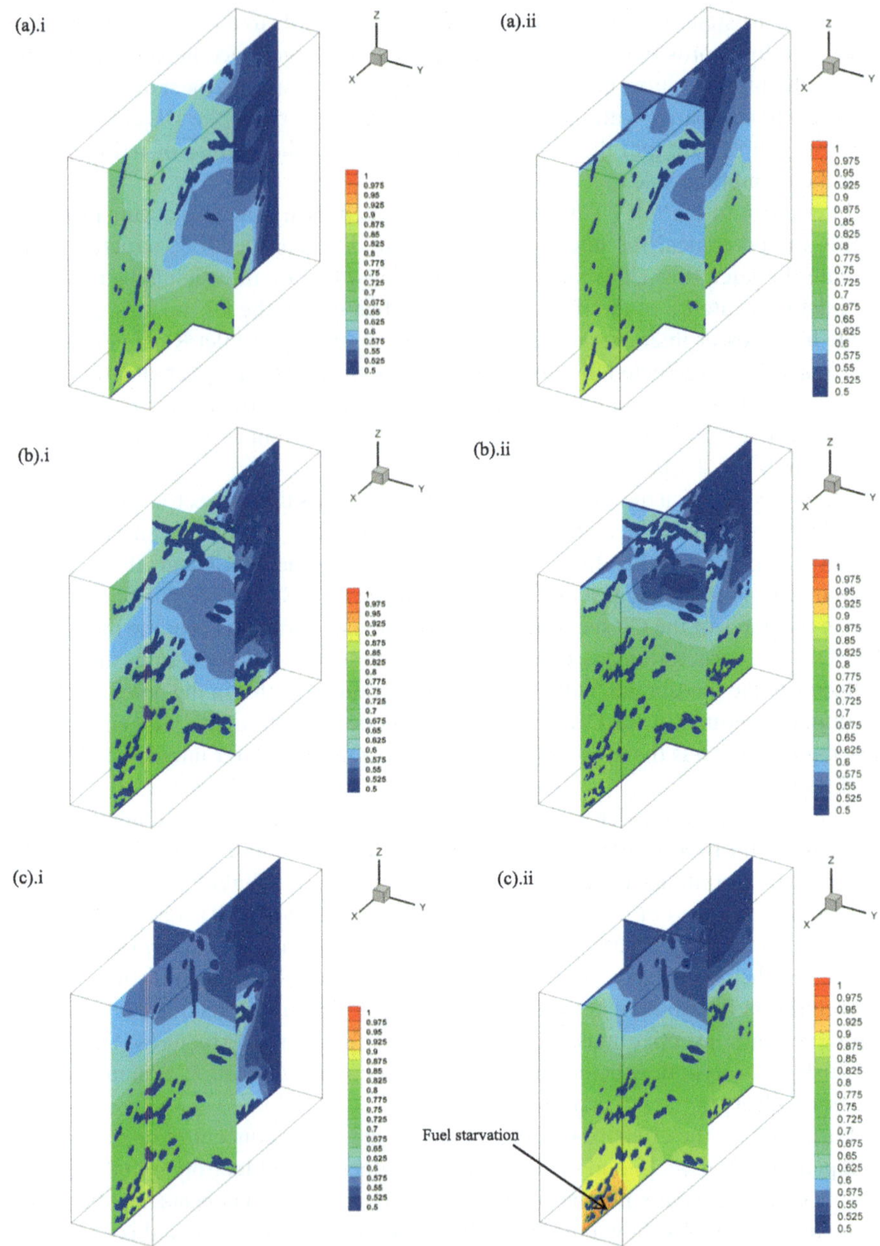

Fig. 4.15 Iso-surfaces of the dimensionless V(III) concentration in different electrodes: **a** GF electrode, **b** CP electrode, **c** DL electrode, (i) FT mode, (ii) IF mode

brane and outlet, as shown in Fig. 4.15c.ii. This risks fuel starvation in these regions, thus limiting the discharging performance for low electrolyte flow rate conditions.

Figure 4.14c shows the transfer current density distribution with different electrodes and feeding modes along the through-plane direction. This figure reveals that the GF electrodes have the highest average transfer current density, while CP electrodes possess the lowest. This suggests that the transfer current density is heavily influenced by the porosity of the electrode. The low porosity CP electrode (red line) provides an abundant surface area for electrochemical reaction, but simultaneously a lower volumetric concentration of reactant V(II), thus leading to a lower transfer current density per unit area during the galvanostatic discharging process. For DL electrodes (blue line), since the CP is placed near the membrane, the average transfer current density decreases along the z direction.

The feeding modes also impact the transfer current density distributions in the different electrode structures. With regard to the electrodes using the FT mode, the transfer current densities exhibit an obvious decrease from the membrane to the current collector, except in the DL case for which the current density is relatively uniform. This result is mainly due to the higher ionic conductivity of the membrane compared to the effective ionic conductivity of the electrolyte, meaning that a higher rate of electron transfer reaction occurs in the region near the membrane. Due to the sluggish convective mass transfer in IF mode, which hinders the electrochemical reaction near the membrane, the transfer current density is relatively evenly distributed along the through-plane direction. For the DL electrode in IF mode, the transfer current density near the current collector is even higher than that near the membrane, due to fuel starvation in the vicinity of the latter.

The pump power losses are depicted in Fig. 4.16a. This figure shows that the pump power losses is determined by the mean porosity of the electrode for the same feeding mode. With regard to the mono-layer electrodes (including GF and CP), the IF mode gives rise to a higher electrolyte flow resistance when compared with the FT mode. As shown in Fig. 4.13, the IF mode leads to a less uniform electrolyte flow pattern, and the seepage resistance along the through-plane direction is higher than that along the through-plane direction of the CP electrode. Conversely, the DL electrode in IF mode has a lower pump power loss compared with the FT mode. This interesting phenomenon may be caused by the inhomogeneity of the flow in IF mode, giving rise to a low electrolyte flux in the CP section, which has a lower permeability. The DL electrode in IF mode has a lower average flow resistance compared with the FT mode.

The electrochemical polarisation losses are shown in Fig. 4.16b, which reveals that the power losses in all cases are primarily caused by the activation overpotential. For all electrodes, the IF mode leads to a higher polarisation loss compared with the FT mode. In particular, for DL electrodes, the total polarisation losses in IF mode are 33.1% higher than in FT mode, due to the more sluggish convective mass transfer near the membrane side. Due to the low inlet flow rate considered in all simulations, the effect of the pump power losses on the operation efficiency of the RFBs is small.

The total power loss (Pt) is dominated by the polarisation losses, as shown in Fig. 4.16(c). Under the current operating conditions, the electrodes in FT mode have

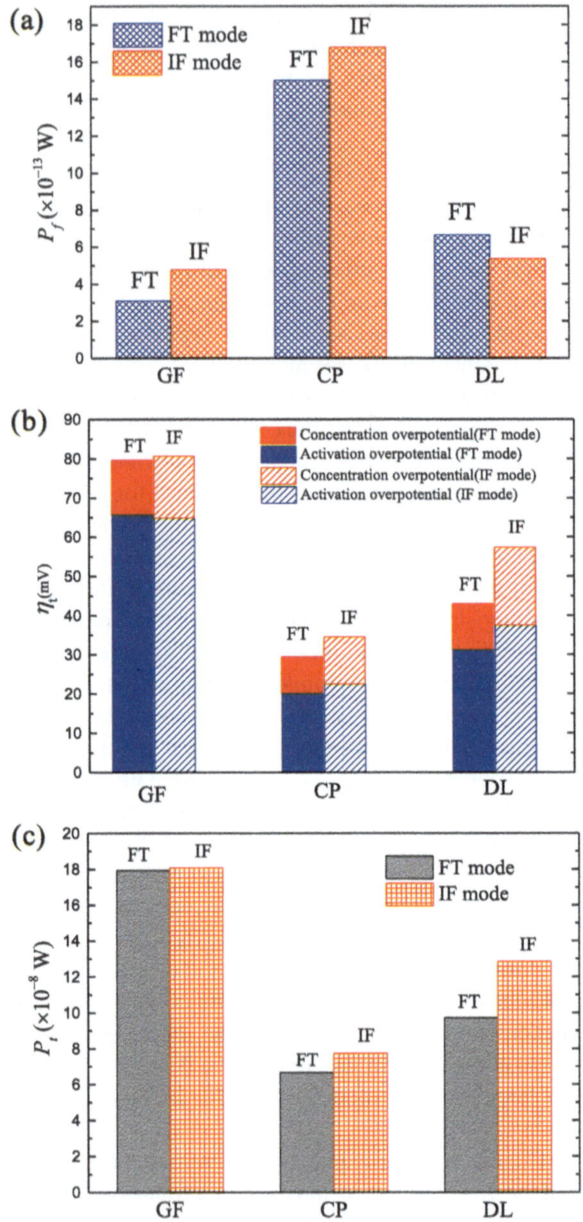

Fig. 4.16 **a** pump power losses, **b** electrochemical polarisation losses and **c** total power losses affected by different electrode structures and feeding modes

a lower total power loss compared with the IF mode. The CP electrode in FT mode achieves the lowest total power loss. Moreover, the feeding modes have the most significant impact on the performance of the RFB with a DL electrode. The IF feeding mode would give rise to a severe inhomogeneity in the electrolyte flow in gradient porous structures, thereby resulting in sluggish convective mass transfer and local fuel starvation near the membrane side, limiting the energy efficiency. Compared with the IF mode, the power loss for the DL electrode in FT mode is 24.3% lower.

4.8 Summary

In this chapter, we provided a detailed description of macroscopic approaches to modelling flow batteries, with full details of the equations that represent the main conservation laws. Additionally, we provided a description of the multiple-relaxation-time (MRT) LBM, which is used for simulating the flow on a mesoscopic or pore scale. As an example, a three-dimensional model of a nonaqueous iron-vanadium redox flow battery with a DES electrolyte was developed, coupling the macroscopic approach with the MRT-LBM to study the transport phenomena in detail and elucidate its influence on the overall performance of the battery. We showed in detail the various steps required to implement the model.

We showed how such a model can be used to study the mass transfer and electrochemical processes within the battery for different flow conditions and electrode structures, including composite electrodes and electrodes with a non-uniform porosity. The differences in the flow velocity under different channel structures were studied and discussed, together with the differences for different types of electrodes. In general, it is found that the porosity distribution has a greater influence on battery performance than the flow regime, due to the effect that the porosity distribution has on mass transport, the specific surface area, the species concentrations and ion transport.

In the next chapter we focus entirely on pore-scale models and additionally incorporate the effects of compression of the electrode, before moving onto data-driven and alternative approaches in Chaps. 6 and 7.

References

1. P. Leung, A.A. Shah, L. Sanz, C. Flox, J.R. Morante, Q. Xu, M.R. Mohamed, C. Ponce de León, F.C. Walsh, Recent developments in organic redox flow batteries: a critical review. J. Power Sources **360**, 243–283 (2017)
2. A.P. Abbott, G. Capper, D.L. Davies, R.K. Rasheed, V. Tambyrajah, Novel solvent properties of choline chloride/urea mixtures. Chem. Commun. (1), 70–71 (2003)
3. D. Zhang, Q. Cai, O.O. Taiwo, V. Yufit, N.P. Brandon, S. Gu, The effect of wetting area in carbon paper electrode on the performance of vanadium redox flow batteries: a three-dimensional lattice boltzmann study. Electrochim. Acta. **283**, 1806–1819 (2018)
4. Duo Zhang, Qiong Cai, Gu. Sai, Three-dimensional lattice-Boltzmann model for liquid water transport and oxygen diffusion in cathode of polymer electrolyte membrane fuel cell with electrochemical reaction. Electrochim. Acta. **262**, 282–296 (2018)
5. Li. Chen, YaLing He, Wen-Quan. Tao, Piotr Zelenay, Rangachary Mukundan, Qinjun Kang, Pore-scale study of multiphase reactive transport in fibrous electrodes of vanadium redox flow batteries. Electrochim. Acta. **248**, 425–439 (2017)
6. D.M. Bernadi, M.W. Verbrugge, AIChE J. **37**, 1151 (1991)
7. D.M. Bernadi, M.W. Verbrugge, A mathematical model of the solid-polymer-electrolyte fuel cell. J. Electrochem. Soc. **139**(9), A2477–A2490 (1992)
8. S. Motupally, A.J. Becker, J.W. Weidner, Diffusion of water in Nafion 115 membranes. J. Electrochem. Soc. **147**(9), A3171–A3177 (2000)
9. A.A. Shah, M.J. Watt-Smith, F.C. Walsh, A dynamic performance model for redox-flow batteries involving soluble species. Electrochim. Acta. **53**(27), 8087–8100 (2008)
10. D. d'Humières, Multiple–relaxation–time lattice boltzmann models in three dimensions. Philos. Trans. R. Soc. Lond. Ser. Math. Phys. Eng. Sci. **360**(1792), 437–451 (2002)
11. Yue-Hong. Qian, Dominique d'Humières, Pierre Lallemand, Lattice BGK models for Navier-Stokes equation. Eur. Lett. **17**(6), 479 (1992)
12. S. Chapman, T.G. Cowling, *The Mathematical Theory of Non-uniform Gases: An Account of the Kinetic Theory of Viscosity, Thermal Conduction and Diffusion in Gases*. Cambridge university press (1990)
13. Shuqi Cui, Ning Hong, Baochang Shi, Zhenhua Chai, Discrete effect on the halfway bounce-back boundary condition of multiple-relaxation-time lattice Boltzmann model for convection-diffusion equations. Phys. Rev. E. **93**(4), 043311 (2016)
14. Li. Chen, Yong-Liang. Feng, Chen-Xi. Song, Lei Chen, Ya-Ling. He, Wen-Quan. Tao, Multi-scale modeling of proton exchange membrane fuel cell by coupling finite volume method and lattice Boltzmann method. Int. J. Heat Mass Transf. **63**, 268–283 (2013)
15. Xin You, Qiang Ye, Trung Van Nguyen, Ping Cheng, 2-d model of a h2/br2 flow battery with flow-through positive electrode. J. Electrochem. Soc. **163**(3), A447 (2015)
16. Xu. Juncai, Qiang Ma, Lei Xing, Huanhuan Li, Puiki Leung, Weiwei Yang, Su. Huaneng, Xu. Qian, Modeling the effect of temperature on performance of an iron-vanadium redox flow battery with deep eutectic solvent (des) electrolyte. J. Power Sources. **449**, 227491 (2020)

Chapter 5
Pore-Scale Modelling of Flow Batteries and Their Components

5.1 Introduction

This chapter overviews *pore-scale* modelling and simulation of the porous electrodes used in VRFBs. The transport phenomena within the electrode are simulated under external compression, coupling fluid motion and solid deformation. The effective transport properties required for model closure of the volume-averaged equations are then computed using various pore-scale models. The workflow of the pore-scale model, including the microstructure reconstruction, explicit dynamics simulation and numerical solution of the transport equations, is explained in detail. The approach is demonstrated with an example and compared with results obtained using experimental data.

The electrode materials used in flow batteries serve multiple functions. They should permit the simultaneous flow of electrons and the electrolyte solution through their solid substrate and pathways of the void space, respectively. They should also have a sufficiently high surface area for the electrochemical reactions to take place. The porous electrode material is often subjected to compression forces to reduce the interfacial contact resistance between the electrode and terminal plates. Therefore, the electrode should also have good mechanical strength and flexibility. Carbon felts (CFs), a fabric-like material made of carbon fibres, are a good choice for porous electrodes for VRFB. The CF is fabricated by weaving graphitised carbon fibres to form a sheet. Unlike the gas diffusion layers (GDLs) commonly employed in proton exchange membrane fuel cells (PEMFCs), the fibres in a CF are usually not glued using binder materials. Catalyst materials can be coated on the CF fibres to facilitate electrochemical reactions.

The transport processes in a CF include convection, diffusion and electromigration of the electrolyte constituents, conduction of heat in the liquid phase, electrical and thermal conduction through the network of carbon fibres, diffusion near the fibre surfaces and charge transfer via electrochemical reactions; see Fig. 5.1. A macroscopic approach can be employed to model these processes. The con-

A. A. Shah et al., *New Paradigms in Flow Battery Modelling*,
Engineering Applications of Computational Methods 16,
https://doi.org/10.1007/978-981-99-2524-7_5

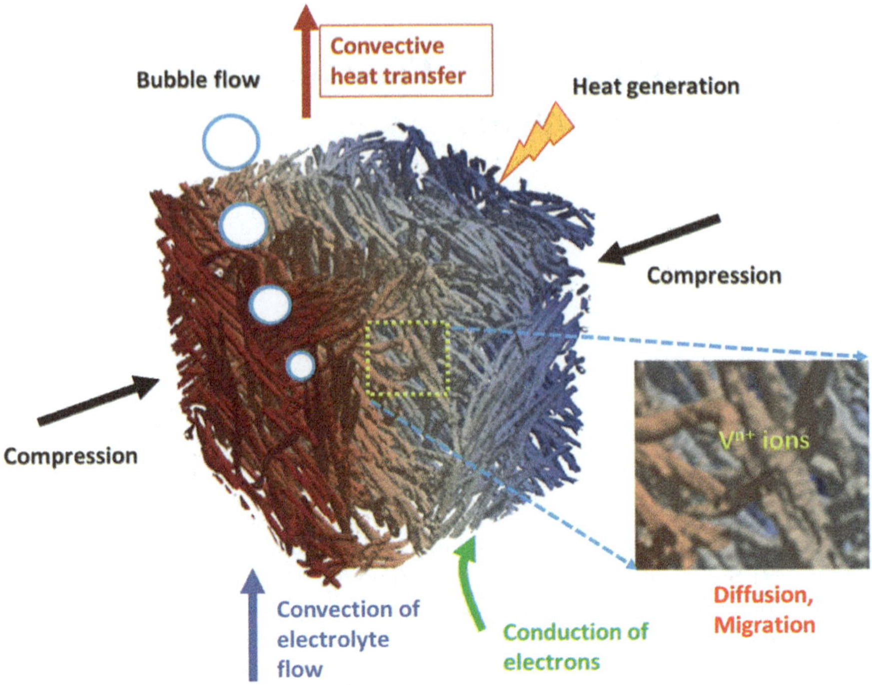

Fig. 5.1 Transport phenomena within a carbon felt representative element volume

servation equations in the entire domain can be solved with appropriate boundary conditions. However, in this approach, the effective transport properties must be obtained a priori. Pore-scale modelling, the focus of this chapter, has been developed in the past few decades to compute these effective properties.

5.2 Pore-Scale Modelling: Averaging Over Space

Before we proceed to discuss the methodology of pore-scale modelling, we need to clarify some terminology. The basic principle of pore-scale modelling is to solve the conservation equations relating to a particular quantity directly within the corresponding phases in which the transport processes occur. A volume-averaging step is then employed to determine the effective transport properties. In some of the literature, pore-scale modelling methodology was termed 'direct numerical simulation' [1]. However, the term direct numerical simulation and its acronym DNS refer to a well-known and extensively used method in computational fluid dynamics for solving the Navier-Stokes equations directly rather than using averaged equations (e.g., RANS) or filtering methods [2]. In this chapter, we therefore avoid the use of the term DNS.

Pore-scale modelling is also referred to as *mesoscopic simulation*, which is generally based on spatial averaging over a group of molecules [3, 4]. Statistical distributions are needed to describe the variables, and these distributions are obtained during the computations. This concept is similar to statistical thermodynamics, which describes physical properties in terms of statistical partition functions. The most popular mesoscopic method is the Lattice-Boltzmann method (LBM). LBM stemmed from the lattice gas automata method, which was developed to simulate fluid flows [5]. Mesoscopic methods can be employed to model a macroscopic, homogeneous domain because they can recover the continuum conservation equations, e.g., the Navier-Stokes equations, when the models are appropriately formulated [4].

Finally, in terms of scales, pore-scale modelling is primarily conducted in a representative element volume (REV) within which sufficient information on the transport processes can be averaged. In this sense, the scale of the domain of interest for the pore-scale model is ideally one order of magnitude greater than the characteristic pore/particle size. This scale is often much larger than atomistic simulation scales such as those in molecular dynamics (MD) [6] but smaller than the component's characteristic length scale.

This chapter aims to introduce pore-scale simulation and demonstrate how this approach can be employed to help understand the couple transport phenomena in porous electrodes for redox flow batteries. We next describe the mathematical formulation, leading to the set of governing equations. The steps required to implement pore-scale simulations, including the microstructure reconstruction and numerical simulation performed in the resulting domain, are then presented, followed by a detailed case study relating to VRFB.

5.3 Transport Phenomena

Multiple transport phenomena occur in the porous electrode during operation of a flow battery. The liquid electrolyte is driven by a pressure gradient and flows through the void space. The flow is convective over the large pore space in the electrode. Near the fibre surface or in narrow pore spaces, it becomes diffusive, and concentration gradients dictate the transport of chemical species. Electrons are conducted through the solid fibres, including the bulk material, and at regions of contact between the fibres and terminal plate surface. Heat is transported by conduction through the fibres and electrolyte, and by convection through the electrolyte. Electrochemical reactions, which require vanadium ions and electrons in a VRFB, occur on the fibre surfaces. The clamping force applied to the electrode causes material deformation, which affects most transport properties and creates local inhomogeneity and anisotropy.

Macroscopic modelling based on conservation laws can be employed to analyse the coupled transport phenomena in the porous electrode. By solving the conservation equations of mass, momentum, charge and energy, one can obtain local distributions of the flow velocity, reactant concentrations, electrical and electrolyte potentials, and temperature.

Macroscopic models are formulated based on the volume-averaging method over REVs. One drawback of this approach is that the accuracy of the model predictions depends on the effective transport properties, such as the diffusivity, permeability, thermal conductivity and electrolyte conductivity. These porous electrode properties depend on the fibre structure, which depends on the external force exerted on the electrode. To address this challenge, pore-scale simulation techniques have recently been developed to compute the effective transport properties directly from reconstructed material microstructures.

5.4 Mathematical Framework

The derivation of the macroscopic model for porous media is based on volume-averaging. We refer to Sect. 3.3 of Chap. 3 for a general introduction to macroscopic models and to Sect. 4.2 of Chap. 4 for a detailed discussion on macroscopic models for flow batteries. The primary variables of interest in the porous electrode are the flow velocity, species concentrations, electrical/electrolyte potentials and temperature. For abnormal operating conditions in which gases evolve, the flow becomes biphasic, and an additional variable related to the liquid or gas saturation can be introduced into the governing equations. Although the complete system of governing equations for porous electrodes can appear complex, some simplifications can be made.

5.4.1 *Multiphase Model and Closure*

When a porous medium is sufficiently dense, the convective flow is dictated by the viscous force of the fibres, and the Navier-Stokes momentum equation can be simplified to yield Darcy's law (3.28)

$$\mathbf{u} = -\frac{k}{\mu}\nabla p \tag{5.1}$$

in which k is the permeability of the medium (m^2), μ is the dynamic viscosity of the fluid (Pa s) and ∇p is the pressure gradient (Nm^{-2}). It should be noted that whether Darcy's law holds for the case of an electrolyte flowing through a CF, which has a high porosity, is subject to further verification.

Charge conservation in the solid fibre phase can be expressed using Ohm's law (3.41)

$$\nabla \cdot (\sigma \nabla \phi_s) = -\nabla \cdot \mathbf{j} \tag{5.2}$$

in which σ is the electric conductivity of the fibre, $\mathbf{j}$ is the total current transferred during reaction and ϕ_s is the electric potential of the solid electrode material.

The movement of charged species (V^{2+} V^{3+}, etc.) is driven by convection, diffusion and electro-migration, approximated by the Nernst-Planck equation (3.27) for a species flux $\mathbf{N}_i$

$$\mathbf{N}_i = -D_i \nabla c_i - \frac{z_i c_i D_i}{RT} F \nabla \phi_m + \mathbf{u} c_i \tag{5.3}$$

in which D_i is the effective diffusion coefficient, z_i is the valence for species i, F is the Faraday constant (96485 C mol^{-1}), R is the universal gas constant, T is the temperature and ϕ_m is the electrolyte potential. Charge conservation in the electrolyte was derived in Sect. 3.3.6 to yield

$$\nabla \cdot \mathbf{j} = \nabla \cdot \left(-\sigma_m \nabla \phi_m - F \sum_i z_i D_i \nabla c_i \right) \tag{5.4}$$

in which $\sigma_m = F^2 \sum_i z_i^2 D_i \nabla c_i / (RT)$ is the conductivity of the electrolyte.

For small current conditions, the heat accompanying electrochemical reactions can be negligible, and the cell can be considered isothermal. However, for high current conditions, the energy conservation equation (3.18) can be used to obtain the temperature field

$$\nabla \cdot \left(\mathbf{u} C_p T \right) + \nabla \cdot (-k \nabla T) + S_T = 0 \tag{5.5}$$

in which C_p is the specific heat of the electrolyte, k is the thermal conductivity of the electrolyte and S_T is the source term, which consists of ohmic heating, entropy change and heat release due to an overpotential. We note that heat transfer occurs not only in the fluid phase but also in the solid fibre phase. In the latter case, the convective term in Eq. (5.5) reduces to zero and the fibre thermal conductivity is used, with ohmic heating as the only source term. In reality, the phases will exchange heat, but such effects will not be considered here.

As discussed in Chap. 4, the presence of a gas phase requires the introduction of a saturation, and corresponding modification of the single-phase equations above. A steady-state equation for the gas saturation can be written as [7]

$$\nabla \cdot (2\rho \mathbf{u}_m s) + \nabla \cdot \left(-\rho_g D_c \nabla s \right) - \nabla \cdot \left(\rho \mathbf{u}_m s^2 \right) = 0 \tag{5.6}$$

in which $\mathbf{u}_m$ and ρ are the volume-averaged velocity and density of the two-phase mixture (4.8), ρ_g is the density of the gas and D_c is the capillary diffusivity. The capillary diffusivity can be expressed as

$$D_c = -\frac{\kappa_g}{\mu_g} \frac{dp_c}{ds} \tag{5.7}$$

in which p_c is the capillary pressure, i.e., the pressure difference between the liquid and vapour phases, κ_g is the gas permeability and μ_g is the dynamic viscosity of the gas.

The derivative of capillary pressure with respect to saturation is a characteristic function of the porous medium in question and can be obtained by pore-scale simulations. The formulation above is a classical model that stemmed from petroleum applications. An underlying assumption of this model is that the porous medium is fairly packed, which is not valid for the CF used in VRFB. We note that the two-phase flow in VRFB is not currently well understood. The flow distribution depends on several factors that include the flow conditions, fibre structure, surface wettability and electric potential distribution, amongst others.

Volume averaging to derive macroscopic equations for systems involving dissimilar phases is rather involved [8], but in essence, following volume-averaging, the original terms of the single-phase equations are retained, with the introduction of additional terms [9] accounting for the interfacial interactions and perturbations. For practical applications, the governing equations take their original forms with the single-phase transport properties replaced by effective transport properties, namely,

$$
\begin{aligned}
\nabla \cdot (\mathbf{u}_m \varphi) + \nabla \cdot \left(-D_\varphi^{eff} \nabla \varphi\right) + S_\varphi = 0 \\
\nabla \cdot \left(\sigma^{eff} \nabla \phi_s\right) = -\nabla \cdot \mathbf{j} \\
\mathbf{N}_i = -D_i^{eff} \nabla c_i - \frac{z_i c_i D_i^{eff}}{RT} F \nabla \phi_m + \mathbf{u}_m c_i \\
\nabla \cdot \mathbf{j} = \nabla \cdot \left(-\sigma_m^{eff} \nabla \phi_m - F \sum_i z_i D_i \nabla c_i\right) \\
\nabla \cdot \left(\mathbf{u}_m C_p T\right) + \nabla \cdot \left(-k^{eff} \nabla T\right) + S_T = 0
\end{aligned}
\tag{5.8}
$$

in which the superscript eff denotes an effective property and we assume a steady state. It should be noted that the governing equation for gas saturation is already expressed in a multiphase format that is averaged over the REV.

Closure of the above volume-averaged equations relies upon knowledge of the effective transport coefficients, including the permeability, diffusivity, electrical and electrolyte conductivities, and thermal conductivity. A classical approach is to relate them to the porosity and tortuosity, as done by Shah et al. [10]. It should be pointed out that many empirical relationships for these properties are based on homogeneous porous media, e.g. sandstone, which are not the same as fibrous materials such as the CF in VRFB. Better estimates can be obtained through pore-scale modelling, which is the focus of this chapter.

5.5 Numerical Procedure for Pore-Scale Simulations

Pore-scale simulations consist of three steps: (1) geometry reconstruction, (2) numerical solution of the field equations with appropriate initial-boundary conditions and (3) post-processing of the results and uplinks to macroscopic models. Details of implementing these steps are discussed in this section.

5.5.1 Geometry Reconstruction

The structure of the porous medium under consideration is needed in order to differentiate between the components in the computational domain. The structure is comprised of void space, in which the fluid flows, and a solid phase, which may consist of different materials. The solid phase can be characterised using advanced microscopy, e.g., computed tomography (CT), scanning electronic microscopy (SEM) and focused ion beam (FIB)-SEM. Alternatively, the geometry of the porous medium can be reconstructed numerically based on defined rules governing the distributions of the materials. The former methods have the advantage of high-fidelity rendering of the actual material, whereas the latter methods are expedient and flexible, making them suitable for parametric studies.

5.5.1.1 X-ray Computed Tomography

X-ray computed tomography (XCT) is a powerful tool for reconstructing objects in 3D. The working principle of this technique is to project an X-ray into the object of interest and collect the attenuated X-ray signal. The object or the X-ray is then rotated at a small angle, and the imaging procedure is repeated until the object has been rotated a full 360°. Each collected image represents the X-ray field after attenuation through the object. The amount of attenuation is proportional to the object's density integrated along the line of sight from the light source to the image sensor. The grayscale values of all the XCT images are then inverted to render the density field of the object in 3D space. These data are then transferred to a 3D matrix for the pore-scale model. We note that the raw images and the 3D reconstructed field are grayscale only, so it is not possible to differentiate between the components and phases of the object.

Additional actions are often needed to determine these components, e.g. with a known porosity or dimensions. In this practice, however, ambiguity arises if the resolution of the XCT imaging equipment is not sufficiently high such that each voxel contains multiple components. Currently, micro-CT equipment can resolve the carbon fibres of the VRFB's CF [11]. A nano-CT may be required for materials with nano-scale structures [12].

5.5.1.2 Focused Ion Beam-Scanning Electron Microscopy

A scanning electron microscope incorporating a Focused Ion Beam (FIB) has been developed to visualise nano-scale material microstructures. The setup usually has an ion beam and an electron beam separated at a fixed angle. The FIB can etch out the object under investigation at the level of tens of nanometres. An SEM image can be taken each time a new specimen cross section is cut, and both steps are repeated until sufficient images are collected to build a 3D model for subsequent pore-scale modelling [13].

5.5.2 Numerical Reconstruction

With a proper understanding of the porous media microstructure, one can numerically reconstruct the 3D model by using a stochastic approach to add components into the domain for visually random materials. This method has been used in constructing GDLs, microporous layers and catalyst layers for PEMFCs. There should be certain constraints, such as volume fractions of each component, to ensure that the reconstructed model structure is compatible with that of the actual material. Similar models can be generated based on the same constraints and used in the pore-scale simulations, with the computed results averaged.

5.5.3 Size of Representative Elementary Volume

The size of the computational domain for the pore-scale models affects the accuracy of the computed transport properties. Ideally, the domain should be as large as one can afford, in order to include a sufficiently large region of dissimilar materials and void space. Domain size dependency can be numerically evaluated by comparing the numerical results of various domain sizes, similar to procedures used for mesh independence.

5.5.4 Pore-Scale Models

Lange et al. [14] developed a pore-scale code based on the finite volume formulation. The code was initially developed to solve the coupled transport processes in a PEMFC catalyst layer. It has been revised for simulations in porous electrodes [11, 15, 16] and will be referred to as PSM in the remainder of this chapter.

The PSM code requires input of the phase structure and the bulk transport properties of the materials within the REV. For the case of CFs for VRFB, the governing equations are discretised and solved with prescribed boundary conditions. For convenience, two Dirichlet boundary conditions are set on two surfaces, while a zero-flux condition is set at the remaining four surfaces of the computational domain. In this way, one can quickly evaluate the effective transport properties by dividing the fluxes obtained from the PSM simulation by the length of the REV surface where the Dirichlet boundary conditions are imposed.

The boundary conditions can be applied to other surfaces to obtain the averaged properties. Some materials are highly anisotropic, e.g., typical GDLs for PEMFCs.

CF materials are less anisotropic than GDLs but their isotropy decreases when compressed, at least to some extent, leading to anisotropic effective transport properties [17].

5.5.5 Multiple Relaxation Time Lattice-Boltzmann Model

The lattice-Boltzmann model, originally stemming from the lattice gas cellular automata (LGCA) method, is a powerful numerical tool for investigating complicated porous media such as GDLs [18]. Full details of the following can be found in Sect. 3.4.3 of Chap. 3. A multiple relaxation time (MRT) LBM model can be employed for multiphase and multicomponent flows with large density ratios using two distribution functions [19, 20]. The governing equations for a two-phase system are the Navier-Stokes equations (3.14) and the Cahn-Hilliard equation (3.69)

$$\begin{aligned} \rho\frac{\mathrm{D}\mathbf{u}}{\mathrm{D}t} = \rho\left(\frac{\partial\mathbf{u}}{\partial t} + \mathbf{u}\cdot\nabla\mathbf{u}\right) &= -\nabla p + \eta\,\nabla^2\mathbf{u} + \frac{1}{3}\eta\,\nabla(\nabla\cdot\mathbf{u}) + F \\ \frac{\partial\varphi}{\partial t} &+ \nabla\cdot(\varphi\mathbf{u}) = M\nabla^2\mu \end{aligned} \tag{5.9}$$

respectively, in which $\mathbf{u}$ is velocity, p is pressure, ρ is density, η is the dynamic viscosity, M is a mobility parameter, φ is an order parameter, μ is the chemical potential and F is an external force.

These equations can be solved on a finite grid using the following two equations [19, 20]

$$\begin{aligned} f_i\,(r + \mathbf{e}_i\delta t,\ t + \delta t) &= f_i(r,t) - Q^{-1}\Lambda_f\left(m_f(r,t) - m_f{}^{eq}(r,t)\right) \\ &\quad + \delta t\left(\mathbf{I} - \frac{1}{2}\mathbf{Q}^{-1}\Lambda_f\mathbf{Q}\right)G_i(r,t) \\ g_i\,(r + \mathbf{e}_i\delta t,\ t + \delta t) &= g_i(r,\ t) - Q^{-1}\Lambda_g\left(m_g(r,t) - m_g{}^{eq}(r,t)\right) \end{aligned} \tag{5.10}$$

in which f_i is the density distribution function and g_i is the order-parameter vector at lattice location r and time t. $\Lambda_{\alpha(\alpha=f,g)}$ is the diagonal relaxation matrix, $\mathbf{I}$ is the identity matrix and $\mathbf{Q}$ is a $q \times q$ matrix which linearly transforms the distribution functions f_i and g_i to the velocity moments m_f and m_g, respectively.

In the case of VRFB, LBM can be employed to study the electrolyte distribution in the CF at various compression ratios. The reconstructed CF electrode sample domain is first discretised for the LB model implementation. All samples were prescribed with a fixed contact angle in the simulation in this chapter.

5.5.6 Solid Mechanics

An important issue in dealing with the CF for VRFB applications is the mechanical compression of the CF. The deformation of the CF fibres and their network will affect the transport pathways for all transport variables. The solution procedure for solid mechanics differs from computational fluid dynamics (CFD) in that the computational domain constantly changes with strain, and the mesh for the domain is required to reflect the volume change. The coupling of computational fluid and solid mechanics falls under the umbrella of flow-structure interaction (FSI).

The motion of the CF fibres can be simulated by an explicit dynamic method using commercial software. A few assumptions are made to simplify the simulation. The fibre materials are assumed to be elastic, homogeneous and isotropic. The kinematic equation that describes the relationship between an external force and stress is written as, in index notation

$$\frac{\partial \sigma_{ij}}{\partial x_j} + \rho f_i = \rho \ddot{u}_i \tag{5.11}$$

in which σ is the stress tensor, ρ is density, f is a body force, u is displacement, dots denote time derivatives and i and j indicate spatial dimensions. The geometric equation that describes the relationship between displacement and strain is

$$\varepsilon_{ij} = \frac{1}{2}\left(u_{i,j} + u_{j,i}\right) \equiv \frac{1}{2}\left(\frac{\partial u_i}{\partial x_j} + \frac{\partial u_j}{\partial x_i}\right) \tag{5.12}$$

in which ε is strain, and the equations describing the relationship between stress and strain can be expressed as

$$\begin{aligned} \sigma_{ij} &= \lambda \varepsilon_{kk} \delta_{ij} + 2\mu \varepsilon_{ij} \\ \sigma_{kk} &= (3\lambda + 2\mu)\varepsilon_{kk} = 3E\varepsilon_{kk} \end{aligned} \tag{5.13}$$

in which λ and μ are the Lame constants of the material, δ_{ij} is the Kronecker delta and E is the bulk elasticity modulus of the material.

Equations (5.11)–(5.13) are solved using the finite-element method (FEM) over the meshed solid material domain. A prescribed displacement rate is applied at the boundaries. The fibres in the CF model move downwards and sideways to accommodate the reduced volume engendered by the external force. Due to the incompatibility between the finite-element and finite-volume methods, real coupling of solid mechanics simulations and CFD cannot be easily carried out. Instead, in PSM simulations, the deformed geometry obtained from the FEM simulation is exported to the PSM for subsequent simulation. The FEM step is used as an alternative to XCT experiments to simulate the compression procedure in order to obtain the deformed CF geometry.

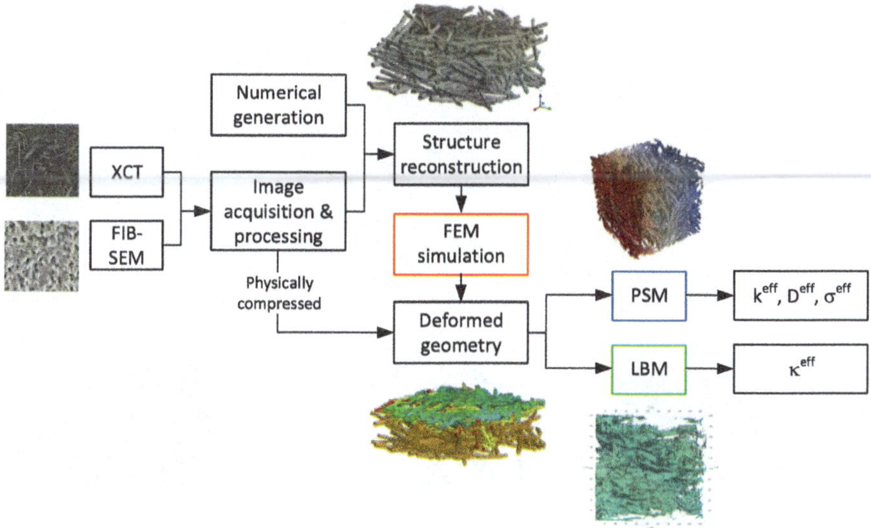

Fig. 5.2 Workflow of microstructure reconstruction and pore-scale simulations

5.6 Results and Discussion

The work in [11, 16, 21] is employed here to demonstrate the workflow of pore-scale modelling for VRFB porous electrodes. Readers are referred to these papers for details omitted in this chapter. Figure 5.2 shows the procedure of typical pore-scale modelling. It starts with image acquisition using advanced microscopy, depending on the length scale of the sample. These images are then processed to generate the 3D geometry of the sample material. If the actual geometry of the sample cannot be obtained experimentally, one can attempt to generate the material structure numerically based on microscopy images and appropriate algorithms, as described above.

The remaining work is to carry out pore-scale simulations of these models. As discussed previously, one can use the PSM code, LBM code or similar tools to compute the transport properties of the models. The computed effective transport properties are then readily applicable to subsequent macroscopic simulations. Details of these steps are discussed below.

5.6.1 Reconstruction

The XCT images of a CF sample are first imported into image processing software. The same procedure is repeated for images of CF samples under different compression ratios. It is worth pointing out that numerical reconstruction using stochastic algorithms cannot be easily implemented due to the curved shape of the CF carbon

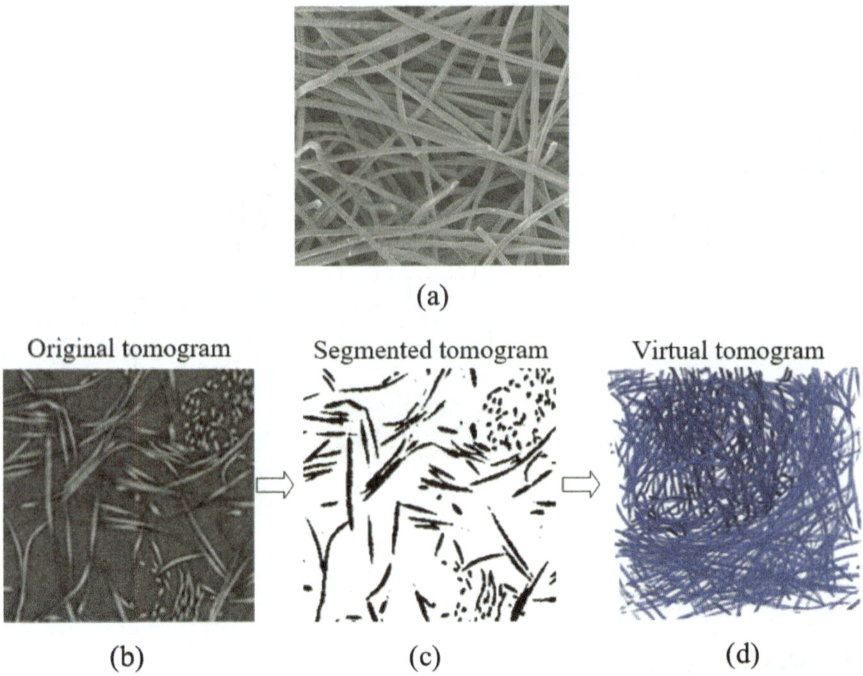

Fig. 5.3 Schematic diagram of the reconstruction process. **a** SEM image of CF sample; **b** Grayscale image with a resolution of 2.44 μm/pixel obtained *via* synchrotron X-ray micro-CT; **c** Binary image obtained to distinguish fibres and pore space; **d** Virtual volume obtained by 3D reconstruction

fibres. The grayscale 3D model of the CF sample is then imported to data visualisation software, where the known porosity of the sample determines the distinction between solid and void. A grayscale value is set to satisfy the given porosity value. The 3D model is then converted to a binary matrix required to run the PSM and LBM codes. Figure 5.3 shows the original CT image, the segmented image based on a given grayscale value and the reconstructed structure.

Before the reconstructed solid model can be ported for pore-scale simulations, the model must be meshed to the format required by the simulation software. Figures 5.4a–d shows the structure of carbon fibres under different compression ratios (CRs). As the CF is compressed, the domain is expected to become more packed by fibres. We point out that the porosity of these cases is relatively high, i.e. 95.2%, 92.6%, 89.5%, and 78.8% for CR = 0, 25, 50 and 75%, respectively.

5.6.2 Explicit Dynamics Simulation of Compression

As with GDLs in PEMFCs, the CF electrode is always compressed to reduce ohmic losses resulting from loose contact among the fibres, and poor contact between the CF

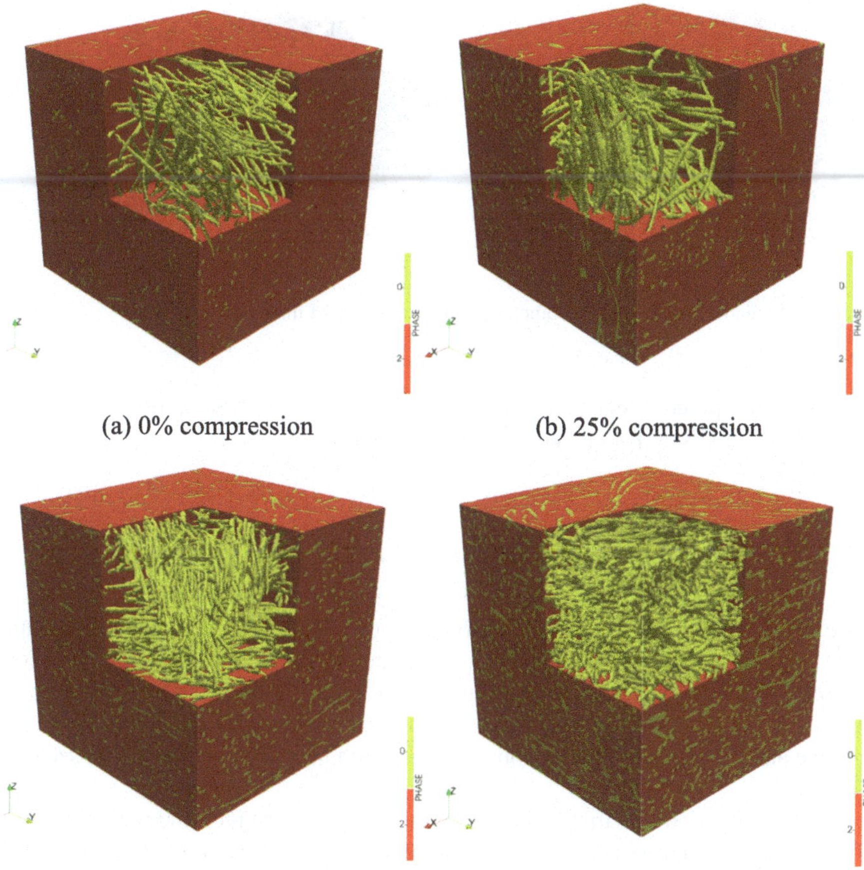

(a) 0% compression (b) 25% compression

(c) 50% compression (d) 75% compression

Fig. 5.4 Electrode morphology at different compression ratios reconstructed

and current terminals. External compression will inherently result in corresponding changes in the effective transport properties. Since a CF is used as a sheet in the VRFB cell, it is convenient to designate two directions when describing the transport and mechanical stress, i.e., the in-plane and the through-plane directions. The former is the two directions parallel to the membrane, whereas the latter is the direction perpendicular to the membrane.

The transport properties in both directions can differ in magnitude depending on how the CF is fabricated and how much compression force is applied to the CF. In an operating VRFB device, a through-plane force is often applied to maintain good electrical conduction through the fibres. Figure 5.5a shows a reconstructed model in which carbon fibres have some degree of entanglement. Figure 5.5b shows the meshed model ready for FEM simulations. Figures 5.6a–c show the displacement

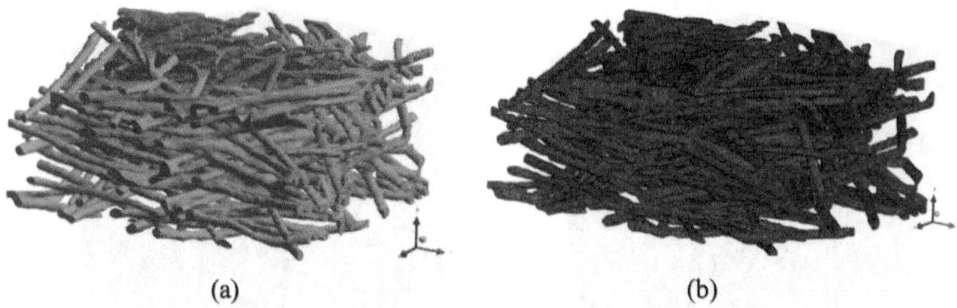

Fig. 5.5 3D rendering of **a** solid geometry and **b** meshed FEM model of carbon felt

in the through-plane direction of the CF material obtained from explicit dynamics simulations for CR = 10, 20 and 30%, respectively.

5.6.3 Computed Effective Transport Properties

Once the deformed geometry of the models at different CRs is obtained, PSM and LBM simulations can be carried out to evaluate the effective transport properties. Figures 5.7a–c show the effective transport properties normalised by the highest values computed using the uncompressed model. The diffusivities for ions in the in-plane and through-plane directions decrease with increased compression; see Fig. 5.7a.

The electric conductivities in both directions (Fig. 5.7b) increase with compression, and the through-plane value is apparently more sensitive to the CR. When the CF is compressed, more contact points among the fibres are formed, and hence the increase in electric conductivity. It should be noted that two different scales are used in Fig. 5.7b, and the CF's in-plane conductivity is lower but close to its through-plane value. The permeability computed using single-phase LBM has the same trend as that reported in the literature, as seen in Fig. 5.7c.

5.6.4 Combining Models at Different Scales

Pore-scale modelling is a pivotal step for the multiscale simulation of materials. Figure 5.8 shows the workflow for a multiscale simulation for porous media such as the CF for VRFB. The properties computed using pore-scale simulations can be used in a macroscopic CFD model to simulate the coupled transport processes at the device or system levels.

These CFD and pore-scale models can be combined into a suite of simulation tools, with various options for the pore-scale and macroscopic CFD approaches,

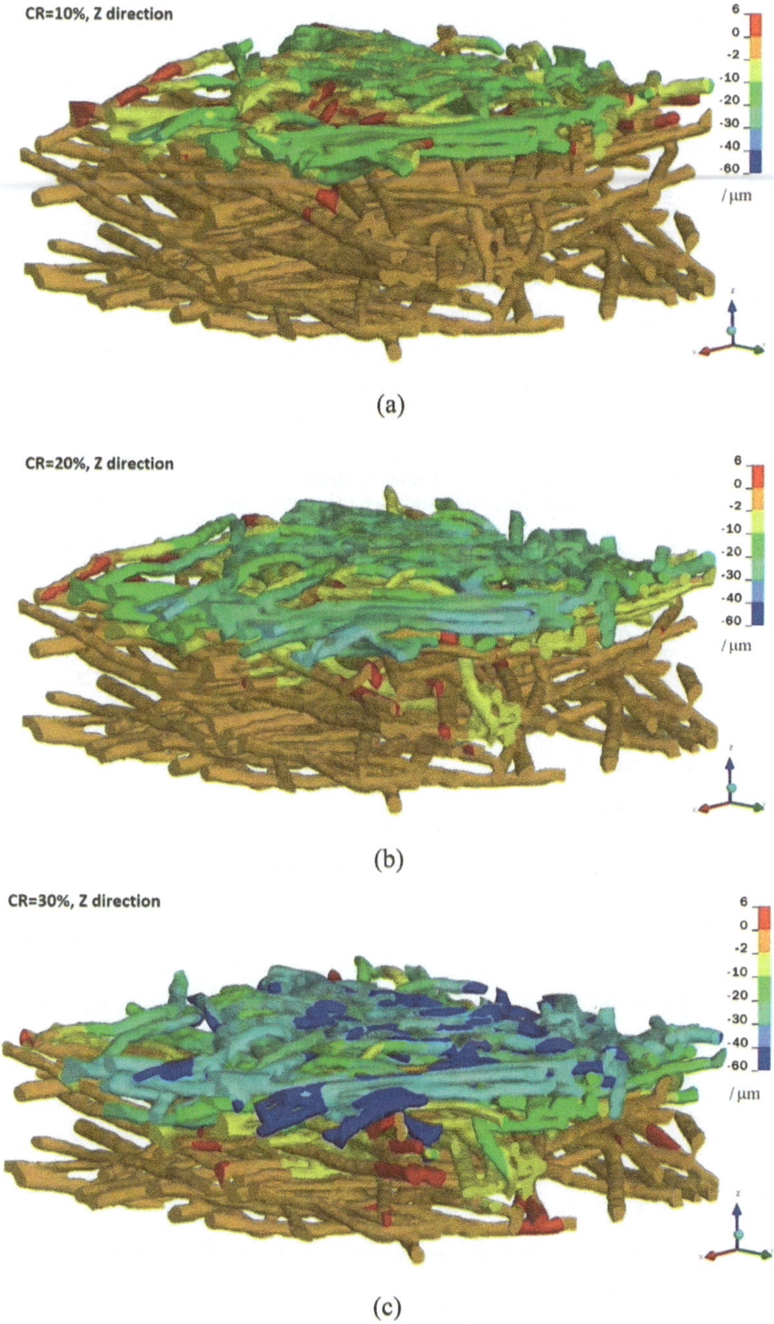

Fig. 5.6 Displacement distributions of carbon felt model in the Z-direction with **a** CR = 10%, **b** CR = 20% and **c** CR = 30%

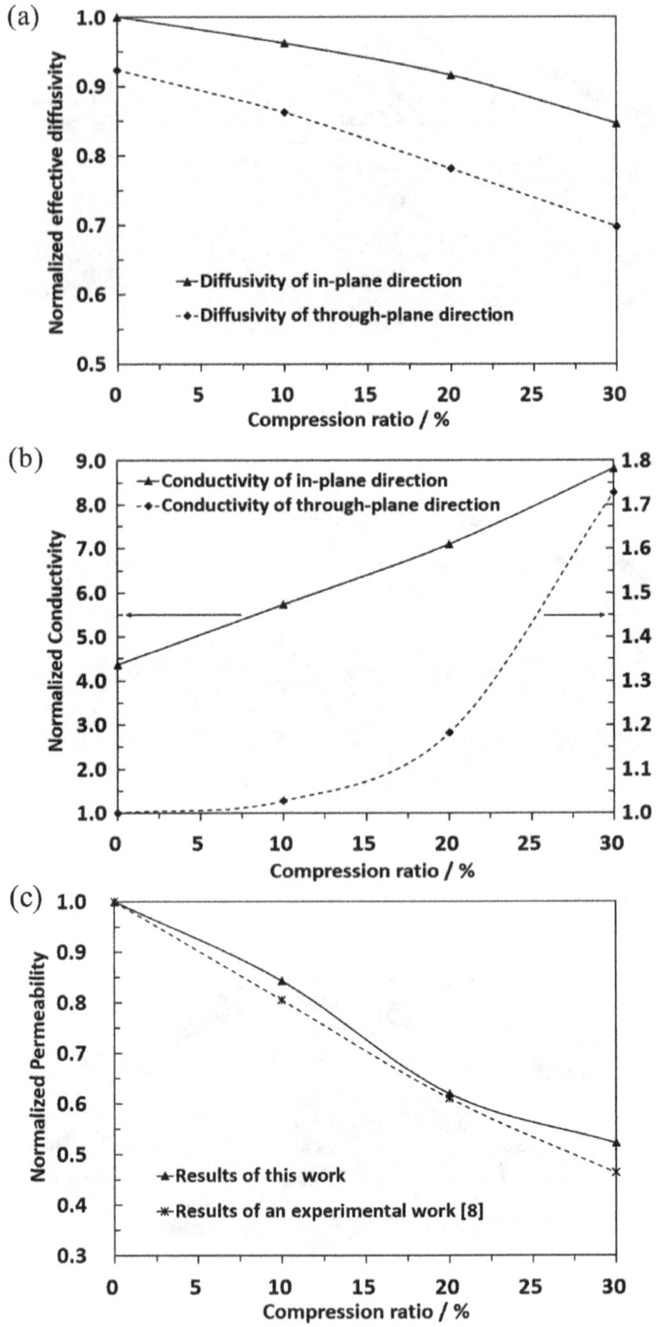

Fig. 5.7 Computed effective transport properties at different CRs. **a** normalised diffusivity of vanadium ion; **b** electronic conductivity; **c** permeability of carbon felt and literature [22]

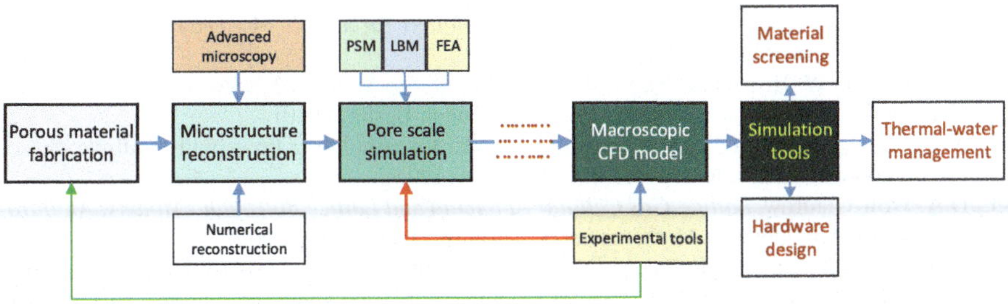

Fig. 5.8 Strategy for multiscale simulation

which engineers can use to assist device or system designs. All of the simulation tools involved should be validated using experimental measurements to improve their fidelity and reliability. Ultimately, one could design a new porous electrode by manipulating the fabrication parameters and producing a CF with properties that optimise VRFB performance and material lifetime.

5.7 Summary

In this chapter we discussed pore-scale modelling and simulation carried out for carbon felt materials in VRFB applications. The methodology introduced here applies to most porous electrode materials such as those in lithium-ion batteries, fuel cells and electrolysers. As advanced microscopy and high-performance computing become increasingly accessible, the approach can be included in the early stage of product design. One can also employ this method to study material degradation using post-mortem characterisation data. Experimental validation is essential to make pore-scale modelling suitable for real-world applications. To this end, there are ample opportunities for future research and development.

Quite apart from these considerations, there are challenges in terms of fully coupling macro- and meso-scopic approaches, namely with regards to the matching of the methods at the interface between the two scales. For porous materials, this is particularly challenging since such an interface is not straightforward to define. Finally, true fluid-structure interaction (FSI) simulations are possible with the methods described in Chap. 3, in particular the arbitrary Eulerian Lagrangian (ALE) approach of Sect. 3.3.9. These more fundamental-level challenges present further opportunities for flow battery researchers.

References

1. P.P. Mukherjee, C.Y. Wang, Direct numerical simulation modeling of bilayer cathode catalyst layers in polymer electrolyte fuel cells. J. Electrochem. Soc. **154**(11), B1121–B1131 (2007)

2. E. Pomraning, *Development of Large Eddy Simulation Turbulence Models*. Ph.D. thesis, University Wisconsin-Madison (2000)
3. K. Binder, D.W. Heermann, *Monte Carlo Simulation in Statistical Physics: An Introduction*, Graduate Texts in Physics (GTP), 5th edn. (Springer, Berlin, 2010)
4. X.W. Shan, H.D. Chen, Lattice boltzmann model for simulating flows with multiple phases and components. Phys. Rev. E **47**(3), 1815–1819 (1993)
5. D.A. Wolf-Gladrow, *Lattice-Gas Cellular Automata and Lattice Boltzmann Models: An Introduction*, Lecture Notes in Mathematics (LNM) (Springer, Berlin, 2000)
6. C.H. Cheng, K. Malek, P.C. Sui, N. Djilali, Effect of pt nano-particle size on the microstructure of pem fuel cell catalyst layers: insights from molecular dynamics simulations. Electrochim. Acta **55**(5), 1588–1597 (2010)
7. S. Mazumder, J.v. Cole, Rigorous 3-d mathematical modeling of pem fuel cells - ii. model predictions with liquid water transport. J. Electrochem. Soc. **150**(11), A1510–A1517 (2003)
8. S. Whitaker, *The Method of Volume Averaging* (Springer, 1999)
9. P.-C. Sui, X. Zhu, N. Djilali, Modeling of pem fuel cell catalyst layers: status and outlook. Electrochem. Energy Rev. pp. 1–39 (2019)
10. A.A. Shah, M.J. Watt-Smith, F.C. Walsh, A dynamic performance model for redox-flow batteries involving soluble species. Electrochim. Acta **53**(27), 8087–8100 (2008)
11. M. Li, N. Bevilacqua, L.K. Zhu, W.L. Leng, K.J. Duan, L.S. Xiao, R. Zeis, P.C. Sui, Mesoscopic modeling and characterization of the porous electrodes for vanadium redox flow batteries. J. Energy Storage **32**, 12 (2020)
12. S. Litster, W.K. Epting, E.A. Wargo, S.R. Kalidindi, E.C. Kumbur, Morphological analyses of polymer electrolyte fuel cell electrodes with nano-scale computed tomography imaging. Fuel Cells **13**(5), 935–945 (2013)
13. R. Singh, A.R. Akhgar, P.C. Sui, K.J. Lange, N. Djilali, Dual-beam fib/sem characterization, statistical reconstruction, and pore scale modeling of a pemfc catalyst layer. J. Electrochem. Soc. **161**(4), F415–F424 (2014)
14. K.J. Lange, P.C. Sui, N. Djilali, Pore scale simulation of transport and electrochemical reactions in reconstructed pemfc catalyst layers. J. Electrochem. Soc. **157**(10), B1434–B1442 (2010)
15. L.J. Zhu, H. Zhang, L.S. Xiao, A. Bazylak, X. Gao, P.C. Sui, Pore-scale modeling of gas diffusion layers: Effects of compression on transport properties. J. Power Sources **496**, 11 (2021)
16. L.S. Xiao, M.J. Luo, L.J. Zhu, K.J. Duan, N. Bevilacqua, L. Eifert, R. Zeis, P.C. Sui, Pore-scale characterization and simulation of porous electrode material for vanadium redox flow battery: effects of compression on transport properties. J. Electrochem. Soc. **167**(11), 10 (2020)
17. H. Zhang, L. Zhu, H.B. Harandi, K. Duan, R. Zeis, P.-C. Sui, P.A. Chuang, Microstructure reconstruction of the gas diffusion layer and analyses of the anisotropic transport properties. Energy Convers. Manag. **241**(11429), 3 (2021)
18. L. Chen, R.Y. Zhang, Q.J. Kang, W.Q. Tao, Pore-scale study of pore-ionomer interfacial reactive transport processes in proton exchange membrane fuel cell catalyst layer. Chem. Eng. J. **391**, 13 (2020)
19. A. Akhgar, *Computational Analysis of Multi-Phase Flow in Porous Media with Application to Fuel Cells*. Ph.D. thesis, University of Victoria (2016)
20. X.D. Niu, T. Munekata, S.A. Hyodo, K. Suga, An investigation of water-gas transport processes in the gas-diffusion-layer of a pem fuel cell by a multiphase multiple-relaxation-time lattice boltzmann model. J. Power Sources **172**(2), 542–552 (2007)
21. L.S. Xiao, M.Q. Bian, L.J. Zhu, K.J. Duan, W.L. Leng, R. Zeis, P.C. Sui, H.C. Zhang, High-density and low-density gas diffusion layers for proton exchange membrane fuel cells: comparison of mechanical and transport properties. Int. J. Hydrog. Energy **47**(53), 22532–22544 (2022)
22. Q. Wang, Z. Qu, Z. Jiang, W. Yang, Experimental study on the performance of a vanadium redox flow battery with non-uniformly compressed carbon felt electrode. Appl. Energy **213**, 293–305 (2018)

Chapter 6
Machine Learning for Flow Battery Systems

6.1 Introduction

We refer to Sect. 3.7 for an introduction to machine learning and the main terminology, as well as an outline of the basic approach to devising a machine learning model. Machine learning can be used in a variety of ways for the design, analysis, optimisation and control of flow batteries, their components and their materials. The majority of likely applications fall within the realm of regression, which is a subset of supervised machine learning. Classification, which is the second major type of supervised machine learning, could also be used for applications such as failure analysis, e.g., predicting modes of failure based on operating and design characteristics.

Surrogate models, as defined in Sect. 3.7, are an obvious application of machine learning to flow batteries, especially for optimisation, sensitivity analysis and uncertainty quantification. Alternatives to machine learning surrogate models are *multi-fidelity* and *reduced-order* models, which are described in Chap. 7. The first of these relies heavily on machine learning, while the latter can be enhanced by machine learning for parameter-dependent and nonlinear problems.

In Sects. 6.2–6.16 we introduce a number of popular classification, regression, *clustering* and *dimension reduction* methods in detail. Many of these methods have been used for the study of battery and fuel cell systems. We cover both linear and nonlinear methods, kernel methods, multivariate models (including tensor-based models) and Bayesian approaches. In Appendix D, we provide a brief outline of gradient-based optimisation, which, as will become apparent early on in the chapter, is an essential element of machine learning in all its forms. Neither semi-supervised nor reinforcement learning will be covered, since there are currently no applications of these approaches to be found in the flow battery literature.

A. A. Shah et al., *New Paradigms in Flow Battery Modelling*,
Engineering Applications of Computational Methods 16,
https://doi.org/10.1007/978-981-99-2524-7_6

6.2 Linear Regression

For univariate training outputs or targets y_n and corresponding design points $\Xi = \{\boldsymbol{\xi}_n\} \subset \mathbb{R}^l$, $n = 1, \ldots, N$, a flexible and general approximating function is available in the form of a *linear combination* of some *known* basis functions $\phi_k(\boldsymbol{\xi})$, $k = 0, \ldots, K-1$, with $\phi_0(\boldsymbol{\xi}) = 1$

$$\begin{aligned}\eta(\boldsymbol{\xi}; \mathbf{w}) &= w_0 + w_1\phi_1(\boldsymbol{\xi}) + \ldots + w_{K-1}\phi_{K-1}(\boldsymbol{\xi}) \\ &= \sum\nolimits_{k=0}^{K-1} w_i\phi_i(\boldsymbol{\xi}) \\ &= (w_0, \ldots, w_{M-1}) \begin{pmatrix} \phi_0(\boldsymbol{\xi}) = 1 \\ \phi_1(\boldsymbol{\xi}) \\ \vdots \\ \phi_{K-1}(\boldsymbol{\xi}) \end{pmatrix} \\ &= \mathbf{w}^T \boldsymbol{\phi}(\boldsymbol{\xi})\end{aligned} \tag{6.1}$$

The coefficients w_k are termed *weights* and $\phi_0(\boldsymbol{\xi}) = 1$ serves to allow for a constant term w_0 in the expansion, known as a *bias*. Clearly, the approximating function is linear in $\mathbf{w}$, although generally nonlinear in $\boldsymbol{\xi}$, which leads to the name *linear regression*. The basic aim is to determine the best values for the parameters $\mathbf{w}$ using the data.

The functions $\phi_k(\boldsymbol{\xi})$ are generally chosen from some formal basis, usually for the space of smooth functions in the relevant domain ($\boldsymbol{\xi}$ space). The most obvious example is a polynomial expansion, e.g., with scalar inputs

$$\eta(\xi; \mathbf{w}) = \sum\nolimits_{k=0}^{K-1} w_k \xi^k = \mathbf{w}^T \boldsymbol{\phi}(\xi) \tag{6.2}$$

In this case, the basis functions are $\phi_0(\xi) = 1$, $\phi_1(\xi) = \xi$, $\phi_2(\xi) = \xi^2, \ldots$, and

$$\boldsymbol{\phi}(\xi) = (1, \xi, \xi^2, \ldots, \xi^{K-1})^T \tag{6.3}$$

There is no restriction on the basis functions chosen. The choice is informed by any prior knowledge of the underlying physical process, or by visualisation or exploratory analysis of the data, which might reveal information about the latent function. On the other hand, the choice could simply be an educated guess. Other popular choices of basis functions include Gaussian basis functions

$$\phi_k(\boldsymbol{\xi}) = \exp\left(-\frac{(\boldsymbol{\xi} - \boldsymbol{\mu}_k)^T(\boldsymbol{\xi} - \boldsymbol{\mu}_k)}{2h^2}\right) = \exp\left(-\frac{\|\boldsymbol{\xi} - \boldsymbol{\mu}_k\|^2}{2h^2}\right) \tag{6.4}$$

in which $\boldsymbol{\mu}_k$ is called a radial basis centre and h is a shape parameter, and sigmoidal basis functions

$$\phi_k(\boldsymbol{\xi}) = \sigma\left(\frac{\|\boldsymbol{\xi} - \boldsymbol{\mu}_k\|^2}{2h^2}\right) \qquad \sigma(a) = \frac{1}{1+e^{-a}} \tag{6.5}$$

with centre $\boldsymbol{\mu}_k$ and shape parameter h. These are both examples of *radial basis functions*, which depend only on the (Euclidean) distance $\|\boldsymbol{\xi} - \boldsymbol{\mu}_k\|$ for some radial basis centre $\boldsymbol{\mu}_k$. The number K of basis functions is user chosen. We return to this choice later. Before moving onto the solution procedure, we highlight some important consequences that follow from this formulation, and which will allow for its extension later using a concept known as kernel substitution.

The vector function $\boldsymbol{\phi}(\boldsymbol{\xi})$ is an example of a *feature map*. It maps the inputs $\boldsymbol{\xi}$ from the design space $\mathcal{X}$ to a *feature space* $\mathcal{F}$, usually of higher dimension than the input space. The coefficients of $\boldsymbol{\xi}$ are called *attributes* and the coefficients of $\boldsymbol{\phi}(\boldsymbol{\xi})$ are called *features*. For example, Let $\boldsymbol{\xi} = \xi$, i.e., a scalar, and define the feature map

$$\boldsymbol{\phi}(\xi) = (1, \xi, \xi^2, \xi^3)^T \tag{6.6}$$

The attribute is ξ, the features are $1, \xi, \xi^2, \xi^3$, $\mathcal{X} \subset \mathbb{R}$ and $\mathcal{F} \subset \mathbb{R}^4$. As a second example, let $\boldsymbol{\xi} = (\xi_1, \xi_2)^T$ and define the feature map

$$\boldsymbol{\phi}(\boldsymbol{\xi}) = (1, \xi_1, \xi_2, \xi_1^2, \xi_2^2, \xi_1\xi_2)^T \tag{6.7}$$

The attributes are ξ_1 and ξ_2, the features are $1, \xi_1, \xi_2, \xi_1^2, \xi_2^2, \xi_1\xi_2$, $\mathcal{X} \subset \mathbb{R}^2$ and $\mathcal{F} \subset \mathbb{R}^6$.

We assume a model of the form (3.186)

$$y = \eta(\boldsymbol{\xi}; \mathbf{w}) + \epsilon \tag{6.8}$$

in which

$$\epsilon|\sigma \sim \mathcal{N}(0, \sigma^2) \tag{6.9}$$

is the error, assumed to be normally distributed (Gaussian) with mean 0 and variance σ^2. The symbol $\mathcal{N}$ denotes a normal distribution and '$\sim$' is to be read as 'distributed according to'. The vertical line '|' is to be read as 'conditioned upon' or 'given'. The latter will often be omitted if the random variables being conditioned upon are obvious from the context. The variance σ^2 is almost always unknown and must be determined along with $\mathbf{w}$. Parameters such as σ and K, which are associated with distributions or higher-level model choices, are called *hyperparameters*.

The (probability) distribution function for a Gaussian random variable Z with realisations z takes the form

$$p(z) = \mathcal{N}(\mu, \sigma^2) = \frac{1}{\sqrt{2\pi\sigma^2}} \exp\left(-\frac{1}{2\sigma^2}(z-\mu)^2\right) \tag{6.10}$$

in which $\mu = \mathbb{E}[Z]$ is the mean or expected value of Z, while $\sigma^2 = \text{var}(Z) = \mathbb{E}[Z^2] - \mathbb{E}[Z]^2$ is its variance, a measure of spread around the mean. Here, $\mathbb{E}[\cdot]$ denotes the expectation operator, defined as

$$\mathbb{E}[f(Z)] = \int_{z \in S} f(z)p(z)dz \tag{6.11}$$

for any function $f(Z)$ of the random variable Z, with *sample space* (the possible values of Z) given by S. A measure-theoretic approach to probability may sometimes be used in the analysis of machine learning methods (and we shall use it occasionally) but for the most part is not necessary.

The probability distribution over a target y at input $\boldsymbol{\xi}$ *given (conditioned on)* $\mathbf{w}$ and σ is normally distributed with mean $\eta(\boldsymbol{\xi}; \mathbf{w})$ and variance σ^2; the sum (6.8) of a constant term $\eta(\boldsymbol{\xi}; \mathbf{w})$ (at a fixed or given $\boldsymbol{\xi}$) and the normally distributed random variable ϵ leads to a normally distributed random variable y with mean $0 + \eta(\boldsymbol{\xi}; \mathbf{w})$ and variance $0 + \sigma^2$

$$p(y|\boldsymbol{\xi}, \mathbf{w}, \beta) = \mathcal{N}(\eta(\boldsymbol{\xi}; \mathbf{w}), \sigma^2) \tag{6.12}$$

It is further assumed that each y_n is drawn independently from this distribution, conditional on knowing $\mathbf{w}$ and σ, which we term *conditional independence*. Defining $\mathbf{y} = (y_1, \ldots, y_N)^T$, the distribution $p(y_1, \ldots, y_N|\Xi, \mathbf{w}, \sigma) = p(\mathbf{y}|\Xi, \mathbf{w}, \sigma)$ over the data is given by

$$\begin{aligned}
\mathcal{L} = p(\mathbf{y}|\Xi, \mathbf{w}, \sigma) &= \prod_{n=1}^{N} p(y_n|\boldsymbol{\xi}_n, \mathbf{w}, \sigma) \\
&= \prod_{n=1}^{N} \mathcal{N}(\mathbf{w}^T \boldsymbol{\phi}(\boldsymbol{\xi}_n), \sigma^2) \\
&= \prod_{n=1}^{N} \sqrt{\frac{1}{2\pi\sigma^2}} \exp\left(-\frac{1}{2\sigma^2}(y_n - \mathbf{w}^T \boldsymbol{\phi}(\boldsymbol{\xi}_n))^2\right) \\
&= \left(\frac{1}{2\pi\sigma^2}\right)^{\frac{N}{2}} \exp\left(-\frac{1}{2\sigma^2}\sum_{n=1}^{N}(y_n - \mathbf{w}^T \boldsymbol{\phi}(\boldsymbol{\xi}_n))^2\right)
\end{aligned} \tag{6.13}$$

$\mathcal{L}$ is called the *likelihood function*, i.e., the probability distribution over the data as a function of the parameters $\mathbf{w}$ and σ.

The *maximum likelihood (ML)* solution is given by maximising $p(\mathbf{y}|\Xi, \mathbf{w}, \sigma)$ over $\mathbf{w}$ and σ. That is, the values of $\mathbf{w}$ and σ that maximise the probability of the data given the assumed model. It is far easier to work with the log likelihood, i.e., the log of the likelihood function

$$\begin{aligned}
\ln \mathcal{L} &= \ln p(\mathbf{y}|\Xi, \mathbf{w}, \sigma) \\
&= \ln\left[\left(\frac{1}{2\pi\sigma^2}\right)^{\frac{N}{2}} \exp\left(-\frac{1}{2\sigma^2}\sum_{n=1}^{N}(y_n - \mathbf{w}^T \boldsymbol{\phi}(\boldsymbol{\xi}_n))^2\right)\right] \\
&= -N \ln \sigma - \frac{N}{2}\ln(2\pi) - \frac{1}{2\sigma^2}\sum_{n=1}^{N}(y_n - \mathbf{w}^T \boldsymbol{\phi}(\boldsymbol{\xi}_n))^2
\end{aligned} \tag{6.14}$$

Maximising the likelihood over $\mathbf{w}$ and σ is the same as maximising the log-likelihood since the log function is monotonically non-decreasing. Maximising $\ln \mathcal{L}$ with respect to (w.r.t.) σ is trivial and leads to its ML estimate ($\widehat{\sigma}$), which is just the sample variance of ϵ

$$\widehat{\sigma}^2 = \frac{1}{N}\sum_{n=1}^{N}(y_n - \mathbf{w}^T\boldsymbol{\phi}(\boldsymbol{\xi}_n))^2 \tag{6.15}$$

This can be calculated once an estimate of $\mathbf{w}$ is available. To obtain the latter requires solving $\nabla_{\mathbf{w}} \ln \mathcal{L} = \mathbf{0}$ in which $\nabla_{\mathbf{w}}$ is the gradient operator with respect to $\mathbf{w}$ and $\mathbf{0}$ is the zero vector. We first note, however, that only the term $-\frac{1}{2\sigma^2}\sum_{n=1}^{N}(y_n - \mathbf{w}^T\boldsymbol{\phi}(\boldsymbol{\xi}_n))^2$ in $\ln \mathcal{L}$ depends on $\mathbf{w}$. Therefore, the maximum likelihood solution is the same as the least squares solution, which is found by minimising the square error

$$\frac{1}{2}\sum_{n=1}^{N}(y_n - \mathbf{w}^T\boldsymbol{\phi}(\boldsymbol{\xi}_n))^2 \tag{6.16}$$

in which the factor of $1/2$ is a convention (it cancels upon taking the gradient). The least squares or ML solution is given by

$$\widehat{\mathbf{w}} = (\boldsymbol{\Phi}^T\boldsymbol{\Phi})^{-1}\boldsymbol{\Phi}^T\mathbf{y} \tag{6.17}$$

in which $(\boldsymbol{\Phi}^T\boldsymbol{\Phi})^{-1}\boldsymbol{\Phi}^T$ is called the *Moore-Penrose pseudoinverse*, and $\boldsymbol{\Phi}$ is called the *design matrix*, corresponding to the design points

$$\boldsymbol{\Phi} = \begin{bmatrix} 1 & \phi_1(\boldsymbol{\xi}_1) & \ldots & \phi_{K-1}(\boldsymbol{\xi}_1) \\ 1 & \phi_1(\boldsymbol{\xi}_2) & \ldots & \phi_{K-1}(\boldsymbol{\xi}_2) \\ \vdots & \vdots & \ddots & \vdots \\ 1 & \phi_1(\boldsymbol{\xi}_N) & \ldots & \phi_{K-1}(\boldsymbol{\xi}_N) \end{bmatrix} \tag{6.18}$$

We note that an alternative notation for the square error is

$$E(\mathbf{w}) = \frac{1}{2}(\mathbf{y} - \boldsymbol{\Phi}\mathbf{w})^T(\mathbf{y} - \boldsymbol{\Phi}\mathbf{w}) = \frac{1}{2}\|\mathbf{y} - \boldsymbol{\Phi}\mathbf{w}\|^2 \tag{6.19}$$

Throughout, $\|\cdot\|$ denotes the standard Euclidean norm unless otherwise specified. Likewise, the likelihood can be written as

$$\begin{aligned} p(\mathbf{y}|\boldsymbol{\Xi}, \mathbf{w}, \sigma) &= \left(\frac{1}{2\pi\sigma^2}\right)^{\frac{N}{2}} \exp\left(-\frac{1}{2\sigma^2}(\mathbf{y} - \boldsymbol{\Phi}\mathbf{w})^T(\mathbf{y} - \boldsymbol{\Phi}\mathbf{w})\right) \\ &= \left(\frac{1}{2\pi\sigma^2}\right)^{\frac{N}{2}} \exp\left(-\frac{1}{2\sigma^2}\|\mathbf{y} - \boldsymbol{\Phi}\mathbf{w}\|^2\right) \\ &= \mathcal{N}(\boldsymbol{\Phi}\mathbf{w}, \sigma^2\mathbf{I}) \end{aligned} \tag{6.20}$$

which is a multi-variate Gaussian distribution over the random vector or multi-variate random variable $\mathbf{y}$, with mean vector $\mathbf{\Phi w}^T$ and covariance matrix $\sigma^2\mathbf{I}$. The diagonal form of the covariance matrix, the components of which are defined by $\text{cov}(y_n, y_m)$, in which $\text{cov}(\cdot, \cdot)$ denotes a covariance between the two arguments, is a consequence of the conditional independence of the targets.

The ML method is elegant and simple to implement but one serious drawback is that it can *overfit*, meaning that if the model complexity is too high (roughly speaking, a high number of basis functions are used), a poor approximation to the latent function is obtained. This problem is particularly acute when there is a high level of noise or the data set is sparse (few data points). An example could be using a tenth-order polynomial function to approximate a linear polynomial latent function. The error on the training data may be low for high model complexities, but between data points the predictions can be highly inaccurate. On the other hand, if the model complexity is too low, the model will *underfit*, e.g., attempting to capture a quartic polynomial latent function with a linear function.

In both cases, the model will not *generalise* well, i.e., fail to accurately predict outputs that were not in the training data. *Generalisation error* is the error measured on an independent set of data that is not used during the training process. In general, some data, called *test data*, is withheld from training in order to assess generalisation, although how much is withheld depends upon how much data is available. One of the choices that we must make in machine learning is the definition of the error, when viewing a problem from a loss (error) minimization perspective (see Chap. 6). The square error is in some sense a natural choice (invariant to the sign of the deviation from the target and equivalent to ML), and the one used most often. There are, however, other choices that could be more suitable for certain types of problems. We shall meet other examples later.

A systematic way of selecting model complexity is *k-fold cross-validation*, in which the data $D = \{\boldsymbol{\xi}_n, y_n\}_{n=1}^N$ is partitioned into k different subsets $D_1, \ldots, D_k$. One of the k subsets D_j is withheld for validation (calculating the error) while the remaining data $D \backslash D_j$ is used for training. This process is repeated with each of the k subsets withheld once, and the errors on each of the subsets are averaged. This procedure can be carried out with different model complexities (e.g., K), with the optimal model being the one with the lowest average error.

Cross-validation, however, is problematic when the volume of data is low or when the *computational complexity* is high. The computational complexity is usually meant in the sense of time and is a measure of how time-intensive a machine learning model is run. There is also a measure of complexity in terms of storage or space, which determines the memory requirements. Alternatives to cross validation that automatically select the correct model complexity (in theory) are as follows:

1. *Regularisation*, in which an extra term is added to a loss function (or sometimes the likelihood) in an optimisation problem, such as least squares. This term penalises high model complexity by encouraging *sparsity*, meaning that it drives some of the weights towards zero, essentially pruning the model by removing terms that make an insignificant contribution.

2. *Bayesian inference*, in which the model parameters are averaged over or integrated out (marginalisation), automatically selecting the correct number of parameters (in theory).

We next discuss the first of these alternatives. The latter approach is discussed later in detail, in the form of Bayesian linear regression and Gaussian process models.

6.3 Regularised Linear Regression

In the ML or least squares estimates of $\mathbf{w}$, we attempt to minimise the square error $E(\mathbf{w})$ given in (6.19). In *regularised* least squares, we instead minimise

$$\widetilde{E}(\mathbf{w}) = E(\mathbf{w}) + \lambda r(\mathbf{w}) \tag{6.21}$$

in which the function $r(\mathbf{w})$ is called a *regularisation* or *penalty* term and λ is called the *regularisation coefficient*. The most common forms of $r(\mathbf{w})$ are

1. $r(\mathbf{w}) = \frac{1}{2}\|\mathbf{w}\|_2^2 = \frac{1}{2}\mathbf{w}^T\mathbf{w}$, leading to *ridge regression*.
2. $r(\mathbf{w}) = \|\mathbf{w}\|_1 = \sum_{i=0}^{K-1}|w_k|$, leading to *least absolute shrinkage and selection operator (LASSO) regression*.

The Manhattan or Taxicab norm $\|\cdot\|_1$ leads to greater sparsity over the standard Euclidean norm, meaning that the values of w_k are driven closer to zero. To see this informally, we note that the penalty term can be removed from the objective function (square error) and instead incorporated as a constraint of the type $\|\mathbf{w}\| \leq k$ or $\|\mathbf{w}\|_1 \leq k$ for some $k \in \mathbb{R}$. For $\mathbf{w} \in \mathbb{R}^2$, the shape of the first is circular, while $\|\mathbf{w}\|_1 \leq k$ is diamond-shaped, forcing values of $\mathbf{w}$ closer to the axes in the plane, thereby driving one or more of the components of $\mathbf{w}$ closer to zero than the constraint $\|\mathbf{w}\| \leq k$.

An advantage of ridge regression (the most widely used) is that it has a closed-form solution $\widetilde{\mathbf{w}}$, readily obtained by taking the gradient of (6.21) and setting it to $\mathbf{0}$

$$\widetilde{\mathbf{w}} = (\mathbf{\Phi}^T\mathbf{\Phi} + \lambda\mathbf{I})^{-1}\mathbf{\Phi}^T\mathbf{y}. \tag{6.22}$$

6.4 Locally Linear and Locally Polynomial Regression

Local linear regression, also known as *locally weighted linear regression* is a non-parametric technique in which we assume that locally (in small regions of input space) the relationship between the output y and the input $\boldsymbol{\xi}$ is linear. Let us focus on the scalar input case $\boldsymbol{\xi} = \xi$ for the purposes of illustration, with data y_n, ξ_n, $n = 1, \ldots, N$. It is convenient to centre the solution around the input value ξ_* at which a prediction is sought, so we can write the model, without any error term, as

$$y = \eta(\xi) = w_0 + w_1(\xi_* - \xi) + \mathcal{O}((\xi_* - \xi)^2) \tag{6.23}$$

This is nothing more than a Taylor expansion of the latent function $\eta(\xi)$ around the point ξ_*, neglecting terms of order $(\xi_* - \xi)^2$ and higher; $\mathcal{O}$ represents the order symbol. By centering around ξ_*, we obtain $\eta(\xi_*) = w_0$, up to higher-order terms, so that the coefficient w_0 is the prediction for any value of ξ_*. At each ξ_*, on the other hand, the value of w_0 is different. Rather than a locally linear fit, we can use quadratic, cubic or higher-order polynomial fits, with the general name *local polynomial regression*.

To determine w_0 and w_1, we minimise a loss function, which is weighted towards those targets y_n such that the corresponding ξ_n are closest to ξ_*. This is achieved by introducing *weights* k_n (not to be confused with the weights w_k) and solving a *weighted least squares* problem

$$\text{argmin}_{w_0,w_1} \sum\nolimits_{n=1}^{N} k_n(y_n - [w_0 + w_1(\xi_* - \xi_n)])^2 \tag{6.24}$$

The resulting w_0 approximates $\eta(\xi_*)$, and is valid only at ξ_*. Defining $\mathbf{w} = (w_0, w_1)^T$, (6.24) is equivalent to

$$\text{argmin}_{\mathbf{w}} \frac{1}{2}(\mathbf{y} - \mathbf{X}\mathbf{w})^T \mathbf{K}(\mathbf{y} - \mathbf{X}\mathbf{w}) \tag{6.25}$$

in which

$$\mathbf{X} = \begin{bmatrix} 1 & \xi_* - \xi_1 \\ 1 & \xi_* - \xi_2 \\ \vdots & \vdots \\ 1 & \xi_* - \xi_N \end{bmatrix} \quad \mathbf{K} = \begin{bmatrix} k_1 & 0 & \dots & 0 \\ 0 & k_2 & \dots & 0 \\ \vdots & \vdots & \ddots & \vdots \\ 0 & 0 & \dots & k_N \end{bmatrix} \tag{6.26}$$

This weighted least square problem has a weight matrix **K** and design matrix **X**. The solution $\mathbf{w}_l$ is readily obtained as

$$\mathbf{w}_l = (\mathbf{X}^T\mathbf{K}\mathbf{X})^{-1}\mathbf{X}^T\mathbf{K}\mathbf{y} \tag{6.27}$$

The first coefficient gives w_0 and therefore the solution.

The key component is the weight matrix, which is defined by the k_n. We can generate the values of k_n using a *kernel function*, which is a function of two arguments, ξ_n and ξ_*. We delay a detailed discussion on kernel functions until Sect. 6.6. There are a number of common kernel functions in use, with the most popular being the *Gaussian kernel:*

$$k_n = k(\xi_n, \xi_*) = \frac{1}{\sqrt{2\pi}} \exp\left(\frac{-(\xi_* - \xi_n)^2}{2h^2}\right) \tag{6.28}$$

in which h is called the *bandwidth*. Clearly, as ξ_* approaches ξ_n, the kernel value (weight) increases. The bandwidth h determines how many points ξ_n make a significant contribution; for a fixed $\xi - \xi_*$, a smaller h leads to a smaller weight.

Locally weighted linear regression is also known as *locally estimated scatterplot smoothing* (LOESS) or locally weighted scatterplot smoothing (LOWESS). Another often used weight function is the *tri-weight* function, which is frequently used in the definition of LOESS/LOWESS

$$k(\xi_*, \xi_n) = W\left(\frac{|\xi_* - \xi_n|}{h}\right)$$
$$W(u) = \begin{cases} (1-u^3)^3 & 0 \le u < 1 \\ 0 & \text{otherwise} \end{cases} \tag{6.29}$$

The most important choice to make is the bandwidth, which is usually done *via* a k-fold cross-validation, especially with $k = N$, which we call *leave one out cross-validation* (LOOCV).

6.5 Bayesian Linear Regression

We now outline an alternative approach to linear regression, from a *Bayesian* perspective. In theory, a Bayesian treatment prevents overfitting and automatically determines model complexity using the training data alone, avoiding costly procedures such as cross-validation. It is particularly appropriate when data sets are small, and additionally in applications requiring a measure of uncertainty in the predictions.

Again, we assume a model $y = \mathbf{w}^T\phi(\boldsymbol{\xi}) + \epsilon$, with $\epsilon|\sigma \sim \mathcal{N}(0, \sigma^2)$. A *prior* probability distribution is placed over the weights $\mathbf{w}$, usually, in the form of a Gaussian

$$p(\mathbf{w}) = \mathcal{N}(\mathbf{m}_0, \mathbf{S}_0) \tag{6.30}$$

in which $\mathbf{m}_0$ and $\mathbf{S}_0$ are, respectively, the mean and covariance matrix. For the purposes of illustration, it is further assumed that we know the noise variance σ^2. Applying *Baye's rule* (a restatement of the laws of conditional probability) yields the following

$$p(\mathbf{w}|\mathbf{y}, \Xi)) = \frac{p(\mathbf{y}|\mathbf{w}, \Xi)p(\mathbf{w})}{p(\mathbf{y}|\Xi)} \tag{6.31}$$

On the left-hand side of this equation is the *posterior* probability distribution over the weights. The numerator on the right-hand side is the product of the data likelihood and the prior, while the denominator is the *marginal* probability distribution or evidence $\int_{\mathbf{w}} p(\mathbf{y}, \mathbf{w}|\Xi)$. The basic principle is to use a prior, informed by some knowledge of the underlying system or simply an educated guess, along with the data to obtain an improved *probability distribution* over the weights. Thus, rather than a point estimate as in ML (see below), we obtain a distribution over the outputs $\mathbf{w}^T\boldsymbol{\phi}(\boldsymbol{\xi})$.

The marginal is usually intractable in Bayesian inference and requires the use of sampling methods. In many cases, however, it is possible to treat it as simply a multiplicative constant. For our purposes, since it does not depend on $\mathbf{w}$, the relationship

$$p(\mathbf{w}|\mathbf{y}, \Xi)) \propto p(\mathbf{y}|\mathbf{w}, \Xi)p(\mathbf{w}) \tag{6.32}$$

will suffice. A special case of the prior (6.30) is

$$p(\mathbf{w}|\beta) = \mathcal{N}(\mathbf{0}, \beta^2\mathbf{I}) \tag{6.33}$$

in which the mean is $\mathbf{0}$ and the covariance is diagonal, i.e., the w_k are not correlated and have a shared variance σ^2, treated as a hyperparameter.

Both the likelihood and the prior are Gaussian, and the product of two Gaussians is again Gaussian. The exponent in the likelihood (6.20) can be expanded as follows:

$$\|\mathbf{y} - \boldsymbol{\Phi}\mathbf{w}\|^2 = (\mathbf{w} - \widehat{\mathbf{w}})^T(\boldsymbol{\Phi}^T\boldsymbol{\Phi})(\mathbf{w} - \widehat{\mathbf{w}}) + (\mathbf{y} - \boldsymbol{\Phi}\widehat{\mathbf{w}})^T(\mathbf{y} - \boldsymbol{\Phi}\widehat{\mathbf{w}}) \tag{6.34}$$

in which $\widehat{\mathbf{w}}$ is the ML solution. The second term on the right-hand side does not depend on $\mathbf{w}$, so that

$$p(\mathbf{y}|\mathbf{w}, \sigma) \propto \exp\left(-\frac{1}{2}(\mathbf{w} - \widehat{\mathbf{w}})^T(\sigma^{-2}\boldsymbol{\Phi}^T\boldsymbol{\Phi})(\mathbf{w} - \widehat{\mathbf{w}})\right) \tag{6.35}$$

Considered as a distribution over $\mathbf{w}$, the right-hand side of (6.35) is equal to

$$\mathcal{N}(\widehat{\mathbf{w}}, \sigma^2(\boldsymbol{\Phi}^T\boldsymbol{\Phi})^{-1}) \tag{6.36}$$

up to a normalising constant that ensures a total probability of 1. The product of this distribution and the prior then leads to the distribution (using a standard result for the product of Gaussians)

$$\begin{gathered} p(\mathbf{w}|\mathbf{y}) = \mathcal{N}(\mathbf{m}, \mathbf{S}) \\ \mathbf{S} = (\sigma^{-2}\boldsymbol{\Phi}^T\boldsymbol{\Phi} + \beta^{-2}\mathbf{I})^{-1}, \quad \mathbf{m} = \sigma^{-2}\mathbf{S}\boldsymbol{\Phi}^T\mathbf{y} \end{gathered} \tag{6.37}$$

Before we outline the procedure required to make predictions, we introduce some important terminology and explain some of the modelling choices available to us.

A *point estimate* of $\mathbf{w}$ is a single-valued estimate, rather than a distribution over $\mathbf{w}$. A point estimate is readily obtained by maximising the posterior $p(\mathbf{w}|\mathbf{y}, \Xi)$ in (6.31). This leads to a *maximum a posteriori (MAP)* solution for $\mathbf{w}$, which is an alternative to the ML point estimate. Since $p(\mathbf{w}|\mathbf{y}, \Xi)) \propto p(\mathbf{y}|\mathbf{w}, \Xi)p(\mathbf{w})$, the MAP problem is the same as maximising

$$\ln p(\mathbf{w}|\mathbf{y}) = -\frac{1}{2\sigma^2}(\mathbf{y} - \mathbf{\Phi w})^T(\mathbf{y} - \mathbf{\Phi w}) - \frac{1}{2\beta^2}\mathbf{w}^T\mathbf{w} + \ldots \tag{6.38}$$

in which $+\ldots$ indicates terms that do not depend on $\mathbf{w}$. Thus, a Gaussian prior leads to ridge regression, which explains why the latter method ameliorates overfitting.

Suppose we wish to predict the output y for some test input $\boldsymbol{\xi}$. We require $p(y|\mathbf{y})$, which we can find by marginalising over $\mathbf{w}$ as follows

$$p(y|\mathbf{y}) = \int p(y|\mathbf{w}, \sigma)p(\mathbf{w}|\mathbf{y})d\mathbf{w} \tag{6.39}$$

Recall that $p(y|\mathbf{w}, \sigma) = \mathcal{N}(\mathbf{w}^T\phi(\xi), \sigma^2)$ and $p(\mathbf{w}|\mathbf{y}) = \mathcal{N}(\mathbf{m}, \mathbf{S})$. The integration in (6.39) can be performed exactly. Given a marginal Gaussian distribution for a random vector $\mathbf{z}$ and a conditional Gaussian distribution for a random vector $\mathbf{u}$ conditioned on $\mathbf{z}$ in the form

$$\begin{aligned} p(\mathbf{z}) &= \mathcal{N}(\boldsymbol{\mu}, \boldsymbol{\Lambda}) \\ p(\mathbf{u}|\mathbf{z}) &= \mathcal{N}(\mathbf{Az} + \mathbf{b}, \mathbf{L}) \end{aligned} \tag{6.40}$$

the marginal distribution of $\mathbf{u}$ is given by

$$p(\mathbf{u}) = \mathcal{N}(\mathbf{A}\boldsymbol{\mu} + \mathbf{b}, \mathbf{L} + \mathbf{A}\boldsymbol{\Lambda}\mathbf{A}^T) \tag{6.41}$$

Setting $\mathbf{u} = y$, $\mathbf{z} = \mathbf{w}$, $\mathbf{A} = \phi(\xi)^T$, $\mathbf{b} = 0$, $\mathbf{L} = \sigma^2$, $\boldsymbol{\mu} = \mathbf{m}$ and $\boldsymbol{\Lambda} = \mathbf{S}$ yields

$$p(y|\mathbf{y}) = \mathcal{N}\left(\phi(\mathbf{x})^T\mathbf{m}, \sigma^2 + \phi(\xi)^T\mathbf{S}\phi(\xi)\right) \tag{6.42}$$

There are two components in the variance: one due to the error ϵ and the second due to the uncertainty in $\mathbf{w}$. As the number of samples N grows, the second term decreases.

In a fully Bayesian treatment, we would introduce prior distributions over the hyperparameters σ and β, and make predictions by marginalising over $\mathbf{w}$ *and* $\{\sigma, \beta\}$. In most cases, exact solutions are not possible since an explicit form for the posterior is not available. Instead, either stochastic sampling techniques such as Markov Chain Monte Carlo (MCMC) [1] or approximate analytical methods such as variational inference [1] can be used.

MCMC is highly computationally expensive and is therefore usually avoided. Variational methods approximate the posterior in terms of some assumed factored distribution and are more scalable, but less accurate than a maximum likelihood, which they approximate. They are used when the likelihood is particularly difficult to optimise, which is the case when there are hidden or latent variables involved. In most cases, a maximum likelihood will suffice.

6.5.1 The Evidence Approximation for Linear Regression

Another approximation for linear regression besides maximum likelihood involves setting the hyperparameters to specific values obtained by maximising the marginal likelihood or evidence $p(\mathbf{y}|\sigma, \beta)$ after integrating over the parameters $\mathbf{w}$. This framework is known in the statistics literature as *empirical Bayes*, type 2 maximum likelihood or generalised maximum likelihood. In the machine learning literature, it is also called the *evidence approximation*.

If we introduce prior distributions over the hyperparameters β and σ, the predictive distribution is obtained by marginalising over $\mathbf{w}$, β and σ, so that

$$p(y|\mathbf{y}) = \int\int\int p(y|\mathbf{w}, \beta)p(\mathbf{w}|\mathbf{y}, \sigma, \beta)p(\sigma, \beta|\mathbf{y})d\mathbf{w}d\sigma d\beta \tag{6.43}$$

We already know that $p(\mathbf{w}|\mathbf{y}, \sigma, \beta) = \mathcal{N}(\mathbf{m}, \mathbf{S})$ (since σ and β are now unknown we are being more explicit in the notation) and $p(y|\mathbf{w}, \sigma) = \mathcal{N}(\mathbf{w}^T\boldsymbol{\phi}(\boldsymbol{\xi}), \sigma^2)$.

If the posterior distribution over the hyperparameters $p(\sigma, \beta|\mathbf{y})$ is sharply peaked around some values $\widehat{\sigma}$ and $\widehat{\beta}$, the predictive distribution $p(y|\mathbf{y})$ can be approximated by marginalising over $\mathbf{w}$ while setting $\sigma = \widehat{\sigma}$ and $\beta = \widehat{\beta}$

$$p(y|\mathbf{y}) \approx p(y|\mathbf{y}, \widehat{\sigma}, \widehat{\beta}) = \int p(y|\mathbf{w}, \sigma)p(\mathbf{w}|\mathbf{y}, \widehat{\alpha}, \widehat{\beta})d\mathbf{w} \tag{6.44}$$

This is the same as the previous result with specific values for σ and β. From Baye's theorem, the posterior distribution for σ and β satisfies

$$p(\sigma, \beta|\mathbf{y}) \propto p(\mathbf{y}|\sigma, \beta)p(\sigma, \beta) \tag{6.45}$$

If the prior $p(\sigma, \beta)$ is relatively flat, then the values of $\widehat{\sigma}$ and $\widehat{\beta}$ are obtained by maximising the marginal likelihood function $p(\mathbf{y}|\sigma, \beta)$. The marginal likelihood function, also called the *evidence function* or *model evidence* can be obtained by integrating over the weight parameters $\mathbf{w}$

$$p(\mathbf{y}|\sigma, \beta) = \int p(\mathbf{y}|\mathbf{w}, \sigma)p(\mathbf{w}|\beta)d\mathbf{w} \tag{6.46}$$

with $p(\mathbf{y}|\mathbf{w}, \beta) = \mathcal{N}(\boldsymbol{\Phi}\mathbf{w}, \sigma^2\mathbf{I})$ and $p(\mathbf{w}|\alpha) = \mathcal{N}(\mathbf{0}, \beta^2\mathbf{I})$. Thus, $p(\mathbf{y}|\sigma, \beta)$ takes the form

$$p(\mathbf{y}|\sigma, \beta) = \left(\frac{1}{2\pi\sigma^2}\right)^{N/2}\left(\frac{1}{2\pi\beta^2}\right)^{M/2}\int \exp\left(\frac{1}{2\sigma^2}\|\mathbf{y} - \boldsymbol{\Phi}\mathbf{w}\|_2^2 + \frac{1}{2\beta^2}\|\mathbf{w}\|_2^2\right)d\mathbf{w} \tag{6.47}$$

Note that the exponent is the regularised sum of squares used in ridge regression. Skipping the algebraic details, the final result is

$$p(\mathbf{y}|\sigma,\beta) = -\frac{M}{2}\ln\beta - \frac{N}{2}\ln\sigma - \frac{\|\mathbf{y}-\mathbf{\Phi m}\|_2^2}{2\sigma^2} + \frac{\|\mathbf{m}\|_2^2}{2\beta^2} - \frac{1}{2}\ln|\mathbf{S}^{-1}| - \frac{N}{2}\ln(2\pi) \tag{6.48}$$

Optimising the log evidence function above over σ and β will yield $\widehat{\sigma}$ and $\widehat{\beta}$. We denote the eigenvalues of $\sigma^{-2}\mathbf{\Phi}^T\mathbf{\Phi}$ by λ_i, i.e.,

$$\lambda_i\mathbf{u}_i = \sigma^{-2}\mathbf{\Phi}^T\mathbf{\Phi}\mathbf{u}_i, \quad i = 1,\ldots,M \tag{6.49}$$

Then $\widehat{\sigma}$ and $\widehat{\beta}$ are given by

$$\widehat{\beta}^{-2} = \frac{\gamma}{\mathbf{m}^T\mathbf{m}}, \quad \widehat{\sigma}^2 = \frac{1}{N-\gamma}\sum\nolimits_{n=1}^{N}\left(y_n - \mathbf{m}^T\boldsymbol{\phi}(\boldsymbol{\xi}_n)\right)^2 \tag{6.50}$$

in which γ is called the *effective number of parameters* and is defined as

$$\gamma = \sum\nolimits_{i=1}^{M}\frac{\lambda_i}{\lambda_i + \widehat{\beta}^{-2}} \tag{6.51}$$

The effective number of parameters essentially determines the number of important parameters in the model. The reader may have noticed an immediate problem: the expressions for γ and $\widehat{\beta}$ depend upon each other. $\mathbf{m}$ also depends on $\widehat{\sigma}$ and $\widehat{\beta}$, while $\widehat{\sigma}$ depends on $\mathbf{m}$ and γ. We therefore require an iterative solution, which proceeds as follows:

1. Find eigenvalues λ_i' of $\mathbf{\Phi}^T\mathbf{\Phi}$ and initialise $\widehat{\sigma}$ and $\widehat{\beta}$.
2. Find $\lambda_i = \widehat{\sigma}^{-2}\lambda_i'$, $\mathbf{m}$ and γ.
3. Use $\mathbf{m}$ and γ to re-estimate $\widehat{\sigma}$ and $\widehat{\beta}$.
4. Return to step 2 until convergence.

Notice that the first step, which finds eigenvalues of $\mathbf{\Phi}^T\mathbf{\Phi}$ rather than $\sigma^{-2}\mathbf{\Phi}^T\mathbf{\Phi}$ requires only one execution.

6.6 Kernel Regression

A powerful nonlinear and non-parametric method can be obtained from the so-called *dual formulation* of linear regression, by employing kernel substitution. In linear regression, we specify a feature map $\boldsymbol{\phi}: \mathcal{X} \to \mathcal{F}$ defined by the basis functions. Linear regression can be rewritten entirely in terms of *dot products* $\boldsymbol{\phi}(\boldsymbol{\xi})^T\boldsymbol{\phi}(\boldsymbol{\xi}')$ of the feature map, for inputs $\boldsymbol{\xi},\boldsymbol{\xi}' \in \mathbb{R}^l$. This alternative formulation is called the dual representation or formulation.

Recall that in ridge regression (6.21) a penalised square error is mimimised, with penalty $r(\mathbf{w}) = \|\mathbf{w}\|_2^2$. Setting the gradient of the objective function to $\mathbf{0}$, it is easy to show that $\mathbf{w} = -\lambda^{-1}\mathbf{\Phi}^T(\mathbf{\Phi w} - \mathbf{y})$, in which we remind the reader that $\mathbf{\Phi}$ is the design matrix, the rows of which are given by $\phi(\xi_n)$, $n = 1, \ldots, N$. We now define a vector $\mathbf{a}$ by $\mathbf{w} = \mathbf{\Phi}^T \mathbf{a}$, so that it satisfies

$$\mathbf{a} = -\frac{(\mathbf{\Phi w} - \mathbf{y})}{\lambda} \tag{6.52}$$

and has coefficients

$$a_n(\mathbf{w}) = -\frac{(\mathbf{w}^T \phi(\xi_n) - y_n)}{\lambda}, \quad n = 1, \ldots, N \tag{6.53}$$

Substituting $\mathbf{w} = \mathbf{\Phi}^T \mathbf{a}$ into the objective function for ridge regression then leads to an alternative formulation

$$\text{argmin}_{\mathbf{a}} \ \frac{1}{2}\mathbf{y}^T\mathbf{y} + \frac{1}{2}\mathbf{a}^T\mathbf{KKa} - \mathbf{a}^T\mathbf{Ky} + \frac{\lambda}{2}\mathbf{a}^T\mathbf{Ka} \tag{6.54}$$

in which $\mathbf{K}$ is the *Gram matrix* or *kernel matrix*, defined as $\mathbf{K} = \mathbf{\Phi\Phi}^T$, with elements

$$K_{nm} = k(\xi_n, \xi_m) = \phi(\xi_n)^T \phi(\xi_m), \quad n, m = 1, \ldots, N \tag{6.55}$$

Notice that $\mathbf{K}$ is symmetric, and that its elements are *dot products of the feature map* $\phi(\xi)$. $k(\xi, \xi') = \phi(\xi)^T\phi(\xi')$ is an example of a *kernel function*, which is a function of two arguments ξ, ξ' satisfying the following properties:

1. Non-negative: $k(\xi, \xi') \geq 0$ for any ξ and ξ'
2. Symmetric: $k(\xi, \xi') = k(\xi', \xi)$

It must also satisfy some technical conditions related to boundedness, which are not important in the present context.

If $k(\xi, \xi')$ depends only on $\xi - \xi'$, i.e., $k(\xi, \xi') = k(\xi - \xi')$, it is called *stationary*, since it can define the covariance function of a wide-sense stationary process. If, moreover, it depends only on $\|\xi - \xi'\|$, it is called *isotropic* or *homogeneous*, since its variation is the same in all directions around ξ'. The most famous example is the Gaussian or *squared exponential* kernel

$$k(\xi, \xi') = \frac{1}{(2\pi)^{1/2}} \exp\left(-\frac{\|\xi - \xi'\|^2}{2\theta^2}\right) \tag{6.56}$$

where θ is a scale factor, bandwidth or correlation length. This is an isotropic kernel. A non-isotropic version called the squared exponential *automatic relevance determination* (SEARD) kernel is given by

$$k(\xi, \xi') = \frac{1}{(2\pi)^{1/2}} \exp\left(-(\xi - \xi)^T \Theta (\xi - \xi)\right) \tag{6.57}$$

in which $\Theta = \text{diag}(\theta_1, \ldots, \theta_l)$ is a diagonal matrix with a different correlation lengths θ_i for each component ξ_i of the input. This allows for the inputs to have different degrees of influence on the kernel values.

Differentiating (6.54) w.r.t. $\mathbf{a}$ and solving yields

$$\mathbf{a} = (\mathbf{K} + \lambda \mathbf{I}_N)^{-1}\mathbf{y} \tag{6.58}$$

in which $\mathbf{I}_N$ is the $N \times N$ identity. Using the definition of the approximating function and the definition of $\mathbf{a}$, we obtain $\eta(\boldsymbol{\xi}; \mathbf{w}) = \boldsymbol{\phi}(\boldsymbol{\xi})^T \boldsymbol{\Phi}^T \mathbf{a}$, whereupon

$$\begin{aligned} \eta(\boldsymbol{\xi}; \mathbf{w}) &= \mathbf{k}(\boldsymbol{\xi})^T (\mathbf{K} + \lambda \mathbf{I}_N)^{-1}\mathbf{y} \\ \mathbf{k}(\boldsymbol{\xi}) &= (k(\boldsymbol{\xi}, \boldsymbol{\xi}_1), \ldots, k(\boldsymbol{\xi}, \boldsymbol{\xi}_N))^T \end{aligned} \tag{6.59}$$

This is called the *dual formulation* for ridge regression, and the solution is given entirely in terms of values of the kernel function $k(\boldsymbol{\xi}, \boldsymbol{\xi}') = \boldsymbol{\phi}(\boldsymbol{\xi})^T \boldsymbol{\phi}(\boldsymbol{\xi}')$. A number of linear models can be reformulated in this way, allowing for use of the *kernel trick* or *kernel substitution*. This consists of replacing the linear kernel $\boldsymbol{\phi}(\boldsymbol{\xi})^T \boldsymbol{\phi}(\boldsymbol{\xi}')$ with a general (equivalent) kernel function $\widehat{k}(\boldsymbol{\xi}, \boldsymbol{\xi}') = \boldsymbol{\psi}(\boldsymbol{\xi})^T \boldsymbol{\psi}(\boldsymbol{\xi}')$, corresponding to a feature map $\boldsymbol{\psi}(\boldsymbol{\xi})$ that is *not* explicitly specified, since only kernel values are needed in (6.59).

The original problem for finding $\mathbf{w}$ required inversion of the $K \times K$ matrix $\boldsymbol{\Phi}^T \boldsymbol{\Phi} + \lambda \mathbf{I}_K$, whereas finding $\mathbf{a}$ requires inversion of the $N \times N$ matrix $\mathbf{K} + \lambda \mathbf{I}_N$, which is more costly since typically $N \gg K$. However, the dual formulation allows us to build more sophisticated models. With appropriate choices of the kernel function, the dimension K of the feature space ($\boldsymbol{\phi}(\boldsymbol{\xi}) \in \mathbb{R}^K$) can be made very high, even infinite, without having to explicitly specify any basis functions (i.e., the feature map). This leads to a richer class of models, potentially able to capture more complex underlying functions.

The reader may have noticed the similarities between kernel regression and local polynomial regression, in that both are weighted linear regressions. In fact, kernel regression corresponds to the case of using a zeroth-order local fit, rather than a linear or higher-order polynomial.

6.7 Univariate Gaussian Process Models

In Gaussian process modelling, a *prior distribution* is placed over the latent function $\eta(\boldsymbol{\xi})$ in the form of a Gaussian process (GP), a particular type of random process. The GP is indexed by the inputs $\boldsymbol{\xi} \in \mathcal{X}$ and can be interpreted in two equivalent ways. For a fixed index $\boldsymbol{\xi}$, the GP defines a random variable, with an associated Gaussian distribution. Realisations of these random variables for all $\boldsymbol{\xi} \in \mathcal{X}$, on the other hand, defines a deterministic function of $\boldsymbol{\xi}$, which is a *realisation* of the GP, also called a *sample path*. GPs have an extremely convenient property that lies at the heart of GP models: the joint distribution of the random variables defined by an arbitrary finite collection of the indices $\boldsymbol{\xi}_1, \ldots, \boldsymbol{\xi}_N$ is a multivariate normal.

We shall use the additive noise assumption, namely $y = \eta(\boldsymbol{\xi}) + \epsilon$. The GP prior over $\eta(\boldsymbol{\xi})$ is

$$\eta(\boldsymbol{\xi})|\boldsymbol{\theta} \sim \mathcal{GP}\left(m(\boldsymbol{\xi}), k(\boldsymbol{\xi}, \boldsymbol{\xi}'|\boldsymbol{\theta})\right) \tag{6.60}$$

in which the first and second arguments are the mean and covariance (or kernel) function, respectively, given by

$$\begin{aligned} m(\boldsymbol{\xi}) &= \mathbb{E}[\eta(\boldsymbol{\xi})] \\ k(\boldsymbol{\xi}, \boldsymbol{\xi}'|\boldsymbol{\theta}) &= \text{cov}\left(\eta(\boldsymbol{\xi}), \eta(\boldsymbol{\xi}')\right) = \mathbb{E}[(\eta(\boldsymbol{\xi}) - m(\boldsymbol{\xi}))(\eta(\boldsymbol{\xi}') - m(\boldsymbol{\xi}'))] \end{aligned} \tag{6.61}$$

$\boldsymbol{\theta}$ is a vector of *hyperparameters* that fully specify the covariance function, as in the kernel (6.56), or its non-isotropic ARD version (6.57). The noise is assumed to be i.i.d.

$$\epsilon \sim \mathcal{GP}\left(0, \sigma^2\delta(\boldsymbol{\xi}, \boldsymbol{\xi}')\right) \tag{6.62}$$

in which $\delta(\boldsymbol{\xi}, \boldsymbol{\xi}')$ is the delta function and σ^2 is the noise variance. The model for the data is then

$$y \sim \mathcal{GP}\left(m(\boldsymbol{\xi}), k(\boldsymbol{\xi}, \boldsymbol{\xi}'|\boldsymbol{\theta}) + \sigma^2\delta(\boldsymbol{\xi}, \boldsymbol{\xi}')\right) \tag{6.63}$$

Strictly speaking, (6.60) represents an approximating function for the true function $\eta(\boldsymbol{\xi})$, but we will not distinguish between the two in the notation. In almost all cases, the hyperparameters $\boldsymbol{\theta}$ are unknown and must be inferred from the data; in fact, this is the main learning task. As in all Bayesian methods, the data is used to update the prior distribution, leading to a predictive *posterior GP distribution* over the outputs, meaning that the mean and covariance functions are updated the prior and posterior GPs specify *distributions over functions* and the latent function can be approximated by the mean function of the posterior GP.

The mean function of the prior is usually assumed to be identically 0, by virtue of centring the data y_n. Constant, polynomial or other functions can also be used, in which case the weights in a basis function expansion also need to be inferred. For the purposes of illustration, it is assumed that $m(\boldsymbol{\xi}) \equiv 0$. The non-zero case will be covered in Sect. 7.4.3 in the next chapter when we discuss GP autoregression for sequential data.

The choice of the kernel function constitutes the main assumption. Different kernel functions will generate different types of functions in terms of smoothness. There are a number of common kernels in use, including the SEARD, the linear kernel, the Matérn set of kernels and polynomial kernels [2]

$$k(\boldsymbol{\xi}, \boldsymbol{\xi}'|\boldsymbol{\theta}) = \theta_0 e^{-r^2}, \quad r^2 = (\boldsymbol{\xi} - \boldsymbol{\xi})^T \Theta (\boldsymbol{\xi} - \boldsymbol{\xi}) \tag{6.64a}$$

$$k(\boldsymbol{\xi}, \boldsymbol{\xi}'|\boldsymbol{\theta}) = \theta_0 r^2 \tag{6.64b}$$

$$k(\boldsymbol{\xi}, \boldsymbol{\xi}'|\boldsymbol{\theta}) = \theta_0 f_\nu(r) \exp\left(-\sqrt{\nu} r\right) \tag{6.64c}$$

$$k(\boldsymbol{\xi}, \boldsymbol{\xi}'|\boldsymbol{\theta}) = (1 + \theta_0 r^2)^p \tag{6.64d}$$

respectively, in which $\Theta = \text{diag}(\theta_1, \ldots, \theta_l), \boldsymbol{\theta} = (\theta_0, \ldots \theta_l)^T, p \in \{0, 1, 2, \ldots\}, \nu \in \{1, 3, 5\}$, and

$$f_1(r) = 1, \quad f_2(r) = 1 + \sqrt{3}r, \quad f_5(r) = 1 + \sqrt{5}r + 5r^2/3 \tag{6.65}$$

Linear combinations of any valid kernels also define valid kernels.

The interchangeability of the terminology 'kernel' and 'covariance' is a consequence of the fact that these types of functions share key properties (positive semidefinite and symmetric). Furthermore, a GP model can be viewed as kernel method, namely a generalisation of Bayesian linear regression in a dual form. In the latter method, the output $\mathbf{w}^T \boldsymbol{\phi}(\boldsymbol{\xi})$ is, formally, a GP index by $\boldsymbol{\xi}$, with zero mean and covariance function

$$k(\boldsymbol{\xi}, \boldsymbol{\xi}') = \beta^2 \boldsymbol{\phi}(\boldsymbol{\xi})^T \boldsymbol{\phi}(\boldsymbol{\xi}') \tag{6.66}$$

In essence, A GP model replaces this linear kernel with an equivalent kernel $k(\boldsymbol{\xi}, \boldsymbol{\xi}'|\boldsymbol{\theta})$.

Defining $\boldsymbol{\eta} = (\eta(\boldsymbol{\xi}_1), \ldots, \eta(\boldsymbol{\xi}_N))^T$, by the properties of a GP, $p(\boldsymbol{\eta}|\boldsymbol{\theta}) = \mathcal{N}(\mathbf{0}, \mathbf{K}(\boldsymbol{\theta}))$, with covariance (kernel) matrix $\mathbf{K}(\boldsymbol{\theta})$ having entries K_{nm}

$$K_{nm}(\boldsymbol{\theta}) = k(\boldsymbol{\xi}_n, \boldsymbol{\xi}_m|\boldsymbol{\theta}), \quad n, m = 1, \ldots, N \tag{6.67}$$

The likelihood $p(\mathbf{y}|\boldsymbol{\theta}, \sigma)$ is the probability distribution over the targets conditioned on the hyperparameters, which is again a multivariate Gaussian

$$p(\mathbf{y}|\boldsymbol{\theta}, \sigma) = \int p(\mathbf{y}|\boldsymbol{\eta}, \sigma)\, p(\boldsymbol{\eta}|\boldsymbol{\theta})\, d\boldsymbol{\eta} = \mathcal{N}(\mathbf{0}, \mathbf{K}(\boldsymbol{\theta}) + \sigma^2 \mathbf{I}) \tag{6.68}$$

The joint distribution $p(\eta(\boldsymbol{\xi}), \mathbf{y}|\boldsymbol{\theta}, \sigma)$ of $\mathbf{y}$ and the latent function value $\eta(\boldsymbol{\xi})$ corresponding to a non-training point $\boldsymbol{\xi} \in \mathcal{X}$ has the distribution $\mathcal{N}(\mathbf{0}, \mathbf{K}'(\boldsymbol{\theta}))$, where

$$\begin{aligned} \mathbf{K}'(\boldsymbol{\theta}) &= \left[\begin{array}{c|c} \mathbf{K}(\boldsymbol{\theta}) + \sigma^2 \mathbf{I} & \mathbf{k}(\boldsymbol{\xi}) \\ \hline \mathbf{k}(\boldsymbol{\xi})^T & k(\boldsymbol{\xi}, \boldsymbol{\xi}|\boldsymbol{\theta}) \end{array}\right] \\ \mathbf{k}(\boldsymbol{\xi}) &= (k(\boldsymbol{\xi}_1, \boldsymbol{\xi}|\boldsymbol{\theta}), \ldots, k(\boldsymbol{\xi}_N, \boldsymbol{\xi}|\boldsymbol{\theta}))^T \end{aligned} \tag{6.69}$$

By conditioning on $\mathbf{y}$, using standard rules for normal distributions, the *conditional predictive distribution* is obtained as [2]

$$
\begin{aligned}
\eta(\boldsymbol{\xi})|\mathbf{y},\boldsymbol{\theta} &\sim \mathcal{GP}\left(m'(\boldsymbol{\xi}|\boldsymbol{\theta}), k'(\boldsymbol{\xi},\boldsymbol{\xi}'|\boldsymbol{\theta})\right)\\
m'(\boldsymbol{\xi}|\boldsymbol{\theta}) &= \mathbf{k}(\boldsymbol{\xi})^T(\mathbf{K}(\boldsymbol{\theta})+\sigma^2\mathbf{I})^{-1}\mathbf{y}\\
k'(\boldsymbol{\xi},\boldsymbol{\xi}'|\boldsymbol{\theta}) &= k(\boldsymbol{\xi},\boldsymbol{\xi}'|\boldsymbol{\theta}) - \mathbf{k}(\boldsymbol{\xi})^T(\mathbf{K}(\boldsymbol{\theta})+\sigma^2\mathbf{I})^{-1}\mathbf{k}(\boldsymbol{\xi}')
\end{aligned}
\tag{6.70}
$$

$\mathbb{E}[\eta(\boldsymbol{\xi})|\mathbf{y},\boldsymbol{\theta}] = m'(\boldsymbol{\xi}|\boldsymbol{\theta})$ is the most likely output at $\boldsymbol{\xi}$, which can be used as a prediction, while the predictive variance in this estimate (a measure of uncertainty) is given by $k'(\boldsymbol{\xi},\boldsymbol{\xi}|\boldsymbol{\theta})$.

For later discussion, a more formal analysis marginalises over $\boldsymbol{\eta}$ in the joint distribution of $\eta(\boldsymbol{\xi})$ and $\boldsymbol{\eta}$

$$
p(\eta(\boldsymbol{\xi})|\mathbf{y},\boldsymbol{\theta},\sigma) = \int p(\eta(\boldsymbol{\xi}),\boldsymbol{\eta}|\mathbf{y},\boldsymbol{\theta},\sigma)d\boldsymbol{\eta} = \int p(\eta(\boldsymbol{\xi})|\boldsymbol{\eta},\boldsymbol{\theta})p(\boldsymbol{\eta}|\mathbf{y},\boldsymbol{\theta},\sigma)d\boldsymbol{\eta}
\tag{6.71}
$$

obtained from $p(\eta(\boldsymbol{\xi})|\boldsymbol{\eta},\boldsymbol{\theta}) = \int p(\eta(\boldsymbol{\xi}),\boldsymbol{\eta}|,\boldsymbol{\theta})d\boldsymbol{\eta}$ and Baye's rule for the posterior over $\boldsymbol{\eta}$

$$
p(\boldsymbol{\eta}|\mathbf{y},\boldsymbol{\theta},\sigma) = \frac{p(\mathbf{y}|\boldsymbol{\eta},\sigma)p(\boldsymbol{\eta}|\boldsymbol{\theta})}{p(\mathbf{y}|\boldsymbol{\theta},\sigma)}
\tag{6.72}
$$

with all integrals being analytically tractable and leading to (6.70) when the noise is Gaussian.

The hyperparameters $\boldsymbol{\theta}$, σ are usually obtained from point estimates, most often a maximum likelihood $\widehat{\boldsymbol{\theta}}$, $\widehat{\sigma}$ [3], and placed inside the posterior predictive distribution. The ML estimate is readily obtained by maximising (numerically) the log-likelihood $\log p(\mathbf{y}|\boldsymbol{\theta},\sigma)$

$$
\widehat{\boldsymbol{\theta}},\widehat{\sigma} = \text{argmax}_{\boldsymbol{\theta},\sigma}\left(-\frac{1}{2}\ln|(\mathbf{K}(\boldsymbol{\theta})+\sigma^2\mathbf{I})| - \frac{1}{2}\mathbf{y}^T(\mathbf{K}(\boldsymbol{\theta})+\sigma^2\mathbf{I})^{-1}\mathbf{y}\right)
\tag{6.73}
$$

6.8 Approximate Inference for Gaussian Process and Other Bayesian Models

Sampling methods are alternatives to point estimates. They are one of a family of approximate inference methods, which are either deterministic or stochastic. The most notable examples of deterministic methods are expectation propagation, Laplace's method and variational inference, while stochastic methods are primarily based on Markov Chain Monte Carlo (MCMC). These approximate methods are required when Bayesian methods employ *non-conjugate* priors, i.e., those which lead to a posterior that is not in the same family of distributions as the prior. In GP models, this occurs when the noise is not a GP, in which case the predictive posterior over the latent function is no longer a GP. Specifically, with non-GP errors, the posterior over latent function values (6.72), the predictive distribution (6.71), and the evidence or marginal likelihood $p(\mathbf{y} \mid \boldsymbol{\theta},\sigma)$ no longer take an analytical form, requiring approximate inference. Moreover, a fully Bayesian inference that introduces priors over the hyperparameters $\boldsymbol{\theta}$ and σ also precludes a closed-form solution.

6.8.1 Laplace's Method

Expectation propagation and Laplace's method look for a Gaussian approximation $q(\boldsymbol{\eta})$ to the posterior over $\boldsymbol{\eta}$ (conditioned on the hyperparameters)

$$p(\boldsymbol{\eta} \mid \mathbf{y}, \boldsymbol{\theta}, \sigma) \approx q(\boldsymbol{\eta}) = \mathcal{N}(\mathbf{m}, \mathbf{C}) \tag{6.74}$$

for some $\mathbf{m}$ and $\mathbf{C}$. Variational inference seeks a similar *variational distribution* $q(\boldsymbol{\eta})$ from some family of distributions that is not necessarily Gaussian. Sampling methods such as MCMC, on the other hand, do not attempt to approximate the distribution but instead take samples from it in an approximate sense. The approximation (6.74) can be used for inference based on Eq. (6.71)

$$p(\eta(\boldsymbol{\xi}) \mid \mathbf{y}, \boldsymbol{\theta}, \sigma) = \int p(\eta(\boldsymbol{\xi}) \mid \boldsymbol{\eta}, \boldsymbol{\theta}) \, \mathcal{N}(\mathbf{m}, \mathbf{C}) \, d\boldsymbol{\eta} \tag{6.75}$$

which is again analytically tractable and leads to a posterior process of the form (6.70)

$$\begin{aligned} \eta(\boldsymbol{\xi}) \mid \mathbf{y}, \boldsymbol{\theta} &\sim \mathcal{GP}\left(m_*(\boldsymbol{\xi}|\boldsymbol{\theta}), k_*(\boldsymbol{\xi}, \boldsymbol{\xi}'|\boldsymbol{\theta})\right) \\ m_*(\boldsymbol{\xi}|\boldsymbol{\theta}) &= \mathbf{k}(\boldsymbol{\xi})^T \mathbf{K}(\boldsymbol{\theta})^{-1} \mathbf{m} \\ k_*(\boldsymbol{\xi}, \boldsymbol{\xi}'|\boldsymbol{\theta}) &= k(\boldsymbol{\xi}, \boldsymbol{\xi}'|\boldsymbol{\theta}) - \mathbf{k}(\boldsymbol{\xi})^T \left(\mathbf{K}(\boldsymbol{\theta})^{-1} - \mathbf{K}(\boldsymbol{\theta})^{-1}\mathbf{C}\mathbf{K}(\boldsymbol{\theta})^{-1}\right) \mathbf{k}(\boldsymbol{\xi}') \end{aligned} \tag{6.76}$$

The solution then boils down to finding suitable values of $\mathbf{m}$ and $\mathbf{C}$. From Eq. (6.72), the log of the posterior is given by

$$\log\, p(\boldsymbol{\eta} \mid \mathbf{y}, \boldsymbol{\theta}, \sigma) = \log\, p(\mathbf{y} \mid \boldsymbol{\eta}, \sigma) + \log\, p(\boldsymbol{\eta} \mid \boldsymbol{\theta}) - \log\, p(\mathbf{y} \mid \boldsymbol{\theta}, \sigma) \tag{6.77}$$

in which the log of the marginal $p(\mathbf{y} \mid \boldsymbol{\theta}, \sigma)$ is independent of $\boldsymbol{\eta}$. In Laplace's method, we define

$$\begin{aligned} \log\, \widetilde{p}(\boldsymbol{\eta} \mid \mathbf{y}, \boldsymbol{\theta}, \sigma) &= \log\, p(\mathbf{y} \mid \boldsymbol{\eta}, \sigma) + \log\, p(\boldsymbol{\eta} \mid \boldsymbol{\theta}) \\ &= \log\, p(\mathbf{y} \mid \boldsymbol{\eta}, \sigma) - \frac{1}{2}\log\, |\mathbf{K}(\boldsymbol{\theta})| - \frac{1}{2}\boldsymbol{\eta}^T \mathbf{K}(\boldsymbol{\theta})^{-1} \boldsymbol{\eta} \end{aligned} \tag{6.78}$$

and consider a second-order Taylor expansion of $\log\, \widetilde{p}(\boldsymbol{\eta} \mid \mathbf{y}, \boldsymbol{\theta}, \sigma)$ around the mode

$$\mathbf{m} = \arg\max_{\boldsymbol{\eta}}\, \log\, \widetilde{p}(\boldsymbol{\eta} \mid \mathbf{y}, \boldsymbol{\theta}, \sigma) \tag{6.79}$$

of the log posterior

$$\log \widetilde{p}(\boldsymbol{\eta} \mid \mathbf{y}, \boldsymbol{\theta}, \sigma) = \log \widetilde{p}(\mathbf{m} \mid \mathbf{y}, \boldsymbol{\theta}, \sigma) + \frac{1}{2}(\mathbf{m} - \boldsymbol{\eta})^T \mathbf{C}^{-1}(\mathbf{m} - \boldsymbol{\eta}) \tag{6.80}$$

$\mathbf{C}^{-1}$ is therefore the Hessian $\nabla_{\eta} \otimes \nabla_{\eta} \log \widetilde{p}(\boldsymbol{\eta} \mid \mathbf{y}, \boldsymbol{\theta}, \sigma)$ evaluated at $\mathbf{m}$

$$\mathbf{C}^{-1} = \nabla_{\eta} \otimes \nabla_{\eta} \log \widetilde{p}(\boldsymbol{\eta} \mid \mathbf{y}, \boldsymbol{\theta}, \sigma)\big|_{\mathbf{m}} = \nabla_{\eta} \otimes \nabla_{\eta} \log p(\mathbf{y} \mid \boldsymbol{\eta}, \sigma)\big|_{\mathbf{m}} - \mathbf{K}(\boldsymbol{\theta})^{-1} \tag{6.81}$$

in which $\nabla_{\eta} \otimes \nabla_{\eta} \log p(\mathbf{y} \mid \boldsymbol{\eta}, \sigma)$ is diagonal, since the likelihood factorises. The mode $\mathbf{m}$ is found iteratively, e.g., using Newton's method. When the posterior is unimodal, Laplace's method can provide accurate results but in other cases its use is problematic.

6.8.2 *Mean Field Variational Inference*

The hyperparameters and latent variables can be lumped together to form a set of unobserved variables $\boldsymbol{\psi} = \{\boldsymbol{\eta}, \boldsymbol{\theta}, \sigma\}$, to which the various approximate-inference methods can be applied for a fully-Bayesian inference. In this section, we describe *mean-field* variational formulations. Although there exist more sophisticated modern (non-mean field) formulations, they will not be covered in this book.

Variational Bayesian inference is typically based on the concept of a Kullback-Leibler (KL) divergence $\mathrm{KL}\left(q(\boldsymbol{\psi}) \,||\, p(\boldsymbol{\psi} \mid \mathbf{y})\right)$ between two distributions, namely, in this case, the true distribution $p(\boldsymbol{\psi} \mid \mathbf{y})$ and its variational approximation $q(\boldsymbol{\psi})$. The KL divergence is a metric that defines a distance between the two distributions, and the optimal variational distribution can therefore be found by minimising the KL divergence with respect to all distributions within some family $\mathcal{Q}$

$$q^{\star}(\boldsymbol{\psi}) = \underset{q \in \mathcal{Q}}{\arg\min}\ \mathrm{KL}\left(q(\boldsymbol{\psi}) \,||\, p(\boldsymbol{\psi} \mid \mathbf{y})\right) \tag{6.82}$$

Moreover, this choice of metric allows for a tractable minimisation. The KL divergence is defined as follows

$$\mathrm{KL}\left(q(\boldsymbol{\psi}) \,||\, p(\boldsymbol{\psi} \mid \mathbf{y})\right) = \int q(\boldsymbol{\psi}) \log \frac{q(\boldsymbol{\psi})}{p(\boldsymbol{\psi} \mid \mathbf{y})} d\boldsymbol{\psi} \tag{6.83}$$

Without knowledge of the marginal $p(\mathbf{y}) = \int p(\mathbf{y} \mid \boldsymbol{\psi}) p(\boldsymbol{\psi}) d\boldsymbol{\psi}$, the KL divergence cannot be minimised. This becomes apparent when we rewrite $\mathrm{KL}\left(q(\boldsymbol{\psi}) \,||\, p(\boldsymbol{\psi} \mid \mathbf{y})\right)$ as follows

$$
\begin{aligned}
\mathrm{KL}\left(q(\psi) \,||\, p(\psi \mid \mathbf{y})\right) &= \mathbb{E}_q\left[\log \frac{q(\psi)}{p(\psi \mid \mathbf{y})}\right] \\
&= \mathbb{E}_q\left[\log q(\psi)\right] - \mathbb{E}_q\left[\log \frac{p(\psi, \mathbf{y})}{p(\mathbf{y})}\right] \\
&= \mathbb{E}_q\left[\log q(\psi)\right] - \mathbb{E}_q\left[\log p(\psi, \mathbf{y})\right] + \mathbb{E}_q\left[\log p(\mathbf{y})\right] \\
&= \mathbb{E}_q\left[\log q(\psi)\right] - \mathbb{E}_q\left[\log p(\psi, \mathbf{y})\right] + \log p(\mathbf{y}) \int q(\psi)\, d\psi \\
&= \mathbb{E}_q\left[\log q(\psi)\right] - \mathbb{E}_q\left[\log p(\psi, \mathbf{y})\right] + \log p(\mathbf{y})
\end{aligned}
\tag{6.84}
$$

in which $\mathbb{E}_q[\cdot]$ denotes an expectation w.r.t. $q(\psi)$ and we use the fact that $\int q(\psi)d\psi = 1$. We notice, however, that $\log p(\mathbf{y})$ is a constant as far as the minimisation over q is concerned. We therefore define the *evidence lower bound* (ELBO)

$$
\begin{aligned}
\mathrm{ELBO}[q] &:= -\mathrm{KL}\left(q(\psi) \,||\, p(\psi \mid \mathbf{y})\right) + \log p(\mathbf{y}) \\
&= \mathbb{E}_q\left[\log p(\psi, \mathbf{y})\right] - \mathbb{E}_q\left[\log q(\psi)\right] \\
&= \mathbb{E}_q\left[\log p(\mathbf{y} \mid \psi)\right] - \mathbb{E}_q\left[\log \left(\frac{q(\psi)}{p(\mathbf{y})}\right)\right] \\
&= \mathbb{E}_q\left[\log p(\mathbf{y} \mid \psi)\right] - \mathrm{KL}\left(q(\psi) \,||\, p(\psi)\right)
\end{aligned}
\tag{6.85}
$$

and instead maximise $\mathrm{ELBO}[q]$ over all $q \in \mathcal{Q}$. The ELBO derives its name from the fact that it is a lower bound on the log marginal likelihood (evidence), which can be seen from its definition and the non-negativity of the KL divergence. The last expression for the ELBO in (6.85) shows that maximising the ELBO yields an approximate distribution $q(\psi)$ that yields a high value of the log likelihood $\log p(\mathbf{y} \mid \psi)$, while penalising distributions $q(\psi)$ that stray far from the prior. The name 'variational' arises from the fact that $\mathrm{ELBO}[q]$ is a functional of q and the problem $\mathrm{argmax}_{q \in \mathcal{Q}} \mathrm{ELBO}[q]$ can be solved using the calculus of variations.

Typically, a mean-field approximation is used, namely that the variational distribution $q(\psi)$ takes a factored form: if we partition ψ into $\psi_1, \ldots, \psi_K$, the mean-field approximation is

$$
q(\psi) = \prod_{i=1}^{K} q_i(\psi_i) \tag{6.86}
$$

with factors $q_i(\psi_i)$ such that each is a well-defined distribution with $\int q_i(\psi_i)\, d\psi_i = 1$. The ELBO becomes

$$
\begin{aligned}
\mathrm{ELBO}[\{q_j\}] &= \mathbb{E}_q\left[\log p(\psi, \mathbf{y})\right] - \mathbb{E}_q\left[\log q(\psi)\right] \\
&= \int \prod_{i=1}^{K} q_i(\psi_i) \log p(\psi, \mathbf{y})\, d\psi - \int \prod_{i=1}^{K} q_i(\psi_i) \log \prod_{i=1}^{K} q_i(\psi_i)\, d\psi
\end{aligned}
\tag{6.87}
$$

We now fix $\psi_i, \forall i \neq j$, and maximise $\mathrm{ELBO}[\{q_j\}]$ w.r.t. ψ_j. Before we perform the maximisation, we define a new functional from the ELBO as follows

$$
\begin{aligned}
\text{ELBO}[\{q_j\}] &= \int \prod_{i=1}^{K} q_i(\psi_i) \log p(\psi, \mathbf{y})\, d\psi - \int \prod_{i=1}^{K} q_i(\psi_i) \log \prod_{i=1}^{K} q_i(\psi_i)\, d\psi \\
&= \int \prod_{i=1}^{K} q_i(\psi_i) \log p(\psi, \mathbf{y})\, d\psi - \int q_j(\psi_j) \log q_j(\psi_j)\, d\psi_j \\
&\qquad - \int \prod_{i\neq j}^{K} q_i(\psi_i) \log \prod_{i\neq j}^{K} q_i(\psi_i)\, \mathrm{d}\psi_{\sim j} \\
&\propto \int \prod_{i=1}^{K} q_i(\psi_i) \log p(\psi, \mathbf{y})\, d\psi - \int q_j(\psi_j) \log q_j(\psi_j)\, d\psi_j \\
&= \int q_j(\psi_j) \left(\int \prod_{i\neq j}^{K} q_i(\psi_i) \log p(\psi, \mathbf{y})\, \mathrm{d}\psi_{\sim j} \right) d\psi_j - \int q_j(\psi_j) \log q_j(\psi_j)\, d\psi_j \\
&= \int q_j(\psi_j)\, \mathbb{E}_{q(\psi_{\sim j})} \left[\log p(\psi, \mathbf{y})\right] d\psi_j - \int q_j(\psi_j) \log q_j(\psi_j)\, d\psi_j
\end{aligned}
\tag{6.88}
$$

in which $\psi_{\sim j}$ denotes all variables in ψ except those in ψ_j. From this functional, we can define a Lagrangian

$$
\begin{aligned}
\mathcal{L}[\{q_j\}] := \int q_j(\psi_j) \left(\mathbb{E}_{q(\psi_{\sim j})} \left[\log p(\psi, \mathbf{y})\right] - \log q_j(\psi_j) \right) d\psi_j \\
- \sum_{i=1}^{K} \lambda_i \left(\int q_i(\psi_i) d\psi_i - 1 \right)
\end{aligned}
\tag{6.89}
$$

in which the second term is a constraint ensuring that all of the q_i are proper densities, with associated Lagrange multipliers λ_i.

Taking the functional derivative of $\mathcal{L}[\{q_j\}]$ w.r.t. q_j then yields

$$
\frac{\delta L}{\delta q_j} = \mathbb{E}_{q(\psi_{\sim j})} \left[\log p(\psi, \mathbf{y})\right] - \log q_j(\psi_j) - 1 - \lambda_j = 0 \tag{6.90}
$$

from which we obtain

$$
q_j(\psi_j) = \frac{1}{Z_j} \exp\left(\mathbb{E}_{q(\psi_{\sim j})} \left[\log p(\psi, \mathbf{y})\right] \right), \quad Z_j = \int e^{\mathbb{E}_{q(\psi_{\sim j})}\left[\log p(\psi, \mathbf{y})\right]} d\psi_j \tag{6.91}
$$

for some normalisation constant Z_j that is typically found by inspection, since $q_j(\psi_j)$ will belong to some known family of distributions (Gaussian, Gamma, etc.). Thus, each of the factors $q_j(\psi_j)$ is defined by (6.91), which leads to a system of consistency equations from $j = 1, \ldots, K$ that together maximise the ELBO (or minimise the KL divergence) under a mean-field assumption. Note, however, that the solution to $q_j(\psi_j)$ depends on the other factors, requiring therefore a self-consistent approach. The term $\mathbb{E}_{q(\psi_{\sim j})} \left[\log p(\psi, \mathbf{y})\right]$ typically simplifies to a function of some fixed hyperparameters associated with the prior distributions placed over the unobserved variables, along with moments of unobservables in $\psi_{\sim j}$. Note also that

$$
p(\psi_j \mid \psi_{\sim j}, \mathbf{y}) = \frac{p(\psi_j, \psi_{\sim j}, \mathbf{y})}{p(\psi_{\sim j}, \mathbf{y})} = \frac{1}{Z} p(\psi, \mathbf{y}) \tag{6.92}
$$

for some normalisation constant Z, so that updates can be performed by replacing $p(\psi, \mathbf{y})$ with the conditional posterior over ψ_j given $\psi_{\sim j}$. Even in the simplest cases, e.g., a univariate Gaussian or linear regression, the derivations are lengthy and cumbersome. We therefore omit any examples of the application of variational inference.

6.8.3 Markov Chain Monte Carlo

From Eq. (6.78), we note that the posterior over the latent variables satisfies

$$p(\boldsymbol{\eta} \mid \mathbf{y}, \boldsymbol{\theta}, \sigma) \propto \widetilde{p}(\boldsymbol{\eta} \mid \mathbf{y}, \boldsymbol{\theta}, \sigma) = p(\mathbf{y} \mid \boldsymbol{\eta}, \sigma)\ p(\boldsymbol{\eta} \mid \boldsymbol{\theta}) \tag{6.93}$$

In this case, and in many other examples of Bayesian modelling, we can sample easily from the likelihood $p(\mathbf{y} \mid \boldsymbol{\eta}, \sigma)$ and the prior $p(\boldsymbol{\eta} \mid \boldsymbol{\theta})$ but the posterior is not in a closed form and/or cannot be used for tractable inference (prediction). In Markov Chain Monte Carlo (MCMC), this issue is avoided altogether by instead generating samples from the posterior $p(\boldsymbol{\eta} \mid \mathbf{y}, \boldsymbol{\theta}, \sigma)$ based only on the tractability of the unnormalised posterior $\widetilde{p}(\boldsymbol{\eta} \mid \mathbf{y}, \boldsymbol{\theta}, \sigma)$. These samples can then be used to yield approximations of properties related to the posterior distribution, especially Monte Carlo approximations of the type

$$\int f(\psi) p(\psi \mid \mathbf{y}) \approx \frac{1}{L} \sum_{i=1}^{L} f(\psi) \tag{6.94}$$

for some function f, and where again $\psi = \{\boldsymbol{\eta}, \boldsymbol{\theta}, \sigma\}$. We note that ordinary Monte Carlo methods also approximate features of a distribution over some random variable X by generating samples $x_1, \ldots, x_L$ from the distribution, but these samples are independent draws. In MCMC, the samples are serially correlated.

We first need to define a Markov chain, and then consider under what conditions we can apply the theory of Markov chains in order to sample from the desired distribution. We further need an easy way to check that the conditions are satisfied and to construct the chain. In particular, we will require that the Markov chain we establish has a stationary distribution (explained below) and that this is the desired distribution.

A Markov process $\{X_t\}$ is defined as a stochastic process (indexed by $t \in \mathcal{T}$) that possesses the *memoryless* property, namely, that the probability of an event depends only on the current state in some probability space common to all X_t, with a defining state space $\mathcal{S}$ (the set of possible values for the realisations or 'states' x_t), i.e., X_{t+1} is conditionally independent of $X_{t-1}, X_{t-2}, \ldots$ given X_t. This is often termed a first-order Markov process, with higher-order processes dependent on X_{t-1} or X_{t-1}, X_{t-2}, etc. In Chap. 7, we will consider such cases when we look at embeddings of time series data. Note also that there is a close connection between Markov chains and graph theory, elements of which (including some spectral theory) are covered in Sect. 6.15.3. Some further theory of Markov chains on continuous state spaces is covered in Sect. 6.15.5.

Another way to write down the Markov property is

$$\begin{aligned} p(x_{t+1} \mid x_t, x_{t-1}, \ldots, x_0) &= p(x_{t+1} \mid x_t) \\ p(x_{t+1}, x_t, x_{t-1}, \ldots, x_0) &= p(x_{t+1} \mid x_t, \ldots, x_0)\, p(x_t \mid x_{t-1}, \ldots, x_0)\, \ldots\, p(x_1 \mid x_0)\, p(x_0) \\ &= p(x_{t+1} \mid x_t)\, p(x_t \mid x_{t-1})\, \ldots\, p(x_1 \mid x_0)\, p(x_0) \end{aligned} \tag{6.95}$$

in which, for compactness, $p(x_{t+1} \mid x_t, x_{t-1}, \ldots, x_0)$ is understood to mean $p(X_{t+1} = x_{t+1} \mid X_t = x_t, X_{t-1} = x_{t-1}, \ldots, X_0 = x_0)$ and so on. Here, $p(\cdot)$ can be either a probability measure or a density. We may use $\mathbb{P}(\cdot)$ for the former if we wish to make the distinction. A Markov chain is often defined as a Markov process on a countable state space $\mathcal{S}$ of cardinality $|\mathcal{S}|$, with $\mathcal{T}$ being either discrete or continuous. The index t is often identified with time, so the terminologies discrete-time and continuous-time Markov chains are usually employed. Many authors, however, define a Markov chain as a discrete-time Markov process with either a countable (discrete) or continuous state space. In perhaps the majority of cases, we deal with a discrete-time, discrete-space Markov process or chain. To keep matters simple, we next discuss some important concepts related to Markov chains by restricting our attention to finite state spaces $\mathcal{S} = \{s_0, \ldots, s_{|\mathcal{S}|}\}$, mentioning generalisations as we go along.

The second of (6.95) shows that we can specify a Markov chain by an initial distribution over x_0, along with the following conditional probabilities

$$T_t(i, j) = p(X_{t+1} = s_j \mid X_t = s_i), \quad s_i, s_j \in \mathcal{S} \tag{6.96}$$

The $T_t(i, j)$ are called (one-step) *transition probabilities*, and if they are independent of t, i.e., $T_t(i, j) = T(i, j)$, we say that the Markov chain is time-homogeneous or simply homogeneous. In the case of a finite state space, we can then place all of the transition probabilities inside a *transition matrix* $\mathbf{T} = [T(i, j)]_{i,j} \in \mathbb{R}^{|\mathcal{S}| \times |\mathcal{S}|}$, in which the rows index the current state $X_t = s_i$ and the columns the next state $X_{t+1} = s_j$. Thus, $\sum_j T(i, j) = 1, \forall i$. These concepts extend to the case of a countably infinite $\mathcal{S}$, whereas a continuous $\mathcal{S}$ requires a transition *kernel density* $T(s_i, s_j) : \mathcal{S} \times \mathcal{S} \to \mathbb{R}$ representing a transition probability *density* (see below for further discussion).

The transition probabilities must satisfy the following property in terms of the marginal probability of X_{t+1} based on the current state

$$p(X_{t+1} = s_j) = \sum_{s_i \in \mathcal{S}} p(X_{t+1} = s_j \mid X_t = s_i)\, p(X_t = s_i) \tag{6.97}$$

Moreover, the probability of going from state $s_i \in \mathcal{S}$ to $s_j \in \mathcal{S}$ in n steps is called the n step transition probability and is given by

$$p(X_{t+n} = s_j \mid X_t = s_i) = \mathbf{T}^n(i, j) \tag{6.98}$$

in which $\mathbf{T}^n(i, j)$ is the (i, j)-th entry of $\mathbf{T}^n$. This is a consequence of the second of (6.95). Note that the trivial relationship $\mathbf{T}^{n+m} = \mathbf{T}^n\mathbf{T}^m$ expresses the *Chapman-Kolomogorov* equation for a homogeneous Markov chain with a countable state space, which states that the conditional probability of going from a state s_i to a state s_j in $n+m$ steps is equal to the sum of the probabilities of going from s_i to some intermediate state s_k in m steps and of going from s_k to s_j in m steps.

Based on (6.96), if we have an initial distribution $\boldsymbol{\pi} = (p(X_0 = s_1), \ldots, p(X_0 = s_{|\mathcal{S}|}))^T$

$$p(X_t) = \boldsymbol{\pi}^T\mathbf{T}^t, \quad p(X_t = s_j) = \sum_{s_i \in \mathcal{S}} p(X_t = s_j \mid X_0 = s_i) = (\boldsymbol{\pi}^T\mathbf{T}^t)(j) \quad (6.99)$$

in which $(\boldsymbol{\pi}^T\mathbf{T}^t)(j)$ is the j-th component of $\boldsymbol{\pi}^T\mathbf{T}^t$. A *stationary* or equilibrium distribution $\boldsymbol{\pi}_* = (\pi_1^*, \ldots, \pi_{|\mathcal{S}|}^*)^T$ of the Markov chain is one for which

$$\sum_{i=1}^{|\mathcal{S}|} \pi_i^* T(i, j) = \pi_i^* \quad \text{or} \quad \boldsymbol{\pi}_*^T\mathbf{T} = \boldsymbol{\pi}_*^T \qquad (6.100)$$

This shows that $\boldsymbol{\pi}_*$ is a left eigenvector of $\mathbf{T}$ with eigenvalue 1. It is also straightforward to show that $p(X_t) = \boldsymbol{\pi}_*$ for any t if $p(X_0) = \boldsymbol{\pi}_*$, so that the X_t are identically distributed, although not independent.

Two states s_i and s_j are said to *communicate*, written $s_i \leftrightarrow s_j$, if it is possible to go from the first to the second and vice versa in a finite number of steps. Mathematically, we can express this as: $\exists n$ such that $\mathbf{T}^n(i, j) > 0$ and $\exists m$ such that $\mathbf{T}^m(j, i) > 0$. $\leftrightarrow$ defines an equivalence relation and it is possible to partition the sample space into disjoint equivalence classes of communicating states. If we are not able to escape a communicating class $\mathcal{C}$, it is said to be *closed*, which is equivalent to $T(i, j) = 0$ if $s_i \in \mathcal{C}$ and $s_j \notin \mathcal{C}$. A special case of a closed class is a singleton set $\{s_j\}$, for which s_j is said to be an *absorbing* state. A Markov chain is termed *irreducible* if $s_i \leftrightarrow s_j\ \forall s_i, s_j \in \mathcal{S}$, in which case the only communicating class is the entire sample space $\mathcal{S}$.

The *period* of a state s_i is defined as $d(s_i) = \text{GCD}\{n : \mathbf{T}^n(i, i) > 0\}$, in which GCD denotes the greatest common divisor. If $s_i \leftrightarrow s_j$, it can be shown that $d(s_i) = d(s_j)$, so that in an irreducible Markov chain all states have the same period, and this common period is called the period of the chain. Furthermore, an irreducible Markov chain is called *aperiodic* if it has a period of one. This means that there is no cycling through states with a finite period.

The Peron-Frobenius theorem states that if $\exists n$ such that $\mathbf{T}^n(ij) > 0\ \forall i, j$, a Markov chain will converge to a unique stationary distribution $\boldsymbol{\pi}_*$. It is generally difficult to establish the required conditions for this theorem. For a finite state-space chain, we can instead demand that it is both irreducible and aperiodic (also called *ergodic*). For countable $\mathcal{S}$, we require an additional constraint. A state s_i is said to be *recurrent* (and otherwise transient) if the probability of returning to s_i infinitely often after starting at s_i is equal to one, or equivalently $\sum_{n=0}^{\infty} \mathbf{T}^n(i, j) = \infty$. Another way to state this is by defining the random variable

$$T_i = \min\{t : X_t = s_i \mid X_0 = s_i\} \tag{6.101}$$

The state s_i is recurrent if $p(T_i < \infty) = 1$; otherwise, it is transient. While this may be the case, there is no guarantee that $\mathbb{E}[T_i] < \infty$, which distinguishes between positive recurrence ($\mathbb{E}[T_i] < \infty$) and null recurrence ($\mathbb{E}[T_i] = \infty$). We require positive recurrence alongside irreducibility and aperiodicity for continuous or countably-infinite state spaces.

Solving for the stationary distribution is in general not an easy task. It is simpler to check for a property termed reversibility, which also allows us to find the stationary distribution. A Markov chain is said to be reversible with respect to a distribution $\boldsymbol{\pi}$ if it satisfies the following *detailed balance* condition

$$\pi_i T(i, j) = \pi_j T(j, i) \tag{6.102}$$

In this case,

$$\sum_{i=1}^{|\mathcal{S}|} \pi_i T(i, j) = \sum_{i=1}^{|\mathcal{S}|} \pi_j T(j, i) = \pi_j \sum_{i=1}^{|\mathcal{S}|} T(j, i) = \pi_j \tag{6.103}$$

and therefore $\boldsymbol{\pi} = \boldsymbol{\pi}_*$ is the unique stationary distribution. We note that detailed balance is a sufficient but not a necessary condition for the stationarity of a distribution with respect to a transition matrix.

Returning to our desired goal in MCMC, we wish to take samples $\psi_1, \psi_2, \ldots$ from the true posterior distribution $p(\psi \mid \mathbf{y})$ by simulating a Markov chain with stationary distribution equal to $p(\psi \mid \mathbf{y})$. We will assume that we only have access to the unnormalised posterior $\tilde{p}(\psi \mid \mathbf{y})$. A random sampling is obtained if the chain is simulated for sufficiently many steps and both the accuracy and efficiency of this process are determined by the rate of convergence to the true distribution (called the mixing time). A chosen number of the first draws from the chain are discarded (usually using convergence diagnostics), since they are not close enough to the true distribution, and this set of draws is called the *burn-in* sample. A further practice of using only each J-th sample after burn-in is called thinning.

In the *Metropolis-Hastings* algorithm, we use a proposal distribution from which it is easy to sample (e.g., an isotropic Gaussian), in common with the non-Markov chain rejection and importance sampling methods. In this case, however, the proposal distribution $q(\psi \mid \psi_t)$ depends on the current state ψ_t, and in contrast to rejection sampling it always generates a sample. Note that for discrete sample spaces, $q(\psi \mid \psi_t)$ is a transition probability, and for continuous sample spaces, it is a transition probability density. The algorithm proceeds as follows:

1. sample $u \sim \mathcal{U}(0, 1)$
2. propose a sample $\psi_* \sim q(\psi \mid \psi_t)$
3. if

$$u \leq A(\psi_t, \psi_*) = \min\left(1, \frac{\tilde{p}(\psi_* \mid \mathbf{y})q(\psi_t \mid \psi_*)}{\tilde{p}(\psi_t \mid \mathbf{y})q(\psi_* \mid \psi_t)}\right) \tag{6.104}$$

set $\psi_{t+1} = \psi_*$
else set $\psi_{t+1} = \psi_t$.

with $A(\psi_t, \psi_*)$ called the acceptance kernel or function. Since the normalising constant cancels in the numerator and denominator, we could equally use p in place of $\widetilde{p}$ in the definition of A. To see that the posterior is the equilibrium distribution of the Markov chain defined by this algorithm, we need only demonstrate detailed balance. We first define a transition kernel $T(\psi_t, \psi_*)$, assuming a continuous sample space $\mathcal{S}$

$$T(\psi_t, \psi_*) = q(\psi_*, \psi_t)A(\psi_t, \psi_*) + \delta_{\psi_t}(\psi_*) \int_{\psi \in \mathcal{S}} \left[1 - A(\psi_t, \psi)\right] q(\psi, \psi_t) d\psi \tag{6.105}$$

in which the bracketed term is the probability of rejecting ψ_*, sampled from the proposal distribution, with $\delta_{\psi_t}(\psi_*)$ being the Dirac measure on $\{\psi_*\}$ (equal to 1 if $\psi_* = \psi_t$ and 0 otherwise). For a discrete $\mathcal{S}$, we replace the integral with a sum over $\psi \in \mathcal{S}$. Using (6.104), for $\psi_{t+1} \neq \psi_t$, in which case $T(\psi_t, \psi_*) = q(\psi_*, \psi_t)A(\psi_t, \psi_*)$, we then obtain

$$p(\psi_t \mid \mathbf{y})q(\psi_* \mid \psi_t)A(\psi_t, \psi_*) = \min\left(p(\psi_t \mid \mathbf{y})q(\psi_* \mid \psi_t), p(\psi_* \mid \mathbf{y})q(\psi_t \mid \psi_*)\right) \tag{6.106}$$

and

$$p(\psi_* \mid \mathbf{y})q(\psi_t \mid \psi_*)A(\psi_*, \psi_t) = \min\left(p(\psi_* \mid \mathbf{y})q(\psi_t \mid \psi_*), p(\psi_t \mid \mathbf{y})q(\psi_* \mid \psi_t)\right) \tag{6.107}$$

which are equal. For $\psi_{t+1} = \psi_t$, detailed balance holds trivially.

A more mathematically rigorous proof shows that

$$\int_{\mathcal{A}} p(\psi \mid \mathbf{y})T(\psi, \mathcal{B})d\psi = \int_{\mathcal{B}} p(\psi \mid \mathbf{y})T(\psi, \mathcal{A})d\psi \tag{6.108}$$

for every measurable $\mathcal{A} \subset \mathcal{S}$ and $\mathcal{B} \subset \mathcal{S}$, which is a continuous analogue of reversibility, with transition kernel

$$T(\psi_t, \mathcal{A}) = \int_{\mathcal{A}} q(\psi_*, \psi_t)A(\psi_t, \psi_*)d\psi_* + \delta_{\psi_t}(\mathcal{A}) \int_{\psi \in \mathcal{S}} \left[1 - A(\psi_t, \psi)\right] q(\psi, \psi_t) d\psi \tag{6.109}$$

that defines the probability $\mathbb{P}\{\psi_{t+1} \in \mathcal{A} \mid \psi_t\} = T(\psi_t, \mathcal{A})$. Thus, the transition function or kernel is a probability measure on the Borel σ-algebra on $\mathcal{S}$. We can also define a kernel density $T(\psi, \psi')$ such that $T(\psi, \mathcal{A}) = \int_{\mathcal{A}} T(\psi, \psi')d\psi'$. Using a symmetric $q(\psi_* \mid \psi_t)$ leads to the *Metropolis* algorithm, with

$$A(\psi_t, \psi_*) = \min\left(1, \frac{\widetilde{p}(\psi_* \mid \mathbf{y})}{\widetilde{p}(\psi_t \mid \mathbf{y})}\right) \tag{6.110}$$

Further imposing $q(\psi_* \mid \psi_t) = q(\|\psi_* - \psi_t\|)$ leads to a *random walk* proposal.

Another often employed sampler is the Gibb's sampler, which cycles through conditional probabilities to update the components of $\boldsymbol{\psi} \in \mathbb{R}^m$

1. initialise: $t = 0$, $\boldsymbol{\psi}_0 = (\psi_i^0, \ldots, \psi_m^0)^T$
2. for $t = 1 : T$
3. for $k = 1 : m$ draw a sample

$$\psi_* \sim p(\psi_k \mid \psi_1^{t+1}, \ldots, \psi_{k-1}^{t+1}, \psi_{k+1}^t, \ldots, \psi_m^t, \mathbf{y}) \tag{6.111}$$

4. set $\psi_1^{t+1} = \psi_*$
5. output $\boldsymbol{\psi}_1, \ldots, \boldsymbol{\psi}_T$

Even though $p(\boldsymbol{\psi} \mid \mathbf{y})$ is known only up to a constant, it is possible to draw from the conditional distributions. Consider the two-stage Gibb's sampler for $p(\mu, \sigma^2 \mid \mathbf{y})$ for a model with $\mathbb{R} \ni y_i \sim \mathcal{N}(\mu, \sigma^2)$ i.i.d., $i = 1, \ldots, N$, and a non-informative prior $p(\mu, \sigma^2) = 1/\sigma^2$. The likelihood and posterior are

$$\begin{aligned} p(\mathbf{y}|\mu, \sigma^2) &= \left(\frac{1}{2\pi\sigma^2}\right)^{N/2} \exp\left(-\frac{1}{2\sigma^2}\sum_i (y_i - \mu)^2\right) \\ p(\mu, \sigma^2|\mathbf{y}) &\propto \left(\frac{1}{2\pi\sigma^2}\right)^{N/2+1} \exp\left(-\frac{1}{2\sigma^2}\sum_i (y_i - \mu)^2\right) \end{aligned} \tag{6.112}$$

The marginal densities are given by $p(\mu \mid \sigma^2, \mathbf{y}) = \mathcal{N}(\bar{y}, \sigma^2/N)$ and $p(\sigma^2 \mid \mu, \mathbf{y}) = \Gamma(N/2, \sum_i (y_i - \mu)^2/2)$.

Consider again a two-stage Gibb's sampler for $p(\boldsymbol{\psi} \mid \mathbf{y})$, in which $\boldsymbol{\psi} = (\psi_1, \psi_2)^T$. The transition kernel density is given by

$$T(\boldsymbol{\psi}, \boldsymbol{\psi}') = p(\psi_2' \mid \psi_1, \mathbf{y})\, p(\psi_1' \mid \psi_2', \mathbf{y}) \tag{6.113}$$

In general, the two-stage Gibbs sampler does not obey detailed balance. To show that $p(\psi_1, \psi_2 \mid \mathbf{y})$ is the stationary distribution of a sampler with this kernel density amounts to showing that $\int T(\boldsymbol{\psi}, \boldsymbol{\psi}') p(\boldsymbol{\psi} \mid \mathbf{y}) d\boldsymbol{\psi} = p(\boldsymbol{\psi}' \mid \mathbf{y})$

$$\begin{aligned} \int_S p(\psi_1, \psi_2 \mid \mathbf{y})\, T(\boldsymbol{\psi}, \boldsymbol{\psi}')\, d\boldsymbol{\psi} &= \int_S p(\psi_1, \psi_2 \mid \mathbf{y})\, p(\psi_2' \mid \psi_1, \mathbf{y})\, p(\psi_1' \mid \psi_2', \mathbf{y})\, d\boldsymbol{\psi} \\ &= \int_S p(\psi_1, \psi_2 \mid \mathbf{y})\, \frac{p(\psi_2', \psi_1 \mid \mathbf{y})}{p(\psi_1 \mid \mathbf{y})}\, \frac{p(\psi_1', \psi_2' \mid \mathbf{y})}{p(\psi_2' \mid \mathbf{y})}\, d\boldsymbol{\psi} \\ &= \int_{\psi_1} p(\psi_1 \mid \mathbf{y}) \frac{p(\psi_2', \psi_1 \mid \mathbf{y})}{p(\psi_1 \mid \mathbf{y})}\, \frac{p(\psi_1', \psi_2' \mid \mathbf{y})}{p(\psi_2' \mid \mathbf{y})}\, d\psi_1 \\ &= \int_{\psi_1} p(\psi_1, \psi_2' \mid \mathbf{y})\, \frac{p(\psi_1', \psi_2' \mid \mathbf{y})}{p(\psi_2' \mid \mathbf{y})}\, d\psi_1 \\ &= p(\psi_1', \psi_2' \mid \mathbf{y}) \end{aligned} \tag{6.114}$$

which can be extended to any number of stages.

There are other popular MCMC sampling methods such as slice sampling and in particular hybrid Monte Carlo. Details can be found in [1].

6.9 Support Vector Regression

The support vector machine (SVM) described later in Sect. 6.13 is a powerful technique for binary classification, based on a *maximum margin* principle, i.e., finding a decision surface and two hyperplanes parallel to this surface such that: (a) the two hyperplanes are of equal perpendicular distance to the surface; (b) the two classes are separated by the parallel hyperplanes; (c) at least one target lies on each hyperplane; and (d) the perpendicular distance between the hyperplanes (or margin) is maximised [1]. It can be turned into a nonlinear method *via* kernel substitution. It can also be extended to multiple classes in a number of different ways that will be detailed in Sect. 6.13.

There exists an equivalent method for regression, termed support vector regression (SVR), in which the goal is to find an approximating function for a scalar latent function $\eta(\boldsymbol{\xi})$, using data $\boldsymbol{\xi}_n \in \mathcal{X}$, $y_n = \eta(\boldsymbol{\xi}_n)$, $i = 1, \ldots, N$. Here, we describe a variant of SVR called ϵ-SVR [1]. The approximating function $\widehat{\eta}(\boldsymbol{\xi})$ of $\eta(\boldsymbol{\xi})$ is assumed to be of the form of a *generalised linear model*

$$\widehat{\eta}(\boldsymbol{\xi}) = \mathbf{w}^T \boldsymbol{\phi}(\boldsymbol{\xi}) + b \tag{6.115}$$

with unknown weight vector $\mathbf{w}$ and bias $b \in \mathbb{R}$, together with a feature map $\boldsymbol{\phi}$ of the inputs. To specify the model parameters, the following problem is solved

$$\text{argmin}_{\mathbf{w},b} \left(\lambda \sum_{n=1}^{N} E_\epsilon(\widehat{\eta}(\boldsymbol{\xi}_n) - y_n) + \frac{1}{2}\|\mathbf{w}\|^2 \right) \tag{6.116}$$

in which $\lambda\|\mathbf{w}\|^2/2$ acts as a penalty term, in which λ is a hyperparameter. The function $E_\epsilon(x)$ in this loss function is called an ϵ-*insensitive error function*. It forces sparsity by setting the error to exactly zero if $\widehat{\eta}(\boldsymbol{\xi}_n)$ is contained within an $\epsilon > 0$ ball of $y_n = \eta(\boldsymbol{\xi}_n)$. We also say that $\widehat{\eta}(\boldsymbol{\xi}_n)$ falls inside an ϵ-*insensitive tube*

$$E_\epsilon(\widehat{\eta}(\boldsymbol{\xi}_n) - y_n) = \begin{cases} 0 & \text{if } |\widehat{\eta}(\boldsymbol{\xi}_n) - y_n| < \epsilon \\ |\widehat{\eta}(\boldsymbol{\xi}_n) - y_n| - \epsilon & \text{otherwise} \end{cases} \tag{6.117}$$

The constant ϵ is a user-chosen hyperparameter. The minimisation problem (6.116) is not solved in its original (*primal*) form, but rather cast in a *dual* from by introducing non-negative slack variables h_n and $\overline{h}_n$, $n = 1, \ldots, N$, which in turn leads to a set of constraints

$$y_n \leq \widehat{\eta}(\boldsymbol{\xi}_n) + \epsilon + h_n, \quad z_n \geq \widehat{\eta}(\boldsymbol{\xi}_n) - \epsilon + \overline{h}_n \tag{6.118}$$

and Lagrange multipliers a_n, $\overline{a}_n$, $n = 1, \ldots, N$. Skipping the mathematical details (more details are provided in Sect. 6.13 on support vector machines), kernel substitution leads to the following dual-constrained optimisation problem:

$$\begin{gathered}
\operatorname{argmin}_{\{a_n,\overline{a}_n\}} \sum_{n,m=1}^{N} (a_n - \overline{a}_n)(a_m - \overline{a}_m)\frac{k(\boldsymbol{\xi}_n, \boldsymbol{\xi}_m)}{2} + \epsilon \sum_{n=1}^{N} (a_n + \overline{a}_n) - \sum_{n=1}^{N} (a_n - \overline{a}_n) y_n \\
0 \le a_n \le \lambda, \quad 0 \le \overline{a}_n \le \lambda \\
\mathbf{w} = \sum\nolimits_{n=1}^{N} (a_n - \overline{a}_n)\boldsymbol{\phi}(\boldsymbol{\xi}_n)
\end{gathered} \tag{6.119}$$

in which $k(\boldsymbol{\xi}_n, \boldsymbol{\xi}_n) = \boldsymbol{\phi}(\boldsymbol{\xi}_n)^T \boldsymbol{\phi}(\boldsymbol{\xi}_n)$ is a kernel function. Predictions can be made for an arbitrary $\boldsymbol{\xi}$ as follows:

$$y = \widehat{\eta}(\boldsymbol{\xi}) = \sum_{n=1}^{N} (a_n - \overline{a}_n) k(\boldsymbol{\xi}, \boldsymbol{\xi}_n) + b \tag{6.120}$$

Only the kernel function is required, which allows for kernel substitution (the feature map $\boldsymbol{\phi}$ need not be specified).

In moving to the dual formulation, the following set of so-called Karush-Kuhn-Tucker (KKT) conditions are introduced, when swapping the order of the optimisation over Lagrange multipliers and the weights/bias (they will be explained in more detail in Sect. 6.13)

$$\begin{gathered}
(\lambda - a_n) h_n = 0 \\
(\lambda - \overline{a}_n) \overline{h}_n = 0 \\
a_n(\epsilon + h_n + \widehat{\eta}(\boldsymbol{\xi}_n) - y_n) = 0 \\
\overline{a}_n(\epsilon + \overline{h}_n - \widehat{\eta}(\boldsymbol{\xi}_n) + y_n) = 0
\end{gathered} \tag{6.121}$$

These conditions ensure that for all $\boldsymbol{\xi}_n$, one of $a_n = 0, \overline{a}_n = 0$ or $a_n = \overline{a}_n = 0$ holds. The last of these options corresponds to the y_n contained inside the ϵ-insensitive tube, which means that only such points, which we call *support vectors*, make a contribution to the predicted value (6.120). The first and third of the KKT conditions (6.121) along with the condition $h_n \ge 0$ requires that for any data $\mathbf{y}_n$ such that $a_n < \lambda$, it must hold that $h_n = 0$. Therefore, the bias is given in terms of such a data point by

$$b = y_n - \epsilon - \sum_{j=1}^{N} (a_j - \overline{a}_j) k(\boldsymbol{\xi}_j, \boldsymbol{\xi}_n) \tag{6.122}$$

employing $\mathbf{w} = \sum_{n=1}^{N} (a_n - \overline{a}_n)\boldsymbol{\phi}(\boldsymbol{\xi}_n)$. A more robust method averages the bias estimates over all $\mathbf{y}_n$ such that $a_n < \lambda$.

6.10 Gaussian Process Models for Multivariate Outputs

GP models can be extended to multivariate outputs $\mathbf{y}_n \in \mathbb{R}^d$ corresponding to a vectorised spatial or spatio-temporal field output $u(\mathbf{x}, t; \boldsymbol{\xi})$, or simply a collection of different outputs. There are a number of ways to extend the univariate framework, falling under the general framework of the *linear model of coregionalisation* (LMC), popularised in geospatial statistics [4, 5]. We consider the noiseless case and define a data matrix as

$$\mathbf{Y} = [\mathbf{y}_1, \ldots, \mathbf{y}_N]^T \in \mathbb{R}^{N \times d} \tag{6.123}$$

collecting the data $\mathbf{y}_n \in \mathbb{R}^d$ as columns, with an associated design $\Xi = \{\boldsymbol{\xi}_n\}_{n=1}^N$.

The basic assumption underlying the LMC is that the coordinates y_j of $\mathbf{y}$ are generated by latent functions $y_j = \eta_j(\boldsymbol{\xi})$ formed by linear combinations of functions that are distributed according to GPs or other random processes

$$\eta_j(\boldsymbol{\xi}) = \sum_{p=1}^{P} a_{j,p} u_p(\boldsymbol{\xi}) \tag{6.124}$$

in which the $u_p(\boldsymbol{\xi})$ are independent GPs with zero-mean and unit variance, and $a_{j,p}$ are non-random coefficients

$$u_p(\boldsymbol{\xi})|\boldsymbol{\theta}_p \sim \mathcal{GP}\left(0, k_p(\boldsymbol{\xi}, \boldsymbol{\xi}'|\boldsymbol{\theta}_p)\right) \tag{6.125}$$

Each $u_p(\boldsymbol{\xi})$ is defined by a kernel function $k_p(\boldsymbol{\xi}, \boldsymbol{\xi}'|\boldsymbol{\theta}_p)$ with associated hyperparameters $\boldsymbol{\theta}_p$.

The formulation (6.124) can be generalised by allowing for groups of the $u_p(\boldsymbol{\xi})$ to possess the same covariance

$$\eta_j(\boldsymbol{\xi}) = \sum_{p=1}^{P} \sum_{i=1}^{R_p} a^i_{j,p} u^i_p(\boldsymbol{\xi}) \tag{6.126}$$

with

$$\operatorname{cov}(u^i_p(\boldsymbol{\xi}), u^{i'}_{p'}(\boldsymbol{\xi}')) = \begin{cases} k_p(\boldsymbol{\xi}, \boldsymbol{\xi}'|\boldsymbol{\theta}_p) & \text{if } p = p' \text{ and } i = i' \\ 0 & \text{otherwise} \end{cases} \tag{6.127}$$

In other words, for each $p = 1, \ldots, P$, $u^i_p(\boldsymbol{\xi}), i = 1, \ldots, R_p$, are independent across i and share the same correlation, with a common kernel $k_p(\boldsymbol{\xi}, \boldsymbol{\xi}'|\boldsymbol{\theta}_p)$. Furthermore, there are P groups of mutually independent functions across p. As a consequence of this independence

$$\begin{aligned} \operatorname{cov}\left(\eta_j(\boldsymbol{\xi}), \eta_{j'}(\boldsymbol{\xi})\right) &= \sum_{p,p'=1}^{P} \sum_{i,i'=1}^{R_p} a^i_{j,p} a^{i'}_{j',p'} k_p(\boldsymbol{\xi}, \boldsymbol{\xi}'|\boldsymbol{\theta}_p) \\ &= \sum_{p=1}^{P} b^p_{d,j'} k_p(\boldsymbol{\xi}, \boldsymbol{\xi}'|\boldsymbol{\theta}_p) \end{aligned} \tag{6.128}$$

in which $b^p_{d,j'} = a^i_{j,p} a^{i'}_{j',p'}$. Defining the multivariate latent function

$$\boldsymbol{\eta}(\boldsymbol{\xi}) = (\eta_j(\boldsymbol{\xi}), \ldots, \eta_j(\boldsymbol{\xi}))^T \tag{6.129}$$

from (6.126), the more familiar form of this model is

$$\boldsymbol{\eta}(\boldsymbol{\xi}) = \sum_{p=1}^{P} \eta_p(\boldsymbol{\xi}) = \sum_{p=1}^{P} \mathbf{A}_p \mathbf{u}_p(\boldsymbol{\xi}) \tag{6.130}$$

with the vector GP processes $\mathbf{u}_p = (u^1_p(\boldsymbol{\xi}), \ldots, u^{R_p}_p(\boldsymbol{\xi}))^T$ and matrices $\mathbf{A}_p = [a^i_{j,p}] \in \mathbb{R}^{d \times R_p}$. The cross-covariance matrix of $\boldsymbol{\eta}(\boldsymbol{\xi})$ can be calculated as follows:

$$\begin{aligned} \text{cov}\left(\boldsymbol{\eta}(\boldsymbol{\xi}), \boldsymbol{\eta}(\boldsymbol{\xi}')\right) &= \sum_{p=1}^{P} \sum_{p'=1}^{P} \mathbf{A}_p \mathbb{E}\left[\mathbf{u}_p(\boldsymbol{\xi}) \mathbf{u}_{p'}(\boldsymbol{\xi})^T\right] \mathbf{A}^T_{p'} \\ &= \sum_{p=1}^{P} \mathbf{B}_p k_p(\boldsymbol{\xi}, \boldsymbol{\xi}' | \boldsymbol{\theta}_p) \end{aligned} \tag{6.131}$$

in which $\mathbf{B}_p = \mathbf{A}_p \mathbf{A}^T_p$ are called *coregionalisation* matrices. This particular covariance structure is called *nested*. The hyperparameters of the model are $\{\boldsymbol{\theta}_p, \mathbf{B}_p\}^P_{p=1}$, which must be learned during training. There are a number of special cases that simplify this model, and which are in practice the most often used. We outline below the two most popular.

6.10.1 Intrinsic Coregionalisation Model

The simplest case, called the *intrinsic coregionalisation model* (ICM) [4, 6] corresponds to $P = 1$, which in turn means that only one GP with kernel $k(\boldsymbol{\xi}, \boldsymbol{\xi}' | \boldsymbol{\theta})$ is used to represent all coordinates of $\mathbf{y} = \boldsymbol{\eta}(\boldsymbol{\xi})$. In this case,

$$\text{cov}\left(\boldsymbol{\eta}(\boldsymbol{\xi}), \boldsymbol{\eta}(\boldsymbol{\xi}')\right) = k(\boldsymbol{\xi}, \boldsymbol{\xi}' | \boldsymbol{\theta}) \otimes \mathbf{B} \tag{6.132}$$

in which $\mathbf{B} = \mathbf{A}\mathbf{A}^T$ is a single coregionalisation matrix and $\otimes$ is the Kronecker product. Since the correlations between coordinates of $\mathbf{y}$ and correlations between inputs $\boldsymbol{\xi}$ are factored, this model is termed *separable*. We can write the prior as

$$\boldsymbol{\eta}(\boldsymbol{\xi}) | \boldsymbol{\theta}, \mathbf{B} \sim \mathcal{GP}\left(\mathbf{0}, k(\boldsymbol{\xi}, \boldsymbol{\xi}' | \boldsymbol{\theta}) \otimes \mathbf{B}\right) \tag{6.133}$$

Note that $k(\boldsymbol{\xi}, \boldsymbol{\xi}' | \boldsymbol{\theta}) \otimes \mathbf{B}$ is a common abuse of notation, and it signifies that with a design Ξ, the covariance would take the form $\mathbf{K} \otimes \mathbf{B}$, in which $\mathbf{K}(\boldsymbol{\theta}) = [k(\boldsymbol{\xi}_n, \boldsymbol{\xi}_m | \boldsymbol{\theta})]$, $n, m = 1, \ldots, N$ is the covariance between inputs. The Kronecker product of two matrices $\mathbf{A} = [a_{ij}] \in \mathbb{R}^{m \times n}$ and $\mathbf{B} \in \mathbb{R}^{p \times q}$ is defined as

$$\mathbf{A} \otimes \mathbf{B} = \begin{bmatrix} a_{11}\mathbf{B} & \cdots & a_{1n}\mathbf{B} \\ \vdots & \ddots & \vdots \\ a_{m1}\mathbf{B} & \cdots & a_{mn}\mathbf{B} \end{bmatrix} \in \mathbb{R}^{mp \times nq} \tag{6.134}$$

In the general model (6.130), each of the $\boldsymbol{\eta}_p(\boldsymbol{\xi}) = \mathbf{A}_p \mathbf{u}_p(\boldsymbol{\xi})$ can now be seen as independent ICM formulations. The simplest intrinsic model sets $\mathbf{B} = \mathbf{I}$, which is the same as assuming that the coordinates of $\boldsymbol{\eta}(\boldsymbol{\xi})$ are identically distributed and conditionally independent given the hyperparameters $\boldsymbol{\theta}$.

A conjugate prior distribution for $\mathbf{B}$ leads to a Student's-t process given the kernel hyperparameters, which can be approximated by a MAP estimate assuming log-logistic priors [6]. Alternatively, a maximum likelihood estimate for $\{\boldsymbol{\theta}, \mathbf{B}\}$ can be employed. It has been demonstrated [7] that there is little difference between these two approaches. The more decisive aspects are the model choices, especially the separability assumption and the choice of prior. Another choice is in the parameterisation of $\mathbf{B}$, which is arbitrary. Commonly, either a Cholesky decomposition [5] or an eigen-decomposition [7] is used.

The log marginal likelihood $\mathcal{L}$ for the model

$$\mathbf{y} = \boldsymbol{\eta}(\boldsymbol{\xi}) + \boldsymbol{\epsilon} \tag{6.135}$$

with the prior (6.133) is

$$\mathcal{L} = \frac{1}{2} \ln |\boldsymbol{\Sigma}| - \frac{1}{2} \mathrm{tr} \left(\mathrm{vec}(\mathbf{Y}) \mathrm{vec}(\mathbf{Y})^T \boldsymbol{\Sigma}^{-1} \right) - \frac{dN}{2} \ln(2\pi) \tag{6.136}$$

in which vec$(\cdot)$ denotes a vectorisation and tr is the trace of a matrix. The covariance matrix is

$$\boldsymbol{\Sigma} = \mathbf{K}(\theta) \otimes \mathbf{B} + \sigma^2 \mathbf{I} \in \mathbb{R}^{dN \times dN} \tag{6.137}$$

in which σ^2 is the variance of the noise

$$\boldsymbol{\epsilon}(\boldsymbol{\xi}) \sim \mathcal{GP}(\mathbf{0}, \sigma^2 \delta(\boldsymbol{\xi}, \boldsymbol{\xi}') \otimes \mathbf{I}) \tag{6.138}$$

Note that we now include the error. A maximum likelihood estimate (MLE) can be used for the hyperparameters $\{\mathbf{B}, \boldsymbol{\theta}\}$, which can be placed in the following posterior distribution for $\boldsymbol{\eta}(\boldsymbol{\xi})$ [2]

$$\begin{aligned} \mathbb{E}[\boldsymbol{\eta}(\boldsymbol{\xi})] &= (\mathbf{B} \otimes \mathbf{k}(\boldsymbol{\xi}))^T \boldsymbol{\Sigma}^{-1} \mathrm{vec}(\mathbf{Y}) \\ \mathrm{var}\,(\boldsymbol{\eta}(\boldsymbol{\xi})) &= k(\boldsymbol{\xi}, \boldsymbol{\xi}|\boldsymbol{\theta})\mathbf{B} - (\mathbf{B} \otimes \mathbf{k}(\boldsymbol{\xi}))^T \boldsymbol{\Sigma}^{-1} (\mathbf{B} \otimes \mathbf{k}(\boldsymbol{\xi})) \\ \mathbf{k}(\boldsymbol{\xi}) &= [k(\boldsymbol{\xi}, \boldsymbol{\xi}_1|\boldsymbol{\theta}), \ldots, k(\boldsymbol{\xi}, \boldsymbol{\xi}_N|\boldsymbol{\theta})]^T \end{aligned} \tag{6.139}$$

6.10.2 Dimensionally Reduced Model

Another special case of the LMC is that of $R_p = 1$, i.e., P independent scalar GPs $u_p(\boldsymbol{\xi})$, for which

$$\boldsymbol{\eta}(\boldsymbol{\xi}) = \sum\nolimits_{p=1}^{P} \mathbf{a}_p u_p(\boldsymbol{\xi}) = \mathbf{A}\mathbf{u}(\boldsymbol{\xi}) \tag{6.140}$$

where the $\mathbf{a}_p$ are the columns of $\mathbf{A}$ and $\mathbf{u} = (u_1(\boldsymbol{\xi}), \ldots, u_P(\boldsymbol{\xi}))^T$. For this model, the cross-covariance matrix becomes

$$\begin{aligned}\operatorname{cov}(\boldsymbol{\eta}(\boldsymbol{\xi}), \boldsymbol{\eta}(\boldsymbol{\xi}')) &= \sum\nolimits_{p=1}^{P} \mathbf{B}_p k_p(\boldsymbol{\xi}, \boldsymbol{\xi}'|\boldsymbol{\theta}_p) \\ &= \mathbf{A}\,\mathrm{diag}(k_1(\boldsymbol{\xi}, \boldsymbol{\xi}'|\boldsymbol{\theta}_1), \ldots, k_P(\boldsymbol{\xi}, \boldsymbol{\xi}'|\boldsymbol{\theta}_P))\,\mathbf{A}^T\end{aligned} \tag{6.141}$$

in which $\mathbf{B}_p = \mathbf{a}_p\mathbf{a}_p^T$.

One particular choice of this model corresponds to $\mathbf{a}_p = \mathbf{v}_p$, in which $\mathbf{v}_p$ are the *principal directions* obtained from a *principal component analysis* (PCA) of the data $\mathbf{Y}$ [8]. The directions are mutually orthogonal and of unit length. Importantly, only $Q \ll d$ of the directions are used, leading to a dimension reduction, namely approximation of the outputs in the Q−dimensional subspace spanned by $\mathbf{v}_1, \ldots, \mathbf{v}_Q$

$$\boldsymbol{\eta}(\boldsymbol{\xi}) = \mathbf{V}_Q\mathbf{u}(\boldsymbol{\xi}) = \sum\nolimits_{p=1}^{Q}, \quad \mathbf{V}_Q = [\mathbf{v}_1, \ldots, \mathbf{v}_Q] u_p(\boldsymbol{\xi})\mathbf{v}_p \tag{6.142}$$

By the properties of PCA, the coefficients u_p are uncorrelated and can therefore be taken to be independent GPs, each with their own covariance function $k_p(\boldsymbol{\xi}, \boldsymbol{\xi}'|\boldsymbol{\theta}_p)$. We note that independent and uncorrelated are equivalent properties for a Gaussian.

In this model, the cross-covariance is

$$\begin{aligned}\operatorname{cov}\left(\boldsymbol{\eta}(\boldsymbol{\xi}), \boldsymbol{\eta}(\boldsymbol{\xi}')\right) &= \mathbf{V}_Q\,\mathrm{diag}[k_p(\boldsymbol{\xi}, \boldsymbol{\xi}'|\boldsymbol{\theta}_p)]\,\mathbf{V}_Q^T \\ &= \sum\nolimits_p k_p(\boldsymbol{\xi}, \boldsymbol{\xi}'|\boldsymbol{\theta}_p)\mathbf{v}_p\mathbf{v}_p^T\end{aligned} \tag{6.143}$$

leading to $\mathbf{B}_p = \mathbf{v}_p\mathbf{v}_p^T$. In contrast to the ICM model, (6.142) is not separable, unless we assume that the u_p are i.i.d.. PCA provides values for the coefficients at the design points: $u_p(\boldsymbol{\xi}_i) = \mathbf{v}_p^T\mathbf{y}_i$, $i = 1, \ldots, N$. Therefore, the coefficients $u_p(\boldsymbol{\xi})$ can be learned independently with Q univariate GP models. PCA is outlined in Sect. 6.14, along with other linear dimension reduction methods.

6.11 Other Approaches to Modelling Random Fields

In order to find a numerical solution for an unknown random function $u(\mathbf{x}, t; \boldsymbol{\xi})$ underlying corresponding data $\mathbf{y}_n \in \mathbb{R}^d$, the field is commonly assumed to take the following separable form:

$$u(\mathbf{x}, t; \boldsymbol{\xi}) = \sum_{r=1}^{R} \psi_r(\mathbf{x}, t)\alpha_r(\boldsymbol{\xi}) \tag{6.144}$$

as $R \to \infty$. This assumption has been widely used in the literature, including in generalised polynomial chaos (gPC) expansions, the stochastic Galerkin method [9–11] and stochastic collocation with gPC [12]. In these spectral methods, the $\{\psi_r(\mathbf{x}, t)\}$ are considered to be deterministic, and are determined from the data (e.g., stochastic collocation) or, as in the stochastic Galerkin approach, from a Galerkin projection onto a basis of random orthogonal functions $\{\alpha_r(\boldsymbol{\xi})\} \subset \mathcal{H}$, where $\mathcal{H}$ is some Hilbert space.

In gPC the $\alpha_r(\boldsymbol{\xi})$ are based on the distributions $\rho_j(\xi_j)$ over $\xi_j \in \mathcal{X}_j$ (if they are known), where $\mathcal{X}_j$ is the design space for the j-th component of $\boldsymbol{\xi}$. The basis is constructed as a tensor product space (see the next section), consisting of the span of tensor products of known orthogonal polynomials in ξ_j (e.g., Hermite polynomials for Gaussian distributions), equipped with the weighted inner product

$$\langle f, g \rangle_{\rho_j} = \int_{\mathcal{X}_j} f g \rho_j d\xi_j \tag{6.145}$$

For a general distribution, these single-parameter polynomials can be constructed *via* the Askey scheme [11] or *via* orthogonalisation. The constructions rely on the assumption that the ξ_j are mutually independent. Then the joint density $\rho(\boldsymbol{\xi}) = \prod_j \rho_j(\xi_j)$ defines a weighted inner product in the Hilbert space $\mathcal{H}$ consisting of products of the polynomials in ξ_j up to some chosen order P.

If $u(\mathbf{x}, t; \boldsymbol{\xi})$ has finite variation, the expansion (6.144) converges in $L^2(\mathcal{X}, \mathcal{B}(\mathcal{X}), \rho(\boldsymbol{\xi})d\boldsymbol{\xi})$, the space of square-integrable equivalence classes defined on the image probability space endowed with the Borel sigma algebra $\mathcal{B}(\mathcal{X})$ on $\mathcal{X}$ for a valid measure $\rho(\boldsymbol{\xi})d\boldsymbol{\xi}$.

Even though convergence of the partial sums (6.144) is guaranteed with Hermite polynomials, it is important to use the optimal polynomials for a given distribution in order to guarantee exponential convergence, ultimately so that the scheme is computationally practical for high-dimensional input spaces. The number of required basis terms R is around $(L + P)!/(L!P!)$, which grows rapidly for large values of P and L. Such methods, therefore, have poor scalability.

From a GP perspective, as discussed earlier, rather than specifying the $u_i(\boldsymbol{\xi})$ in expansion (6.144), they are treated as latent functions, while the $\psi_r(\mathbf{x}, t)$ are considered to be basis functions [8, 13, 14]. We can place independent GP priors over each latent process

$$u_r(\boldsymbol{\xi}) \sim \mathcal{GP}\left(0, k_r(\boldsymbol{\xi}, \boldsymbol{\xi}')\right) \tag{6.146}$$

with mean function 0 and covariance (kernel) function $k_r(\boldsymbol{\xi}, \boldsymbol{\xi}')$ (for simplicity, we do not include the hyperparameters $\boldsymbol{\theta}$ in the notation). The joint prior is

$$\mathbf{u}(\boldsymbol{\xi}) \sim \mathcal{GP}\left(\mathbf{0}, \text{diag}(k_1(\boldsymbol{\xi}, \boldsymbol{\xi}') \cdots, k_R(\boldsymbol{\xi}, \boldsymbol{\xi}')\right) \tag{6.147}$$

where $\mathbf{u}(\boldsymbol{\xi}) = (u_1(\boldsymbol{\xi}), \ldots, u_R(\boldsymbol{\xi}))^T$. Expansion (6.144) clearly defines a GP

$$u(\mathbf{x}, t; \boldsymbol{\xi}) \sim \mathcal{GP}\left(0, \sum\nolimits_{r=1}^{R} \psi_r(\mathbf{x}, t) \otimes \psi_r(\mathbf{x}', t') \otimes k_r(\boldsymbol{\xi}, \boldsymbol{\xi}')\right) \tag{6.148}$$

in which $\psi_r(\mathbf{x}, t) \otimes \psi_r(\mathbf{x}', t') \otimes k_r(\boldsymbol{\xi}, \boldsymbol{\xi}')$ can be interpreted as an element of a tensor product space by defining suitable spaces for $\psi_r(\mathbf{x}, t)$ and $k_r(\boldsymbol{\xi}, \boldsymbol{\xi}')$ over the real numbers (Sect. 6.11.1). A finite-dimensional formulation within the GP framework is obviously necessary, and this is achieved as described in Sect. 3.7.1 (Eqs. (3.181) and (3.182)), leading to data Ξ, $\mathbf{Y}$, and the model

$$\boldsymbol{\eta}(\boldsymbol{\xi}) = \sum\nolimits_{r=1}^{R} \boldsymbol{\psi}_r u_r(\boldsymbol{\xi}) \tag{6.149}$$

in which $u_r(\boldsymbol{\xi}) \sim \mathcal{GP}\left(0, k_r(\boldsymbol{\xi}, \boldsymbol{\xi}')\right)$ as before and $\boldsymbol{\psi}_r \in \mathbb{R}^d$ is now a basis vector in a Euclidean space, the components of which correspond to values of $\psi_r(\mathbf{x}, t)$ at the specified spatio-temporal grid locations. The GP model is

$$\mathbf{y} = \boldsymbol{\eta}(\boldsymbol{\xi}) + \boldsymbol{\epsilon} \sim \mathcal{GP}\left(\mathbf{0}, \sum\nolimits_{r=1}^{R} \boldsymbol{\psi}_r \boldsymbol{\psi}_r^T \otimes k_r(\boldsymbol{\xi}, \boldsymbol{\xi}') + \delta(\boldsymbol{\xi}, \boldsymbol{\xi}')\sigma^2 \mathbf{I}\right) \tag{6.150}$$

which is the discrete form of (6.148) with added noise. This is the variant of the LMC discussed in Sect. 6.10.2.

For certain types of problems, we may run into computational difficulties, namely when the size of d is very high. One can imagine, e.g., a 3D problem on a $100 \times 100 \times 100$ grid with 10 times steps, leading to $d = 10^7$. Performing matrix inversions (on the covariance matrix) and storing the components soon becomes problematic. In such cases, it may be more convenient to assume that the correlations in the various dimensions (spatial, temporal and parametric) can be factored. This leads to tensor-based formulations, which utilise various decompositions of tensors to reduce the computational burden. We first introduce the concept of a tensor before discussing the implementation of tensor-based formulations for regression. This material will also be required for Sect. 6.14.3 on tensor decompositions.

6.11.1 Tensors and Multi-arrays

A *tensor* of *order* K is an element of a tensor product space, written

$$V_1 \otimes V_2 \otimes \ldots \otimes V_K \quad \text{or} \quad \bigotimes_{k=1}^{K} V_k \tag{6.151}$$

for arbitrary underlying vector spaces V_k defined over $\mathbb{R}$. Any number of the V_k can be replaced by their duals $V_k^* = \mathrm{Hom}(V_k, \mathbb{R})$, the set of all homomorphisms from V_k to $\mathbb{R}$, with the convention that the dual spaces appear first in the ordering. We are concerned with the case $V_k = \mathbb{R}^{p_k}$, for some $p_k \in \mathbb{N}$. An *elementary* (or pure) element of the tensor product space can be written as $\mathbf{a}_1 \otimes \mathbf{a}_2 \otimes \ldots \otimes \mathbf{a}_K$, for vectors $\mathbf{a}_k \in \mathbb{R}^{p_k}$. In general, however, elements of the tensor product space (denoted by calligraphic letters) are finite linear combinations of elementary tensors

$$\mathcal{A} = \sum_{j=1}^{r} \mathbf{a}_1^j \otimes \mathbf{a}_2^j \otimes \ldots \otimes \mathbf{a}_K^j \tag{6.152}$$

for some $\mathbf{a}_k^j \in \mathbb{R}^{p_k}$, in which r is called the *rank* of the tensor. The $\otimes$ operation is associative and satisfies the following multilinearity law for any $k \in \{1, \ldots, K\}$

$$\mathbf{a}_1 \otimes \ldots \otimes (\lambda \mathbf{a}_k + \mathbf{c}) \otimes \ldots \otimes \mathbf{a}_K = \lambda \mathbf{a}_1 \otimes \ldots \otimes \mathbf{a}_k \otimes \ldots \otimes \mathbf{a}_K + \mathbf{a}_1 \otimes \ldots \otimes \mathbf{c} \otimes \ldots \otimes \mathbf{a}_K$$

for all vectors $\mathbf{a}_k \in \mathbb{R}^{p_k}$, $\mathbf{c} \in \mathbb{R}^{p_k}$ and any $\lambda \in \mathbb{R}$. Moreover, λ can be placed in any of the factors in the first tensor on the r.h.s., without changing the tensor. These rules can be used to define an equivalence relation on formal linear combinations of elementary tensors.

Since $\mathbb{R}^{p_k}$ is an inner product space, there is a natural isomorphism $\mathbb{R}^{p_k} \to \mathbb{R}^{p_k *}$ given by $\mathbf{a}_k \mapsto \mathbf{a}_k^*(\cdot) := \langle \mathbf{a}_k, \cdot \rangle$, using the standard inner product $\langle \cdot, \cdot \rangle : \mathbb{R}^{p_k} \times \mathbb{R}^{p_k} \mapsto \mathbb{R}$. By the Riesz representation theorem, every element of $\mathbb{R}^{p_k *}$ is of this form. Thus, using the standard basis $\{\mathbf{e}_{i_k}^k\}_{i_k=1}^{p_k}$, for a given a vector

$$\mathbf{a} = \sum_{i_k} a_{k,i_k} \mathbf{e}_{i_k}^k \tag{6.153}$$

the corresponding dual vector is

$$\mathbf{a}_k^* = \sum_{i_k} a_{k,i_k} \mathbf{e}_{i_k}^{k*} \tag{6.154}$$

written in the corresponding standard dual basis $\{\mathbf{e}_{i_k}^{k*}\}_{i_k=1}^{p_k}$. Here $\mathbf{e}_{i_k}^{k*}(\cdot) \in \mathbb{R}^{p_k *}$ is the action of left multiplication of the argument by $(\mathbf{e}_{i_k}^k)^T$. Therefore, each $\mathbf{a}_k^* \in \mathbb{R}^{p_k *}$ written in the standard dual basis is equivalent to left multiplication by $\mathbf{a}_k^T$. Since the double dual satisfies $\mathbb{R}^{p_k **} \cong \mathbb{R}^{p_k}$, with $\cong$ denoting topological equivalence (homoemorphic spaces), every element $\mathbf{a}_k \in \mathbb{R}^{p_k}$ can be identified with the linear map $\mathbb{R}^{p_k *} \to \mathbb{R} : \mathbf{b}_k^* \mapsto \mathbf{b}_k^*(\mathbf{a}_k)$.

Based on the tensor product rules defined above, and using the standard dual basis for each $\mathbb{R}^{p_k}$, a basis for $\bigotimes_{k=1}^K \mathbb{R}^{p_k}$ is given by

$$\{\mathbf{e}_{i_1}^1 \otimes \ldots \otimes \mathbf{e}_{i_K}^K\}_{\mathbf{i} \in I}, \quad \mathbf{i} = (i_1, \ldots, i_K) \tag{6.155}$$

with values in the set $I = \{\mathbf{i} : i_k \in \{1, \ldots, p_k\}\}$. The dimension of the space is therefore $\prod_{k=1}^K p_k$. In terms of this basis, we may write

$$\mathbf{a}_1^j \otimes \ldots \otimes \mathbf{a}_K^j = \sum_{\mathbf{i}} a_{\mathbf{i}}^j \mathbf{e}_{i_1}^1 \otimes \ldots \otimes \mathbf{e}_{i_K}^K \quad (6.156)$$

where the *tensor components* are given by $a_{\mathbf{i}}^j := a_{1,i_1}^j \ldots a_{K,i_K}^j$, in which a_{k,i_k}^j is component i_k of $\mathbf{a}_k^j \in \mathbb{R}^{p_k}$. A general tensor has the form

$$\mathcal{A} = \sum_{\mathbf{i}} a_{\mathbf{i}} \mathbf{e}_{i_1}^1 \otimes \ldots \otimes \mathbf{e}_{i_K}^K, \quad a_{\mathbf{i}} = \sum_j a_{\mathbf{i}}^j \quad (6.157)$$

Scalar multiplication and vector addition over the vector space can be defined by identifying $\bigotimes_{k=1}^K \mathbb{R}^{p_k}$ with a space of multilinear functionals: any tensor $\mathcal{A} \in \bigotimes_{k=1}^K \mathbb{R}^{p_k}$ defines a multilinear functional $f : \mathbb{R}^{p_1 *} \times \ldots \times \mathbb{R}^{p_K *} \to \mathbb{R}$ as follows

$$f(\mathbf{b}_1^*, \ldots, \mathbf{b}_K^*) = \sum_{\mathbf{i}} a_{\mathbf{i}} \mathbf{b}_1^* \left(\mathbf{e}_{i_1}^1\right) \mathbf{b}_2^* \left(\mathbf{e}_{i_2}^2\right) \ldots \mathbf{b}_K^* \left(\mathbf{e}_{i_K}^K\right) \quad (6.158)$$

using the identification $\mathbb{R}^{p_k **} \cong \mathbb{R}^{p_k}$. From this equation, it is clear that $f(\mathbf{e}_{i_1}^{1*}, \ldots, \mathbf{e}_{i_K}^{K*})$ are the components $a_{\mathbf{i}}$ of $\mathcal{A}$. Multilinearlity follows from the linearity of $\mathbf{b}_k^* \in \mathbb{R}^{p_k *}$.

In the context of machine learning, $\bigotimes_{k=1}^K \mathbb{R}^{p_k}$ is usually identified with the space of *hypermatrices* or *multidimensional arrays* (also called multi-arrays) $\mathbb{R}^{p_1 \times \ldots \times p_K}$, by virtue of the fact that $\bigotimes_{k=1}^K \mathbb{R}^{p_k} \cong \mathbb{R}^{p_1 \times \ldots \times p_K}$. The obvious way to establish an isomorphism is to introduce a basis (as above) and set the entries of the hypermatrix to be the tensor components, with addition and scalar multiplication for the vector space of hypermatrices defined entry-wise.

Vectors $\mathbf{a}_k$ considered as component vectors can be used to construct order-K hypermatrices *via* the *outer product*, which possesses the same properties as the tensor product operation. For vectors $\mathbf{a} \in \mathbb{R}^n$, $\mathbf{b} \in \mathbb{R}^m$ with elements a_i and b_j, the outer product $\mathbf{a} \circ \mathbf{b}$ can be written in a number of equivalent ways

$$\mathbf{a} \circ \mathbf{b} = \mathbf{a}\mathbf{b}^T = \mathbf{a} \otimes \mathbf{b}^T \in \mathbb{R}^{n \times m} \quad (6.159)$$

in which the i, j-th entry is $a_i b_j$. For example, $\mathbf{a}_1^j \circ \ldots \circ \mathbf{a}_K^j = (a_{\mathbf{i}}^j)_{\mathbf{i} \in I}$, which is identified with the elementary tensor $\mathbf{a}_1^j \otimes \ldots \otimes \mathbf{a}_K^j = \sum_{\mathbf{i}} a_{\mathbf{i}}^j \mathbf{e}_{i_1}^1 \otimes \ldots \otimes \mathbf{e}_{i_K}^K$ defined earlier. The unfolding operations defined below lead to equivalent new hypermatrices *via* isomorphisms, e.g., $\bigotimes_{k=1}^K \mathbb{R}^{p_k} \cong \mathbb{R}^{p_1 \times \prod_{k=1}^K p_k}$. No distinction is made between the tensor and its corresponding hypermatrix representation and both are referred to as order$-K$ tensors.

The Kronecker product (6.134), also denoted $\otimes$, is likewise a tensor product specialised to matrices. There are two more important matrix products, the first of which is the *Hadamard product* $\odot$ of two $m \times n$ matrices $\mathbf{A} = (a_{ij})$ and $\mathbf{B} = (b_{ij})$

$$(\mathbf{A} \odot \mathbf{B})_{ij} = (a)_{ij} (b)_{ij} \quad (6.160)$$

which is nothing other than an element-wise multiplication. The second important product is the *Khatri-Rao product* $*$ of two matrices $\mathbf{A} = [\mathbf{a}_1 \dots \mathbf{a}_n] \in \mathbb{R}^{m\times n}$ and $\mathbf{B} = [\mathbf{b}_1 \dots \mathbf{b}_n] \in \mathbb{R}^{p\times n}$

$$\mathbf{A} * \mathbf{B} = [\mathrm{a}_1 \otimes \mathrm{b}_1 \;\; \mathrm{a}_2 \otimes \mathrm{b}_2 \;\; \dots \;\; \mathrm{a}_k \otimes \mathrm{b}_k] \in \mathbb{R}^{mp\times n} \tag{6.161}$$

in which $\mathrm{a}_i \otimes \mathrm{b}_i \in \mathbb{R}^{mp}$ should not to be confused with $\mathrm{a}_i \circ \mathrm{b}_i = \mathrm{a}_i \otimes \mathrm{b}_i^T \in \mathbb{R}^{m\times p}$.

The dimensions of the multidimensional array are referred to as *modes* or legs. The *mode−k vectors* (or fibres) of an order−K tensor $\mathcal{A}$ interpreted as a hypermatrix are column vectors formed by fixing all but the k−th index. *Vectorisation* of $\mathcal{A}$ is a simple reorganisation of the tensor by stacking the mode−1 fibres into a column vector, and is denoted vec($\mathcal{A}$). This allows us to define a natural inner product $\langle\cdot,\cdot\rangle$ for tensors of the same size, and also to define the *Frobenius norm* $\|\cdot\|_F$

$$\langle\mathcal{A},\mathcal{B}\rangle = \langle\mathrm{vec}(\mathcal{A}), \mathrm{vec}(\mathcal{B})\rangle, \quad \|\mathcal{A}\|_F = \langle\mathcal{A},\mathcal{A}\rangle^{\frac{1}{2}} \tag{6.162}$$

Note that the Frobenius norm, which consists of a sum of the element-wise products of the tensors, is a natural extension of the standard Euclidean norm. It applies to any two tensors, including matrices.

The mode−k product of $\mathcal{A}$ with a matrix $\mathbf{D} \in \mathbb{R}^{J\times p_k}$ having elements d_{ij} is a K−order tensor, denoted $\mathcal{A} \times_k \mathbf{D}$, with elements defined as follows

$$(\mathcal{A} \times_k \mathbf{D})_{i_1\dots i_{k-1} j i_{k+1}\dots i_K} = \sum\nolimits_{i_k=1}^{p_k} a_{i_1 i_2\dots i_K} d_{j i_k} \tag{6.163}$$

The resulting tensor $\mathcal{B} = \mathcal{A} \times_k \mathbf{D}$ is of size $p_1 \times p_{k-1} \times J \times p_{k+1} \times \dots \times p_K$. It is equivalent to pre-multiplying each mode−k fibre by $\mathbf{D}$. The equivalent matrix form is

$$\mathbf{B}_{(k)} = \mathbf{D}\mathbf{P}_{(k)}, \quad \mathbf{B}_{(k)} \in \mathbb{R}^{J\times \prod_{j\neq k} p_j}, \; \mathbf{P}_{(k)} \in \mathbb{R}^{p_k\times \prod_{j\neq k} p_j} \tag{6.164}$$

where $\mathbf{B}_{(k)}$ and $\mathbf{P}_{(k)}$ are the mode−k unfoldings (matricisations) of $\mathcal{B}$ and $\mathcal{A}$, respectively, which are formed by arranging the mode-k fibres as columns in a matrix. An important property of the n-mode product is that for distinct modes $k_1 \neq k_2$

$$\mathcal{A} \times_{k_1} \mathbf{D}_1 \times_{k_2} \mathbf{D}_2 = \mathcal{A} \times_{k_2} \mathbf{D}_2 \times_{k_1} \mathbf{D}_1 \tag{6.165}$$

while for $k_1 = k_2 = k$, $\mathcal{A} \times_k \mathbf{D}_1 \times_k \mathbf{D}_2 = \mathcal{A} \times_k (\mathbf{D}_2\mathbf{D}_1)$.

The mode−k (vector) product of a K−order tensor $\mathcal{A}$ with a vector $\mathbf{v} = (v_1, \dots, v_K)^T \in \mathbb{R}^{p_k}$ is denoted $\mathcal{A} \bar{\times}_k \mathbf{v}^T$ and is defined element-wise as follows

$$(\mathcal{A} \bar{\times}_k \mathbf{v}^T)_{i_1\dots i_{k-1} i_{k+1}\dots i_K} = \sum\nolimits_{i_k=1}^{p_k} a_{i_1 i_2\dots i_K} v_{i_k} \tag{6.166}$$

This is equivalent to taking the inner product of each mode−k fibre with the vector $\mathbf{v}$ and thus results in a tensor of order $K-1$ and size $p_1 \times \dots \times p_{k-1} \times p_{k+1} \times \dots \times p_K$.

For any $\mathcal{A} \in \mathbb{R}^{p_1 \times \ldots \times p_K}$, there exists $r \in \mathbb{N}$ and $\mathbf{b}_i^k \in \mathbb{R}^{p_k}$ such that [15]

$$\mathcal{A} = \sum_{i=1}^{r} \mathbf{b}_i^1 \circ \mathbf{b}_i^2 \circ \ldots \circ \mathbf{b}_i^K \tag{6.167}$$

The smallest r for which such a representation holds is called the *canonical polyadic (CP) rank* of $\mathcal{A}$. The representation (6.167) is referred to as the *CP decomposition* of $\mathcal{A}$ and can be written as

$$\mathcal{A} = [\![\mathbf{B}_1, \ldots, \mathbf{B}_K]\!] = \sum_{i=1}^{r} \mathbf{b}_i^1 \circ \mathbf{b}_i^2 \circ \ldots \circ \mathbf{b}_i^K \tag{6.168}$$

in which the $\mathbf{B}_i = [\mathbf{b}_i^1 \ldots \mathbf{b}_i^r]$ are called *factors*. The operator $[\![\cdot]\!]$ is called the *Kruskal operator*. This decomposition can form the basis for methods to compress the tensor $\mathcal{A}$ by finding low CP-rank approximations.

Another decomposition that is frequently used for compressing tensors is the *Tucker decomposition* [15]. For any $\mathcal{A} \in \mathbb{R}^{p_1 \times \ldots \times p_K}$, the Tucker decomposition is defined as

$$\begin{aligned} \mathcal{A} = \mathcal{G} \times_1 \mathbf{B}_1 \times_2 \ldots \times_K \mathbf{B}_K &= \sum_{r_1=1}^{n_1} \cdots \sum_{r_K=1}^{n_K} g_{r_1 \ldots K} \mathbf{b}_1^{r_1} \circ \ldots \circ \mathbf{b}_K^{r_K} \\ &:= [\![\mathcal{G}; \mathbf{B}_1, \ldots, \mathbf{B}_K]\!] \end{aligned} \tag{6.169}$$

in which $\mathcal{G} \in \mathbb{R}^{n_1 \times \ldots \times n_K}$ is called the *core tensor* and $\mathbf{B}_k = [\mathbf{b}_k^1 \ldots \mathbf{b}_k^{n_k}] \in \mathbb{R}^{p_k \times n_K}$ are again called factors. The tuple $(n_1, \ldots, n_K)$ is called the *multilinear rank* of $\mathcal{A}$. Equivalently,

$$\begin{aligned} \mathbf{A}_{(k)} &= \mathbf{B}_k \mathbf{G}_{(k)} (\mathbf{B}_K \otimes \ldots \otimes \mathbf{B}_{k+1} \otimes \mathbf{B}_{k-1} \otimes \ldots \otimes \mathbf{B}_1)^T \\ \mathrm{vec}(\mathcal{A}) &= (\mathbf{B}_K \otimes \ldots \otimes \mathbf{B}_1) \mathrm{vec}(\mathcal{G}) \end{aligned} \tag{6.170}$$

A tensor-variate normal for a random tensor $\mathcal{A}$ is denoted

$$\mathcal{A} \sim \mathcal{TN}_{p_1, \ldots, p_K}(\mathcal{M}, \mathbf{K}_1, \ldots, \mathbf{K}_K) \tag{6.171}$$

where the subscripts indicate the dimensions of the modes, the first argument is the mean tensor $\mathcal{M} = \mathbb{E}[\mathcal{W}]$ and the remaining arguments are the covariance matrices corresponding to each mode; that is, the entries of $\mathbf{K}_k(i_k, i_k')$ are the covariances between elements $a_{\mathbf{i}}$ and $a_{\mathbf{i}'}$ with $\mathbf{i} = (i_1, \ldots, i_k, \ldots, i_K)$ and $\mathbf{i}' = (i_1, \ldots, i_k', \ldots, i_K)$. The defining property of a tensor variate normal distribution is that

$$\mathrm{vec}(\mathcal{A}) \sim \mathcal{N}\left(\mathrm{vec}(\mathcal{M}), \bigotimes_{k=K}^{1} \mathbf{K}_k(i_k, i_k')\right) \tag{6.172}$$

Note that matrices are second-order tensors while vectors are first-order tensors, but we will use the notation $\mathcal{TN}$ only for order 2 and higher.

6.11.2 Tensor-Variate Gaussian Process Models

Suppose we have a spatio-temporal field $u(\mathbf{x}, t; \boldsymbol{\xi})$ parameterised by $\boldsymbol{\xi}$. The output on a spatial grid can be organised for different times into an order-4 tensor (hyper-matrix) format $\mathcal{Y}_n$ such that $\text{vec}(\mathcal{Y}_n) = \mathbf{y}_n \in \mathbb{R}^d$, or $\text{vec}(\mathcal{Y}) = \mathbf{y}$ for a general $\mathbf{y}$. The ordering of the modes can be taken as the spatial coordinates x_1, x_2, x_3, followed by time. Then

$$\mathcal{Y} = (y_{\mathbf{i}})_{\mathbf{i}\in I} \in \mathbb{R}^{d_1\times d_2\times d_3\times d_4}, \quad \mathbf{i} = (i_1, \ldots, i_4), \quad I = \{\mathbf{i} : i_k \in \{1, \ldots, d_k\}\} \tag{6.173}$$

d_1, d_2, and d_3 are the numbers of points in the first, second and third spatial dimensions, while d_4 is the number of snapshots in time, with $d_1 d_2 d_3 d_4 = d$.

A component-wise model that separates parameter covariances from spatio-temporal location covariances is

$$y_{\mathbf{i}} = g_{\mathbf{i}}(\boldsymbol{\xi}) + \epsilon, \quad g_{\mathbf{i}}(\boldsymbol{\xi}) \sim \mathcal{GP}\left(0, k(\boldsymbol{\xi}, \boldsymbol{\xi}'|\boldsymbol{\theta})c(\mathbf{i}, \mathbf{i}')\right) \tag{6.174}$$

in which $k(\mathbf{i}, \mathbf{i}') = \text{cov}(g_{\mathbf{i}}(\boldsymbol{\xi}), g_{\mathbf{i}'}(\boldsymbol{\xi}))$ and $k(\boldsymbol{\xi}, \boldsymbol{\xi}'|\boldsymbol{\theta})$ is the covariance function across parameters. $\epsilon \sim \mathcal{GP}(0, \tau^{-1}\delta(\boldsymbol{\xi}, \boldsymbol{\xi}'))$ is an error with precision τ (inverse variance). If we further assume that values in each of the coordinates (in space and time) are uncorrelated, we obtain a *separable* form for $k(\mathbf{i}, \mathbf{i}')$

$$k(\mathbf{i}, \mathbf{i}') = k^{(1)}(i_1, i_1')k^{(2)}(i_2, i_2')k^{(3)}(i_3, i_3')k^{(4)}(i_4, i_4') \tag{6.175}$$

for kernel functions $k^{(j)}$ that define correlations across space and time. This is a natural structure in view of the anisotropy of many physical processes. This separability leads to a model with the prior

$$\mathcal{Y} \sim \mathcal{TN}_{d_1,\ldots,d_4}(\mathcal{O}|\mathcal{M}, \mathbf{K}_1, \ldots, \mathbf{K}_4, k(\boldsymbol{\xi}, \boldsymbol{\xi}'|\boldsymbol{\theta})) + \mathcal{E}(\boldsymbol{\xi}) \tag{6.176}$$

in which $\mathbf{K}_k = [k^{(k)}(i_k, i_k')]_{i_k, i_k'} \in \mathbb{R}^{d_k\times d_k}$ and we assume a zero-tensor mean, $\mathbb{E}[\mathcal{Y}] = \mathcal{O}$. The homogeneous error term satisfies

$$\mathcal{E}(\boldsymbol{\xi}) \sim \mathcal{TN}_{d_1,\ldots,d_4}(0, \mathbf{I}, \ldots, \mathbf{I}, \tau^{-1}\delta(\boldsymbol{\xi}, \boldsymbol{\xi}')) \tag{6.177}$$

such that $\text{vec}(\mathcal{E}(\boldsymbol{\xi})) = \boldsymbol{\epsilon}(\boldsymbol{\xi})$. Note that we place $k(\boldsymbol{\xi}, \boldsymbol{\xi}'|\boldsymbol{\theta})$ and the covariance for the error $\tau^{-1}\delta(\boldsymbol{\xi}, \boldsymbol{\xi}')$ as the last argument, which is similar to the notation in (6.133).

An equivalent formulation is

$$\mathbf{y} = \text{vec}(\mathcal{Y}) \sim \mathcal{GP}\left(\mathbf{0}, k(\boldsymbol{\xi}, \boldsymbol{\xi}'|\boldsymbol{\theta}) \otimes \bigotimes_{k=4}^{1} \mathbf{K}_k(i_k, i_k') + \tau^{-1}\delta(\boldsymbol{\xi}, \boldsymbol{\xi}')\mathbf{I}\right) \tag{6.178}$$

which is precisely the ICM multivariate model with solution (6.139). Defining an order−5 tensor $\widehat{\mathcal{Y}} \in \mathbb{R}^{d_1 \times d_2 \times d_3 \times d_4 \times N}$ that collects all of the $\mathcal{Y}_n$, the likelihood function is

$$\begin{aligned}\log p(\widehat{\mathcal{Y}}|\Xi) &= \log p(\text{vec}(\widehat{\mathcal{Y}})|\Xi) \\ &= -\frac{1}{2}\log|\mathbf{S}| - \frac{1}{2}\text{vec}(\widehat{\mathcal{Y}})\mathbf{S}^{-1}\text{vec}(\widehat{\mathcal{Y}}) - \frac{Nd}{2}\log(2\pi)\end{aligned} \tag{6.179}$$

with kernel matrix

$$\mathbf{S} = \mathbf{K}_5 \otimes \mathbf{K}_4 \otimes \mathbf{K}_3 \otimes \mathbf{K}_2 \otimes \mathbf{K}_4 + \tau^{-1}\mathbf{I} \tag{6.180}$$

in which $\mathbf{K}_5 \in \mathbb{R}^{N \times N}$ is the covariance matrix across parameters with entries $k(\boldsymbol{\xi}_n, \boldsymbol{\xi}_m|\boldsymbol{\theta})$. We can now optimise $\mathcal{L}$ w.r.t. $\mathbf{K}_k$ and $\boldsymbol{\theta}$.

This does not, however, appear to have gained us any particular advantage. On the other hand, tensor-product structure of the kernel $\mathbf{S}$ in (6.180) arising from the tensor-variate formulation (6.176), underlying which are the assumptions on the correlation structure offers large computational savings. The major challenge is in the calculation of the inverse and determinant of $\mathbf{S} \in \mathbb{R}^{Nd \times Nd}$, with up to $\mathcal{O}(N^3 d^3)$ complexity. Below, we explain how to reduce this complexity to $\mathcal{O}\big(Nd(\sum_{k=1}^{4} d_k + N)\big)$, following the working of [16] closely. Later we look at different types of (non-GP-based) tensor variate models in Sect. 6.11.3.

Each of the covariance matrices $\mathbf{K}_i$ is symmetric, so that it has a spectral decomposition

$$\mathbf{K}_i = \mathbf{U}_k \text{diag}(\boldsymbol{\lambda}_k)\mathbf{U}_k^T \tag{6.181}$$

in which the columns of $\mathbf{U}_k \in \mathbb{R}^{d_k \times d_k}$ are eigenvectors of $\mathbf{K}_k$ and $\text{diag}(\boldsymbol{\lambda}_k) \in \mathbb{R}^{d_k \times d_k}$ is a diagonal matrix, with entries equal to the eigenvalues of $\mathbf{K}_k$, for $k = 1, \ldots, 4$. For $k = 5$, we replace d_k with N. The diagonal elements are contained in the vector $\boldsymbol{\lambda}_k$. Repeated application of the property $(\mathbf{A}_1 \otimes \mathbf{A}_2)(\mathbf{A}_3 \otimes \mathbf{A}_4) = (\mathbf{A}_1\mathbf{A}_2) \otimes (\mathbf{A}_2\mathbf{A}_4)$ for any matrices $\mathbf{A}_i$ leads to

$$\begin{aligned}\mathbf{S} &= \bigotimes_{k=1}^{K} \mathbf{K}_k + \tau^{-1}\mathbf{I} \\ &= \bigotimes_{k=1}^{K} (\mathbf{U}_k \text{diag}(\boldsymbol{\lambda}_k)\mathbf{U}_k^T) + \tau^{-1}\mathbf{I} \\ &= \bigotimes_{k=1}^{K} \mathbf{U}_k \left[\text{diag}\left(\bigotimes_{k=1}^{K} \boldsymbol{\lambda}_k\right) + \tau^{-1}\mathbf{I}\right] \bigotimes_{k=1}^{K} \mathbf{U}_k^T \\ &:= \mathbf{P}(\Lambda + \tau^{-1}\mathbf{I})\mathbf{P}^T\end{aligned} \tag{6.182}$$

in which $K = 5$. Note that $\otimes_{k=1}^{K} \boldsymbol{\lambda}_k$ is a vector of size $1 \times Nd$ so that $\Lambda \in \mathbb{R}^{Nd \times Nd}$. The matrix $\mathbf{P}$ is orthogonal by virtue of the fact that the $\mathbf{U}_k$ are orthogonal, so that the noise term can be taken inside the brackets. From (6.182) we obtain

$$\mathbf{S}^{-1} = \mathbf{P}(\Lambda + \tau^{-1}\mathbf{I})^{-1}\mathbf{P}^T \tag{6.183}$$

The likelihood (6.179) can now be evaluated efficiently. The term $\log |\mathbf{S}|$ can be decomposed as follows

$$\log |\mathbf{S}| = \log |\mathbf{P}(\Lambda + \tau^{-1}\mathbf{I})\mathbf{P}^T| = \log |\mathbf{P}^T\mathbf{P}(\Lambda + \tau^{-1}\mathbf{I})| = \log |\Lambda + \tau^{-1}| \tag{6.184}$$

and since $\Lambda + \tau^{-1}\mathbf{I}$ is diagonal, it has only Nd non-zero elements. We can reshape $\Lambda + \tau^{-1}\mathbf{I}$ into a tensor $\mathcal{A} \in \mathbb{R}^{d_1 \times \ldots \times d_4 \times N}$ as follows:

$$\mathcal{A} = [\![\boldsymbol{\lambda}_1, \ldots, \boldsymbol{\lambda}_K]\!] + \tau^{-1}\mathbb{1} = \boldsymbol{\lambda}_1 \circ \ldots \circ \boldsymbol{\lambda}_K + \tau^{-1}\mathbb{1}, \tag{6.185}$$

in which $\mathbb{1}$ is a tensor of all ones. $\log |\Lambda + \tau^{-1}|$ can be computed by summing the logs of all elements of $\mathcal{A}$, with $\mathcal{O}(Nd)$ complexity.

The term $\text{vec}(\widehat{\mathcal{Y}})^T\mathbf{S}^{-1}\text{vec}(\widehat{\mathcal{Y}})$ has a decomposition

$$\begin{aligned}
\text{vec}(\widehat{\mathcal{Y}})^T\mathbf{S}^{-1}\text{vec}(\widehat{\mathcal{Y}}) &= \left(\text{vec}(\widehat{\mathcal{Y}})\mathbf{S}^{-\frac{1}{2}}\right)\left(\mathbf{S}^{-\frac{1}{2}}\text{vec}(\widehat{\mathcal{Y}})\right) \\
&= \left(\text{vec}(\widehat{\mathcal{Y}})\mathbf{P}^T(\Lambda + \tau^{-1})^{-\frac{1}{2}}\mathbf{P}\right)\left(\mathbf{P}(\Lambda + \tau^{-1})^{-\frac{1}{2}}\mathbf{P}^T\text{vec}(\widehat{\mathcal{Y}})\right) \\
&:= \mathbf{b}^T\mathbf{b}
\end{aligned} \tag{6.186}$$

using the fact that the spectral decomposition of $\mathbf{S}^{-\frac{1}{2}}$ is $\mathbf{S}^{-\frac{1}{2}} = \mathbf{P}(\Lambda + \tau^{-1}\mathbf{I})^{-\frac{1}{2}}\mathbf{P}^T$. For a tensor $\mathcal{M}$ and L matrices $\mathbf{A}_1, \ldots, \mathbf{A}_L$, it holds that

$$\text{vec}(\mathcal{M} \times_1 \mathbf{A}_1 \ldots \times_K \mathbf{A}_L) = (\mathbf{A}_1 \otimes \ldots \otimes \mathbf{A}_L)\text{vec}(\mathcal{M}) \tag{6.187}$$

so that

$$\mathbf{b} = \text{vec}\left(\mathcal{T} \times_1 \mathbf{U}_1 \ldots \times_K \mathbf{U}_K\right) \tag{6.188}$$

$$\mathcal{T} = \widehat{\mathcal{Y}} \times_1 \mathbf{U}_1^T \ldots \times_K \mathbf{U}_K^T \odot \mathcal{A}^{-\frac{1}{2}} \tag{6.189}$$

in which $(\cdot)^{-\frac{1}{2}}$ is applied element-wise, i.e., to the diagonal of $(\Lambda + \tau^{-1}\mathbf{I})^{-\frac{1}{2}}$. This leads to a computational complexity for (6.189) of $\mathcal{O}\big(Nd(\sum_{k=1}^{K-1} d_k + N)\big)$.

Optimising the likelihood also requires access to the derivatives with respect to the parameters. These are calculated by computing the derivative w.r.t. $\mathbf{S}$ and applying the chain rule. We use the relationships

$$\partial \ln |\mathbf{S}| = \text{tr}(\mathbf{S}^{-1}\partial\mathbf{S}) \tag{6.190}$$

$$\partial\mathbf{S}^{-1} = -\mathbf{S}^{-1}\,(\partial\mathbf{S})\,\mathbf{S}^{-1} \tag{6.191}$$

the first being valid for a symmetric matrix $\mathbf{S}$. For the likelihood, therefore,

$$\partial\mathcal{L} = -\frac{1}{2}\text{tr}(\mathbf{S}^{-1}\partial\mathbf{S}) + \frac{1}{2}\text{vec}(\widehat{\mathcal{Y}})\mathbf{S}^{-1}\left(\partial\mathbf{S}\right)\mathbf{S}^{-1}\text{vec}(\widehat{\mathcal{Y}}) \tag{6.192}$$

The derivative w.r.t. τ is calculated in straightforward manner

$$\frac{\partial\mathcal{L}}{\partial\tau} = \frac{1}{2}\tau^{-2}\big(\text{tr}(\mathbf{S}^{-1}) - \text{vec}(\widehat{\mathcal{Y}})\mathbf{S}^{-1}\mathbf{S}^{-1}\text{vec}(\widehat{\mathcal{Y}})\big) \tag{6.193}$$

By the cyclic permutation properties of a trace

$$\text{tr}(\mathbf{S}^{-1}) = \text{tr}\big(\mathbf{P}(\Lambda + \tau^{-1}\mathbf{I})^{-1}\mathbf{P}^T\big) = \text{tr}\big(\mathbf{P}^T\mathbf{P}(\Lambda + \tau^{-1}\mathbf{I})^{-1}\big) = \text{tr}\big((\Lambda + \tau^{-1}\mathbf{I})^{-1}\big) \tag{6.194}$$

This requires a sum over the $\mathcal{O}(Nd)$ elements in $\mathcal{A}^{-1}$, while the computation of $S^{-1}\text{vec}(\widehat{\mathcal{Y}})$ has a complexity of $\mathcal{O}\big(Nd(\sum_{k=1}^{K-1} d_k + N)\big)$.

The kernel parameters $\boldsymbol{\theta}$ and other parameters are contained in the kernel matrices. To compute derivatives w.r.t. these parameters, the derivative with respect to each kernel matrix is first computed, followed by application of the chain rule. Taking $\mathbf{K}_1$ as an example and using the fact that $\mathbf{U}_k^T\mathbf{U}_k = \mathbf{I}$

$$\begin{aligned}
\nabla\mathbf{S} &= \partial\mathbf{K}_1 \otimes \mathbf{K}_2 \otimes \ldots \otimes \mathbf{K}_K \\
&= \mathbf{U}_1(\mathbf{U}_1^T\partial\mathbf{K}_1\mathbf{U}_1)\mathbf{U}_1^T \otimes \bigotimes_{k=2}^{K}(\mathbf{U}_k\text{diag}(\boldsymbol{\lambda}_i)\mathbf{U}_k^T) \\
&= \mathbf{P}\mathbf{U}_1^T\partial\mathbf{K}_1\mathbf{U}_1 \otimes \left(\bigotimes_{k=2}^{K}\text{diag}(\boldsymbol{\lambda}_k)\right)\mathbf{P}^T
\end{aligned} \tag{6.195}$$

Therefore, by the trace properties

$$\begin{aligned}
\text{tr}(\mathbf{S}^{-1}\partial\mathbf{S}) &= \text{tr}\left((\Lambda + \tau^{-1}\mathbf{I})^{-1}(\mathbf{U}_1^T\partial\mathbf{K}_1\mathbf{U}_1) \otimes \bigotimes_{k=2}^{K}\text{diag}(\boldsymbol{\lambda}_k)\right) \\
&= \left(\text{diag}^T(\mathbf{U}_1^T\partial\mathbf{K}_1\mathbf{U}_1)\bigotimes_{k=2}^{K}\boldsymbol{\lambda}_k^T\right)\text{vec}(\mathcal{A}^{-1}) \\
&= \text{vec}\left(\mathcal{A}^{-1} \times_1 \text{diag}^T(\mathbf{U}_1^T\partial\mathbf{K}_1\mathbf{U}_1) \times_2 \boldsymbol{\lambda}_2^T \ldots \times_K \boldsymbol{\lambda}_K^T)\right) \\
&= \text{vec}\left([\![\mathcal{A}^{-1}; \text{diag}^T(\mathbf{U}_1^T\partial\mathbf{K}_1\mathbf{U}_1), \boldsymbol{\lambda}_2^T \ldots, \boldsymbol{\lambda}_K^T]\!]\right) \\
&= \text{tr}\left(\mathbf{U}_1\text{diag}(\mathcal{A}^{-1} \times_2 \boldsymbol{\lambda}_2^T \ldots \times_K \boldsymbol{\lambda}_K^T)\mathbf{U}_1^T\partial\mathbf{K}_1\right)
\end{aligned} \tag{6.196}$$

The time complexity of (6.196) is $\mathcal{O}(Nd)$. The term $\text{vec}(\widehat{\mathcal{Y}})\mathbf{S}^{-1}\left(\partial\mathbf{S}\right)\mathbf{S}^{-1}\text{vec}(\widehat{\mathcal{Y}})$ can be treated similarly [16] to show that the complexity is $\mathcal{O}\big(Nd(\sum_{k=1}^{K-1} d_k + N)\big)$.

6.11.3 Tensor Linear Regression

Most tensor variate models do not involve GPs, and most deal with inputs in high-dimensional spaces, rather than high-dimensional output spaces. These arise naturally in a number of important applications. For example, in medical imaging there are numerous data sets of ultra-high dimension, such as those from magnetic resonance imaging (MRI) and positron emission tomography (PET). In the context of flow batteries, computed tomography scans could be the motivation. Finding relationships between general inputs and tensor outputs or between tensor inputs and general outputs presents an enormous challenge. Classical techniques to reduce dimensionality, such as principal component analysis (to be detailed later), can lead to the loss of vital information.

To overcome these issues, Zhou et al. [17] introduced the following linear tensor regression model for multi-array inputs $\mathcal{X} \in \mathbb{R}^{p_1 \times \ldots \times p_K}$ and outputs $y \in \mathbb{R}$

$$y = (\boldsymbol{\beta}_K \otimes, \ldots, \otimes \boldsymbol{\beta}_1)^T \text{vec}(\mathcal{X}) + \epsilon \tag{6.197}$$

for parameters $\boldsymbol{\beta}_k \in \mathbb{R}^{p_k}$ and noise ϵ. This model was motivated by the generalised linear model (GLM) $y = \boldsymbol{\beta}_1^T \mathbf{X} \boldsymbol{\beta}_2$ for matrix valued inputs $\mathbf{X}$, in which the bilinear form can be expressed as $y = (\boldsymbol{\beta}_2 \otimes \boldsymbol{\beta}_1)\text{vec}(\mathcal{X})$. A GLM assumes that the data arises from an exponential distribution and generalises linear regression by passing the linear part through a link or activation function (Equation (6.115) in the next section defines a GLM in more detail).

The model (6.197) can be written as

$$y = \langle \mathcal{B}, \mathcal{X} \rangle + \epsilon = \langle \mathcal{B}, \mathcal{X} \rangle + \epsilon \tag{6.198}$$

for a tensor-variate weight $\mathcal{B} \in \mathbb{R}^{p_1 \times \ldots \times p_K}$ such that

$$\text{vec}(\mathcal{B}) = \text{vec}(\boldsymbol{\beta}_1 \circ \ldots \circ \boldsymbol{\beta}_K) = \boldsymbol{\beta}_1 * \ldots * \boldsymbol{\beta}_K = \boldsymbol{\beta}_1 \otimes \ldots \otimes \boldsymbol{\beta}_K \tag{6.199}$$

It can be generalised to a CP-r rank model as follows

$$\begin{aligned} y = \langle \mathcal{B}, \mathcal{X} \rangle + \epsilon &= \left\langle [\![\mathbf{B}_1, \ldots, \mathbf{B}_K]\!], \mathcal{X} \right\rangle + \epsilon \\ &= \left\langle \sum\nolimits_{i=1}^{r} \boldsymbol{\beta}_1^i \circ \cdots \circ \boldsymbol{\beta}_K^i, \mathcal{X} \right\rangle + \epsilon \\ &= \langle (\mathbf{B}_K * \ldots * \mathbf{B}_1)\mathbf{1}, \mathcal{X} \rangle + \epsilon \end{aligned} \tag{6.200}$$

in which $\mathbf{B}_k = [\boldsymbol{\beta}_k^i \ldots \boldsymbol{\beta}_k^r] \in \mathbb{R}^{p_k \times r}$ and $\mathbf{1}$ is a vector of ones. The second expression follows from a well-known result on CP decompositions. Another important results is that if $\mathcal{B}$ admits a CP-r decomposition, then

$$\mathbf{B}_{(k)} = \mathbf{B}_k (\mathbf{B}_K * \ldots * \mathbf{B}_{k+1} * \mathbf{B}_{k-1} * \ldots * \mathbf{B}_1)^T \tag{6.201}$$

A solution is obtained by using some low rank approximation of $\mathcal{B}$. We note that the CP decomposition is often abbreviated CANDECOMP and is also called the parallel factors (PARAFAC) decomposition. Imposing the above structure reduces the complexity massively from $\prod_k p_k$ to $r\sum_k p_k$.

An alternative is to use a Tucker decomposition rather than a CP decomposition, leading to the model [18]

$$\begin{aligned} y &= \left\langle [\![\mathcal{G}; \mathbf{B}_1, \ldots, \mathbf{B}_K]\!], \mathcal{X} \right\rangle + \epsilon \\ &= \langle \mathcal{G} \times_1 \mathbf{B}_1 \times_2 \ldots \times_K \mathbf{B}_K, \mathcal{X} \rangle + \epsilon \end{aligned} \tag{6.202}$$

for the core tensor $\mathcal{G} \in \mathbb{R}^{n_1 \times \ldots \times n_K}$ and factors $\mathbf{B}_k \in \mathbb{R}^{p_k \times n_K}$.

Assumptions are required regarding the noise in order to solve for the $\{\mathbf{B}_k\}$ in the CP model or $\mathcal{G}$ and $\{\mathbf{B}_k\}$ in the Tucker model, either by minimising a loss function or by maximising a likelihood function. In [17], the authors used a generic exponential distribution to define a log likelihood $\mathcal{L}(\{\mathbf{B}_k\})$, which is minimised using a block relaxation algorithm [19] for updating the $\{\mathbf{B}_k\}$. This algorithm cycles through and updates the $\{\mathbf{B}_k\}$ in turn, as well as other parameters that appear in extended versions of the model. By virtue of the fact that (see (6.201))

$$\left\langle \sum_{i=1}^{r} \beta_1^i \circ \cdots \circ \beta_K^i, \mathcal{X} \right\rangle = \langle \mathbf{B}_k, \mathbf{X}_{(d)}(\mathbf{B}_K * \ldots * \mathbf{B}_{k+1} * \mathbf{B}_{k-1} * \ldots * \mathbf{B}_1) \rangle \tag{6.203}$$

the updates of the $\mathbf{B}_k$ involve Rp_k parameters, leading to a sequence of relatively low-dimensional optimisations.

The Tucker decomposition model can also be solved using the same algorithm, cycling through the core tensor and factors in turn. When updating the $\mathbf{B}_k$, the inner product can be written as

$$\begin{aligned} \langle \mathcal{B}, \mathcal{X} \rangle &= \langle \mathbf{U}_{(k)}, \mathbf{X}_{(k)} \rangle \\ &= \langle \mathbf{B}_k \mathbf{G}_{(k)} (\mathbf{B}_K \otimes \cdots \otimes \mathbf{B}_{k+1} \otimes \mathbf{B}_{k-1} \otimes \cdots \otimes \mathbf{B}_1)^T, \mathbf{X}_{(k)} \rangle \\ &= \langle \mathbf{B}_k, \mathbf{X}_{(k)} (\mathbf{B}_K \otimes \cdots \otimes \mathbf{B}_{k+1} \otimes \mathbf{B}_{k-1} \otimes \cdots \otimes \mathbf{B}_1) \mathbf{G}_{(k)}^T \rangle \end{aligned} \tag{6.204}$$

with $p_k R$ parameters. When updating the core tensor, the problem can be written as

$$\begin{aligned} \langle \mathcal{B}, \mathcal{X} \rangle &= \langle \mathrm{vec}(\mathcal{B}), \mathrm{vec}(\mathcal{X}) \rangle \\ &= \langle (\mathbf{B}_K \otimes \cdots \otimes \mathbf{B}_1) \mathrm{vec}(\mathcal{G}), \mathrm{vec}(\mathcal{X}) \rangle \\ &= \langle \mathrm{vec}(\mathcal{G}), (\mathbf{B}_K \otimes \cdots \otimes \mathbf{B}_1)^{\mathrm{T}} \mathrm{vec}(\mathcal{X}) \rangle. \end{aligned} \tag{6.205}$$

containing $\prod_k n_k$ parameters. Both approaches can be solved using a loss function approach, e.g.,

$$\mathrm{argmin}_{\{\mathbf{B}_k\}, \mathcal{G}, \lambda} \left(y - \left\langle [\![\mathcal{G}; \mathbf{B}_1, \ldots, \mathbf{B}_K]\!], \mathcal{X} \right\rangle \right)^2 + \lambda r(\{\mathbf{B}_k\}, \mathcal{G}) \tag{6.206}$$

in which $\lambda r(\{\mathbf{B}_k\}, \mathcal{G})$ is a penalty term that encourages sparsity in the parameters (in this case $\{\mathbf{B}_k\}$ and $\mathcal{G}$), with a regularisation constant λ. In fact, in both models mentioned above, a penalty term was added directly to the likelihood function.

6.12 Neural Networks and Deep Learning for Regression and Classification

Neural networks (ANNs) and especially deep learning versions of ANNs have become extremely popular in recent years, largely due to their superiority for certain types of problems, as well as the availability of open-source codes and functions on platforms such as TensorFlow and PyTorch, frequently with accompanying tutorials. This is not the case for Gaussian process models, which also involve a steeper learning curve.

Deep learning is superior for applications in which large data sets are available, especially in pattern recognition and natural language processing in which the tasks are usually classification, object detection, sequence prediction and feature extraction or representation learning. For regression problems, the benefits of deep learning are not as clear-cut, and much more work is required to assess its accuracy in comparison to alternatives such as GP models, SVR and kernel regression.

We point out that time-dependent or sequence problems are somewhat different from pure regression problems since they involve *forecasting*, or *extrapolation* rather than *interpolation*. While such problems can be posed as supervised machine learning problems *via* an embedding of the data, it must be borne in mind that the training data in this case is restricted to only a portion of the design space (time up to the present or ordered indexed up to the current), whereas predictions are required in the future or for the next index in the ordering. This is much more challenging because an historical pattern is not necessarily an accurate predictor of future patterns. No design can access the portion of design space (future) that is of interest, in contrast to standard regression in which there is in principle no restriction on the design. The recurrent networks described below were motivated by such problems. Further discussion on this issue will be provided in Sect. 7.4 of Chap. 7.

An ANN is a mapping $\widehat{\eta}(\boldsymbol{\xi})$ or $\widehat{\boldsymbol{\eta}}(\boldsymbol{\xi})$ of the inputs, which is used as an approximating function for the latent function $\eta(\boldsymbol{\xi})$ or $\boldsymbol{\eta}(\boldsymbol{\xi})$. One of the great advantages of ANNs is that no great effort or modifications are required to treat vector-valued outputs, although the computational costs increase. Throughout, it is assumed that the noise is zero, so that $y = \eta(\boldsymbol{\xi})$ or $\mathbf{y} = \boldsymbol{\eta}(\boldsymbol{\xi})$. Noise can be fully incorporated into ANN models, although this is rarely the case given the associated computational cost of a fully Bayesian approach. Partial Bayesian approaches are possible but we shall not cover the details. Typically, overfitting is ameliorated using a specialised training procedure and regularisation, to be discussed below.

6.12.1 Multi-layer Perceptron

A multi-layer perceptron (MLP) is a type of *feed-forward* network in that information is fed in one direction (forwards) from an input layer to an output layer, with intervening *hidden layers*. To each layer are associated *nodes* or *neurons*, which are nothing more than placeholders for certain mathematical operations. In the input layer are l neurons, one for each of the inputs in $\boldsymbol{\xi} \in \mathbb{R}^l$, and the output layer contains d neurons, one for each of the outputs in $\mathbf{y} \in \mathbb{R}^d$. We present the details for the vector case, with the scalar case obtained trivially by setting $d = 1$.

There are one or more hidden layers between the input and output layers. We start by considering a network with one hidden layer, which contains a specified number of neurons. The hidden layer acts on linear combinations of the inputs, which are contained in a vector $\mathbf{a}_n$ found by an affine transformation of the input $\boldsymbol{\xi}_n$

$$\mathbf{a}_n = \mathbf{W}_1 \boldsymbol{\xi}_n + \mathbf{b}_1 \tag{6.207}$$

in which the (i, j)-th entry of $\mathbf{W}_1 \in \mathbb{R}^{K \times l}$ is a weight connecting the j-th input to the i-th hidden neuron, and $\mathbf{b}_1 \in \mathbb{R}^K$ is a vector of biases. Neuron k in the hidden layer therefore receives a linear combination of the inputs with a bias included, equal to the k-th component a_k of $\mathbf{a}_n \in \mathbb{R}^K$. These components are called *activations*, and the number K of hidden neurons is a user-chosen hyperparameter. The weights and biases are unknown, and are the quantities that require estimation during the learning phase. The activations are subjected to a nonlinear transformation using an *activation function* $g(\cdot)$, which is applied component-wise to produce a vector $\mathbf{o}_n \in \mathbb{R}^K$

$$\mathbf{o}_n = g(\mathbf{a}_n) \tag{6.208}$$

The hyperbolic tangent (tanh), sigmoid σ and rectified linear unit ReLU

$$\begin{aligned} f(a) &= \frac{e^a - e^{-a}}{e^a + e^{-a}} \\ f(a) &= \frac{1}{1 + e^{-a}} \\ f(a) &= \begin{cases} a & a > 0 \\ 0 & a \leq 0 \end{cases} \end{aligned} \tag{6.209}$$

respectively, are the most common activation functions.

If there are no further hidden layers, the final output is calculated *via* another affine transformation of $\mathbf{o}_n$ followed by a nonlinear transformation using an activation function $g(\cdot)$, which need not be the same as that in (6.208)

$$\widehat{\boldsymbol{\eta}}(\boldsymbol{\xi}) = g\left(\mathbf{W}_y \mathbf{o}_n + \mathbf{b}_y\right) \tag{6.210}$$

in which the (i, j)-th entry of $\mathbf{W}_y \in \mathbb{R}^{d\times K}$ is a weight connecting the j-th hidden neuron to the i-th output neuron, and $\mathbf{b}_y \in \mathbb{R}^d$ is a vector of biases. For regression problems, $g(\cdot)$ is the identity map, while, for classification problems, it is usually the logistic function (binary case) or softmax function (multi-class case). In classification, the outputs (one for each class) are interpreted as *class probabilities* and an input or *pattern* $\boldsymbol{\xi}$ is assigned to the class for which the corresponding class probability is highest. Notice that

$$\widehat{\boldsymbol{\eta}}(\boldsymbol{\xi}) = g\left(\mathbf{W}_y\mathbf{o}_n + \mathbf{b}_y\right) = g\left(\mathbf{W}_y g(\mathbf{W}_1\boldsymbol{\xi}_n + \mathbf{b}_1) + \mathbf{b}_y\right)$$

which can be expanded to reveal a linear combination of functions of $\boldsymbol{\xi}$. Thus, an MLP is akin to a basis function expansion.

If there is a second hidden layer, the output $\mathbf{o}_n \in \mathbb{R}^K$ from the first hidden layer is used as the input for the next hidden layer. The operations (6.207) and (6.208) are repeated with an unknown weight matrix $\mathbf{W}_2 \in \mathbb{R}^{J\times K}$ and bias $\mathbf{b}_2 \in \mathbb{R}^J$, for J user-chosen neurons in the second hidden layer, to produce an output $\mathbf{o}_{2n} \in \mathbb{R}^J$. The process can be repeated with more hidden layers and the final output passed through an output layer as described above.

The hidden layers are considered to be *dense* or *fully connected*, i.e., each neuron in the hidden layer is connected to each neuron in the preceding layer *via* a weight. This leads to a large number of weights and therefore a challenging learning task. With one hidden layer, the network is termed *shallow*, whereas for two or more hidden layers it is considered a *deep network* (DNN).

A form of regularisation called *dropout* is often employed to lower model complexity (reduce the number of parameters). It is applied to outputs $\mathbf{o}$ of the hidden layers and can also be used to map the input to a low-dimensional feature space. Dropout consists of taking samples $z_j \in \{0, 1\}$, $j = 0, \ldots, k$, from a Bernoulli distribution, $p(z) = p^z(1-p)^{1-z}$, for some $p \in (0, 1)$, followed by the transformation

$$\mathbf{o} \mapsto \operatorname{diag}(z_1, \ldots, z_k)\mathbf{o} \tag{6.211}$$

This transformation randomly assigns certain of the hidden layer outputs to zero, i.e., removes some of the hidden layer neurons from consideration.

Learning consists of approximating the weights based on minimising a loss function, usually the square error

$$E(\mathbf{w}) = \sum\nolimits_{n=1}^{N} E_n(\mathbf{w}) = \frac{1}{2}\sum\nolimits_{n=1}^{N} \|\widehat{\boldsymbol{\eta}}(\boldsymbol{\xi}_n) - \mathbf{y}_n\|^2 \tag{6.212}$$

or the square error with a penalty term, e.g., $E(\mathbf{w}) + \frac{\lambda}{2}\|\mathbf{w}\|^2$, for a regularisation parameter λ. Here $\mathbf{w}$ is a vectorisation of all weights and biases, i.e., all entries of $\mathbf{W}_1$, $\mathbf{W}_2$, $\mathbf{W}_y$, etc. A standard gradient based optimisation *update rule* is employed (see Sect. D.1), namely

$$\mathbf{w}_{i+1} = \mathbf{w}_i + \alpha_i\mathbf{d}_i, \quad i = 0, 2, \ldots \tag{6.213}$$

with usually $\mathbf{d}_i = -\nabla_{\mathbf{w}} E(\mathbf{w}_i)$ as the search direction, namely *steepest descent* (detailed in Sect. D.1), in which $\nabla_{\mathbf{w}}$ is the gradient operator w.r.t. $\mathbf{w}$. The algorithm is usually initialised by setting $\mathbf{w}_0$ to the zero vector, a vector of ones multiplied by a small number or a vector of small random numbers according to various distributions, e.g., a normal. The *step length* or *learning rate* α_i is usually adaptive, i.e., it decreases as learning proceeds in order to prevent oscillations.

Typically, not all of the data is used in the update rule (6.213), i.e., it is not *batch learning*. Instead, online or *stochastic gradient descent* (SGD) is implemented, in which only a randomly selected subset of the data is employed [20]

$$\mathbf{d}_i = -\sum\nolimits_{n \in \mathcal{M}_i} \nabla_{\mathbf{w}} E_n(\mathbf{w}_i) \tag{6.214}$$

for some *mini-batch* $\mathcal{M}_i$ that is randomly selected (without replacement) from the data set at each iteration i. The extreme case is one data point at each iteration $\mathcal{M}_i = \{\boldsymbol{\xi}_{n_i}\}$, but normally some larger fraction of the data is used. SGD can massively reduce the computational burden, leading to faster convergence. It is almost always combined with a particularly convenient method for calculating the gradients, called *backpropagation*. Backpropagation is simply the application of the chain rule to evaluate the components $\partial E_n / \partial w_j(\mathbf{w}_i)$ of $\nabla_{\mathbf{w}} E_n(\mathbf{w}_i)$.

An additional regularisation term called *momentum* is often included, leading to a weighted sum of the current and past updates

$$\mathbf{w}_{i+1} = \mathbf{w}_i - \alpha_i \sum\nolimits_{n \in \mathcal{M}_i} \nabla_{\mathbf{w}} E_n(\mathbf{w}_i) + \mu(\mathbf{w}_i - \mathbf{w}_{i-1}), \quad i = 0, 2, \ldots \tag{6.215}$$

in which μ is a *momentum rate*. State-of-the-art methods are refined versions of SGD with backpropagation, such as Adaptive Moment Estimation (Adam), which uses an adaptive learning rate and momentum [21] calculated from decaying averages of the first two moments of the gradient. An *epoch* refers to one pass through the entire set of data in the iterative learning procedure (6.213) or (6.215). One epoch is rarely sufficient to reach convergence; most algorithms require at least a few tens of epochs to complete the training.

It is common to split the data into training, *hold out* (or validation) and test sets. The square or other error is calculated on the hold-out set as training proceeds, with the training performed using only the training set. *Early stopping* refers to the cessation of training when the error on the hold-out set begins to increase, even though it may continue to decrease on the training set. This helps to prevent overfitting. Once training is complete, the true generalisation error can be assessed on the test set. Other techniques such as cross-validation and random data shuffling can also be employed during training.

6.12.2 Convolutional Networks

Convolutional networks (CNNs) contain so-called *convolutional layers*, in which the operations can be regarded as regularised versions of those in dense layers. With fewer parameters, such layers are less prone to overfitting. The input $\boldsymbol{\xi}$ undergoes a transformation *via* a *kernel, filter* or *window* in the first convolutional layer. The kernel is a multidimensional array (or multi-array) in a space of lower dimension that of the input space. Inputs can be of any size, and a typical input for these types of networks is an image, described by a two- or three-dimensional array of pixels.

Consider a three-dimensional multi-array input $\mathbf{L} \in \mathbb{R}^{H\times W\times D}$ having entries $L_{i,j,k}$, and a kernel $\mathbf{K} \in \mathbb{R}^{F\times F\times D}$ having entries $K_{i,j,k}$. More precisely we should say order-3 multi-array, as in section 6.11.1, but we shall not be strict with the terminology. The channel dimension D of the two multi-arrays must agree and the entries of the kernel are hyperparameters to be learned. A convolutional layer operation is denoted $*$ (not to be confused with the Khatri-Rao product) and the output is $\mathbf{A} = \mathbf{L} * \mathbf{K}$, the u, v-th component of which is

$$A_{u,v} = \sum_{i=1}^{H}\sum_{j=1}^{W}\sum_{p=1}^{D} K_{i,j,p} L_{u+i-1,v+j-1,p} \tag{6.216}$$

if we set the stride to 1 (see below). For a 1D input $\boldsymbol{\xi} \in \mathbb{R}^l$ with entries ξ_j, the convolution operation with a kernel $\mathbf{k} = (k_1, \ldots, k_f)^T \in \mathbb{R}^f$, $f < l$, yields $\mathbf{a} = \boldsymbol{\xi} * \mathbf{k}$, with entries

$$a_j = \sum_{i=1}^{f} \xi_{j+i-1+s-1} k_i, \quad j = 1, \ldots \lfloor \frac{l-f}{s} \rfloor + 1 \tag{6.217}$$

in which s is called the *stride* of the convolutional operation and $\lfloor\cdot\rfloor$ denotes the floor operation. It should be pointed out that this is not a convolution operation in the strict mathematical sense of the convolution of two continuous or discrete functions.

It is easier to understand this operation from the following description: the kernel scans f-length segments of the input of f-length and the dot product of these segments with the kernel are taken (sum of element-wise products considering both to be vectors) to form the a_j. The first scan to produce a_1 starts with the first f entries of the input, while the second scan to produce a_2 is taken with the entries $s+1, \ldots, s+f$, and the k-th scan to produce a_k is with the entires $(k-1)s+1, \ldots, (k-1)s+f$. The stride s is the number of entries by which the kernel is shifted between each scan.

When the input is 2D (or 3D), the kernel is 2D (or 3D) and the procedure is the same, except that the dot product becomes a Hadamard product for 2D (3D) arrays and the kernel must be shifted along more than one dimension. *Padding* adds zero-valued elements to the input in the first and last p entries in each dimension. This serves to increase the size of $\mathbf{a}$ to $\lfloor \frac{m+2p-f}{s} \rfloor + 1$. Usually, multiple kernels are employed and the outputs are stacked along the second dimension of a 2D array in the 1D input case.

To each kernel scan we can associate a neuron in an equivalent hidden layer, with corresponding activation a_j. This activation is a linear combination of only a subset of the inputs, so this interpretation means that the hidden layer is not fully connected. This lack of fully connectedness should in theory ameliorate overfitting when compared to an MLP. The main advantage, however, is in dealing with inputs in high-dimensional spaces (e.g. images), for which the convolution layer essentially finds a feature mapping of a specified size. This mapping may then be subjected to further refinement and size reduction as described below.

$\mathbf{a} = \boldsymbol{\xi} * \mathbf{h}$ is passed through an activation function $g(\cdot)$ (applied entry-wise) to yield $\mathbf{o} = g(\mathbf{a})$. A type of regularisation often used in CNNs is *pooling* or *down-sampling*. A window of size n is passed over segments of $\mathbf{a}$ of length n and the n elements of the segments are combined into a single value, such as the maximum or average (max and average pooling, respectively). This can be performed on $\mathbf{a}$, with an activation function subsequently applied, or the activation function can be applied before applying the pooling operation on $\mathbf{o}$.

The output $\mathbf{o} = g(\mathbf{a})$, possibly after pooling, can form the input for a second convolutional layer. The number of convolutional layers is user-chosen, with care taken to balance model complexity with the complexity of the underlying problem and the volume of data available. The output of the last convolutional layer $\mathbf{o}_f = g(\mathbf{a}_f)$ is then fed to a dense layer to yield $\widehat{\eta}(\boldsymbol{\xi}) = g(\mathbf{W}_y \mathbf{o}_f + \mathbf{b}_y)$, with weight matrix $\mathbf{W}_y$, bias $\mathbf{b}_y$ and activation function $g(\cdot)$ (identify map for regression). As with the MLP, there may be multiple dense layers, with the output from a dense layer forming the input of the next dense layer. Training to learn the kernel hyperparameters and weights in the dense layers is based on a variant of SGD with backpropagation.

6.12.3 Recurrent Networks

To deal with the special case of sequences, e.g., a series of measurements in time or words in a sentence, recurrent networks (RNNs) are more appropriate than MLPs and CNNs. The sequence values can be scalars or vectors, e.g., a one-hot encoding of a word in a sentence. We can consider a scalar sequence represented as the vector $\boldsymbol{\xi} \in \mathbb{R}^l$, with elements that we shall now label ξ_t, $t = 1, \ldots, d$. Here, the suggestive index t usually denotes a strict ordering (such as time), although this does not have to be the case; $\boldsymbol{\xi}$ can be any vector-valued input. We shall call the index t time, in agreement with the usual terminology. As usual, we have an output $\mathbf{y} = \widehat{\eta}(\boldsymbol{\xi}) \in \mathbb{R}^d$ corresponding to the input $\boldsymbol{\xi}$. In many problems, the components of $\mathbf{y}$ are a continuation of a sequence represented by $\boldsymbol{\xi}$, although again this need not be the case.

In simple terms, an RNN computes an output $\mathbf{y}$ through a sequence of *hidden state vectors* $\mathbf{h}_t \in \mathbb{R}^H$, for hidden state-space dimension H (a hyperparameter), *via* the following iterative process, for $t = 1, \ldots, l$

$$\begin{aligned} \mathbf{h}_t &= \mathcal{H}\left(\mathbf{W}_{xh}\xi_t + \mathbf{W}_{hh}\mathbf{h}_{t-1} + \mathbf{b}_h\right) \\ \mathbf{y} &= \mathbf{W}_{hy}\mathbf{h}_m + \mathbf{b}_y \end{aligned} \tag{6.218}$$

in which $\mathbf{W}_{xh} \in \mathbb{R}^{H\times 1}$, $\mathbf{W}_{hh} \in \mathbb{R}^{H\times H}$, $\mathbf{W}_{hy} \in \mathbb{R}^{h\times H}$, $\mathbf{b}_h \in \mathbb{R}^H$ and $\mathbf{b}_y \in \mathbb{R}^h$ are weights and biases. The nonlinear transformation $\mathcal{H}$ can take different forms. The hidden state can be initialised as $\mathbf{h}_0 = \mathbf{0}$. There are a number of variants, but usually only the last state vector $\mathbf{h}_l$ is used as an input to a dense layer. This dense layer may be the output layer as in (6.218) or there may be one or more dense hidden layers before a final output layer. Alternatively, the hidden states for all t may be combined to produce the output or even a sequence of outputs $\mathbf{y}_t$, one from each $\mathbf{h}_t$. More recurrent layers are added by using $\mathbf{h}_1, \ldots, \mathbf{h}_l$ as inputs for the next layer, modifying the weight and bias sizes. Alternatively, only $\mathbf{h}_l$ is used as the input for the second recurrent layer at each time step.

Early versions of RNNs used standard activation functions for $\mathcal{H}$, which led to instability (so-called exploding or vanishing gradients). To ameliorate such issues, several variants were introduced, especially the *Long short-term memory* (LSTM) network [22] and the *Gated Recurrent Unit* (GRU) [23]. In the LSTM the following set of operations define $\mathcal{H}$

$$\begin{aligned} \mathbf{s}_t^f &= \sigma\left(\mathbf{W}_{\xi h}^f\xi_t + \mathbf{W}_{hh}^f\mathbf{h}_{t-1} + \mathbf{b}_f\right) \\ \mathbf{s}_t^i &= \sigma\left(\mathbf{W}_{\xi h}^i\xi_t + \mathbf{W}_{hh}^i\mathbf{h}_{t-1} + \mathbf{b}_i\right) \\ \mathbf{s}_t^o &= \sigma\left(\mathbf{W}_{\xi h}^o\xi_t + \mathbf{W}_{hh}^o\mathbf{h}_{t-1} + \mathbf{b}_o\right) \\ \widetilde{\mathbf{c}}_t &= \tanh(\mathbf{W}_{\xi h}^{\widetilde{c}}\xi_t + \mathbf{W}_{hh}^{\widetilde{c}}\mathbf{h}_{t-1} + \mathbf{b}_{\widetilde{c}}) \\ \mathbf{c}_t &= \mathbf{s}_t^f \odot \mathbf{c}_{t-1} + \mathbf{s}_t^i \odot \widetilde{\mathbf{c}}_t \\ \mathbf{h}_t &= \mathbf{s}_t^o \odot \tanh(\mathbf{c}_t) \end{aligned} \tag{6.219}$$

in which $\odot$ is the Hadamard product. $\mathbf{c}_t \in \mathbb{R}^H$ is called a *cell state*, which acts as a memory, retaining important information between time steps. $\mathbf{h}_t \in \mathbb{R}^H$ is the hidden state, while $\mathbf{s}_t^i$, $\mathbf{s}_t^o$ and $\mathbf{s}_t^f \in \mathbb{R}^H$ are called input, output and forget states. A new candidate is proposed for the cell state using $\widetilde{\mathbf{c}}_t \in \mathbb{R}^H$, while unnecessary information is discarded via $\mathbf{s}_t^f$. $\mathbf{W}_{\xi h}^j \in \mathbb{R}^{H\times 1}$, $\mathbf{W}_{hh}^j \in \mathbb{R}^{H\times H}$, $\mathbf{b}_j \in \mathbb{R}^H$, $j \in \{i, o, f, \widetilde{c}\}$, are weights and biases to be learned. The cell and hidden states are normally set to zero to initialise the algorithm.

A variant of the LSTM is the Gated Recurrent Unit (GRU) [23], which has only reset and update gates, with $\mathcal{H}$ defined by

$$\begin{aligned} \mathbf{z}_t &= \sigma(\mathbf{W}_{\xi h}^z\xi_t + \mathbf{W}_{hh}^z\mathbf{h}_{t-1} + \mathbf{b}_z) \\ \mathbf{r}_t &= \sigma(\mathbf{W}_{\xi h}^r\xi_t + \mathbf{W}_{hh}^r\mathbf{h}_{t-1} + \mathbf{b}_r) \\ \widehat{\mathbf{h}}_t &= \tanh(\mathbf{W}_{\xi h}^{\hat{h}}\xi_t + \mathbf{W}_{hh}^{\hat{h}}(\mathbf{r}_t \odot \mathbf{h}_{t-1}) + \mathbf{b}_{\hat{h}}) \\ \mathbf{h}_t &= \mathbf{z}_t \odot \widehat{\mathbf{h}}_t + (1 - \mathbf{z}_t) \odot \mathbf{h}_{t-1} \end{aligned} \tag{6.220}$$

in which $\mathbf{h}_t$ is the hidden state, $\widehat{\mathbf{h}}_t$ acts as a candidate activation, and $\mathbf{z}_t$ and $\mathbf{r}_t$ are the update and reset gate states, respectively. $\mathbf{W}^j_{xh} \in \mathbb{R}^{H\times 1}$, $\mathbf{W}^j_{hh} \in \mathbb{R}^{H\times H}$ and $\mathbf{b}_j \in \mathbb{R}^H$, $j \in \{z, r, \widehat{h}\}$, are the weights and biases, while H is the hidden state-space dimension.

Training to learn all of the weights and biases is a variant of backpropagation with SGD, called backpropgation through time (BPTT) [24]. An unfolded network treats the time steps as separate layers and therefore defines an equivalent feedforward network with l hidden layers. BPTT propagates through these layers. A slight modification to ensure stability is termed *truncated BPTT*. It does not use all timesteps on the backward pass to obtain the gradients and is useful when the sequence length l is very long, usually a few hundreds or more.

6.12.4 Bi-directional Recurrent Networks

There are a number of more sophisticated networks that can be used for particular types of problems. In *bi-directional RNNs* (bi-RNNs) [25] sequential data is simultaneously passed in forward and backward directions *via* connected layers. Both layers produce outputs that are transformed to yields the final output. There are now forward and backward hidden state vectors $\overrightarrow{\mathbf{h}}_t$ and $\overleftarrow{\mathbf{h}}_t$, together with an output $\mathbf{y}$, calculated by iteration as follows, from $t = 1, \ldots, l$

$$\begin{aligned}
\overrightarrow{\mathbf{h}}_t &= \mathcal{H}\left(\overrightarrow{\mathbf{W}}_{\xi h}\, \xi_t + \overrightarrow{\mathbf{W}}_{hh}\overrightarrow{\mathbf{h}}_{t-1} + \overrightarrow{\mathbf{b}}_h\right)\\
\overleftarrow{\mathbf{h}}_t &= \mathcal{H}\left(\overleftarrow{\mathbf{W}}_{\xi h}\, \xi_t + \overleftarrow{\mathbf{W}}_{hh}\overleftarrow{\mathbf{h}}_{t-1} + \overleftarrow{\mathbf{b}}_h\right)\\
\mathbf{y} &= \overrightarrow{\mathbf{W}}_{hy}\overrightarrow{\mathbf{h}}_l + \overleftarrow{\mathbf{W}}_{hy}\overleftarrow{\mathbf{h}}_1 + \mathbf{b}_y
\end{aligned} \tag{6.221}$$

in which $\overrightarrow{\mathbf{W}}$, $\overleftarrow{\mathbf{W}}$, $\overrightarrow{\mathbf{b}}$, $\overleftarrow{\mathbf{b}}$ and $\mathbf{b}$ are weights and biases. Alternatively, a sequence of outputs $\mathbf{y}_t$ can be obtained from the sequences $\overrightarrow{\mathbf{h}}_t$ and $\overleftarrow{\mathbf{h}}_t$. A forward hidden layer passes inputs ξ_t and a hidden state $\overrightarrow{\mathbf{h}}_{t-1}$ from $t = 1, \ldots, l$, while the backward layer passes ξ_t and a hidden state $\overleftarrow{\mathbf{h}}_{t-1}$ from $t = l, \ldots, 1$.

The hidden states at each t are usually concatenated, summed together, averaged or multiplied using a Hadamard product to form $g(\overrightarrow{\mathbf{h}}_t, \overleftarrow{\mathbf{h}}_t)$, where $g(\cdot, \cdot)$ represents one of these operations. Each of the $g(\overrightarrow{\mathbf{h}}_t, \overleftarrow{\mathbf{h}}_t)$ can be used as inputs to a second bi-RNN layer. Following the final layer, an output $\mathbf{y}$ is usually produced by concatenating the last forward and backward hidden states and passing the result through a dense layer, as in (6.221). The RNN is usually an LSTM, although a GRU or other RNNs can also be used.

6.12.5 Encoder-Decoder Models

An *encoder-decoder* network also employs two RNNs, in which the first discovers a representation or encoding of the network input, while the decoder decodes this representation to predict an output. It is also called a *sequence-to-sequence* model since it maps input sequences to output sequences. Again, it is designed for sequence data. An encoder-decoder model performs the following operations, from $t = 1, \ldots, l$ and $t' = 1, \ldots, d$

$$\begin{aligned} \mathbf{h}_t^e &= \mathcal{H}^e\left(\mathbf{W}_{\xi h}^e \xi_t + \mathbf{W}_{hh}^e \mathbf{h}_{t-1}^e + \mathbf{b}_h^e\right) \\ \mathbf{v} &= q(\mathbf{h}_1^e, \ldots, \mathbf{h}_l^e) \\ \mathbf{h}_{t'}^d &= \mathcal{H}^d\left(\mathbf{W}_{vh}^d \mathbf{v} + \mathbf{W}_{hh}^d \mathbf{h}_{t'-1}^d + \mathbf{W}_{yh}^d y_{t'-1} + \mathbf{b}_h^d\right) \\ y_{t'} &= \mathbf{W}_{hy}^d \mathbf{h}_{t'}^d + b_y^d \end{aligned} \tag{6.222}$$

in which we now have an encoder hidden state $\mathbf{h}_t^e$ and a decoder hidden state $\mathbf{h}_{t'}^d$, with transformations $\mathcal{H}^k$ (usually LSTM or GRU), and weights and biases $\mathbf{W}^k$ and $\mathbf{b}^k \in \mathbb{R}^H$, $k \in \{e, d\}$. The vector $\mathbf{v}$ is called a context vector and it is a function q of the encoder hidden states, with typically $\mathbf{v} = \mathbf{h}_l^e$. The hidden state output from the encoder initialises the hidden state of the decoder, $\mathbf{h}_0^d = \mathbf{h}_l^e$. If an LSTM is used, the cell state for the decoder is initialised in the same way. Alternatively, some function of the entire sequence of hidden states, e.g., an average, is used for initialisation.

There are several further choices to be made in terms of the decoder input. The most common method uses $y_{t'-1}$ and the hidden state $\mathbf{h}_{t'-1}^d$, with either the $\mathbf{v}$ at each time step, as in (6.222) [26], or with $\mathbf{v}$ only at $t' = 1$ [27], in which case $\mathbf{W}_{vh}^d = \mathbf{0}$ for $t' = 2, \ldots, d$. The real (given) value of $y_{t'-1}$ can also be used in place of the predicted value when only the prediction of y_d is required, which is referred to as *teacher forcing*. The unknown value of y_0 is specified by a special token value in order to initialise the decoder. Alternatively, only the hidden state and $\mathbf{v}$ are used as inputs, in which case $\mathbf{W}_{yh} = \mathbf{0}$.

The encoder can also be replaced by a bi-RNN, with the context vector given as a function of the concatenated forward and backward encoder hidden states $\overrightarrow{\mathbf{h}}_{e,t}$ and $\overleftarrow{\mathbf{h}}_{e,t}$, i.e., $\mathbf{v} = q(\{[\overrightarrow{\mathbf{h}}_{e,t}; \overleftarrow{\mathbf{h}}_{e,t}]\}_t)$. When using an LSTM, the decoder cell and hidden states are initialised by similar concatenations, requiring the decoder to have a hidden state-space dimension of $2H$.

6.12.6 The Attention Mechanism

The context vector is of fixed length, which can engender an information bottleneck, since it is the sole link between the encoder and decoder. To overcome this issue,

Bahdanau et al. [28] developed the concept of *attention*. During the decoder process, attention is used to focus on the most relevant information emerging from the encoder. At each time t', the context vector is given by a linear combination of the encoder hidden states

$$\mathbf{v}_{t'} = \sum_{t=1}^{m} \alpha_{t,t'} \mathbf{h}_t^e \tag{6.223}$$

in which $\alpha_{t,t'}$ are referred to as *attention weights*, defined as

$$\begin{aligned} \alpha_{t,t'} &= \text{softmax}(e_{t,t'}) = \frac{\exp(e_{t,t'})}{\sum_{k=1}^{l} \exp(e_{k,t'})} \\ e_{t,t'} &= \text{score}(\mathbf{h}_t^e, \mathbf{h}_{t'-1}^d) \end{aligned} \tag{6.224}$$

This is essentially an attempt to pair decoder states $\mathbf{h}_{t'-1}^d$ with relevant encoder states $\mathbf{h}_t^e$ *via* a similarity score. The similarity score in (6.224) was defined by a dense layer in the original formulation [28]

$$\text{score}(\mathbf{h}_t^e, \mathbf{h}_{t'-1}^d) = \mathbf{v}_a^T \tanh(\mathbf{W}_d \mathbf{h}_{t'-1}^d + \mathbf{W}_e \mathbf{h}_t^e) \tag{6.225}$$

in which the $\mathbf{W}_j$ and $\mathbf{v}_a$ are parameters to be learned. The following simpler score functions were later proposed [29]

$$\text{score}(\mathbf{h}_t^e, \mathbf{h}_{t'-1}^d) = \begin{cases} (\mathbf{h}_{t'-1}^d)^T \mathbf{W}_l \mathbf{h}_t^e & \text{Luong } et\ al. [310] \\ (\mathbf{h}_{t'-1}^d)^T \mathbf{h}_t^e & \text{Luong } et\ al. [310] \end{cases} \tag{6.226}$$

where again $\mathbf{W}_l$ is a parameter to be learned.

6.13 Linear Discriminant Classification and Support Vector Machines

Despite the advent of very powerful deep learning methods for classification, there are a number of methods that are still able to compete with or even outperform these methods in many problems. Chief amongst these is the *support vector machine* (SVM), the classification analogue of support vector regression introduced in Sect. 6.9. Consider data $\boldsymbol{\xi}_n \in C_k$, $n = 1, \ldots, N$, belonging to one of K classes C_k, $k = 1, \ldots, K$. It is important to point out that by the notation $\boldsymbol{\xi}_n \in C_k$ we are assigning $\boldsymbol{\xi}$ to a class or label, although $\boldsymbol{\xi}$ is a numerical representation of the object to which it assigned, living in some Euclidean space $\boldsymbol{\xi}_n \in \mathbb{R}^l$.

To explain the concepts behind SVM, we first briefly review linear classification methods based on *linear discriminant functions*. In such methods we search for a linear function $z(\boldsymbol{\xi})$ of the patterns $\boldsymbol{\xi}$

$$z(\boldsymbol{\xi}) = \mathbf{w}^T \boldsymbol{\xi} + w_0 \tag{6.227}$$

in which $\mathbf{w}$ is a set of weights and w_0 is a bias, both of which are to be learned from the data, e.g., using a least squares fit. $z(\boldsymbol{\xi})$ is called a linear discriminant (function). Outputs can be defined either as $y_n \in \{-1, 1\}$ in the binary case of $K = 2$ or by a *1-of-K-coding* otherwise: if $\boldsymbol{\xi}_n \in \mathcal{C}_k$ we can define a corresponding target by

$$\mathbf{y}_n = (0, \dots, 1, \dots, 0)^T \tag{6.228}$$

in which the 1 is placed in the k-th entry. The discriminant function separates the input space $\mathbb{R}^l$ into two regions, one with $z(\boldsymbol{\xi}) > 0$ and one with $z(\boldsymbol{\xi}) < 0$, by the hyperplane

$$0 = z(\boldsymbol{\xi}) = \mathbf{w}^T \boldsymbol{\xi} + w_0 \tag{6.229}$$

In binary classification, $z(\boldsymbol{\xi}) > 0$ would correspond to class $\mathcal{C}_1$ and $z(\boldsymbol{\xi}) < 0$ would correspond to class $\mathcal{C}_2$. The hyperlane is called a *decision surface* or *decision boundary*. For $K > 2$ classes we define a single K-class discriminant function $\mathbf{z}(\boldsymbol{\xi}) = (z_1, \dots, z_K)^T$ comprising K linear functions of the form

$$z_k(\boldsymbol{\xi}) = \mathbf{w}_k^T \boldsymbol{\xi} + w_{k0} \tag{6.230}$$

with weights and biases for each function. We assign a point $\boldsymbol{\xi}$ to $\mathcal{C}_k$ if

$$z_k(\boldsymbol{\xi}) > z_j(\boldsymbol{\xi}) \quad \text{for all } j \neq k \tag{6.231}$$

On the decision boundary between class $\mathcal{C}_i$ and class $\mathcal{C}_j$ we must have $z_i(\boldsymbol{\xi}) = z_j(\boldsymbol{\xi})$, so that

$$(\mathbf{w}_i - \mathbf{w}_j)^T \boldsymbol{\xi} + (w_{i0} - w_{j0}) = 0 \tag{6.232}$$

which is again an $(l - 1)$-dimensional hyperplane. The classes are separated by these hyperplanes.

A simple method to find the weights and biases is to minimise the square error on the targets. With the following definitions

$$\widetilde{\boldsymbol{\xi}} = \begin{pmatrix} 1 \\ \uparrow \\ \boldsymbol{\xi} \\ \downarrow \end{pmatrix} \quad \widetilde{\mathbf{w}}_k = \begin{pmatrix} w_{0k} \\ \uparrow \\ \mathbf{w}_k \\ \downarrow \end{pmatrix} \quad \mathbf{W} = \begin{pmatrix} \uparrow & \dots & \uparrow \\ \widetilde{\mathbf{w}}_1 & \dots & \widetilde{\mathbf{w}}_K \\ \downarrow & \dots & \downarrow \end{pmatrix} \quad \mathbf{z}(\boldsymbol{\xi}) = \begin{pmatrix} z_1(\boldsymbol{\xi}) \\ \vdots \\ z_K(\boldsymbol{\xi}) \end{pmatrix} \tag{6.233}$$

we obtain the compact form

$$\mathbf{z}(\boldsymbol{\xi}) = \mathbf{W}^T \widetilde{\boldsymbol{\xi}} \tag{6.234}$$

$\widetilde{\boldsymbol{\xi}}$ and $\widetilde{\mathbf{w}}_k$ are called *augmented* inputs and weight vectors, and serve to absorb the biases into a compact representation (linear rather than affine). From this point onwards we redefine the inputs and weights as follows

$$\boldsymbol{\xi} \mapsto (1, \boldsymbol{\xi})^T, \quad \mathbf{w}_k \mapsto (w_{0k}, \mathbf{w}_k)^T \tag{6.235}$$

Once we have learned $\mathbf{W}$, a new pattern $\boldsymbol{\xi}$ is assigned to the class $\mathcal{C}_k$ such that $k = \text{argmax}_j z_j(\boldsymbol{\xi}) = \mathbf{w}_j^T \boldsymbol{\xi}$. $\mathbf{W}$ can be determined by minimising the square error on the targets, defined as

$$\begin{aligned} E_D(\mathbf{W}) &= \frac{1}{2}\sum_{n=1}^{N} (\mathbf{W}^T \boldsymbol{\xi}_n - \mathbf{y}_n)^T (\mathbf{W}^T \boldsymbol{\xi}_n - \mathbf{y}_n) \\ &= \frac{1}{2}\text{tr}\left((\mathbf{XW} - \mathbf{Y})^T (\mathbf{XW} - \mathbf{Y})\right) \end{aligned} \tag{6.236}$$

in which $\mathbf{X} \in \mathbb{R}^{(l+1)\times N}$ and $\mathbf{Y} \in \mathbb{R}^{K\times N}$ collect the augmented inputs and 1-of-K outputs $\mathbf{y}_n$ as columns, respectively. Using basic rules for transposes, expanding the term inside the trace and using the properties of the derivatives of a trace, we obtain

$$\mathbf{W} = \left(\mathbf{X}^T\mathbf{X}\right)^{-1} \mathbf{X}^T\mathbf{Y} \tag{6.237}$$

in which $\left(\mathbf{X}^T\mathbf{X}\right)^{-1} \mathbf{X}^T$ is the pseudoinverse. The least-squares approach gives an exact closed-form solution for the discriminant function parameters, but it can suffers from some severe problems. The square error function penalises predictions that are 'too correct', i.e., they lie a long way on the correct side of the decision boundary. This is not surprising because the method corresponds to maximum likelihood under the assumption of a Gaussian conditional distribution, while discrete target vectors are far from Gaussian.

Returning to the two class case $\mathcal{C}_1$ and $\mathcal{C}_2$, with outputs $y_n \in \{-1, 1\}$, in both cases we have

$$y_n z(\boldsymbol{\xi}_n) = y_n(\mathbf{w}^T \boldsymbol{\xi}_n + w_0) > 0 \tag{6.238}$$

We can introduce a useful notation to represent this classifier

$$f(\boldsymbol{\xi}; \mathbf{w}) = \text{sign}(z(\boldsymbol{\xi})) = \text{sign}(\mathbf{w}^T \boldsymbol{\xi} + w_0) = \begin{cases} 1 & \text{if } \mathbf{w}^T \boldsymbol{\xi} + w_0 > 0 \\ -1 & \text{if } \mathbf{w}^T \boldsymbol{\xi} + w_0 < 0 \end{cases} \tag{6.239}$$

In SVM we attempt to find a decision boundary that leads to an optimal separation of the classes. What we mean by 'optimal' is defined in terms of the *margin*, introduced in Sect. 6.9: the smallest perpendicular distance between the decision boundary $z(\boldsymbol{\xi}) = 0$ and any of the patterns $\boldsymbol{\xi}_n$. SVM essentially attempts to maximise the margin. We note that $\mathbf{w}$ is orthogonal to the hyperplane: let $\boldsymbol{\xi}_A$ and $\boldsymbol{\xi}_B$ be two points on the hyperplane, then

$$z(\boldsymbol{\xi}_A) - z(\boldsymbol{\xi}_B) = \mathbf{w}^T (\boldsymbol{\xi}_A - \boldsymbol{\xi}_B) = 0 \tag{6.240}$$

which shows that $\mathbf{w}$ is orthogonal to $\boldsymbol{\xi}_A - \boldsymbol{\xi}_B$, which lies in the hyperplane. Let $\boldsymbol{\xi}_p$ be the orthogonal projection of a point $\boldsymbol{\xi}$ onto the decision surface.

Then $\boldsymbol{\xi} = \boldsymbol{\xi}_p + (\boldsymbol{\xi} - \boldsymbol{\xi}_p)$, where $\boldsymbol{\xi} - \boldsymbol{\xi}_p$ is parallel to $\mathbf{w}$, so that $\boldsymbol{\xi} = \boldsymbol{\xi}_p + b\mathbf{w}/\|\mathbf{w}\|$, for some $b \in \mathbb{R}$. Therefore

$$\begin{aligned} z(\boldsymbol{\xi}) &= \mathbf{w}^T\boldsymbol{\xi} + w_0 \\ &= \mathbf{w}^T\boldsymbol{\xi}_p + w_0 + \frac{b\mathbf{w}^T\mathbf{w}}{\|\mathbf{w}\|} \end{aligned} \tag{6.241}$$

which shows that $z(\boldsymbol{\xi}) = b\|\mathbf{w}\|$. $z(\boldsymbol{\xi})/\|\mathbf{w}\|$ therefore defines the perpendicular distance of a point $\boldsymbol{\xi}$ to the decision surface $z(\boldsymbol{\xi}) = 0$ and if all points are correctly classified, $|z(\boldsymbol{\xi})| = y_n z(\boldsymbol{\xi}_n)$. Thus the magnitude of the perpendicular distance of $\boldsymbol{\xi}_n$ to the decision surface is given by (6.238) as

$$\frac{y_n z(\boldsymbol{\xi}_n)}{\|\mathbf{w}\|} = \frac{y_n(\mathbf{w}^T\boldsymbol{\xi}_n + w_0)}{\|\mathbf{w}\|} \tag{6.242}$$

The *maximum margin* solution searches for the parameters $\mathbf{w}$ and w_0 such that the perpendicular distance to the closest point $\boldsymbol{\xi}_n$ on *either side* of the decision surface is maximised

$$\operatorname{argmax}_{\mathbf{w},w_0}\left\{\min_n \frac{y_n(\mathbf{w}^T\boldsymbol{\xi}_n + w_0)}{\|\mathbf{w}\|}\right\} = \operatorname{argmax}_{\mathbf{w},w_0}\left\{\frac{1}{\|\mathbf{w}\|}\min_n(y_n(\mathbf{w}^T\boldsymbol{\xi}_n + w_0)\right\} \tag{6.243}$$

in which we take $\|\mathbf{w}\|$ outside the min because it does not depend on n. This optimisation problem is not easy to solve so we shall transform it into an equivalent *dual* problem, to be explained below.

We first notice that the perpendicular distance of any point $\boldsymbol{\xi}_n$ to the decision surface does not change if we rescale $\mathbf{w}$ and w_0 by any constant k, i.e., $\mathbf{w} \mapsto k\mathbf{w}$ and $w_0 \mapsto kw_0$. We therefore (arbitrarily) set

$$y_n(\mathbf{w}^T\boldsymbol{\xi}_n + w_0) = 1 \tag{6.244}$$

for *the point* $\boldsymbol{\xi}_n$ *closest to the decision surface*. The maximum margin solution will then split the input space into two regions, separated by the maximum margin decision surface. The points closest to this surface lie on two parallel hyperplanes of equal distance $1/\|\mathbf{w}\|$ from the surface

$$w_0 + \mathbf{w}^T\boldsymbol{\xi} = -1 \quad \text{and} \quad w_0 + \mathbf{w}^T\boldsymbol{\xi} = 1 \tag{6.245}$$

Points $\boldsymbol{\xi}_n \in C_1$ satisfy $w_0 + \mathbf{w}^T\boldsymbol{\xi}_n \geq 1$ and points $\boldsymbol{\xi}_n \in C_2$ will satisfy $w_0 + \mathbf{w}^T\boldsymbol{\xi}_n \leq -1$, while all points will satisfy

$$y_n(w_0 + \mathbf{w}^T\boldsymbol{\xi}_n) \geq 1 \tag{6.246}$$

The width m separating the two hyperplanes in (6.245) is the margin. We call this a *hard margin* because all points are correctly classified. To reiterate, the basic strategy is to find the decision surface that maximises the margin m, and we wish to do this in an indirect manner that will alleviate numerical issues. Maximising the margin $m = 2/\|\mathbf{w}\|$ is the same as minimising $\|\mathbf{w}\|$ or $\|\mathbf{w}\|^2$, which are linear and quadratic functions, much more suited to optimisation problems.

We also have *constraints* in the form

$$y_n(w_0 + \mathbf{w}^T \boldsymbol{\xi}_n) \geq 1 \tag{6.247}$$

so we can instead solve the following *constrained quadratic* problem

$$\begin{gathered} \text{argmin}_{\mathbf{w}} \frac{1}{2}\|\mathbf{w}\|^2 \\ \text{subject to } y_n(\mathbf{w}^T \boldsymbol{\xi}_n + w_0) \geq 1 \quad \forall n \end{gathered} \tag{6.248}$$

This is called the hard margin *primal problem*. Given $\mathbf{w}$ we can classify a new input $\boldsymbol{\xi}$ by calculating $\text{sign}(\mathbf{w}^T \boldsymbol{\xi} + w_0)$. Problem (6.248) can be solved in its original form, but a more efficient solution is obtained by transforming it to a dual form, which also leads to nonlinear SVM by using kernel substitution. The constrained problem above can be transformed into an equivalent unconstrained problem by introducing Lagrange multipliers

$$\begin{aligned} &\text{argmin}_{\mathbf{w}, w_0, \{\lambda\}} \mathcal{L}(\mathbf{w}, w_0, \{\lambda_n\}) \\ &= \text{argmin}_{\mathbf{w}, w_0, \{\lambda\}} \left(\frac{1}{2}\|\mathbf{w}\|^2 - \sum_{n=1}^{N} \lambda_n \left\{ (y_n(\mathbf{w}^T \boldsymbol{\xi}_n + w_0) - 1 \right\} \right) \end{aligned} \tag{6.249}$$

in which $\mathcal{L}(\mathbf{w}, w_0, \{\lambda_n\})$ is the Lagrangian and λ_n are the Lagrange multipliers, related to the constraints.

Values of $\mathbf{w}$, w_0 for which the constraints $y_n(w_0 + \mathbf{w}^T \boldsymbol{\xi}_n) \geq 1$ are satisfied for all n are called *feasible* value. We shall denote the feasible set of values by $\mathcal{W}$, and set $f(\mathbf{w}) = \|\mathbf{w}\|^2/2$ to simplify the notation. At each set of feasible $\mathbf{w}, w_0 \in \mathcal{W}$ values, $f(\mathbf{w})$ satisfies

$$f(\mathbf{w}) = \max_{\{\lambda_n \geq 0\}} \mathcal{L}(\mathbf{w}, w_0, \{\lambda_n\}) \tag{6.250}$$

and the maximum is attained *if and only if*

$$\lambda_n \geq 0 \quad \text{and} \quad \lambda_n \left\{ (y_n(\mathbf{w}^T \boldsymbol{\xi}_n + w_0) - 1 \right\} = 0 \tag{6.251}$$

To see this, notice that for $\mathbf{w}, w_0 \in \mathcal{W}$

$$\mathcal{L}(\mathbf{w}, w_0, \{\lambda_n\}) = f(\mathbf{w}) - \sum_{n=1}^{N} \lambda_n \left\{ (y_n(\mathbf{w}^T \boldsymbol{\xi}_n + w_0) - 1 \right\} \leq f(\mathbf{w}) \tag{6.252}$$

since the constraints are non-negative, so that the sum is non-negative if $\lambda_n \geq 0$. Thus, the Lagrangian is bounded above by $f(\mathbf{w})$ for $\mathbf{w}, w_0 \in \mathcal{W}$. In order to obtain equality, $\mathcal{L}(\mathbf{w}, w_0, \{\lambda_n\}) = f(\mathbf{w})$, we need to maximise $\mathcal{L}(\mathbf{w}, w_0, \{\lambda_n\})$ over $\{\lambda_n\}$, and this maximum only occurs when $\lambda_n \left\{(y_n(\mathbf{w}^T \boldsymbol{\xi}_n + w_0) - 1\right\} = 0$. Therefore, either $\lambda_n = 0$ *or* $y_n(\mathbf{w}^T \boldsymbol{\xi}_n + w_0) - 1 = 0$.

The minimum value of $f(\mathbf{w})$, labelled f^*, can now be written as

$$f^* = \min_{\mathbf{w}, w_0} \max_{\{\lambda_n \geq 0\}} \mathcal{L}(\mathbf{w}, w_0, \{\lambda_n\}) \tag{6.253}$$

Consider feasible values $\mathbf{w}, w_0 \in \mathcal{W}$: from the previous result, $f(\mathbf{w}) = \max_{\{\lambda_n \geq 0\}} \mathcal{L}(\mathbf{w}, w_0, \{\lambda_n\})$, so that

$$f^* = \min_{\mathbf{w}, w_0 \in \mathcal{W}} f(\mathbf{w}) = \min_{\mathbf{w}, w_0 \in \mathcal{W}} \max_{\{\lambda_n \geq 0\}} \mathcal{L}(\mathbf{w}, w_0, \{\lambda_n\}) \tag{6.254}$$

Next consider a non-feasible $\mathbf{w}, w_0 \notin \mathcal{W}$; then

$$\begin{aligned} &\max_{\{\lambda_n \geq 0\}} \mathcal{L}(\mathbf{w}, w_0, \{\lambda_n\}) \\ &= \max_{\{\lambda_n \geq 0\}} \left(\frac{1}{2}\|\mathbf{w}\|^2 - \sum_{n=1}^{N} \lambda_n \left\{(y_n(\mathbf{w}^T \boldsymbol{\xi}_n + w_0) - 1\right\} \right) = \infty \end{aligned} \tag{6.255}$$

since the λ_n are unbounded. Therefore, as claimed, f^* is the minimum of $\max_{\{\lambda_n \geq 0\}} \mathcal{L}(\mathbf{w}, w_0, \{\lambda_n\})$, which is clearly when $\mathbf{w}, w_0 \in \mathcal{W}$.

We also require $\lambda_n \geq 0$ and $\lambda_n \left\{(y_n(\mathbf{w}^T \boldsymbol{\xi}_n + w_0) - 1\right\} = 0$ for all n. The last two constraints are called *Karush-Kuhn-Tucker (KKT) conditions* [30]. The principle behind the *dual problem* is to swap the order of the minimum and maximum, i.e.,

$$\begin{aligned} &\text{find} \quad \max_{\{\lambda_n \geq 0\}} \min_{\mathbf{w}, w_0} \mathcal{L}(\mathbf{w}, w_0, \{\lambda_n\}) \\ &\text{rather than} \quad \min_{\mathbf{w}, w_0} \max_{\{\lambda_n \geq 0\}} \mathcal{L}(\mathbf{w}, w_0, \{\lambda_n\}) \end{aligned} \tag{6.256}$$

Under certain necessary conditions these two problems are the same, which we assume is the case. Performing the minimisation over $\mathbf{w}, w_0$ leads to

$$\mathbf{w} = \sum_{n=1}^{N} \lambda_n y_n \boldsymbol{\xi}_n, \quad \sum_{n=1}^{N} \lambda_n y_n = 0 \tag{6.257}$$

From the definition of the Lagrangian

$$\begin{aligned} \mathcal{L}(\mathbf{w}, w_0, \{\lambda_n\}) &= f(\mathbf{w}) - \textstyle\sum_{n=1}^{N} \lambda_n \left\{(y_n(\mathbf{w}^T \boldsymbol{\xi}_n + w_0) - 1\right\} \\ &= f(\mathbf{w}) - \sum_{m=1}^{N} \sum_{n=1}^{N} \lambda_m \lambda_n y_m y_n \boldsymbol{\xi}_m^T \boldsymbol{\xi}_n + \sum_{n=1}^{N} \lambda_n \end{aligned} \tag{6.258}$$

We can now write down the dual problem for SVM as follows

$$\operatorname{argmax}_{\{\lambda_n\}} \left(f(\mathbf{w}) - \sum_{m=1}^{N} \sum_{n=1}^{N} \lambda_m \lambda_n y_m y_n \boldsymbol{\xi}_m^T \boldsymbol{\xi}_n + \sum_{n=1}^{N} \lambda_n \right)$$
$$\lambda_n \geq 0, \quad \forall n \tag{6.259}$$
$$\sum_{n=1}^{N} \lambda_n y_n = 0$$

Note that $\mathbf{w} = \sum_{n=1}^{N} \lambda_n y_n \boldsymbol{\xi}_n$ is given in terms of the $\{\lambda_n\}$. The form of (6.259) with regards to the *dot product* $\boldsymbol{\xi}_m^T \boldsymbol{\xi}_n$ allows for use of kernel substitution in order to define a nonlinear version. We simply replace $\boldsymbol{\xi}_m^T \boldsymbol{\xi}_n$ by a kernel function $k(\boldsymbol{\xi}_m, \boldsymbol{\xi}_n)$, implicitly defining a feature map $\boldsymbol{\phi}(\boldsymbol{\xi})$.

To classify a new input $\boldsymbol{\xi}$ we again calculate $\operatorname{sign}(\mathbf{w}^T \boldsymbol{\xi} + w_0)$, using expression (6.257) for $\mathbf{w}$

$$w_0 + \mathbf{w}^T \boldsymbol{\xi} = w_0 + \sum_{n=1}^{N} \lambda_n y_n \boldsymbol{\xi}_n^T \boldsymbol{\xi}$$
$$w_0 + \mathbf{w}^T \boldsymbol{\xi} = w_0 + \sum_{n=1}^{N} \lambda_n y_n k(\boldsymbol{\xi}, \boldsymbol{\xi}_n) \tag{6.260}$$

for the original and kernel versions, respectively. Recall that from the KKT conditions, for every data point $\boldsymbol{\xi}_n$ either $\lambda_n = 0$ or $y_n(\mathbf{w}^T \boldsymbol{\xi}_n + w_0) = 1$. Any data point for which $\lambda_n = 0$ will not appear in the sums above. The remaining data points $\boldsymbol{\xi}_m$, satisfying $y_m(\mathbf{w}^T \boldsymbol{\xi}_m + w_0) = 1$, are called the *support vectors*, These vectors are precisely those that lie on the hyperplanes (6.245).

Once the SVM is trained, we only require support vectors for making predictions. Having obtained $\mathbf{w}$, for any support vector $\boldsymbol{\xi}_m$

$$y_m(\mathbf{w}^T \boldsymbol{\xi}_m + w_0) = \begin{cases} y_m \left(w_0 + \sum_{n=1}^{N} \lambda_n y_n \boldsymbol{\xi}_n^T \boldsymbol{\xi}_m \right) = 1 & \text{original} \\ y_m \left(w_0 + \sum_{n=1}^{N} \lambda_n y_n k(\boldsymbol{\xi}_n, \boldsymbol{\xi}_m) \right) = 1 & \text{kernelised} \end{cases} \tag{6.261}$$

from which we can calculate the bias w_0. A more stable method is to average over the values of w_0 from all of the support vectors.

SVMs can be extended to multiple classes, although this is not straightforward and there is no entirely consistent method. One simple method is called *one-versus-all (OVA)*, for data $\boldsymbol{\xi}_n$, $n = 1, \ldots, N$, in K classes C_k. It proceeds as follows:

1. **for** $k = 1, \ldots, K$
    1. Set $y_n = 1$ for all $\boldsymbol{\xi}_n$ in C_k, and for the rest of the $\boldsymbol{\xi}_n$ set $y_n = -1$
    2. Train a binary SVM $z_k(\boldsymbol{\xi}) = \mathbf{w}_k^T \boldsymbol{\xi} + w_{k0}$ using this data
2. **end for**
3. Classify a point $\boldsymbol{\xi}$ according to the which of the discriminant functions $z_k(\boldsymbol{\xi})$ gives the highest value, i.e., $\boldsymbol{\xi} \in C_i$ where $i = \operatorname{argmax}_k z_k(\boldsymbol{\xi})$

This method requires K separate binary classifiers. An alternative method is the *all-versus-all (AVA)* or *one-versus-one* method, requiring $K(K-1)/2$ separate binary classifiers, and it proceeds as follows, using a *voting strategy*

1. **for** $k, m = 1, \ldots, K$ (without repetition of indices k, m)
    1. Set $y_n = 1$ for all $\boldsymbol{\xi}_n$ in C_k and set $y_n = -1$ for all $\boldsymbol{\xi}_n$ in C_m
    2. Train a binary SVM $z_{km}(\boldsymbol{\xi}) = \mathbf{w}_{km}^T \boldsymbol{\xi} + w_{km0}$ using this data
2. **end for**
3. For all $k, m = 1, \ldots, K$, if $\text{sign}(z_{km}(\boldsymbol{\xi})) = 1$ add one to the vote for class C_k otherwise add one to the vote for class C_m
4. The class with the highest number of votes wins and $\boldsymbol{\xi}$ is assigned to this class.

6.14 Linear Dimension Reduction

Dimension reduction methods seek either to compress the size of a data set or to find hidden patterns and relationships (representation learning). It could be, e.g., that amongst many variables in a vector-valued data point, some are redundant or of low importance. More common is that the data points have an *alternative representation*, which is given by some transformation of the points or change of variables. This transformation or change of variables can have very useful (equivalent) properties: it can locate the *intrinsic* dimension of the subspace in which the data lives or it can reveal new *explanatory* variables that are linear on nonlinear combinations of the original attributes. This allows for more compact representations or further analysis in terms of these more meaningful (in some sense) variables.

Linear methods look for a *linear vector subspace* of the original space in which the data resides (at least approximately) in order to define the above transformation. Alternatively, these methods transform the data in a linear or affine manner. In non-linear methods (also called *manifold learning*), such a linear transformation or linear subspace for achieving the desired goal simply does not exist. Instead, a nonlinear transformation can be found, as detailed in the next section. Here we outline the classical linear methods. It should be stressed that dimension reduction is an example of unsupervised learning, in which the data points have no corresponding input or output.

6.14.1 Principal Component Analysis and the Singular Value Decomposition

The data $\mathbf{y}_n$ can be treated as realisations of a random vector $\mathbf{y}$, which generally has an unknown distribution. Principal component analysis (PCA) constitutes a linear transformation of a data set $\mathbf{y}_n$, $n = 1, \ldots, N$

$$\mathbf{z}_n = \mathbf{V}^T \mathbf{y}_n \quad \text{or} \quad \mathbf{y}_n = \sum_{j=1}^{d} z_{jn} \mathbf{v}_j = \mathbf{V} \mathbf{z}_n \tag{6.262}$$

where $\mathbf{V} = [\mathbf{v}_1 \dots \mathbf{v}_d] \in \mathbb{R}^{d\times d}$. The columns of $\mathbf{V}$ are called *principal directions* and are orthonormal. The components z_{jn} of $\mathbf{y}_n$ in the orthonormal basis $\{\mathbf{v}_j\}_{j=1}^d$ for $\mathbb{R}^d$ are termed principal components. It is straightforward to show *via* orthogonal projection that $z_{jn} = \mathbf{v}_j^T\mathbf{y}_n$. The definition of PCA from a probabilistic perspective is that the principal components are mutually uncorrelated across j, and are ordered, together with the $\mathbf{v}_j$, such that their variances are non-increasing in j.

We centre the data and make the assumption $\mathbb{E}[\mathbf{y}] = \mathbf{0}$, so that $\text{cov}(\mathbf{y}) = \mathbb{E}\left[\mathbf{y}\mathbf{y}^T\right]$. Consider the first principal component $\mathbf{y}^T\mathbf{v}$ for an arbitrary $\mathbf{y}$. To find the first principal direction $\mathbf{v}$ we maximise the variance of this component

$$\begin{aligned}\text{var}(\mathbf{y}^T\mathbf{v}) &= \mathbb{E}\left[\left(\mathbf{v}^T\mathbf{y} - \mathbb{E}\left[\mathbf{v}^T\mathbf{y}\right]\right)^2\right]\\ &= \mathbb{E}\left[\left(\mathbf{v}^T\mathbf{y}\right)\left(\mathbf{v}^T\mathbf{x}\right)^T\right]\\ &= \mathbf{u}^T\mathbb{E}\left[\mathbf{y}\mathbf{y}^T\right]\mathbf{v}\\ &= \mathbf{v}^T\text{cov}(\mathbf{y})\mathbf{v}\end{aligned} \tag{6.263}$$

Therefore, we need to solve the constrained optimisation problem

$$\text{argmax}_{\mathbf{v}}\ \mathbf{v}^T\text{cov}(\mathbf{x})\mathbf{u} \quad \text{subject to } \|\mathbf{v}\|^2 = 1 \tag{6.264}$$

for which we introduce a Lagrange multiplier λ and instead solve

$$\text{argmax}_{\mathbf{v},\lambda}\ \left(\mathbf{v}^T\text{cov}(\mathbf{x})\mathbf{v} - \lambda(\mathbf{v}^T\mathbf{v} - 1)\right) \tag{6.265}$$

The solution satisfies

$$\text{cov}(\mathbf{y})\mathbf{v} = \lambda\mathbf{v} \tag{6.266}$$

which is an eigenvalue problem for the population covariance matrix $\text{cov}(\mathbf{y})$, providing d eigenvector-eigenvalue pairs. Clearly, the population covariance matrix is inaccessible, which necessitates an approximation in the form of the empirical or *sample covariance matrix*.

$$\mathbf{\Sigma} = \frac{1}{N-1}\mathbf{Y}\mathbf{Y}^T, \quad \mathbf{Y} = [\mathbf{y}_1 \dots \mathbf{y}_N] \in \mathbb{R}^{d\times N} \tag{6.267}$$

so that

$$\mathbf{\Sigma}\mathbf{v}_j = \lambda_j\mathbf{v}_j, \quad j = 1, \dots, d \tag{6.268}$$

We use the same notation for the eigenpairs although they are different from those of $\text{cov}(\mathbf{y})$. The eigenvectors λ_j are equal to the variance of the z_{qm} in the corresponding principal direction, which is seen from

$$\text{var}(\mathbf{y}^T\mathbf{v}_j) = \mathbf{v}_j^T\text{cov}(\mathbf{y})\mathbf{v}_j \approx \mathbf{v}_j^T\mathbf{\Sigma}\mathbf{v}_j = \mathbf{v}_j^T\lambda_j\mathbf{v}_j = \lambda_j \tag{6.269}$$

Note that $\text{cov}(\mathbf{y}^T\mathbf{v}_j, \mathbf{y}^T\mathbf{v}_i) = \lambda_i \mathbf{v}_j^T \mathbf{v}_i = 0$ for $i \neq j$, so that the components are mutually uncorrelated.

Once (6.268) is solved, the eigenvalues are ordered, together with the corresponding eigenvectors, according to $\lambda_j \geq \lambda_{j+1}$. There is usually a rapid decline in the variances of the components with j, allowing for accurate subspace approximations. A 'total variance' can be defined as $\sum_{q=1}^{d} \lambda_q$. To perform a dimension reduction, a required fraction of the variance ε is specified, and the value of Q such that

$$\frac{\sum_{j=1}^{Q} \lambda_j}{\sum_{j=1}^{d} \lambda_j} \geq \varepsilon \tag{6.270}$$

is calculated. An approximation of each $\mathbf{y}_n$ can then be made in the $Q-$dimensional subspace given by linear combinations of (spanned by) the directions $\{\mathbf{v}_j\}_{j=1}^{Q} \subset \mathbb{R}^d$

$$\mathbf{y}_n \approx \sum_{j=1}^{Q} z_{jn} \mathbf{v}_j = \mathbf{V}_Q \mathbf{z}_{Qn} \tag{6.271}$$

in which $\mathbf{V}_Q = [\mathbf{v}_1, \ldots, \mathbf{v}_Q]$ and $\mathbf{z}_{Qn} = (z_{1n}, \ldots, z_{Qn})^T$. Of all possible Q-dimensional subspace approximations of the data, the lowest square error is achieved with the basis $\{\mathbf{v}_j\}_{q=1}^{Q}$. This is a second interpretation of PCA and can also be used to derive the basis and coefficients.

In the context of a supervised learning problem based on a dimensionality reduction, the data point $\mathbf{y}_n$ is associated with a design point $\boldsymbol{\xi}_n$. In this case, the coefficients z_{jn} are functions of $\boldsymbol{\xi}_n$. For any general input $\boldsymbol{\xi}$, we may then form the $Q-$dimensional subspace approximation

$$\mathbf{y} = \boldsymbol{\eta}(\boldsymbol{\xi}) \approx \sum_{j=1}^{Q} z_j(\boldsymbol{\xi}) \mathbf{v}_j = \mathbf{V}_Q \mathbf{z}(\boldsymbol{\xi}) \tag{6.272}$$

in which $\mathbf{z}(\boldsymbol{\xi}) = (z_1(\boldsymbol{\xi}), \ldots, z_Q(\boldsymbol{\xi}))^T$ can be treated as a vector of uncorrelated random variables or random processes.

In practise, PCA is usually implemented not by solving the eigenvalue problem for the covariance matrix but by a *singular value decomposition* (SVD) of the data matrix $\mathbf{Y}$. Since $\boldsymbol{\Sigma}$ is symmetric, it can be *diagonalised*, i.e., written as

$$\boldsymbol{\Sigma} = \mathbf{V}\mathbf{L}\mathbf{V}^T \tag{6.273}$$

Each column of $\mathbf{V}$ is an eigenvector $\mathbf{v}_j$ of $\boldsymbol{\Sigma}$, while $\mathbf{L}$ is a diagonal matrix, with entries λ_j. On the other hand, a SVD of $\mathbf{Y}$ is a decomposition of the form

$$\mathbf{Y} = \mathbf{V}\mathbf{S}\mathbf{U}^T \tag{6.274}$$

in which both $\mathbf{V} \in \mathbb{R}^{d \times d}$ and $\mathbf{U} \in \mathbb{R}^{N \times N}$ are *unitary* matrices, i.e., the columns are unit length vectors and are mutually orthogonal, so that $\mathbf{U}^T\mathbf{U} = \mathbf{V}^T\mathbf{V} = \mathbf{I}$. The column vectors of $\mathbf{V}$ ($\mathbf{U}$) are called *left (right) singular vectors* of $\mathbf{Y}$. The matrix $\mathbf{S} \in \mathbb{R}^{d \times N}$ is again diagonal, with entries s_j, called *singular values*. By definition

$$\boldsymbol{\Sigma} = \frac{\mathbf{Y}\mathbf{Y}^T}{N-1} = \mathbf{V}\frac{\mathbf{S}^2}{N-1}\mathbf{V}^T \tag{6.275}$$

Comparing to $\boldsymbol{\Sigma} = \mathbf{V}\mathbf{L}\mathbf{V}^T$, it is obvious that the left singular vectors are the eigenvectors of $\boldsymbol{\Sigma}$ and that the non-zero singular values satisfy $s_j^2/(N-1) = \lambda_j$. Therefore, we can obtain the principal directions and eigenvalues from the SVD of $\mathbf{Y}$ rather than an eigendecomposition of $\boldsymbol{\Sigma}$. The primary reason for using a SVD is numerical stability and accuracy, even though the eigendecomposition can be less computationally costly. Moreover, SVD can be performed as a *reduced SVD*, in which not all principal directions and eigenvalues are obtained, making it highly efficient.

6.14.2 Multidimensional Scaling

Metric Multidimensional scaling is another linear method in its original form [31], consisting of an embedding of the data such that *dissimilarities* d_{nm} between points $\mathbf{y}_n$ and $\mathbf{y}_m$ in the original spaced are related to Euclidean distances $\widetilde{d}_{nm}$ between points in a low-dimensional approximating subspace. *Classical scaling* chooses the Euclidean distance as the dissimilarity measure and is isometric, i.e., it sets $\widetilde{d}_{nm} = d_{nm}$ in a least squares approximate sense.

We first define a dissimilarity matrix $\mathbf{D} = [d_{nm}]$, i.e., the matrix of dissimilarities between untransformed points $\mathbf{y}_n \in \mathbb{R}^d$, $n = 1 \ldots, N$. By the definition of classical scaling, the transformed (embedded) points $\mathbf{z}_n$ are obtained from

$$\widetilde{d}_{nm} := \|\mathbf{z}_n - \mathbf{z}_m\| = d_{nm} \tag{6.276}$$

Defining $\mathbf{Z}$ to be the matrix with rows given by $\mathbf{z}_n$, we obtained a centred version from

$$\widehat{\mathbf{Z}} = \mathbf{H}\mathbf{Z} = [\widehat{\mathbf{z}}_1 \ldots \widehat{\mathbf{z}}_N]^T, \quad \mathbf{H} = \mathbf{I} - \frac{1}{N}\mathbf{1}\mathbf{1}^T \tag{6.277}$$

in which $\mathbf{H}$ is called a centering matrix and $\widehat{\mathbf{z}}_m = \mathbf{z}_m - \overline{\mathbf{z}}$, with $\overline{\mathbf{z}} = \sum_{i=1}^N \mathbf{z}_n$ denoting the mean. The centering is performed as a constraint, since (6.276) does not have a unique solution; any solutions $\mathbf{z}_n$ will generate a family of solutions $\mathbf{z}_n + k$, $k \in \mathbb{R}$, since the distances between these points are not affected by a constant. By virtue of the simple relationship

$$d_{nm}^2 = (\mathbf{y}_n - \mathbf{y}_m)^T(\mathbf{y}_n - \mathbf{y}_m) \tag{6.278}$$

we can write the centred kernel matrix in the original space, $\mathbf{K} = (\mathbf{H}\mathbf{Y})^T(\mathbf{H}\mathbf{Y})$, in terms of the distance matrix $\mathbf{D}$ and set this equal to the kernel matrix in the embedded space in order to satisfy (6.276)

$$\mathbf{K} = -\frac{1}{2}\mathbf{H}(\mathbf{D} \odot \mathbf{D})\mathbf{H} = \widehat{\mathbf{Z}}\widehat{\mathbf{Z}}^T \tag{6.279}$$

The Hadamard product $\mathbf{D} \odot \mathbf{D}$ gives a matrix of the square Euclidean distances d_{nm}^2, while applying $\mathbf{H}$ on the left and right leads to a matrix in which the means of both the column and row vectors are zero.

$\mathbf{K}$ has a spectral decomposition $\mathbf{K} = \mathbf{V}\mathbf{\Lambda}\mathbf{V}^T$, in which

$$\mathbf{\Lambda} = \text{diag}(\lambda_1, \ldots \lambda_d) \in \mathbb{R}^{N\times N} \quad \mathbf{U} = [\mathbf{u}_1, \ldots \mathbf{u}_d] \in \mathbb{R}^{N\times N} \tag{6.280}$$

As in PCA, the eigenvalues λ_i, $i = 1, \ldots, N$, are arranged such that they are non-increasing, and the eigenvectors $\mathbf{u}_i \in \mathbb{R}^N$ have unit length. If $d < N$, there are $N - d$ zero eigenvalues. Consequently, the following representation of the data is obtained

$$\widehat{\mathbf{Z}} = \mathbf{U}\mathbf{\Lambda}^{1/2} \in \mathbb{R}^{N\times N} \tag{6.281}$$

A reduction in dimensionality is achieved by restricting the representation to the first $Q < N$ directions $\mathbf{u}_i$ having the largest associated eigenvalues, that is

$$\widehat{\mathbf{Z}}_Q = \mathbf{U}_Q\mathbf{\Lambda}_Q^{1/2}, \quad \mathbf{V}_Q = [\mathbf{u}_1, \ldots \mathbf{u}_Q], \quad \mathbf{\Lambda}_Q = \text{diag}(\lambda_1, \ldots \lambda_Q) \tag{6.282}$$

The low-dimensional approximations $\widehat{\mathbf{z}}_{Qn}$ are simply the rows of $\widehat{\mathbf{Z}}_Q$.

Classical scaling is in fact equivalent to PCA. To see this, we first note the similarities between the eigenproblems for $\mathbf{K}$ and the sample covariance matrix. We recall that $\mathbf{K}\mathbf{u}_i = (\mathbf{HY})^T(\mathbf{HY})\mathbf{u}_i = \lambda_i\mathbf{u}_i$, so that

$$\left[(\mathbf{HY})(\mathbf{HY})^T\right](\mathbf{HY})\mathbf{u}_i = \lambda_i(\mathbf{HY})\mathbf{u}_i \tag{6.283}$$

Thus, the eigenvalues λ_i of the scaled covariance matrix $\mathbf{HY}(\mathbf{HY})^T$, assuming now that the data is not centred, are identical to those of $\mathbf{K}$, while the eigenvectors $\mathbf{v}_i$ of the scaled covariance matrix satisfy $\mathbf{v}_i = \mathbf{HY}\mathbf{u}_i$. This leads to

$$(\mathbf{HY})^T\mathbf{v}_i = (\mathbf{HY})^T\mathbf{HY}\mathbf{u}_i = \lambda_i\mathbf{u}_i \tag{6.284}$$

Thus, the projection of the centred data matrix onto the (scaled) principal directions leads again to the classical scaling solution (6.281).

In the more general case, a metric other than the Euclidean metric is used to define the dissimilarities, and the problem is posed as

$$\min_{\mathbf{z}_n} \ \text{Stress} := \left\| \mathbf{D} - \|\mathbf{z}_n - \mathbf{z}_m\| \right\|^2 \tag{6.285}$$

PCA and MDS are linear dimension reduction techniques, meaning that they seeks a linear subspace of the ambient space $\mathbb{R}^d$ that can well approximate the data points. Nonlinear methods attempt instead to find a nonlinear subspace, in cases where no linear subspace can be found. There are various such methods, such as kernel PCA [32], local linear embedding [33], Isomap [34] and diffusion maps [35], which are discussed in the Sect. 6.15. We first outline an extension of PCA and SVD to tensor-variate data.

6.14.3 Reduced Rank Tensor Decompositions

It is often the case that data is most naturally expressed in its 2D or higher-dimensional format, such as data from SEM/TEM images, flow visualisation experiments or 3D computed tomography. The question is whether there exist version of PCA or SVD in such cases. The answer is to the positive, although such reduced-rank *decompositions* (as discussed in Sect. 6.11.1) may not be unique.

The CP and Tucker decompositions (6.168) and (6.169) were used in Sect. 6.11.1 to solve tensor linear regression problems. They are repeated here for convenience. For a $\mathcal{A} \in \mathbb{R}^{p_1 \times \ldots \times p_K}$, the CP-$r$ rank decomposition of $\mathcal{A}$ is given by

$$\mathcal{A} = [\![\mathbf{B}_1, \ldots, \mathbf{B}_K]\!] = \sum\nolimits_{i=1}^{r} \mathbf{b}_i^1 \circ \mathbf{b}_i^2 \circ \ldots \circ \mathbf{b}_i^K \tag{6.286}$$

in which $\mathbf{B}_i = [\mathbf{b}_i^1 \ldots \mathbf{b}_i^r] \in \mathbb{R}^{p_k \times r}$ are the factors and $r \in \mathbb{N}$, while

$$\mathcal{A} = [\![\mathcal{G}; \mathbf{B}_1, \ldots, \mathbf{B}_K]\!] = \sum\nolimits_{r_1=1}^{n_1} \cdots \sum\nolimits_{r_K=1}^{n_K} g_{r_1 \ldots K} \mathbf{b}_1^{r_1} \circ \ldots \circ \mathbf{b}_K^{r_K} \tag{6.287}$$

is the Tucker decomposition for a core tensor $\mathcal{G} \in \mathbb{R}^{n_1 \times \ldots \times n_K}$ and factors $\mathbf{B}_k = [\mathbf{b}_k^1 \ldots \mathbf{b}_k^{n_k}] \in \mathbb{R}^{p_k \times n_K}$, with multilinear rank $(n_1, \ldots, n_K)$, $n_k \in \mathbb{N}$.

Storing these tensors using the factors and the core clearly leads to a reduction in the number of components required, e.g., for the CP decomposition the number is reduced from $\prod p_k$ to $r \sum p_k$. The CP decomposition is unique up to certain scaling and permutation operations, provided the tensor is of order 3 and above (matrix decompositions are not unique). The problem to be solved in order to compute a CP decomposition is

$$\text{argmin}_{\{\mathbf{B}_k\}} \left(\|\mathcal{X} - [\![\mathbf{B}_1, \ldots, \mathbf{B}_K]\!]\| = \langle \mathcal{X}, [\![\mathbf{B}_1, \ldots, \mathbf{B}_K]\!] \rangle^{\frac{1}{2}} \right) \tag{6.288}$$

The workhorse method for solving this problem is known as *alternating least squares* (ALS), in which each factor but the n-th is in turn fixed to solve for $\mathbf{A}_n$, and the process is repeated until convergence. From (6.201), the mode-k matricised version of (6.286) is

$$\mathbf{A}_{(k)} = \mathbf{B}_k (\mathbf{B}_K * \ldots * \mathbf{B}_{k+1} * \mathbf{B}_{k-1} * \ldots * \mathbf{B}_1)^T \tag{6.289}$$

so that ALS consists of

1. Initialise $\{\mathbf{B}_k\}$
2. **for** $k = 1 : K$
3. Solve

$$\min_{\mathbf{B}_k} \|\mathbf{A}_{(k)} - \mathbf{B}_k(\mathbf{B}_K * \ldots * \mathbf{B}_{k+1} * \mathbf{B}_{k-1} * \ldots * \mathbf{B}_1)^T\| \tag{6.290}$$

4. **end for**
5. Return to 2 if $\|(\mathbf{B}_k)_{\text{new}} - (\mathbf{B}_k)_{\text{old}}\|_F > \varepsilon$ for threshold ε or until max iterations reached
6. **Return** $\{\mathbf{B}_k\}$

Problem (6.291) is a classical least squares minimisation with solution

$$\mathbf{B}_k = \mathbf{A}_{(k)} \left[\mathbf{B}_k(\mathbf{B}_K * \ldots * \mathbf{B}_{k+1} * \mathbf{B}_{k-1} * \ldots * \mathbf{B}_1)^T\right]^\dagger \tag{6.291}$$

in which $\dagger$ denotes a Moore-Penrose right pseudo inverse. We mention that there are a number of alternative schemes to ALS, although few have been shown to work better. Additionally, the selection of r is a main concern. The procedure above is repeated for different $r = 1, 2, \ldots$ until the best fit is found. There is no systematic way of determining r *a-priori*.

The problem to be solved in the case of a Tucker decomposition is

$$\text{argmin}_{\mathcal{G},\{\mathbf{B}_k\}} \left(\|\mathcal{X} - [\![\mathcal{G}; \mathbf{B}_1, \ldots, \mathbf{B}_K]\!]\| = \langle \mathcal{X}, [\![\mathcal{G}; \mathbf{B}_1, \ldots, \mathbf{B}_K]\!]\rangle^{\frac{1}{2}}\right) \tag{6.292}$$

in which the $\mathbf{B}_k$ are assumed to be column-wise orthogonal. By virtue of (6.170)

$$\text{argmin}_{\mathcal{G},\{\mathbf{B}_k\}} \|\text{vec}(\mathcal{X}) - (\mathbf{B}_K \otimes \ldots \otimes \mathbf{B}_1)\text{vec}(\mathcal{G})\| \tag{6.293}$$

so that the core tensor must satisfy

$$\mathcal{G} = \mathcal{X} \times_1 \mathbf{B}_1^T \times_2 \mathbf{B}_2^T \times_3 \ldots \times_K \mathbf{B}_K^T \tag{6.294}$$

There are a several ways of solving this problem, with the simplest called the Tucker1 or higher-order SVD, iteratively finding independently the components in each mode that best capture the variations in the data, along with updating the core tensor. This method can also be used to truncate the rank n_k for one or more modes by virtue of the SVD properties, in which case it is called truncated HOSVD. The truncated HOSVD is not optimal in a least-squares sense, so an ALS version called *higher-order orthogonal iteration* (HOOI) is usually preferred.

Based on (6.294) we obtain

$$\begin{aligned}
&\|\mathcal{X} - [\![\mathcal{G}; \mathbf{B}_1, \ldots, \mathbf{B}_K]\!]\|^2 \\
&= \|\mathcal{X}\|^2 - 2\langle \mathcal{X}, [\![\mathcal{G}; \mathbf{B}_1, \ldots, \mathbf{B}_K]\!]\rangle + \|[\![\mathcal{G}; \mathbf{B}_1, \ldots, \mathbf{B}_K]\!]\|^2 \\
&= \|\mathcal{X}\|^2 - 2\langle \mathcal{X} \times_1 \mathbf{B}_1^T \times_2 \mathbf{B}_2^T \times_3 \ldots \times_K \mathbf{B}_K^T, \mathcal{G}\rangle + \|\mathcal{G}\|^2 \\
&= \|\mathcal{X}\|^2 - 2\langle \mathcal{G}, \mathcal{G}\rangle + \|\mathcal{G}\|^2 \\
&= \|\mathcal{X}\|^2 - \|\mathcal{G}\|^2 \\
&= \|\mathcal{X}\|^2 - \|\mathcal{X} \times_1 \mathbf{B}_1^T \times_2 \mathbf{B}_2^T \times_3 \ldots \times_K \mathbf{B}_K^T\|^2
\end{aligned} \tag{6.295}$$

We also used the fact that the mode-k multiplication commutes with respect to the inner product, namely

$$\langle \mathcal{A}, \mathcal{B} \times_k \mathbf{A}\rangle = \langle \mathcal{A} \times_k \mathbf{A}^T, \mathcal{B}\rangle \tag{6.296}$$

for tensors $\mathcal{A}, \mathcal{B}$ and an appropriately sized matrix $\mathbf{A}$. Moreover, we used the property that for any tensor $\mathcal{A}$

$$\|\mathcal{A}\| = \|\mathcal{A} \times_k \mathbf{U}\| \tag{6.297}$$

for any orthogonal matrix $\mathbf{U}$.

The objective function in (6.295) can first be simplified by ignoring $\|\mathcal{X}\|^2$, which is a constant. The remaining term, which must now be maximised, can be rewritten using (6.170), so that the ALS problem becomes, for each $k = 1, \ldots, K$, to iteratively solve

$$\begin{aligned}
\text{argmax}_{\mathbf{B}_k} \left(\|\mathcal{X} \times_1 \mathbf{B}_1^T \times_2 \mathbf{B}_2^T \times_3 \ldots \times_K \mathbf{B}_K^T\|^2 = \|\mathbf{B}_k \mathbf{W}\|^2\right) \\
\mathbf{W} = \mathbf{X}_{(k)} \left(\mathbf{B}_K \otimes \ldots \otimes \mathbf{B}_{k+1} \otimes \mathbf{B}_{k-1} \otimes \ldots \otimes \mathbf{B}_1\right)
\end{aligned} \tag{6.298}$$

The solution is given in terms of the SVD of the matrix $\mathbf{W}$, namely $\mathbf{B}_k$ is defined by the leading n_k singular vectors of $\mathbf{W}$.

6.15 Manifold Learning and Nonlinear Dimension Reduction

A $Q-$dimensional manifold $\mathcal{M} \subset \mathbb{R}^d$ embedded in $\mathbb{R}^d$ is defined as a subspace of $\mathbb{R}^d$ in which all points in $\mathcal{M}$ can be parameterised by Q independent variables. Any parameterisation is referred to as a *coordinate system* or a chart, with possibly multiple coordinate systems required to characterise the entire manifold using overlapping patches, each possessing a system of non-unique coordinates.

More precisely, a smooth $Q-$manifold is a topological space $\mathcal{M}$ with a *maximal open cover* $\{U_\alpha\}_{\alpha\in\Gamma}$, for some index set Γ, comprising coordinate neighbourhoods or patches U_α, and a corresponding set of homeomorphisms (the coordinate charts)

$$\phi_\alpha : U_\alpha \to \phi_\alpha(U_\alpha) \subset \mathbb{R}^Q \tag{6.299}$$

These charts map onto open subsets $\phi_\alpha(U_\alpha) \subset \mathbb{R}^Q$ with the property that

$$\phi_\alpha(U_\alpha \cap U_\beta) \quad \text{and} \quad \phi_\beta(U_\alpha \cap U_\beta) \tag{6.300}$$

are $\mathbb{R}^Q$-open, in which case it is said that ϕ_α and ϕ_β are *compatible*. The transition maps $\phi_\beta \circ \phi_\alpha^{-1}$ that define a coordinate change are diffeomorphic for all $\alpha, \beta \in \Gamma$.

Our data lies in some set $\mathcal{M}$, which could be a manifold of dimension $Q \ll d$. We can therefore attempt to represent the data $\mathbf{y} \in \mathcal{M}$ by points $\mathbf{z}_Q$ in a feature or latent space $\mathcal{F}_Q \subset \mathbb{R}^Q$ *via* a smooth, and unknown function $\mathbf{f}_Q^{-1} : \mathcal{F}_Q \to \mathcal{M}$, with $\mathbf{y} = \mathbf{f}_Q^{-1}(\mathbf{z}_Q)$. Manifold learning refers to the problem of approximating $\mathbf{f}_Q^{-1}$ along with its inverse $\mathbf{f}_Q$ based on a given data set. Nonlinear dimensionality reduction (in fact all dimension reduction), on the other hand, refers to the equivalent problem of finding the latent space representation $\mathbf{z}_Q \in \mathcal{F}_Q$.

The intrinsic representation of points $\mathbf{y} \in \mathcal{M}$ using Q parameters is generally not easy to find, and even if it were possible to find, such a representation would not be especially convenient. The equivalent representation $\mathbf{z}_Q \in \mathcal{F}_Q$, on the other hand, uses a standard Euclidean basis, and is therefore highly convenient. Most, if not all nonlinear dimension reduction methods can be viewed as embeddings, relying on an appropriate non-Euclidean measure of distance (metric) between the data points in the ambient space. This metric is then preserved in the embedding of the data in a low-dimensional Euclidean (latent or feature) space, in which distances are measured using the Euclidean metric. Different methods use different metrics, arguing in one way or another that the metric used is the most natural. They can also be viewed as feature mappings, devised in such a way that dimension reduction in the form of a linear subspace approximation is possible in the feature space, while it is not possible in the original space.

6.15.1 Kernel Principal Component Analysis

Kernel PCA (kPCA) [32] essentially performs PCA in a feature space $\mathcal{F}'$ by mapping points $\mathbf{y}_n \in \mathbb{R}^d, n = 1, \ldots, N$ in an original space using a feature map $\boldsymbol{\phi} : \mathbb{R}^d \to \mathcal{F}'$. It is a *kernel method*, meaning that the feature map is (generally) implicitly specified through a kernel function. We first define the sample covariance matrix $\mathbf{C}_{\mathcal{F}'}$ in the feature space and solve the eigenvalue problem

$$\mathbf{C}_{\mathcal{F}'}\mathbf{w} = \left(\frac{1}{N} \sum_{n=1}^{N} \widetilde{\boldsymbol{\phi}}(\mathbf{y}_n) \left(\widetilde{\boldsymbol{\phi}}(\mathbf{y}_n) \right)^T \right) \mathbf{w} = \lambda \mathbf{w} \tag{6.301}$$

where

$$\widetilde{\boldsymbol{\phi}}(\mathbf{y}_n) = \boldsymbol{\phi}(\mathbf{y}_n) - \overline{\boldsymbol{\phi}}, \quad \overline{\boldsymbol{\phi}} = \frac{1}{N} \sum_{n=1}^{N} \boldsymbol{\phi}(\mathbf{y}_n) \tag{6.302}$$

is data point i after centering in the feature space. The feature map $\boldsymbol{\phi}(\cdot)$ is defined by a kernel $k(\mathbf{y}_n, \mathbf{y}_m) = \boldsymbol{\phi}(\mathbf{y}_n)^T\boldsymbol{\phi}(\mathbf{y}_m)$, e.g., the Gaussian kernel (6.64). A corresponding kernel matrix is given by $\mathbf{K} = [K_{nm}]$, in which $K_{nm} = k(\mathbf{y}_n, \mathbf{y}_m)$. A centred kernel $\widetilde{k}(\mathbf{y}_n, \mathbf{y}_m) = \widetilde{\boldsymbol{\phi}}(\mathbf{y}_n)^T\widetilde{\boldsymbol{\phi}}(\mathbf{y}_m)$ leads to a centred matrix $\widetilde{\mathbf{K}} = [\widetilde{K}_{ij}]$ with entries $\widetilde{K}_{ij} = \widetilde{\boldsymbol{\phi}}(\mathbf{y}_n)^T\widetilde{\boldsymbol{\phi}}(\mathbf{y}_m)$. The relationship between the two is given by

$$\widetilde{\mathbf{K}} = \mathbf{HKH} \tag{6.303}$$

in which $\mathbf{H}$ is the previously defined centering matrix.

According to Eq. (6.301), the eigenvectors $\mathbf{w}$ of the covariance matrix are linear combinations of $\widetilde{\boldsymbol{\phi}}(\mathbf{y}_n)$, that is $\mathbf{w} = \sum_{n=1}^{N} \alpha_n\widetilde{\boldsymbol{\phi}}(\mathbf{y}_n)$, with some coefficients α_n. Inserting this expression into (6.301) and multiplying on the left by $\widetilde{\boldsymbol{\phi}}(\mathbf{y}_n)^T$, leads to an equivalent eigenvalue problem

$$\widetilde{\mathbf{K}}\boldsymbol{\alpha} = m\lambda\boldsymbol{\alpha} \tag{6.304}$$

in which $\boldsymbol{\alpha} = (\alpha_1, \ldots, \alpha_N)^T$. This problem can be solved for the orthonormal $\boldsymbol{\alpha}_n$, after which they are rescaled according to $\boldsymbol{\alpha}_n \mapsto \boldsymbol{\alpha}_n/\sqrt{\lambda_n} = \widetilde{\boldsymbol{\alpha}}_n$, resulting in orthonormal eigenvectors

$$\widetilde{\mathbf{w}}_i = \sum_{j=1}^{N} \widetilde{\alpha}_{ji}\widetilde{\boldsymbol{\phi}}(\mathbf{y}_j), \quad i = 1, \ldots, N \tag{6.305}$$

in which $\widetilde{\alpha}_{ji} = \alpha_{ji}/\sqrt{\lambda_n}$ and α_{ji} are respectively the j-th components of $\widetilde{\boldsymbol{\alpha}}_i$ and $\boldsymbol{\alpha}_i$. There are in fact $\min(\dim \mathscr{F}', N)$ basis vectors $\widetilde{\mathbf{w}}_i$, but in the general case $\dim \mathscr{F}' > N$. The expansion of a point $\widetilde{\boldsymbol{\phi}}(\mathbf{y}_n)$ in feature space in terms of the basis $\{\widetilde{\mathbf{w}}_i\}_{i=1}^{N} \subset \mathcal{F}'$ is

$$\widetilde{\boldsymbol{\phi}}(\mathbf{y}_n) = \sum_{i=1}^{N} z_i(\mathbf{y}_n)\widetilde{\mathbf{w}}_i \tag{6.306}$$

in which the coefficients are given by

$$\begin{aligned} z_i(\mathbf{y}_n) = \widetilde{\mathbf{w}}_i^T\widetilde{\boldsymbol{\phi}}(\mathbf{y}_n) &= \sum_{l=1}^{N} \widetilde{\alpha}_{li}\widetilde{\boldsymbol{\phi}}(\mathbf{y}_l)^T\widetilde{\boldsymbol{\phi}}(\mathbf{y}_n) \\ &= \sum_{l=1}^{N} \widetilde{\alpha}_{li}\widetilde{K}_{ln} \\ &= \widetilde{\boldsymbol{\alpha}}_i^T\widetilde{\mathbf{k}}_n \\ &= \widetilde{\boldsymbol{\alpha}}_i^T\mathbf{H}(\mathbf{k}_n - \mathbf{K1}), \quad \forall i = 1, \ldots, N \end{aligned} \tag{6.307}$$

where

$$\mathbf{k}_n = (K_{1n}, \ldots, K_{Nn})^T, \quad \widetilde{\mathbf{k}}_n = (\widetilde{K}_{1n}, \ldots, \widetilde{K}_{Nn})^T \tag{6.308}$$

This allows us to define the following quantity

$$\mathbf{z}(\mathbf{y}_n) = (z_1(\mathbf{y}_n), \ldots, z_N(\mathbf{y}_n))^T \tag{6.309}$$

in which $z_i(\mathbf{y}_n)$, $i = 1, \ldots, N$, are defined in (6.307). It is important to note that while the coefficients in (6.306) are computed according to these formulae, the basis vectors $\widetilde{\mathbf{w}}_i$ in (6.305) are unknown since the map $\widetilde{\phi}(\cdot)$ is not explicitly specified.

As in PCA, the eigenvalues are arranged such that $\lambda_i < \lambda_{i-1}$, $i = 2, \ldots, N$, and are equal to the variances along the directions $\widetilde{\mathbf{w}}_i$, while the coefficients of a mapped point in the basis $\{\widetilde{\mathbf{w}}_i\}_{i=1}^N$ are uncorrelated. Moreover, a Q-dimensional subspace approximation of mapped points $\widetilde{\phi}(\mathbf{y}_n)$ can be obtained as in PCA, to define a latent mapping $\mathbf{f}_Q : \mathbb{R}^d \to \mathcal{F}_Q \subset \mathbb{R}^Q$, with latent space $\mathcal{F}_Q$, as follows

$$\mathbf{z}_Q = \mathbf{f}_Q(\mathbf{y}_n) = (z_1(\mathbf{y}_n), \ldots, z_Q(\mathbf{y}_n))^T \in \mathcal{F}_Q \tag{6.310}$$

for which we may use the standard Euclidean basis for convenience and without loss of generality. The error in the projection

$$\widetilde{\phi}_Q(\mathbf{y}_n) = \sum\nolimits_{i=1}^{Q} z_i(\mathbf{y}_n)\widetilde{\mathbf{w}}_i \tag{6.311}$$

of $\widetilde{\phi}(\mathbf{y}_n)$ onto $\mathcal{F}'_Q = \text{span}(\widetilde{\mathbf{w}}_1, \ldots, \widetilde{\mathbf{w}}_Q)$, with the coefficient vector (6.310), is bounded by

$$\|\widetilde{\phi}_Q(\mathbf{y}_n) - \widetilde{\phi}(\mathbf{y}_n)\|^2 = \sum\nolimits_{i=Q+1}^{N} \lambda_i^2 \tag{6.312}$$

using the standard Euclidean norm. Note that we use $\mathcal{F}_Q$ for the space of coordinates, as opposed to the space $\mathcal{F}'_Q \subset \mathcal{F}'$.

6.15.2 *Isomap*

Isomap [36] extends MDS by using a geodesic distance to measure dissimilarities. Essentially, the classical MDS method is applied to a dissimilarity matrix with these geodesic distances, resulting in an isometric embedding of the data and a simultaneous dimension reduction [36].

Given data $\mathbf{y}_n \in \mathbb{R}^d$, $n = 1 \ldots, N$, the method first selects neighbourhood points based on the Euclidean distance. This can be achieved in different ways. The first main method selects as neighbours those points lying within an ϵ ball, while another uses the M (neighbourhood number) closest points. The matrix $\mathbf{D} = [d_{ij}]$ of dissimilarities is then constructed. Distances between neighbours are defined by Euclidean distances, while distances between non-neighbours are set to the shortest path distances through the neighbouring points.

Finally, the method applies classical scaling on $\mathbf{K} = -(1/2)\mathbf{H}(\mathbf{D} \odot \mathbf{D})\mathbf{H}$ (as described earlier) to produce the low-dimensional representations $\widehat{\mathbf{z}}_{Qn}$, with the rows of (6.282) defining the map $\mathbf{f}_Q(\mathbf{y}_n) : \mathbb{R}^d \to \mathcal{F}_Q \subset \mathbb{R}^Q$, with latent space $\mathcal{F}_Q$.

The method is very simple, and has an intimate connection to kPCA, explained in [37–39]. The matrix $\mathbf{D}$ can be viewed as defining distances between mapped points in a feature space $\mathcal{F}$, namely

$$d_{ij}^2 = (\boldsymbol{\phi}(\mathbf{y}_n) - \boldsymbol{\phi}(\mathbf{y}_m))^T(\boldsymbol{\phi}(\mathbf{y}_n - \boldsymbol{\phi}(\mathbf{y}_m)) \tag{6.313}$$

in which $\boldsymbol{\phi} : \mathbb{R}^d \to \mathcal{F}$ is the feature map, with an associated kernel $k'(\mathbf{y}_n, \mathbf{y}_m) = \boldsymbol{\phi}(\mathbf{y}_n)^T\boldsymbol{\phi}(\mathbf{y}_m)$. For isotropic kernels, with the scaling $k'(\mathbf{y}_n, \mathbf{y}_n) = 1$

$$d_{nm}^2 = 2 - 2k'(\mathbf{y}_n, \mathbf{y}_m) \quad \text{or} \quad -\frac{1}{2}(\mathbf{D} \odot \mathbf{D}) = \mathbf{K}' - \mathbf{1}\mathbf{1}^T \tag{6.314}$$

where $\mathbf{K}'$ is a kernel matrix, with a centred version

$$\mathbf{K} = -\frac{1}{2}\mathbf{H}(\mathbf{D} \odot \mathbf{D})\mathbf{H} = \mathbf{H}\mathbf{K}'\mathbf{H} \tag{6.315}$$

In classical MDS, this kernel matrix (obtained from a dissimilarity matrix) essentially finds the components in a representation of data points in $\mathcal{F}'$, with the basis comprising eigenvectors of the centred sample covariance matrix $\mathbf{C}_{\mathcal{F}'}$ in (6.301), which is identical to kPCA.

One issue faced in Isomap is that there is no guarantee of a positive definite kernel matrix, which, theoretically, is required to ensure that the low-dimensional embedding space is Euclidean and to ensure the existence of a feature space. When a matrix $\mathbf{D}$ generates a positive semi-definite kernel matrix, it is referred to as having an *exact Euclidean representation*. To avoid non-Euclidean $\mathbf{D}$, Choi and Choi developed the *kernel Isomap* method, which exhibits greater robustness [38].

6.15.3 Diffusion Maps

Diffusion maps are based on mapping data $\mathbf{y}_n \in \mathbb{R}^d$, $n = 1, \ldots, N$ to a subset $\mathcal{D}_t$ of $\mathbb{R}^N$ known as the *diffusion space*, and subsequently performing a dimensional reduction [35, 40]. In actual fact, there is a family of such spaces indexed by a parameter t. The mapping is achieved *via* an embedding in the diffusion space, such that a *diffusion distance* defined between points in the original space is preserved between points in the embedded space. The method is motivated by graph theory, by first identifying data $\mathbf{y}_n$ with nodes on a graph. A Markov chain is then generated by defining a *connectivity* (or a 'kernel') between the points or nodes (see Sect. 6.8.3 for details on Markov chains and Markov processes on continuous state spaces).

Starting with an undirected, connected graph $\mathcal{G}$ with a vertex set $\{\mathbf{y}_1, \ldots, \mathbf{y}_N\}$, we define edge weights using a kernel function $k(\mathbf{y}_n, \mathbf{y}_m)$, such as the Gaussian kernel encountered earlier. We construct a diffusion process on $\mathcal{G}$ [41] using a normalised connectivity or adjacency matrix, defined as $\mathbf{K} = [k_{nm}]$, in which $k_{nm} = k(\mathbf{y}_n, \mathbf{y}_m)$.

The so-called *degree matrix* is given by

$$\mathbf{D} = \text{diag}(d_1, \ldots, d_N), \quad d_n = \sum\nolimits_m k_{nm} \tag{6.316}$$

while the $N \times N$ *diffusion matrix* is given as

$$\mathbf{P} = \mathbf{D}^{-1}\mathbf{K} \tag{6.317}$$

This matrix $\mathbf{P} = [P_{nm}]$ is in fact a Markov matrix, with P_{nm} being a transition probability $p(\mathbf{y}_n, \mathbf{y}_m)$ of going from a point or node $\mathbf{y}_n$ on the graph to the node $\mathbf{y}_m$ during a random walk. We can also define t step transition probabilities $p_t(\mathbf{y}_n, \mathbf{y}_m)$ in going from node $\mathbf{y}_n$ to $\mathbf{y}_m$ in $t \in \mathbb{N}$ steps as the (n, m)-th entries of the matrix $\mathbf{P}^t = \mathbf{P} \times \cdots \times \mathbf{P}$, formed by the t-th power of $\mathbf{P}$.

One of the key properties of the graph $\mathcal{G}$ is that it is connected, so that the Markov chain $\mathbf{P}$ is ergodic. It therefore has a unique stationary distribution $\boldsymbol{\pi}$ such that $\pi_n = d_n / \sum_m d_m$ [35]. The matrix

$$\mathbf{P}' = \mathbf{D}^{-1/2}\mathbf{K}\mathbf{D}^{1/2} \tag{6.318}$$

is symmetric and its eigenvalues γ_i' are identical to those of $\mathbf{P}$. Performing a spectral decomposition on this matrix yields

$$\mathbf{P}' = \mathbf{S}\boldsymbol{\Gamma}'\mathbf{S}^T \tag{6.319}$$

in which $\mathbf{S}$ is an orthogonal matrix with columns given by the eigenvectors $\mathbf{s}_n$, $n = 1, \ldots, N$, of $\mathbf{P}'$, while $\boldsymbol{\Gamma}' = \text{diag}(\gamma_1', \ldots, \gamma_N')$. As is by now familiar, we arrange the eigenvalues so that $1 = \gamma_1' > \cdots > \gamma_N'$. The first eigenvector $\mathbf{s}_1$ has elements $\sqrt{\pi_n}$ [42].

The spectral decomposition of $\mathbf{P}$ on the other hand is

$$\mathbf{P} = \mathbf{Q}\boldsymbol{\Gamma}'\mathbf{Q}^{-1} \tag{6.320}$$

in which $\mathbf{Q} = \mathbf{D}^{-1/2}\mathbf{S}$. The eigenvectors of $\mathbf{P}$ can be found to the left and to the right, and are given by

$$\mathbf{r}_n = \mathbf{D}^{-1/2}\mathbf{s}_n, \quad \mathbf{l}_n = \mathbf{D}^{1/2}\mathbf{s}_n \tag{6.321}$$

respectively, satisfying a bi-orthogonality property, namely, $\mathbf{l}_n^T \mathbf{r}_m = \delta_{nm}$, where δ_{nm} is the Kronecker delta. Moreover

$$\mathbf{l}_1 = \boldsymbol{\pi}\sqrt{\sum\nolimits_n d_n}, \quad \mathbf{r}_1 = \frac{\mathbf{1}^T}{\sqrt{\sum_n d_n}} \tag{6.322}$$

Due to the orthogonality of $\mathbf{S}$, we obtain

$$\mathbf{P}^t = \mathbf{Q}\boldsymbol{\Gamma}^t\mathbf{Q}^{-1}, \quad \mathbf{P}^t = \sum\nolimits_{n=1}^{N} (\gamma_n')^t \mathbf{r}_n \mathbf{l}_n^T \tag{6.323}$$

Denoting by $\mathbf{p}_n^t$ the n-th row of $\mathbf{P}^t$, we have

$$\mathbf{p}_n^t = (p_t(\mathbf{y}_n, \mathbf{y}_1), \ldots, p_t(\mathbf{y}_n, \mathbf{y}_N))^T = \sum_{m=1}^{N} (\gamma_m')^t r_{nm} \mathbf{l}_m \tag{6.324}$$

in which r_{nm} denotes the n-th coordinate of $\mathbf{r}_m$. The quantity $\mathbf{p}_n^t$ is a probability mass function, with component $m = 1, \ldots, N$, being the probability of reaching node $\mathbf{y}_m$ after a random walk of t steps on $\mathcal{G}$ starting from the node $\mathbf{y}_n$.

Diffusion maps $\mathbf{f}^t : \mathbb{R}^d \to \mathcal{D}^{(t)} \subset \mathbb{R}^N$ map the $\mathbf{y}_n$ to a diffusion space $\mathcal{D}^{(t)}$ using [35, 40]

$$\mathbf{z}^t = \mathbf{f}^t(\mathbf{y}_n) = \left((\gamma_1')^t r_{n1}, \ldots, (\gamma_N')^t r_{nN}\right)^T \tag{6.325}$$

These components are nothing more than the coefficients of $\mathbf{p}_n^t$ in terms of the basis $\{\mathbf{l}_m\}_{m=1}^N$. The parameter t is free to be chosen by the user. Defining a diffusion distance D_t (in the original space in $\mathbb{R}^d$) by [35]

$$D_t(\mathbf{y}_n, \mathbf{y}_m) = \left((\mathbf{p}_n^t - \mathbf{p}_m^t)^T \mathbf{D}^{-1} (\mathbf{p}_n^t - \mathbf{p}_m^t)\right)^{1/2} \tag{6.326}$$

it can be shown that diffusion maps are an embedding of the data in the spaces $\mathcal{D}^{(t)}$ such that [35, 40, 43]

$$||\mathbf{f}^t(\mathbf{y}_n) - \mathbf{f}^t(\mathbf{y}_m)|| = D_t(\mathbf{y}_n, \mathbf{y}_m) \tag{6.327}$$

in which $||\cdot||$ is the standard Euclidean norm. This is a direct consequence of the the bi-orthogonality property of $\mathbf{l}_n$ and $\mathbf{s}_n$. By virtue of the decay in the eigenvalues, from (6.325) we can develop Q-dimensional approximations via mapping $\mathbf{f}_Q^t : \mathbb{R}^d \to \mathcal{F}_Q^t \subset \mathbb{R}^Q$, with latent spaces $\mathcal{F}_Q^t$, as follows:

$$\mathbf{z}_Q^t = \mathbf{f}_Q^t(\mathbf{y}_n) = ((\gamma_1')^t r_{n1}, \ldots, (\gamma_Q')^t r_{nQ})^T \in \mathcal{F}_Q^t \tag{6.328}$$

The value of Q is determined by some criterion related to the eigenvalues, such as the largest index j satisfying $|(\gamma_j')^t| > \upsilon|(\gamma_2')^t|$, for some υ [35]. The parameter t has to be fixed. It changes the diffusion distance, and therefore changes the diffusion map itself. For increasing t, the diffusion distances decrease for two fixed points because the rows of $\mathbf{P}^t$ will approach the stationary distribution as t increases (see Eq. (6.324)). More often than not, t is selected as 1.

6.15.4 Local Tangent Space Alignment

Suppose $\mathcal{A} = \{(U_\alpha, \phi_\alpha)\}_{\alpha\in\Gamma}$ is an *atlas* on $\mathcal{M}$, i.e., $\{U_\alpha\}_{\alpha\in\Gamma}$ covers $\mathcal{M}$ and $\{\phi_\alpha\}_{\alpha\in\Gamma}$ are pairwise compatible. Smooth curves $\gamma_0, \gamma_1 : \mathbb{R} \to \mathcal{M}$ parameterised by t are called $\mathbf{y}$-*equivalent* if $\gamma_0(0) = \gamma_1(0) = \mathbf{y}$ for each $\alpha \in \Gamma$ with $\mathbf{y} \in U_\alpha$, and, moreover

$$\left.\frac{d}{dt}\right|_{t=0} \phi_\alpha(\gamma_0(t)) = \left.\frac{d}{dt}\right|_{t=0} \phi_\alpha(\gamma_1(t)) \tag{6.329}$$

This defines an equivalence relation and the equivalence class of such a smooth curve γ satisfying $\gamma(0) = \mathbf{y}$ is labelled $[\gamma]_\mathbf{y}$. The *tangent space* $T_\mathbf{y}\mathcal{M}$ of $\mathcal{M}$ at a point $\mathbf{y} \in \mathcal{M}$ is defined as the set of equivalence classes

$$T_\mathbf{y}\mathcal{M} = \{[\gamma]_\mathbf{y} : \gamma(0) = \mathbf{y}\} \tag{6.330}$$

This space is a Q-dimensional linear subspace of $\mathbb{R}^d$, which is evident by the identification of $T_\mathbf{y}\mathcal{Y}$ with the set consisting of all derivations at the point $\mathbf{y}$, namely all linear maps from $C^\infty(\mathcal{M})$ to $\mathbb{R}$ satisfying the derivation property.

The space $\mathcal{M} \ni \mathbf{y}$ is assumed to be a Q-dimensional manifold embedded in the space $\mathbb{R}^d$ and there is assumed to be a smooth function $\mathbf{f}_Q^{-1} : \mathcal{F}_Q \to \mathcal{M}$ from some feature or latent space $\mathcal{F}_Q \subset \mathbb{R}^Q$ to $\mathcal{M}$. Each point $\mathbf{y}_n$ in a data set can be approximated using a basis for $T_{\mathbf{y}_n}\mathcal{Y}$ and the approximation can be used to find Q-dimensional representations in a *global coordinate system*, by using a procedure that aligns these tangent spaces *via* local affine transformations [44]. This is thc basis of local tangent space alignment (LTSA) [44]. Implicit is the assumption that a single chart (homeomorphism) $\mathbf{f}_Q$ exists.

By the above assumptions

$$\mathbf{y} = \mathbf{f}_Q^{-1}(\mathbf{z}) = \left(f_1^{-1}(\mathbf{z}), \ldots, f_d^{-1}(\mathbf{z})\right)^T \tag{6.331}$$

in which $\mathbf{z} = (z_1, \ldots, z_Q)^T \in \mathcal{F}_Q$ is the latent representation of $\mathbf{y}$. If $\mathbf{f}_Q^{-1}$ is a smooth function, it admits a first-order Taylor series expansion around $\mathbf{z}$, within a small neighbourhood $\Omega(\mathbf{z})$

$$\mathbf{f}_Q^{-1}(\widehat{\mathbf{z}}) = \mathbf{f}_Q^{-1}(\mathbf{z}) + \mathbf{J}(\mathbf{z})(\widehat{\mathbf{z}} - \mathbf{z}) + \mathcal{O}(\|\widehat{\mathbf{z}} - \mathbf{z}\|^2), \quad \forall \widehat{\mathbf{z}} \in \Omega(\mathbf{z}) \tag{6.332}$$

in which $\mathbf{J}(\mathbf{z}) \in \mathbb{R}^{d\times Q}$ is the Jacobian of $\mathbf{f}_Q^{-1}$ evaluated $\mathbf{z}$, with element i, j equal to $\partial f_i^{-1}/\partial z_j$.

The columns of $\mathbf{J}$ form a basis for $T_\mathbf{y}\mathcal{M}$, the tangent space of $\mathcal{M}$ at $\mathbf{y} = \mathbf{f}_Q^{-1}(\mathbf{z})$. The quantity $\widehat{\mathbf{z}} - \mathbf{z}$ then provides the components of $\mathbf{f}\,(\widehat{\mathbf{z}})$ in an affine subspace $\mathbf{f}_Q^{-1}(\mathbf{z}) + T_\mathbf{y}\mathcal{M}$. The Jacobian $\mathbf{J}$ is unknown without knowing $\mathbf{f}_Q^{-1}$. To overcome this problem, we write $T_\mathbf{y}\mathcal{M}$ in terms of some matrix $\mathbf{Q}_\mathbf{z}$ whose columns are orthonormal and form a basis for $T_\mathbf{y}\mathcal{M}$

$$\mathbf{J}(\mathbf{z})\,(\widehat{\mathbf{z}} - \mathbf{z}) = \mathbf{Q}_\mathbf{z}\pi_\mathbf{z}^* \tag{6.333}$$

in which the quantity

$$\pi_\mathbf{z}^* = \mathbf{Q}_\mathbf{z}^T\mathbf{J}(\mathbf{z})\,(\widehat{\mathbf{z}} - \mathbf{z}) := \mathbf{P}_\mathbf{z}\,(\widehat{\mathbf{z}} - \mathbf{z}) \tag{6.334}$$

is also not known since it depends on $\mathbf{f}_Q^{-1}$. Note that the above relationship follows from the orthogonality of $\mathbf{Q}_\mathbf{z}$. However, if we combine (6.333) and (6.332), it is possible to approximate $\pi_\mathbf{z}^*$ as an orthogonal projection of $\mathbf{f}(\widehat{\mathbf{z}}) - \mathbf{f}(\mathbf{z})$ onto $T_\mathbf{y}\mathcal{M}$

$$\boldsymbol{\pi}_\mathbf{z} \equiv \mathbf{Q}_\mathbf{z}^T \left(\mathbf{f}_Q^{-1}(\widehat{\mathbf{z}}) - \mathbf{f}_Q^{-1}(\mathbf{z})\right) = \boldsymbol{\pi}_\mathbf{z}^* + \mathcal{O}\left(\|\widehat{\mathbf{z}} - \mathbf{z}\|^2\right) \tag{6.335}$$

if $\mathbf{Q}_\mathbf{z}$ for each $\mathbf{z}$ is known ($\mathbf{f}_Q^{-1}(\cdot)$ are the known data points). The first-order Taylor expansion shows that $\mathbf{z}$ satisfies

$$\int d\mathbf{z} \int_{\Omega(\mathbf{z})} \|\mathbf{P}_\mathbf{z}(\widehat{\mathbf{z}} - \mathbf{z}) - \boldsymbol{\pi}_\mathbf{z}\| d\widehat{\mathbf{z}} \approx 0 \tag{6.336}$$

in some neighbourhood $\Omega(\mathbf{z})$. A natural way to approximate the *global coordinate* is to find the value of $\mathbf{z}$ and the affine transformation $\mathbf{P}_\mathbf{z}$ that minimise the left-hand side of (6.336). However, a simpler approach is to use a linear alignment, which is explained below.

Provided the Jacobian matrix has full column rank, and therefore $\mathbf{P}$ is invertible, it is possible to find the following local affine transformation based on the loss function in (6.336)

$$\widehat{\mathbf{z}} - \mathbf{z} \approx \mathbf{P}_\mathbf{z}^{-1}\boldsymbol{\pi}_\mathbf{z} \equiv \mathbf{L}_\mathbf{z}\boldsymbol{\pi}_\mathbf{z} \tag{6.337}$$

The affine transformation $\mathbf{L}_\mathbf{z}$ should align the local coordinate with the global coordinate $\widehat{\mathbf{z}} - \mathbf{z}$ at the point $\mathbf{y} = \mathbf{f}_Q^{-1}(\widehat{\mathbf{z}})$. The global coordinate $\mathbf{z}$ and the local affine transformation $\mathbf{L}_\mathbf{z}$ can then be found by minimisation of

$$\int d\mathbf{z} \int_{\Omega(\mathbf{z})} \|\widehat{\mathbf{z}} - \mathbf{z} - \mathbf{L}_\mathbf{z}\boldsymbol{\pi}_\mathbf{z}\| d\widehat{\mathbf{z}} \tag{6.338}$$

However, $\mathbf{Q}_\mathbf{z}$ is still unknown for all of the tangent spaces. If we have a data set $\mathbf{y}_n, n = 1, \ldots, N$ containing noise ϵ_n the usual additive model is

$$\mathbf{y}_n = \mathbf{f}_Q^{-1}(\mathbf{z}_n) + \epsilon_n. \tag{6.339}$$

For a given $\mathbf{y}_n$, define $\mathbf{Y}_n = [\mathbf{y}_{n_1} \ldots \mathbf{y}_{n_P}]$ to be a matrix of the P nearest neighbours, inclusive of $\mathbf{y}_n$, measuring the distances using the Euclidean metric. Of all Q-dimensional local affine subspace approximations for $\{\mathbf{y}_{n_k}\}$, the best in a least-squares sense is given by the solution to

$$\arg\min_{\mathbf{y},\boldsymbol{\Pi},\mathbf{Q}} \sum_{k=1}^{P} \|\mathbf{y}_{n_k} - (\mathbf{y} + \mathbf{Q}\boldsymbol{\pi}_k)\|_2^2 = \arg\min_{\mathbf{y},\boldsymbol{\Pi},\mathbf{Q}} \|\mathbf{Y}_n - \left(\mathbf{y}\mathbf{1}^T + \mathbf{Q}\boldsymbol{\Pi}\right)\|_2^2 \tag{6.340}$$

where the Q columns of $\mathbf{Q}$ are orthonormal, and the matrix $\boldsymbol{\Pi}$ is defined as $\boldsymbol{\Pi} = [\boldsymbol{\pi}_1 \ldots \boldsymbol{\pi}_P]$. By SVD (we omit the details) the solution to (6.340) is such that $\mathbf{y} = \bar{\mathbf{y}}_n$, the mean of $\{\mathbf{y}_{n_k}\}_k$, while $\mathbf{Q} = \mathbf{Q}_n$, a matrix with columns given by the Q left singular vectors of $\mathbf{Y}_n\left(\mathbf{I} - \mathbf{1}\mathbf{1}^T/P\right)$ associated with the Q largest singular values. $\boldsymbol{\Pi}$ is equal to $\mathbf{\Pi}_n$, defined as

$$\mathbf{\Pi}_n = \mathbf{Q}_n^T \mathbf{Y}_n \left(\mathbf{I} - \frac{1}{P} \mathbf{1}\mathbf{1}^T \right) = \left[\boldsymbol{\pi}_1^{(i)}, \ldots, \boldsymbol{\pi}_K^{(i)} \right] \tag{6.341}$$

in which

$$\boldsymbol{\pi}_k^{(n)} = \mathbf{Q}_n^T \left(\mathbf{y}_{n_k} - \bar{\mathbf{y}}_n \right) \tag{6.342}$$

We therefore obtain

$$\mathbf{y}_{n_k} = \bar{\mathbf{y}}_n + \mathbf{Q}_n \boldsymbol{\pi}_k^{(n)} + \varphi_k^{(n)} \tag{6.343}$$

in which

$$\varphi_k^{(n)} = \left(\mathbf{I} - \mathbf{Q}_n \mathbf{Q}_n^T \right) \left(\mathbf{y}_{n_k} - \bar{\mathbf{y}}_n \right) \tag{6.344}$$

is a reconstruction error.

We now wish to find global coordinates $\mathbf{z} = [\mathbf{z}_1 \ldots \mathbf{z}_N] \in \mathbb{R}^{Q \times N}$ associated with $\mathbf{y}_n$, given the local coordinates $\boldsymbol{\pi}_k^{(n)}$, which characterise the local geometry. To achieve this, the global coordinates $\mathbf{z}_{n_k}$ are chosen to satisfy the following conditions in order to be consistent with the local geometry contained in the $\boldsymbol{\pi}_k^{(n)}$

$$\begin{aligned} \mathbf{z}_{n_k} &= \bar{\mathbf{z}}_n + \mathbf{L}_n \boldsymbol{\pi}_k^{(n)} + \epsilon_k^{(n)}, \quad k = 1, \ldots, P, \; n = 1, \ldots, N \\ \mathbf{Z}_n &= \frac{1}{P} \mathbf{Z}_n \mathbf{1}\mathbf{1}^T + \mathbf{L}_n \mathbf{\Pi}_n + \mathbf{E}_n \end{aligned} \tag{6.345}$$

in which $\bar{\mathbf{z}}_n$ is defined as the mean of $\{\mathbf{z}_{n_k}\}$, and

$$\begin{aligned} \mathbf{Z}_n &= [\mathbf{z}_{n_1} \ldots \mathbf{z}_{n_P}] \\ \mathbf{E}_n &= [\epsilon_1^{(n)} \ldots \epsilon_P^{(n)}] = \mathbf{Z}_n \left(\mathbf{I} - \frac{1}{P} \mathbf{1}\mathbf{1}^T \right) - \mathbf{L}_n \mathbf{\Pi}_n \end{aligned} \tag{6.346}$$

The latent points and affine transformations $\mathbf{L}_n$ are found by minimising $\|\mathbf{E}_n\|_F$, in which $|| \cdot ||_F$ is the Frobenius norm defined in (6.162). The solution is

$$\begin{aligned} \mathbf{L}_n &= \mathbf{Z}_n \left(\mathbf{I} - \frac{1}{P} \mathbf{1}\mathbf{1}^T \right) \mathbf{\Pi}_n^\dagger \\ \mathbf{E}_n &= \mathbf{Z}_n \left(\mathbf{I} - \frac{1}{P} \mathbf{1}\mathbf{1}^T \right) (\mathbf{I} - \mathbf{\Pi}_n^\dagger \mathbf{\Pi}_n) \end{aligned} \tag{6.347}$$

in which $\mathbf{\Pi}_n^\dagger$ is the Moore-Penrose pseudo inverse of $\mathbf{\Pi}_n$. A selection matrix $\mathbf{S}_n \in \mathbb{R}^{N \times P}$ can be defined such that $\mathbf{Z}\mathbf{S}_n = \mathbf{Z}_n$ and the global coordinates are found by minimising the overall reconstruction error as follows, imposing $\mathbf{Z}^T\mathbf{Z} = \mathbf{I}$ to ensure that the solutions are unique

$$\underset{\mathbf{Z}:\mathbf{Z}^T\mathbf{Z}=\mathbf{I}}{\arg\min} \sum_n \|\mathbf{E}_n\|_F^2 = \underset{\mathbf{Z}:\mathbf{Z}^T\mathbf{Z}=\mathbf{I}}{\arg\min} \|\mathbf{ZSW}\|_F^2, \tag{6.348}$$

in which

$$\begin{aligned} \mathbf{S} &= [\mathbf{S}_1 \dots \mathbf{S}_N] \\ \mathbf{W} &= \text{diag}\,(\mathbf{W}_1, \dots, \mathbf{W}_N) \\ \mathbf{W}_n &= \left(\mathbf{I} - \frac{1}{P}\mathbf{1}\mathbf{1}^T\right)(\mathbf{I} - \boldsymbol{\Pi}_n^{\dagger}\boldsymbol{\Pi}_n), \quad n = 1, \dots, N \end{aligned} \tag{6.349}$$

In these expressions, $\mathbf{1}$ is an eigenvector corresponding to a zero eigenvalue of the matrix

$$\mathbf{B} \equiv \mathbf{SWW}^T\mathbf{S}^T \in \mathbb{R}^{N\times N} \tag{6.350}$$

Arranging the eigenvalues such that they are non-decreasing leads to the optimal $\mathbf{z}$ as

$$\mathbf{z}' = [\boldsymbol{\zeta}_2 \dots \boldsymbol{\zeta}_{Q+1}]^T \tag{6.351}$$

in which $\boldsymbol{\zeta}_j \in \mathbb{R}^N$ are the first $Q+1$ eigenvectors of $\mathbf{B}$, excluding that corresponding to a zero eigenvalue. An approximation of the map $\mathbf{f}_Q : \mathbb{R}^d \to \mathcal{F}_Q \subset \mathbb{R}^Q$ based on the data is then given by the n-th column $\mathbf{z}'_{n,:}$ of $\mathbf{z}'$

$$\mathbf{z}_n = \mathbf{f}_Q(\mathbf{y}_n) = \mathbf{z}'_{n,:} \tag{6.352}$$

A fixed number of neighbours assumes a degree of smoothness for the manifold and using an equal number of neighbours for each of the tangent spaces suggests a global smoothness property. Both of these assumptions can lead to inaccuracies in the predictions. A number of adaptive algorithms have been developed to avoid these assumptions [45–47], as well as to increase robustness when the data has a high degree of noise [48].

The *out-of-sample* problem refers to finding the coordinates of a new point that was not included in the training. Generally, this requires a fresh run of the algorithm with a new data set incorporating the new point, which obviously leads to a high cost if new points are regularly made available. This is an issue faced with all manifold learning methods and solutions to this out-of-sample problem are of great interest, with a solution for the LTSA method developed in [49].

6.15.5 The Inverse Mapping Problem in Manifold Learning

The problem of mapping a point back to the original space is called the *pre-image* or inverse-mapping problem, i.e., finding $\mathbf{f}_Q^{-1} : \mathcal{F}_Q \to \mathcal{M} \subset \mathbb{R}^d$, in which $\mathbf{f}_Q$ is any one of the maps obtained in the previous sections. For all methods introduced above, such inverses can be approximated. In kPCA, a least-squares solution is given in [50, 51], while a fixed-point iterative method can be found in [52], although both suffer from numerical instabilities if d exceeds N. In the case of diffusion maps, optimisation methods for low-dimensional problems have been proposed [53, 54]. When Q and d are large, however, these methods are unstable and time-consuming. We outline later a more scalable method developed in [13].

Given the coordinates of a point in feature space, corresponding to a point $\mathbf{y}$ in the original space, a general framework for solving the pre-image problem involves a weighted average of neighbouring points of $\mathbf{y}$, with the neighbourhood $\mathbf{y}_j$, $j \in \mathcal{J} \subseteq \{1, 2, \ldots, N\}$, defined according to some criterion

$$\mathbf{y} = \sum\nolimits_{j \in \mathcal{J}} \vartheta(\mathbf{y}_j)\mathbf{y}_j \tag{6.353}$$

in which $\vartheta(\mathbf{y}_j)$ are the weights. The data points can be used to define the neighbouring points since the components of these points in the feature space are known.

A general method defines the weights as functions of the distances $d_{n,*}$, between the point $\mathbf{y}$ and $\mathbf{y}_n$, $n = 1, \ldots, N$, assuming for the moment that these distances can be computed. In a local linear interpolation [55, 56] (not to be confused with locally linear interpolation)

$$\vartheta(\mathbf{y}_n) = \frac{d_{n,*}^{-1}}{\sum_{j=1}^{N} d_{j,*}^{-1}} \tag{6.354}$$

The index set $\mathcal{J}$ is then defined as those points in the complement of a ball of predefined size ϵ, i.e.,

$$\{n \in \{1, \ldots, N\} | \vartheta(\mathbf{y}_n) \geq \epsilon\} \tag{6.355}$$

or a predefined number of points N' with the largest values of $\vartheta(\mathbf{y}_n)$. We may generalise (6.354) using an isotropic kernel $\chi(\mathbf{y}, \mathbf{y}') = \chi(||\mathbf{y} - \mathbf{y}'||)$, e.g., a Gaussian $\chi(\mathbf{y}, \mathbf{y}') = \exp(-||\mathbf{y} - \mathbf{y}'||^2)$, to define weights as follows [57]:

$$\vartheta(\mathbf{y}_n) = \frac{\chi(\mathbf{y}, \mathbf{y}_n)}{\sum_{j=1}^{N} \chi(\mathbf{y}, \mathbf{y}_j)} = \frac{\chi(d_{n,*})}{\sum_{j=1}^{N} \chi(d_{j,*})} \tag{6.356}$$

In order to find $\mathbf{y}$, therefore, we need access to the $d_{n,*}$, $i = 1, \ldots, N$. We shall see below how to obtain these distances for the various methods.

kPCA Inverse Mapping

In the case of kPCA, the mapped points

$$\Phi = [\phi(\mathbf{y}_1), \ldots, \phi(\mathbf{y}_N)] \tag{6.357}$$

can be subjected to the centering transformation $\widetilde{\Phi} = \Phi\mathbf{H}$ to obtain the following representations of the basis vectors (6.305)

$$\widetilde{\mathbf{w}}_i = \sum\nolimits_{j=1}^{N} \widetilde{\alpha}_{ji}\widetilde{\phi}(\mathbf{y}_j) = \widetilde{\Phi}\widetilde{\alpha}_i = \Phi\mathbf{H}\widetilde{\alpha}_i \tag{6.358}$$

in which the $\widetilde{\alpha}_i$ are known from the kPCA procedure. The projection (6.311) of a general point $\phi(\mathbf{y}) \in \mathcal{F}$ (not centred) onto the first of the Q basis vectors is

$$\begin{aligned}\phi_Q(\mathbf{y}) = \sum_{i=1}^{Q} z_i\widetilde{\mathbf{w}}_i + \overline{\phi} &= \sum_{i=1}^{Q} z_i\Phi\mathbf{H}\widetilde{\alpha}_i + \Phi\mathbf{1} \\ &= \Phi\left(\mathbf{H}[\widetilde{\alpha}_1\ldots,\widetilde{\alpha}_Q]\mathbf{z}_Q + \mathbf{1}\right) \\ &:= \Phi\boldsymbol{\tau}\end{aligned} \tag{6.359}$$

The distance $\widetilde{d}_{n,*}$ between a point $\phi(\mathbf{y}_n)$ and the point $\phi(\mathbf{y})$ is

$$\widetilde{d}^2_{n,*} = \phi(\mathbf{y})^T\phi(\mathbf{y}) + \phi(\mathbf{y}_n)^T\phi(\mathbf{y}_n) - 2\phi(\mathbf{y})^T\phi(\mathbf{y}_n) \tag{6.360}$$

Setting $\phi(\mathbf{y}) \approx \phi_Q(\mathbf{y})$ and using (6.359) in (6.360), noting that $\Phi^T\Phi = \mathbf{K}$ and $\mathbf{k}_n = \Phi^T\phi(\mathbf{y}_n)$, then shows that

$$\widetilde{d}^2_{n,*} \approx \boldsymbol{\tau}^T\mathbf{K}\boldsymbol{\tau} + k(\mathbf{y}_n, \mathbf{y}_n) - 2\boldsymbol{\tau}^T\mathbf{k}_n \tag{6.361}$$

with $\boldsymbol{\tau}$ given by (6.359).

For normalised isotropic kernels satisfying $k(\mathbf{y}', \mathbf{y}') = 1$, Eq. (6.360) yields

$$\widetilde{d}^2_{n,*} = 2 - 2k(\mathbf{y}_n, \mathbf{y}) \tag{6.362}$$

Equating this result with (6.361) then provides the value of $k(\mathbf{y}_n, \mathbf{y})$. Employing a Gaussian kernel $k(\mathbf{y}_n, \mathbf{y}) = \exp\left(-\|\mathbf{y}_n - \mathbf{y}\|^2/\theta^2\right)$, for a correlation length θ, we therefore find the distances as follows:

$$d^2_{n,*} = \|\mathbf{y}_n - \mathbf{y}\|^2 = -\theta^2 \ln k(\mathbf{y}_n, \mathbf{y}) \tag{6.363}$$

A similar procedure can be carried out for other kernel functions [58] such as the polynomial kernel $k_n(\mathbf{y}, \mathbf{y}') = \left(\mathbf{y}^T\mathbf{y}' + c\right)^n$, $c \in \mathbb{R}$, $n = \mathbb{N}$.

Isomap Inverse Mapping

In the case of Isomap, the distances between $\widehat{\mathbf{z}}_{nQ}$, $n = 1, \ldots, N$, are computable and are equal to the geodesic distances between $d_{i,*}$ between the corresponding $\mathbf{y}_n$. In a neighbourhood of a point $\mathbf{y}$, the geodesic distances are approximately Euclidean distances. Therefore, according to these distances, we can simply choose some neighbourhood of points closest to $\mathbf{y}$ in order to implement (6.354) or (6.356). The natural way to select the neighbourhood points is to use the neighbourhood defined in the Isomap procedure.

Diffusion Map Inverse Mapping

For diffusion maps we first set $t = 1$ in order to illustrate the procedure in [13]. Taking the limit $N \to \infty$, when using a Gaussian kernel the Markov chain with transition matrix $\mathbf{P}$ approaches a Markov chain on $\mathcal{M} \subset \mathbb{R}^d$, i.e., the continuous state space in which the data resides [35, 40, 43, 59, 60]. The time step for this discrete-time Markov Chain is equal to the scale factor θ^2. We denote by μ a probability measure on $\mathcal{M}$, which defines the density of points. For instance a uniform density would correspond to the Lebesgue measure.

Taking the limit $N \to \infty$, a one-step transition kernel from $\mathbf{y}' \in \mathcal{M}$ to $\mathbf{y} \in \mathcal{M}$ for the above Markov chain can be defined by

$$p(\mathbf{y}', \mathbf{y}) = \frac{k(\mathbf{y}, \mathbf{y}')}{d(\mathbf{y}')}, \quad d(\mathbf{y}') = \int_{\mathcal{M}} k(\mathbf{y}, \mathbf{y}') d\mu(\mathbf{y}) \tag{6.364}$$

in which $d(\mathbf{y}')$ normalises the probability. The kernel $p(\mathbf{y}', \mathbf{y})$ generalises $\mathbf{P}$ to the continuous case. Let $\varphi(\mathbf{y})$ be a probability distribution; its evolution is determined by a so-called Markov operator $\mathcal{L}$, also called the *forward transfer operator* or the propagator, given by [43, 59]

$$\mathcal{L}\varphi(\mathbf{y}) = \int_{\mathcal{M}} p(\mathbf{y}', \mathbf{y})\varphi(\mathbf{y}') d\mu(\mathbf{y}'), \quad \forall \varphi(\mathbf{y}') \in L^2(\mathcal{M}, \mu) \tag{6.365}$$

After t steps, the distribution is found by applying this operator t times, i.e., $\mathcal{L}^t \varphi = \mathcal{L} \circ \mathcal{L} \circ \cdots \circ \mathcal{L}\varphi$. In the finite state space case discussed before, left multiplication of $\mathbf{P}$ is the analogue of $\mathcal{L}$. The $L^2(\mathcal{M}, \mu)$ inner product is defined by

$$\langle \varphi_1, \varphi_2 \rangle = \int_{\mathcal{M}} \varphi_1(\mathbf{y})\varphi_2(\mathbf{y}) d\mu(\mathbf{y}), \quad \forall \varphi_1, \varphi_2 \in L^2(\mathcal{M}, \mu) \tag{6.366}$$

and the *backward transfer operator* is equal to the adjoint of $\mathcal{L}$ under this inner product [43, 59]

$$\begin{gathered} \mathcal{R}\varphi(\mathbf{y}) = \int_{\mathcal{M}} p(\mathbf{y}, \mathbf{y}')\varphi(\mathbf{y}') d\mu(\mathbf{y}') \\ \langle \mathcal{L}\varphi_1, \varphi_2 \rangle = \langle \varphi_1, \mathcal{R}\varphi_2 \rangle \end{gathered} \tag{6.367}$$

The meaning of this operator is that when $\varphi(\mathbf{y})$ is a function on $\mathcal{M}$, its mean value following a single step of a random walk starting at $\mathbf{y}$ is given by $\mathcal{R}\varphi(\mathbf{y})$, while $\mathcal{R}^t\varphi$ is the mean after t steps of the walk. Complementary to $\mathcal{L}$, $\mathcal{R}$ is the analogue of right multiplication of $\mathbf{P}$.

We can also define a symmetric transition kernel by

$$p_s(\mathbf{y}', \mathbf{y}) = \frac{k(\mathbf{y}, \mathbf{y}')}{\sqrt{d(\mathbf{y}')}\sqrt{d(\mathbf{y})}} \tag{6.368}$$

which defines a self-adjoint, compact operator $\mathcal{S}$ as follows [35, 40]

$$\mathcal{S}\varphi(\mathbf{y}) = \int_{\mathcal{M}} p_s(\mathbf{y}, \mathbf{y}')\varphi(\mathbf{y}')d\mu(\mathbf{y}')$$
$$\langle \mathcal{S}\varphi_1, \varphi_2 \rangle = \langle \varphi_1, \mathcal{S}\varphi_2 \rangle, \quad \forall \varphi_1, \varphi_2 \in L^2(\mathcal{M}, \mu) \tag{6.369}$$

$\mathcal{S}$ is analogue of the finite-dimensional operator $\mathbf{P}' = \mathbf{D}^{-1/2}\mathbf{K}\mathbf{D}^{1/2}$. Compact, self-adjoint operators have a well-known spectral theory, from which we can deduce that $\mathcal{S}$ has an eigendecomposition $\mathcal{S}s_i = \gamma_i s_i$, $i \in \mathbb{N}$, with positive eigenvalues $1 = \gamma_1 > \gamma_2 > \cdots$. The eigenfunctions of $\mathcal{S}$ are orthonormal and from a basis for $L^2(\mathcal{M}, \mu)$. In addition, $p_s(\mathbf{y}, \mathbf{y}')$ can be expanded as

$$p_s(\mathbf{y}, \mathbf{y}') = \sum\nolimits_{i=1}^{\infty} \gamma_i s_i(\mathbf{y}) s_i(\mathbf{y}') \tag{6.370}$$

The eigenvalues of $\mathcal{L}$, $\mathcal{R}$ and $\mathcal{S}$ are identical, while the eigenfunctions of $\mathcal{L}$ and $\mathcal{R}$ are

$$l_i = s_i(\mathbf{y})\sqrt{d(\mathbf{y})}, \quad r_i = \frac{s_i(\mathbf{y})}{\sqrt{d(\mathbf{y})}}, \quad i \in \mathbb{N} \tag{6.371}$$

respectively. These results follow from the fact that $\mathcal{S}$ arises from the conjugation of the kernel $p(\mathbf{y}', \mathbf{y})$ with $\sqrt{d(\mathbf{y})}$.

By virtue of (6.370), together with (6.371), it follows that [35]

$$p(\mathbf{y}, \mathbf{y}') = \sum\nolimits_{i=1}^{\infty} \gamma_i r_i(\mathbf{y}) l_i(\mathbf{y}') \tag{6.372}$$

The probabilities $p_t(\mathbf{y}_j, \mathbf{y}_n)$ of (6.324) (entries of $\mathbf{P}^t$) are the analogues of the transition kernel $p_t(\mathbf{y}, \mathbf{y}')$ of $\mathcal{R}^t = \mathcal{R} \circ \cdots \circ \mathcal{R}$, which can be expanded as

$$p_t(\mathbf{y}, \mathbf{y}') = \sum\nolimits_{i=1}^{\infty} \gamma_i^t r_i(\mathbf{y}) l_i(\mathbf{y}') \tag{6.373}$$

For a fixed $\mathbf{y} \in \mathcal{M}$, this is a continuous analogue (as a function of $\mathbf{y}' \in \mathcal{M}$) of the vector $\mathbf{p}_j^t$ in (6.324), with $\mathbf{y} = \mathbf{y}_j$ and $\mathbf{y}' \in \{\mathbf{y}_1, \ldots, \mathbf{y}_N\}$, being a finite set of states that can be accessed from $\mathbf{y}_j$. Similarly, the set of basis vectors $\{\mathbf{l}_i\}_{i=1}^{m}$ have the continuous analogues $\{l_i\}_{i=1}^{\infty}$ (functions defined on $\mathcal{M}$), while the i-th coordinate $(\gamma_i')^t r_{ji}$ is equivalent to the function $\gamma_i^t r_i$ evaluated at $\mathbf{y} \in \mathcal{M}$, the starting point. From the ordering property of the eigenvalues, it is possible to restrict the expansion (6.373) to the eigenfunctions $\{l_i\}_{i=1}^{Q}$, for some Q, mirroring the finite-dimensional case.

The diffusion distance has a continuous analogue given by [35]

$$D_t^2(\mathbf{y}_1, \mathbf{y}_2) = ||p_t(\mathbf{y}_1, \mathbf{y}') - p_t(\mathbf{y}_2, \mathbf{y}')||_{1/d}^2 = \sum\nolimits_{i=1}^{\infty} \gamma_i^{2t} [r_i(\mathbf{y}_1) - r_i(\mathbf{y}_2)]^2 \tag{6.374}$$

in which the norm is defined by

$$\|\varphi\|_{1/d}^2 = \langle\varphi,\varphi\rangle_{1/d} := \int_{\mathbf{y}'\in\mathcal{M}} \frac{|\varphi(\mathbf{y}')|^2}{d(\mathbf{y}')} d\mu(\mathbf{y}'), \quad \forall\{\varphi : \|\varphi\|_{1/d} < \infty\} \tag{6.375}$$

The second equality in (6.374) is a direct consequence of the orthonormality of $\{l_i\}_{i=1}^{\infty}$ w.r.t. $\langle\cdot,\cdot\rangle_{1/d}$. The diffusion maps $\mathbf{f}^t : \mathbb{R}^d \supset \mathcal{M} \to \mathcal{D}^{(t)} \subset \mathbb{R}^N$ can therefore be generalised to maps $\mathbf{f}^t : \mathcal{M} \to \mathcal{D}^{(t)} \subset \ell^2$ on the whole of $\mathcal{M}$

$$\mathbf{f}^t(\mathbf{y}) = (\gamma_1^t r_1(\mathbf{y}), \gamma_2^t r_2(\mathbf{y}), \ldots) \tag{6.376}$$

in which ℓ^2 is the space of sequences $(x_1, x_2, \ldots)$

$$\ell^2 = \{(x_1, x_2 \ldots) : \sum_{j=1}^{\infty} x_j^2 < \infty\} \tag{6.377}$$

Restricting (6.373) to the Q eigenfunctions corresponding to the largest Q eigenvalues, the latent maps $\mathbf{f}_Q^t : \mathcal{M} \to \mathcal{D}_Q^{(t)} \subset \mathbb{R}^Q$ are defined by

$$\mathbf{f}_Q^t(\mathbf{y}) = (\gamma_1^t r_1(\mathbf{y}), \ldots, \gamma_Q^t r_Q(\mathbf{y}))^T \in \mathcal{D}_Q^{(t)} \tag{6.378}$$

In the finite-dimensional case, given training data $\{\mathbf{y}_n\}_{n=1}^N$ and a new point $\mathbf{y}$ for which we have a representation $\mathbf{f}_Q^t(\mathbf{y})$, the kernel matrix $\mathbf{K}$, degree matrix $\mathbf{D}$ and Markov matrix $\mathbf{P}$ in terms of the training data can be augmented to include $\mathbf{y}$, producing matrices $\underline{\mathbf{K}}$, $\underline{\mathbf{D}}$ and $\underline{\mathbf{P}} = \underline{\mathbf{D}}^{-1}\underline{\mathbf{K}}$

$$\underline{\mathbf{K}} = \left[\begin{array}{c|c} \mathbf{K} & (k(\mathbf{y}_1,\mathbf{y}), \ldots, k(\mathbf{y}_N,\mathbf{y}))^T \\ \hline (k(\mathbf{y}_1,\mathbf{y}), \ldots, k(\mathbf{y}_N,\mathbf{y})) & k(\mathbf{y},\mathbf{y}) \end{array}\right] \tag{6.379}$$

$$\underline{\mathbf{D}} = \left[\begin{array}{c|c} \widehat{\mathbf{D}} & 0 \\ \hline 0 & k(\mathbf{y},\mathbf{y}) + \sum_j k(\mathbf{y}_j,\mathbf{y}) \end{array}\right] \tag{6.380}$$

$$\underline{\mathbf{P}} = \left[\begin{array}{c|c} \widehat{\mathbf{D}}^{-1}\mathbf{K} & \widehat{\mathbf{D}}^{-1}(k(\mathbf{y}_1,\mathbf{y}), \ldots, k(\mathbf{y}_N,\mathbf{y}))^T \\ \hline \dfrac{(k(\mathbf{y}_1,\mathbf{y}), \ldots, k(\mathbf{y}_N,\mathbf{y}))}{k(\mathbf{y},\mathbf{y}) + \sum_j k(\mathbf{y}_j,\mathbf{y})} & \dfrac{k(\mathbf{y},\mathbf{y})}{k(\mathbf{y},\mathbf{y}) + \sum_j k(\mathbf{y}_j,\mathbf{y})} \end{array}\right] \tag{6.381}$$

in which

$$\widehat{\mathbf{D}} = \mathbf{D} + \mathrm{diag}(k(\mathbf{y}_1,\mathbf{y}), \ldots, k(\mathbf{y}_N,\mathbf{y})) \tag{6.382}$$

We write $\underline{\mathbf{p}}_{N+1}$ to denote row $(N+1)$ of the matrix $\underline{\mathbf{P}}$. The i-th element of $\underline{\mathbf{p}}_{N+1}$ is equal to the transition probability of going from the point $\mathbf{y}$ to the point $\mathbf{y}_i$, $i = 1, \ldots, N$, while the last element is the probability of going from $\mathbf{y}$ to itself.

From the continuous state space analysis, the n-th element of $\underline{\mathbf{p}}_{N+1}$ approximates

$$p(\mathbf{y},\mathbf{y}') = \sum_{j=1}^{\infty} \gamma_j r_j(\mathbf{y}) l_i(\mathbf{y}') \tag{6.383}$$

for a finite set of points $\{\mathbf{y}_n\}_{n=1}^N$, with $\mathbf{y}$ fixed, $\mathbf{y}' = \mathbf{y}_n$, and last element $\mathbf{y}' = \mathbf{y}$. This yields

$$\begin{aligned}\underline{\mathbf{p}}_{N+1} &\approx \sum_{j=1}^{\infty} \gamma_j r_j(\mathbf{y})(l_j(\mathbf{y}_1), \ldots, l_j(\mathbf{y}_N), l_j(\mathbf{y}))^T \\ &\approx \sum_{j=1}^{Q} \gamma_j r_j(\mathbf{y})(l_j(\mathbf{y}_1), \ldots, l_j(\mathbf{y}_N), l_j(\mathbf{y}))^T\end{aligned} \tag{6.384}$$

by virtue of the ordering of the eigenvalues γ_i. The eigenvectors $\mathbf{l}_j$ are empirical versions of l_j and $l_j(\mathbf{y}_i)$, $i = 1, \ldots, N$, is equivalent to the i-th entry l_{ij} of $\mathbf{l}_j$. The coordinates of $\mathbf{y}$ in the diffusion space satisfy

$$\mathbf{f}_Q(\mathbf{y}) = (\gamma_1' r_1(\mathbf{y}), \ldots, \gamma_Q' r_Q(\mathbf{y}))^T \tag{6.385}$$

and are known. Therefore, component i of $\underline{\mathbf{p}}_{N+1}$, denoted $\underline{p}_{N+1,i}$, has an approximation

$$\underline{p}_{N+1,i} \approx \sum_{j=1}^{Q} \gamma_j' r_j(\mathbf{y}) l_{ij}, \quad i = 1, \ldots, N \tag{6.386}$$

Note that we only have the empirical eigenvalues γ_i' rather than γ_i, which we use to approximate the latter. Equating (6.386) with the corresponding entry in (6.381) yields

$$\sum_{j=1}^{Q} \gamma_j' r_j(\mathbf{y}) l_{nj} = \frac{k(\mathbf{y}_n, \mathbf{y})}{k(\mathbf{y}, \mathbf{y}) + \sum_{j=1}^{N} k(\mathbf{y}_j, \mathbf{y})}, \quad n = 1, \ldots, N \tag{6.387}$$

In the case of the Gaussian kernel, $k(\mathbf{y}, \mathbf{y}) = 1$, and solution of the N equations (6.387) yields values of $k(\mathbf{y}_n, \mathbf{y})$, $m = 1, \ldots, N$. The distances $d_{i,*}$ can then be extracted from these kernel values as in the case of kPCA, e.g., $d_{n,*}^2 = -\theta^2 \ln k(\mathbf{y}_n, \mathbf{y})$.

LTSA Inverse Mapping

To map a point $\mathbf{z} \in \mathcal{F}_Q$ in the latent space to a point $\mathbf{y} \in \mathbb{R}^d$ in the original space is relatively straightforward for LTSA. Defining $\mathbf{z}_k$ to be the nearest neighbour of the point $\mathbf{z}$, from (6.345) we obtain

$$\boldsymbol{\pi}_*^{(k)} = \mathbf{L}_k^{-1} (\mathbf{z} - \bar{\mathbf{z}}_k) - \mathbf{L}_k^{-1} \epsilon_*^{(k)} \tag{6.388}$$

while from (6.343) we can define

$$\mathbf{y} = \bar{\mathbf{y}}_k + \mathbf{Q}_k \boldsymbol{\pi}_*^{(k)} + \varphi_*^{(k)} \tag{6.389}$$

From these relationships we can develop an approximate pre-image $\mathbf{f}_Q^{-1} : \mathcal{F}_Q \to \mathcal{M} \subset \mathbb{R}^d$ as follows:

$$\begin{aligned}\mathbf{y} = \mathbf{f}_Q^{-1}(\mathbf{z}) &= \bar{\mathbf{y}}_k + \mathbf{Q}_k \left(\mathbf{L}_k^{-1} (\mathbf{z} - \bar{\mathbf{z}}_k) - \mathbf{L}_k^{-1} \epsilon_*^{(k)}\right) + \varphi_*^{(k)} \\ &= \bar{\mathbf{y}}_k + \mathbf{Q}_k \mathbf{L}_k^{-1} (\mathbf{z} - \bar{\mathbf{z}}_k) + \mathbf{e}\end{aligned} \tag{6.390}$$

in which $k = \arg\min_n \|\mathbf{z} - \mathbf{z}_n\|$ and

$$\mathbf{e} = -\mathbf{Q}_k \mathbf{L}_k^{-1} \epsilon_*^{(k)} + \varphi_*^{(k)} \tag{6.391}$$

includes all of the error terms.

6.15.6 A General Framework for Gaussian Process Latent Variable Models and Dual Probabilistic PCA

Another popular method for finding low-dimensional representations of (or extracting features from) data is the *Gaussian process latent variable model* (GPLVM) [61]. Probabilistic versions of PCA, called *probabilistic PCA* (PPCA) [62] and dual PPCA are related to this method. In this section we show how GPLVM and dual PPCA arise naturally from a generalised linear model, in which kernel substitution can be used to define a nonlinear method that extends GPLVM and dual PPCA.

Given data $\mathbf{y}_n \in \mathbb{R}^d$, $n = 1, \ldots, N$, we assume that there exist corresponding latent or feature representations $\mathbf{z}_n \in \mathcal{F}_Q \subset \mathbb{R}^Q$. We then define a nonlinear feature mapping $\phi(\mathbf{z}) : \mathcal{F}_Q \to \mathcal{F}$ that maps each $\mathbf{z}$ in the latent space onto another f−dimensional feature space $\mathcal{F} \subset \mathbb{R}^f$, where $f = \infty$ is allowed. We then assume the following generalised linear model

$$\mathbf{y} = \mathbf{W}\phi(\mathbf{z}) + \boldsymbol{\epsilon} \tag{6.392}$$

in which $\epsilon \sim \mathcal{N}(\mathbf{0}, \sigma^2\mathbf{I})$ is i.i.d. noise (across different values of $\mathbf{z}$) and $\mathbf{W} = (w_{ij}) \in \mathbb{R}^{d\times f}$ is a matrix of coefficients that defines the transformation. $\mathbf{y}$ is assumed to be a nonlinear function of the latent variable $\mathbf{z}$, in which the i−th attribute takes the form $y_i = \sum_{j=1}^{f} w_{ij}\phi_j(\mathbf{z})$ for features $\phi_j(\mathbf{z})$ of $\phi(\mathbf{z})$. In the limit $f \to \infty$, we obtain an expansion of each attribute in terms of a basis $\{\phi_j(\mathbf{z})\}_{j=1}^{\infty}$. A general matrix Gaussian prior is placed over $\mathbf{W}$

$$\mathbf{W} \sim \mathcal{MN}_{d,f}(\mathbf{0}, \mathbf{K}_d, \mathbf{K}_f) = \frac{\exp\left(-\frac{1}{2}\mathrm{tr}[\mathbf{K}_f^{-1}\mathbf{W}^T\mathbf{K}_d^{-1}\mathbf{W}]\right)}{(2\pi)^{df/2}|\mathbf{K}_d|^{f/2}|\mathbf{K}_f|^{d/2}} \tag{6.393}$$

in which $\mathbf{K}_f \in \mathbb{R}^{f\times f}$ and $\mathbf{K}_d \in \mathbb{R}^{d\times d}$ are the row and column covariance matrices, respectively. Working with a high-dimensional explicitly-defined feature space is clearly impractical. We can instead integrate out $\mathbf{W}$ and introduce a kernel function to obtain a GP marginal likelihood, avoiding an explicit specification of ϕ and $\mathbf{K}_f$, and leading to a probabilistic model for $\mathbf{y}$ conditioned on $\mathbf{z}$.

The prior (6.393) can equivalently be written in a vectorised form

$$\mathrm{vec}(\mathbf{W}) \sim \mathcal{N}(\mathbf{0}, \mathbf{K}_f \otimes \mathbf{K}_d) = \frac{\exp\left(-\frac{1}{2}\mathrm{vec}(\mathbf{W})^T(\mathbf{K}_f \otimes \mathbf{K}_d)^{-1}\mathbf{W})\right)}{(2\pi)^{df/2}|\mathbf{K}_d|^{f/2}|\mathbf{K}_f|^{d/2}} \tag{6.394}$$

Marginalising over $\mathbf{W}$ yields

$$\begin{aligned} p\left(\mathbf{y}_n|\mathbf{z}_n\right) &= \int p(\mathbf{y}_n|\mathbf{W},\mathbf{z}_n))p(\text{vec}(\mathbf{W}))\,d\text{vec}(\mathbf{W}) \\ &= \int \mathcal{N}\left(\mathbf{W}\boldsymbol{\phi}(\mathbf{z}_n),\sigma^2\mathbf{I}\right)\mathcal{N}(\mathbf{0},\mathbf{K}_f\otimes\mathbf{K}_d)\,d\text{vec}(\mathbf{W}) \\ &= \int \mathcal{N}\left((\boldsymbol{\phi}(\mathbf{z}_n)\otimes\mathbf{I})^T\text{vec}(\mathbf{W}),\sigma^2\mathbf{I}\right)\mathcal{N}(\mathbf{0},\mathbf{K}_f\otimes\mathbf{K}_d)\,d\text{vec}(\mathbf{W}) \\ &= \mathcal{N}\left(\mathbf{0},(\boldsymbol{\phi}(\mathbf{z}_n)\otimes\mathbf{I})^T(\mathbf{K}_f\otimes\mathbf{K}_d)(\boldsymbol{\phi}(\mathbf{z}_n)\otimes\mathbf{I})+\sigma^2\mathbf{I}\right) \\ &= \mathcal{N}\left(\mathbf{0},\boldsymbol{\phi}^T(\mathbf{z}_n)\mathbf{K}_f\boldsymbol{\phi}(\mathbf{z}_n)\mathbf{K}_d+\sigma^2\mathbf{I}\right) \end{aligned} \tag{6.395}$$

Since $\mathbf{K}_f$ is symmetric and positive semidefinite (PSD) by definition, it possesses a unique PSD symmetric square root $\sqrt{\mathbf{K}_f}$. Hence, the term $\boldsymbol{\phi}^T(\mathbf{z}_n)\mathbf{K}_f\boldsymbol{\phi}(\mathbf{z}_n)$ defines an inner product in the feature space $\mathcal{F}$ as follows

$$\boldsymbol{\phi}^T(\mathbf{z}_n)\mathbf{K}_f\boldsymbol{\phi}(\mathbf{z}'_n) = \langle\widetilde{\boldsymbol{\phi}}(\mathbf{z}_n),\widetilde{\boldsymbol{\phi}}(\mathbf{z}'_n)\rangle := \langle\boldsymbol{\phi}(\mathbf{z}_n),\boldsymbol{\phi}(\mathbf{z}'_n)\rangle_{\mathbf{K}_f} \tag{6.396}$$

in which $\langle\cdot,\cdot\rangle$ denotes the standard inner product and $\widetilde{\boldsymbol{\phi}}(\mathbf{z}_n)=\sqrt{\mathbf{K}_f}\boldsymbol{\phi}(\mathbf{z}_n)$. Equation (6.395) defines a multivariate GP with a kernel given by the inner product $\langle\cdot,\cdot\rangle_{\mathbf{K}_f}$. We can now employ kernel substitution to replace the kernel in (6.396) with a general kernel $k(\mathbf{z}_n,\mathbf{z}'_n)$ to obtain the model

$$\mathbf{y}|\mathbf{z}\sim\mathcal{GP}(\mathbf{0},k(\mathbf{z},\mathbf{z}')\otimes\mathbf{K}_d+\delta(\mathbf{z},\mathbf{z}')\otimes\sigma^2\mathbf{I}). \tag{6.397}$$

Proceeding with the marginal distribution (6.397), which is of the type seen in Sect. 6.10, will lead to an optimisation over the covariance $\mathbf{K}\otimes\mathbf{K}_d$, in which $\mathbf{K}$ is the covariance matrix across different values of $\mathbf{z}$. Since $\mathbf{K}_d$ is PSD, the correlations can be modelled indirectly using a full-rank or low-rank Cholesky decomposition as in [63]. The number of hyperparameters to infer for these approaches is still, however, $\mathcal{O}(N^2d^2)$ while the matrix inversions are $\mathcal{O}(N^3d^3)$. We can instead use the ideas in Sect. 6.11.2 to break the dependence on d while retaining a rich spatial covariance model. That is, we reorganise $\mathbf{y}$ into its natural order−3 or higher tensor (hypermatrix) format $\mathcal{A}_n$ such that $\text{vec}(\mathcal{A}_n)=\mathbf{y}_n$, or $\text{vec}(\mathcal{A})=\mathbf{y}$ for a general $\mathbf{y}$. For example, if $\mathbf{y}$ is a vectorised spatio-temporal field, the ordering of the modes follows some ordering of the spatial coordinates and time.

We illustrate the ideas with a 3D random field at $d_1\times d_2\times d_3$ locations ($d_1d_2d_3=d$), for which $\mathcal{A}_n=(a_{\mathbf{i}}^{(n)})_{\mathbf{i}\in I}\in\mathbb{R}^{d_1\times d_2\times d_3}$, where $\mathbf{i}=(i_1,i_2,i_3)\in I$, $I=\{\mathbf{i}:i_j\in\{1,\ldots,d_j\},\,j\in\{1,2,3\}\}$. A component-wise model can be written as

$$a_{\mathbf{i}}=f_{\mathbf{i}}(\mathbf{z}_n)+\epsilon,\quad f_{\mathbf{i}}(\mathbf{z})\sim\mathcal{GP}\left(0,k(\mathbf{z},\mathbf{z}')k_d(\mathbf{i},\mathbf{i}')\right) \tag{6.398}$$

in which $\epsilon\sim\mathcal{N}(0,\sigma^2)$ (across $\mathbf{z}$) and $k_d(\mathbf{i},\mathbf{i}')=\text{cov}(f_{\mathbf{i}}(\mathbf{z}),f_{\mathbf{i}'}(\mathbf{z}))$ is the spatial covariance between any indices $\mathbf{i}\in I$ and $\mathbf{i}'=(i'_1,i'_2,i'_3)'\in I$. A linearly separable structure for k_d as in Sect. 6.11.2 then leads to the form

$$k_d(\mathbf{i}, \mathbf{i}') = k_d^{(1)}(i_1, i_1')k_d^{(2)}(i_2, i_2')k_d^{(3)}(i_3, i_3') \tag{6.399}$$

in which $k_d^{(j)}$ are kernel functions across each mode. Model (6.392) can be written in the form

$$\mathcal{A} = \mathcal{W} \bar{\times}_1 \phi(\mathbf{z})^T + \mathcal{E} \tag{6.400}$$

for some tensor $\mathcal{W} \in \mathbb{R}^{f \times d_1 \times d_2 \times d_3}$ and a tensor valued error $\mathcal{E}$, which contains i.i.d. entries $\epsilon \sim \mathcal{N}(0, \sigma^2)$ along the super-diagonal with all other entries equal to 0. Vectorisation leads to

$$\mathbf{y} = \mathbf{W}_{(1)}^T \phi(\mathbf{z}) + \boldsymbol{\epsilon} \tag{6.401}$$

which is the same as model (6.392) if $\mathbf{W} = \mathbf{W}_{(1)}^T$, where $\mathbf{W}_{(1)} \in \mathbb{R}^{f \times d}$ is the mode−1 unfolding of $\mathcal{W}$. The prior (6.393) is equivalent to the tensor normal prior

$$\mathcal{W} \sim \mathcal{TN}_{f,d_1,d_2,d_3}(\mathcal{O}, \mathbf{K}_f, \mathbf{K}_d^{(1)}, \mathbf{K}_d^{(2)}, \mathbf{K}_d^{(3)}) \tag{6.402}$$

in which the covariance matrices $\mathbf{K}_d^{(j)} \in \mathbb{R}^{d_j \times d_j}$ have entries $k_d^{(j)}(i_j, i_j')$. Equivalently

$$\mathbf{W}_{(1)}^T \sim \mathcal{MN}_{l,f}(\mathbf{W}_{(1)}^T | \mathbf{0}, \mathbf{K}_d^{(3)} \otimes \mathbf{K}_d^{(2)} \otimes \mathbf{K}_d^{(1)}, \mathbf{K}_f) \tag{6.403}$$

Thus, (6.397) can now be written as

$$\mathbf{y}|\mathbf{z} \sim \mathcal{GP}(\mathbf{0}, k(\mathbf{z}, \mathbf{z}') \otimes \mathbf{K}_d^{(3)} \otimes \mathbf{K}_d^{(2)} \otimes \mathbf{K}_d^{(1)} + \delta(\mathbf{z}, \mathbf{z}') \otimes \sigma^2\mathbf{I}). \tag{6.404}$$

If we consider the feature map to be the identity, $\phi(\mathbf{z}_n) = \mathbf{z}_n$, the model (6.404) takes the form

$$\mathbf{y}_n = \mathbf{W}_{(1)}^T \mathbf{z}_n + \boldsymbol{\epsilon}_n \tag{6.405}$$

in which $\mathbf{W} = \mathbf{W}_{(1)}^T$. Using a prior $\mathbf{W} \sim \prod_{i=1}^l \mathcal{N}(\mathbf{0}, \mathbf{I})$, i.e, a product of independent priors over the rows $\mathbf{r}_i$, $i = 1, \ldots, d$, of $\mathbf{W}$, leads to *dual PPCA*. In PPCA, we instead place i.i.d. priors over the latent variables in the form $\mathbf{z}_n \sim \mathcal{N}(\mathbf{0}, \mathbf{I})$ to obtain a marginal distribution

$$p(\mathbf{y}_n \mid \mathbf{W}) = \int p(\mathbf{y}_n \mid \mathbf{z}_n, \mathbf{W}) p(\mathbf{z}_n \mid \mathbf{W}) d\mathbf{z}_n = \mathcal{N}(\mathbf{0}, \mathbf{W}\mathbf{W}^T + \sigma^2\mathbf{I}) \tag{6.406}$$

Defining $\mathbf{Y} = [\mathbf{y}_1 \ldots \mathbf{y}_N]$, we then optimise the marginal log likelihood $\mathcal{L}(\mathbf{W})$ with respect to $\mathbf{W}$

$$\mathcal{L}(\mathbf{W}) = -\frac{Nd}{2} \ln 2\pi - \frac{N}{2} \ln |\mathbf{W}\mathbf{W}^T + \sigma^2\mathbf{I}| - \frac{1}{2} \mathrm{tr}\left((\mathbf{W}\mathbf{W}^T + \sigma^2\mathbf{I})^{-1}\mathbf{Y}\mathbf{Y}^T\right) \tag{6.407}$$

with the maximum likelihood estimate of $\mathbf{W}$ being equivalent to standard PCA, up to scaling and rotation.

In dual, PPCA on the other hand, the prior is placed over the rows of $\mathbf{W}$ as above. The marginalised likelihood of the data $\{\mathbf{y}_n\}$ is a product of d independent Gaussian distributions over $\mathbf{y}_{:,i} \in \mathbb{R}^N$, defined as vectors of the i−th components (attributes) of the $\mathbf{y}_n$

$$p(\mathbf{Y} \mid \mathbf{Z}) = \prod_{i=1}^{l} \mathcal{N}(\mathbf{0}, \mathbf{Z}\mathbf{Z}^T + \sigma^2\mathbf{I}), \quad \mathbf{Z} = [\mathbf{z}_1 \ldots \mathbf{z}_N]^T \tag{6.408}$$

in which $\mathbf{Z}$ is the design matrix, while $\mathbf{Z}\mathbf{Z}^T$ is the Gram (kernel) matrix. Replacing the linear kernel $\mathbf{Z}\mathbf{Z}^T$ with an equivalent kernel $\mathbf{K} = [k(\mathbf{z}_n, \mathbf{z}_m)]_{n,m}$ for some kernel $k(\mathbf{z}, \mathbf{z}')$ leads to GPLVM [61] and is equivalent to placing i.i.d. GP priors indexed by the latent variable over each attribute of the $\mathbf{y}_n$

$$\mathbf{y}|\mathbf{z} \sim \mathcal{GP}(\mathbf{0}, (k(\mathbf{z}, \mathbf{z}') + \delta(\mathbf{z}, \mathbf{z}')\sigma^2) \otimes \mathbf{I}) \tag{6.409}$$

This is a special case of (6.397), obtained by neglecting spatial correlations. The log likelihood is

$$\mathcal{L}(\mathbf{Z}) = -\frac{Nd}{2} \ln 2\pi - \frac{d}{2} \ln |\mathbf{K} + \sigma^2\mathbf{I}| - \frac{1}{2} \mathrm{tr}\left((\mathbf{K} + \sigma^2\mathbf{I})^{-1}\mathbf{Y}^T\mathbf{Y}\right) \tag{6.410}$$

and optimisation over the latent variable is equivalent to PPCA and PCA for a linear kernel. For nonlinear kernels, the optimisation over the latent variables $\mathbf{Z}$ is achieved through the chain rule, first taking the derivative with respect to $\mathbf{C} = \mathbf{K} + \sigma^2\mathbf{I}$

$$\frac{\partial \mathcal{L}}{\partial \mathbf{C}} = \mathbf{C}^{-1}\mathbf{Y}^T\mathbf{Y}\mathbf{C}^{-1} - d\mathbf{C}^{-1} \tag{6.411}$$

and subsequently with respect to the components of $\mathbf{Z}$.

The more general form (6.397) incorporates spatial correlations and is especially appropriate for data that has a natural tensor structure. Defining an order−4 tensor $\mathcal{Y} \in \mathbb{R}^{N \times d_1 \times d_2 \times d_3}$ that collects all of the $\mathcal{A}_n$ along the first mode, the likelihood of the observations is given by

$$\mathcal{L} = \log p(\mathrm{vec}(\mathcal{Y})|\mathbf{Z}) = -\frac{d}{2} \log |\mathbf{\Sigma}_\xi| - \frac{1}{2} \mathrm{vec}(\mathcal{Y})^T \mathbf{\Sigma}_\xi^{-1} \mathrm{vec}(\mathcal{Y}) - \frac{Nd}{2} \log(2\pi) \tag{6.412}$$

in which $\mathbf{\Sigma}_\xi = \mathbf{K}_l^{(3)} \otimes \mathbf{K}_l^{(2)} \otimes \mathbf{K}_l^{(1)} \otimes \mathbf{K} + \sigma^2\mathbf{I}$ is the kernel matrix with $\mathbf{K} = [k(\mathbf{z}_n, \mathbf{z}_m)]_{n,m}$. The major challenge is in the calculation of the inverse and determinant of $\mathbf{\Sigma} \in \mathbb{R}^{Nd \times Nd}$, for which we may take advantage of the Kronecker product structure to reduce the computational complexity from $\mathcal{O}(N^3d^3)$ to $\mathcal{O}(Nd(d_1 + d_2 + d_3 + N))$, as in Sect. 6.11.2.

6.16 K-means and K-Medoids Clustering

We now consider a classical method for *clustering*, in which again we have only inputs $\mathbf{y}_n$, $n = 1, \ldots, N$, so that this method is unsupervised. The task this time is to find *similar groupings* of the data points. This may be done as a form of exploratory analysis to visualise relationships between attributes of the data points, or it may be used as way to group similar data points for a supervised learning analysis.

The most standard method is simple to implement and is called *K-means clustering*. It starts by picking a value of K, which is the number of clusters C_k, $k = 1, \ldots, K$, and the goal is to assign each data point $\mathbf{y}_n$ to a cluster. To each cluster we assign a *centre* $\boldsymbol{\mu}_k$. We are going to pick these centres and assign each data point such that the *sum of square distances of each data point* $\mathbf{y}_n$ *to its cluster centre* $\boldsymbol{\mu}_k$ *is minimised.* For each $\mathbf{y}_n$, we introduce a 1-of-K representation $\mathbf{r}_n$ of its cluster membership, i.e., if $\mathbf{y}_n \in C_k$, the k-th entry is 1 and the other entries are 0

$$\mathbf{r}_n = (0, \ldots, 0, 1, 0, \ldots, 0)^T \in \mathbb{R}^K \tag{6.413}$$

We denote the coefficients of $\mathbf{r}_n \in \{0, 1\}$ by r_{nk}, $k = 1, \ldots, K$.

We can now write down an objective function to minimise, i.e., the sum of square distances of each $\mathbf{y}_n$ to its cluster centre $\boldsymbol{\mu}_k$

$$\begin{aligned} J &= \sum\nolimits_{k=1}^{K} r_{1k} \|\mathbf{y}_1 - \boldsymbol{\mu}_k\|^2 + \ldots + \sum\nolimits_{k=1}^{K} r_{Nk} \|\mathbf{y}_N - \boldsymbol{\mu}_k\|^2 \\ &:= J_1 + \ldots + J_N \end{aligned} \tag{6.414}$$

J is also called a *distortion measure* and we minimise J over r_{nk} and $\boldsymbol{\mu}_k$ iteratively. The algorithm is as follows:

1. make an initial guess for the $\boldsymbol{\mu}_k$
2. minimise over r_{nk} keeping $\boldsymbol{\mu}_k$ fixed
3. minimise over $\boldsymbol{\mu}_k$ keeping r_{nk} fixed
4. repeat steps (2) and (3) until convergence

In step 1 we differentiate J w.r.t. r_{nk} with fixed $\boldsymbol{\mu}_k$'s yielding $\|\mathbf{y}_n - \boldsymbol{\mu}_k\|^2$, so the minimisation is independent for each n

$$\min_{r_{nk}} J_n = \min_{r_{nk}} \sum\nolimits_{k=1}^{K} r_{nk} \|\mathbf{y}_n - \boldsymbol{\mu}_k\|^2 \quad \text{such that } r_{nk} \in \{0, 1\} \tag{6.415}$$

The solution is straightforward: pick $r_{nk} = 1$ for $k = \text{argmin}_j \|\mathbf{y}_n - \boldsymbol{\mu}_j\|^2$ and otherwise set $r_{nk} = 0$, i.e, choose that k such that $\mathbf{y}_n$ is closest to $\boldsymbol{\mu}_k$. For the second minimisation we calculate $\nabla_{\boldsymbol{\mu}_k} J$ w.r.t. each $\boldsymbol{\mu}_k$, yielding

$$-2r_{1k}(\mathbf{y}_1 - \boldsymbol{\mu}_k) - \ldots - 2r_{Nk}(\mathbf{y}_N - \boldsymbol{\mu}_k) = -2\sum\nolimits_{n=1}^{N} r_{nk}(\mathbf{y}_n - \boldsymbol{\mu}_k) = 0 \tag{6.416}$$

so that

$$\mu_k = \frac{\sum_{n=1}^{N} r_{nk} \mathbf{y}_n}{\sum_{n=1}^{N} r_{nk}} \tag{6.417}$$

The denominator is the total number of points in cluster k so that $\boldsymbol{\mu}_k$ is the mean of all points in this cluster, hence the name 'K means'.

To generalise the method, we can replace the distance $\|\mathbf{y}_1 - \boldsymbol{\mu}_k\|$ with any other distance measure, for example the Manhattan norm. Another generalisation is the K-*medoid* method, in which we cluster in the same way but each centre $\boldsymbol{\mu}_k$ is one of the data points $\mathbf{y}_n$ and is referred to as a *medoid* or prototype. Moreover, we can use a general *dissimilarity measure* $\nu(\mathbf{y}_n, \boldsymbol{\mu}_k)$ in place of $\|\mathbf{y}_1 - \boldsymbol{\mu}_k\|$ in the distortion measure J

$$J = \sum_{n=1}^{N} \sum_{k=1}^{K} r_{nk} \nu(\mathbf{y}_n, \boldsymbol{\mu}_k) \tag{6.418}$$

The K-medoids algorithm proceeds as follows:

1. Make an initial guess for the K cluster medoids $\boldsymbol{\mu}_k \in \{\mathbf{y}_n\}$
2. For each $\mathbf{y}_n$ find the cluster medoid $\boldsymbol{\mu}_k$ with the smallest dissimilarity $\nu(\mathbf{y}_n, \boldsymbol{\mu}_k)$ and assign $\mathbf{y}_n \in C_k$, i.e., $k = \text{argmin}_j \nu(\mathbf{y}_n, \boldsymbol{\mu}_j)$
3. For each medoid $\boldsymbol{\mu}_k$

   - For each $\mathbf{y}_n$ in the cluster C_k swap $\boldsymbol{\mu}_k$ with $\mathbf{y}_n$ and calculate the *total dissimilarity*

     $$\sum_{\mathbf{y}_m \in C_k} \nu(\mathbf{y}_m, \boldsymbol{\mu}_k)$$

     in which $\boldsymbol{\mu}_k$ is now $\mathbf{y}_n$. Select as the new medoid the $\mathbf{y}_n$ with the smallest total dissimilarity

4. Return to step 2 until convergence

As in the K-means algorithm, the number of clusters K has to be decided *a-priori*, which is not a trivial task. The K-means algorithm is a special case of a method that models the data using a mixture (linear combination) of Gaussians and uses an iterative algorithm called *Expectation Maximisation* to approximately maximise the likelihood (actually, a lower bound). This method is able to automatically select the number of clusters, al least in theory.

The dissimilarity measure can be any number of measures, e.g., the Manhattan norm or any *Minkowski norm*

$$\nu(\mathbf{y}_n, \boldsymbol{\mu}_k) = \left(\sum_{i=1}^{D} |y_{ni} - \mu_{ki}|^p\right)^{\frac{1}{p}} \tag{6.419}$$

in which y_{ni} is the i-th component of $\mathbf{y}_n$ and μ_{ki} is the i-th component of $\boldsymbol{\mu}_k$. The Manhattan norm corresponds to $p = 1$ and the standard Euclidean corresponds to $p = 2$.

6.17 Machine Learning-Assisted Macroscopic Modelling

In recent years, there has been a growth in the applications of machine learning to the study of flow batteries, at various scales from the macroscopic [64] to mesoscopic [65] to nanoscale [66]. In this and the next two sections, we illustrate the potential of machining learning and its capabilities with selected examples, starting with the macroscopic case.

There are, of course, many more potential applications and studies that could be covered, but rather than survey the entire literature and engage in a high-level discussion, our approach is to provide an in-depth analysis and discussion of selected examples, so that readers are of sufficient information to attempt their own implementations. Other applications follow a similar pattern in terms of machine learning, with the slight exception of time-series problems, covered in Chap. 7.

In [64], Shah and co-workers developed surrogate models for the 2D, dynamic all-vanadium model in [67]. This model includes conservation of mass, charge and momentum inside a domain comprising the positive and negative porous electrodes, the membrane and the electrolyte reservoirs (see Sect. 4.2 of Chap. 4 for the corresponding equations and details). The species considered were V(II), V(III), V(IV), V(V), H_2O and H^+. Transport was assumed to occur *via* diffusion, convection and electro-migration, and separate balances for the species were considered for the reservoirs. Charge balances in the electrolyte were derived from charge conservation taking into account the individual species fluxes and local electroneutrality.

For the surrogate model, the authors used as the inputs

$$\boldsymbol{\xi} = \xi = (L_m, c_{3,0}, u)^T \tag{6.420}$$

in which L_m is the thickness of the membrane, $c_{3,0}$ is the initial concentration of V(III) and u is the flow rate of the electrolyte. Design points were selected using Latin hypercube sampling (LHS) to obtain $N = 100$ design points $\boldsymbol{\xi}_i$, with the final design $\Xi = \{\boldsymbol{\xi}_i\}$. The inputs were constrained such that

$$c_{3,0} \in [750, 1500]\ \text{mol m}^{-3}, \quad L_m \in [50, 200]\mu\text{m}, \quad u \in [0.1, 1]\ \text{cm s}^{-1} \tag{6.421}$$

For each input, voltage charge-discharge curves were simulated using a charge/discharge current of 10 A. The open-circuit voltage values of 1.4 V and 1.7 V were used to define the end of discharge and charge, respectively. The cell voltage was recorded at 200 time instances for each charge-discharge cycle, yielding vectorised values of voltage $\mathbf{y}_i \in \mathbb{R}^d$, $d = 200$, corresponding to $\boldsymbol{\xi}_i$.

Moreover, the time interval Δ_i between each of the 200 time instances was recorded. Only this time interval, corresponding to $\boldsymbol{\xi}_i$ is required to define all times, since they are equally spaced and for each $\boldsymbol{\xi}_i$ there are 200 instances. That is, the time sequence for $\boldsymbol{\xi}_i$ is

$$t_n^i = (n-1)\Delta_i, \quad n = 1, \ldots, 200 \tag{6.422}$$

Table 6.1 RMSE values for different methods as the number of training points M is increased, against 20 test points

M	GP	SVR	MLP	DMLP	CNN
20	0.4615	0.5273	1.2259	1.4671	0.9868
40	0.3732	0.4135	0.9118	1.1043	0.7173
60	0.3309	0.3455	0.6918	0.9921	0.4386
80	0.3214	0.3345	0.4680	0.6810	0.3653

The voltage values can be collected in a matrix $\mathbf{Y} \in \mathbb{R}^{d\times N}$. Importantly, in this first example the explicit times were not taken into account, or more precisely, a state-of-charge was used in lieu of time.

Several machine learning methods were considered for the mapping $\mathbf{y}_i = \boldsymbol{\eta}(\boldsymbol{\xi}) + \boldsymbol{\epsilon}$, both with and without the error term $\boldsymbol{\epsilon}$, using the data $\{\Xi, \mathbf{Y}\}$. A dimensionally reduced GP model and an equivalent SVR were implemented (see Sects. 6.10.2, 6.7 and 6.9), along with shallow and deep MLPs and CNNs (Sects. 6.12.1 and 6.12.2). The dimension reduction for the GP and SVR models was performed with a singular value decomposition (Sect. 6.14.1) on the data $\mathbf{Y}$, with six principal components selected to capture at least 99.9 % of the total variance. Twenty of the data points were used for testing while the remainder were available for training.

The authors used a $\nu = 5$ Matérn kernel (6.64) for the GP model and a Gaussian kernel (6.56) for SVR, with the hyperparameters optimised using a grid search combined with cross-validation. The shallow MLP used 1024 hidden layer neurons and a ReLU activation function (6.209), while the deep MLP used 2 hidden layers with 512 and 32 neurons, alongside a ReLU activation function. The CNN used 1024 kernels of size 2 and stride 1 and a fully connected layer with 512 neurons and a ReLU function.

The root mean square error (RMSE) between a prediction $\mathbf{y}_j^p$ and a test point $\mathbf{y}_j$ is defined as

$$RMSE = \sqrt{\sum_{j=1}^{M'} \frac{1}{M'} \|\mathbf{y}_j^p - \mathbf{y}_j\|} \tag{6.423}$$

in which $M' = 20$ is the number of test points. (6.423) has an obvious equivalent for the scalar case.

The RMSE values obtained using the various methods are shown in Table 6.1 for an increasing number of training points M. The SVR and GP models are clearly superior to the networks, with the GP model performing particularly well. The MLPs lead to high errors, while the CNN shows significant improvement as M is increased. This is as expected and is quite typical of the comparison between networks and GP or SVR models. The networks contain a high number of parameters, and thus require larger training point numbers to achieve a similar level of accuracy.

The GP and SVR models contain $\mathcal{O}(l)$ hyperparameters, whereas the shallow MLP contains $1024 \times (l + 1)$ weights and biases between the input and hidden layer, with

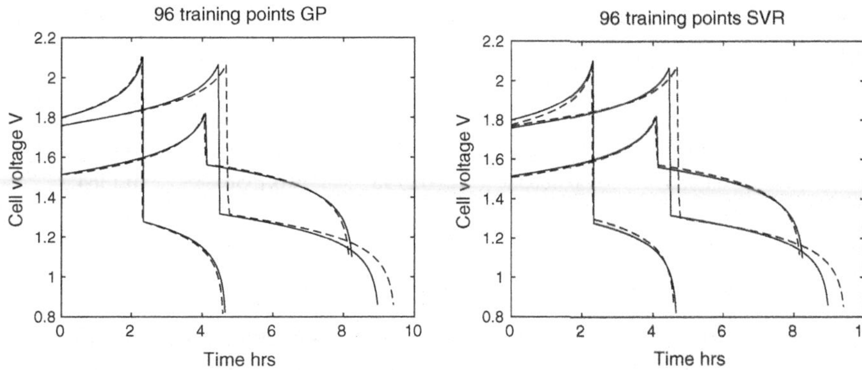

Fig. 6.1 Example predictions of the charge-discharge curves and time sequences using the GP and SVR approaches for 3 of the test points and 96 training points. The solid curves are the test and the dashed curves are the predictions

a further $(1024 + 1) \times d$ weights between the hidden and output layer. This leads to a high model variance, meaning that the results can change quite markedly depending on the weight and bias initialisation.

As a second example, the authors used the input

$$\boldsymbol{\xi} = (L_m, c_{3,0}, u, I_{app}, V_R, k_\sigma, \epsilon)^T \tag{6.424}$$

in which $I_{app} \in [2.5, 30]$ A is the load current, $V_R \in [0.1, 1]$ dm^3 is the volume of the reservoir, $\epsilon \in [0.4, 0.95]$ is the porosity of the electrode, and $k_\sigma \in [0.5, 4]$ S m^{-1} is a constant that defines the membrane conductivity. They again used LHS to generate 156 design points $\boldsymbol{\xi}_i$. The other inputs were now in the ranges $c_{3,0} \in [500, 3000]$ mol m^{-3}, $L_m \in [50, 350]$ μm and $u \in [0.1, 3]$ cm s^{-1}. In this case, the time intervals $\Delta_i \in \mathbb{R}$ were also targets, along with the vectorised voltage data $\mathbf{y}_i \in \mathbb{R}^d$, defined as before.

The time interval data can be treated with the model $\Delta_i = \eta(\boldsymbol{\xi}) + \epsilon$, with noise ϵ and latent function η. The authors used a variety of methods to approximate the latent function, with data $\{\Delta_i, \boldsymbol{\xi}_i\}$. Learning of the time interval and voltage sequence was conducted independently, which ignores correlations between the two and may thus be sub-optimal. The same method was used for approximating both $\eta(\boldsymbol{\xi})$ and $\boldsymbol{\eta}(\boldsymbol{\xi})$, meaning the univariate and multivariate versions of the method, with the dimensionally reduced GP and SVR methods as described above for the multivariate case. 28 of the data points were reserved for testing, while the maximum number of training points was 128. Examples of the predictions (3 of the 28) are shown in Fig. 6.1 for both the GP and SVR approaches, using 96 training points.

Measuring the error in this case is not straightforward, e.g., by using the RMSE. The authors defined it is as the integral of the magnitude of the difference between the prediction and test, which is equal to the area enclosed by the prediction and test curves in Fig. 6.1. Unfortunately, the supports of the test and prediction as functions

Table 6.2 Values of error (A) for different methods as a function of the number of training points M, against 28 test points

M	GP	SVR	MLP	DMLP	CNN
32	0.1120	0.1244	1.2097	0.9771	0.7752
64	0.1006	0.1106	0.4818	0.3298	0.2487
96	0.0586	0.0603	0.2198	0.1556	0.1364
128	0.0450	0.0442	0.2610	0.1089	0.1032

of time are not the same. The authors therefore calculated the area A for only that portion of the charge-discharge duration that is shared by both curves, that is, up to $t = \min(t_f^p, t_f^t)$, in which t_f^p is the end time for the prediction curve and t_f^t is the end time for the test curve. This involved (since the times are different) interpolating the predicted values $\mathbf{y}_j^p$ at the times corresponding to the test $\mathbf{y}_j$ in the case of $t_f^p > t_f^t$, or else interpolating the test values at the predicted time sequence if $t_f^p < t_f^t$.

The values of A obtained for each of the methods are shown in Table 6.2 for an increasing number of training point M. The GP and SVR models were again shown to be superior, for the reasons provided earlier.

6.18 Machine Learning-Assisted Mesoscopic Models

Mesoscopic models (see Chaps. 4 and 5 and Sect. 3.4) are used to bridging the gap between the macroscopic and atomistic scales, providing detailed analyses of pore scale phenomena, including mass, momentum, heat and charge transport or transfer processes. From the results, certain material properties such as a hydraulic permeability can be extracted.

Recently, Wan et al. [65] used machine learning to discover optimal electrode structures for RFBs, based on a dataset containing the specific surface area and hydraulic permeability of 2275 fibrous structures, calculated using a lattice Boltzmann model (Sect. 3.4.3) with microstructures generated using stochastic reconstruction. Four inputs related to the microstructure (fibre diameter, electrode porosity, in-plane orientation factor, through-plane orientation factor) were selected to learn the specific area and permeability relationship, using linear regression, an MLP and a random forest (RF) method. High accuracy results were achieved on the surface area, while the errors on the hydraulic permeability were somewhat higher (around 10%), with the MLP performing best overall.

The authors then combined the MLP model with a genetic algorithm to conduct a multi-objective optimisation over the inputs to find electrodes with higher specific areas and higher permeabilities. They found that the fibre diameter and porosity of promising candidates exhibited a triangle-like joint distribution, and that fibre diameters of around 5 μm with aligned arrangements were preferred. Figure 6.2 shows the

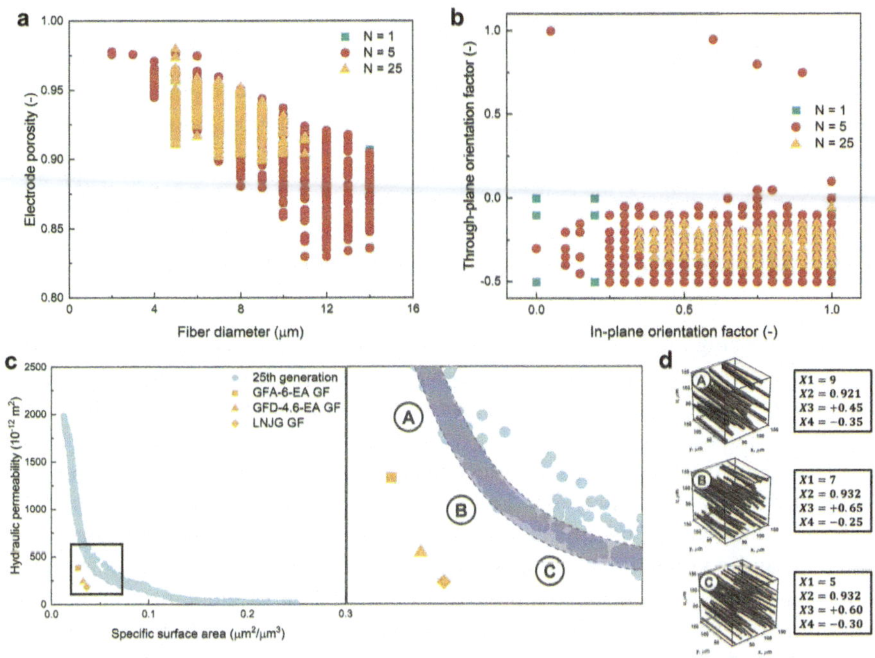

Fig. 6.2 Structural optimisation of electrodes for RFBs based on an MLP using data from a lattice-Boltzmann model combined with stochastic reconstruction to create the porous structure. **a**, **b** show the distributions of the four structural inputs from 95 candidates selected from the 1st generation, 573 from the 5th generation, and 714 from the 25th generation of the genetic algorithm. **c** shows the 714 twenty-fifth-generation candidates on the Pareto front, divided into 3 regions, A, B and C. **d** shows randomly selected examples of the structures with one from each of the regions, together with the structural (input) parameters. Reprinted with permission (License Number 5482980892153) from [65]. Copyright 2023 Elsevier

results of the optimisation. Figure 6.2a and b show the distributions of the four structural inputs from 95 candidates selected from the 1st generation, 573 from the 5th generation, and 714 from the 25th generation of the genetic algorithm. As the generation number increases, the fibre diameter and electrode porosity begin to exhibit the triangle-like joint distribution, with ranges of 5–10 μm and 0.9–0.96, respectively. From these results, it was concluded that fine fibres (around 5 μm diameter) are feasible for RFB electrodes. Figure 6.2b shows that the in-plane and through-plane orientation factors should be in the ranges 0.4–1.0 and 0.4–0.2 respectively, suggesting that slightly aligned arrangements should be used.

The 714 twenty-fifth-generation candidates on the Pareto front shown in Fig. 6.2c were considered to be solutions to the structural optimisation problem. The majority of the solutions are distributed along the purple curve. The shaded region can be divided into three regions, A, B and C, with solutions in each region improving the electrode structure with respect to different requirements. For example, to increase the specific surface area, solutions from region C could be selected, with up to ca. 80% increase, without reducing the permeability. Finally, Fig. 6.2d illustrates randomly selected examples of the structures with one from each of the regions, together with the structural (input) parameters. All of the examples can be seen to have an aligned structure.

6.19 Machine Learning Models for Material Properties

Machine learning has been used extensively for screening and designing new active redox species, especially organic molecules [66]. The primary goal is to find a balance between the redox potential, the solubility in a given medium and the stability of the species.

Electronic-structure (or quantum-mechanical) methods (Sect. 3.6) such as density functional theory (DFT) play a major role in determining material properties, either directly or indirectly. These techniques form a hierarchy, in terms of the associated accuracy and computational burden. In general, they are highly time-consuming, which has led to great efforts to replace all or part of their formulation with machine learning surrogates, especially for large-scale screening of materials. In fact, the range and number of applications of machine learning in this area is quite staggering.

The most direct approach is to learn a mapping between the output of interest, e.g., an atomisation energy, and a suitable characterisation of the molecule, which forms the input for supervised machine learning [68, 69]. The resulting model can also be used inside *ab-initio* molecular dynamics (MD) simulations [70, 71]. Another approach is to find a mapping between the charge density and the various contributions to the total system energy, in addition to maps between the charge density and the external potential [71, 72]. Often the motivation is to approximate the kinetic energy functional $T_{KS}[\rho]$ (3.167) in order to accelerate the calculations in Kohn-Sham DFT [73, 74] (Sect. 3.6.4). There are, however, challenges in terms of calculating the functional derivatives. An alternative approach instead uses the density of states [75, 76].

At the heart of DFT and related methods is the exchange-correlation functional (Sects. 3.6.4 and 3.6.5). It is no surprise, therefore, that machine learning has been used to replace, or find corrections to this functional in order to facilitate more accurate and/or faster approximations [77]. One of the great challenges with this and the other approaches is transferability across systems, so that these approaches are usually developed for a particular type of system or molecule.

A critical component of any machine learning-assisted electronic-structure method is the characterisation of the input, namely a good numerical descriptor of the molecule. A broad range of methods have been explored. Examples include Coulomb matrices [78, 79] or their eigenvalues [68, 80], the bag of bonds method [81], generalised symmetry functions [82], smooth overlap of atomic positions (SOAP) [83], and molecular fingerprints [84]. Enormous variations in the accuracy are observed with different descriptors [84, 85], and choosing a good descriptor is almost certainly more important than the choice of machine learning method.

To predict certain key properties of molecules such as the solubility of redox pairs, the a well-established approach is to find a mapping between the solubility and a molecular descriptor using a purely machine-learning approach with a given data set. This is known as a *quantitative structure-activity relationship* (QSAR) model, and it does not generally involve any physics based simulations. As with the machine learning-assisted electronic structure methods, key to QSAR models is the descriptor. Apart from those mentioned above, numerous other descriptors have been developed, including descriptors extracted from quantum mechanical models. Below, we briefly introduce QSAR models. We then present a number of case studies that use QSAR models and machine learning-assisted electronic-structure methods to calculate vital properties of organic molecules, intended for application as the active species in organic redox flow batteries (ORFBs) (Chap. 2).

6.19.1 Introduction to Quantitative Structure-Activity Relationship Models

Quantitative structure-activity relationship (QSAR) or quantitative structure-property relationship (QSPR) models attempt to predict certain fundamental properties of a chemical compound, such as the activity, solubility or toxicity, from molecular descriptors. Typically, the compounds are organic and the application area is pharmacological or medicinal. Descriptors can be any information deemed essential to the prediction of the target variable, and can relate to the constituents of the molecule, its electronic structure, its mechanical structure and geometry, its hydrophobicity, and its quantum-chemical properties such as the highest or lowest unoccupied molecular orbitals (HOMO and LUMO) [86]. These descriptors are usually employed in combinations.

Denoting the combination of descriptors $\boldsymbol{\xi}$ and the target property (to be predicted) y, the basic assumption of QSAR models is that a relationship of the form

$$y = \eta(\boldsymbol{\xi}) \tag{6.425}$$

exists, in which $\eta(\cdot)$ is a function of the descriptors. This (assumed) relationship between the input $\boldsymbol{\xi}$ and output y is precisely the type of relationship that machine learning seeks to approximate. One only requires examples, i.e., data of the type $y_n, \boldsymbol{\xi}_n, n = 1, \ldots, N$, and a proposed statistical model such as that above, or more generally

$$y = \eta(\boldsymbol{\xi}) + \epsilon \tag{6.426}$$

with noise ϵ and latent function $\eta(\cdot)$.

Molecular descriptors are at the heart of QSAR models, and a good model is one in which the descriptor provides a good causal link to the target [87]. Descriptors can be broadly classified into three groups [88]. The simplest is 1D, and includes atom counts, molecular weights, numbers of functional groups, numbers of single, aromatic and other bond types, and other properties related solely to the molecular formulae. 2D descriptors are based on the 2D molecular structure. They can also use atom counts, but unlike 1D descriptors, counting is done for different types of atoms (considering such things as hybridisation status). Bond information can again be included in such descriptors. There is a vast array of other 2D descriptors that include van der Waals volumes, polarisabilities, 2D autocorrelations, connectivity and cyclicity indices, topological distance indices, adjacency matrices and their eigenvalues, and molecular walk counts based on graphs, to name but a few.

3D methods are those that use descriptors derived from a spatial representation of the molecule, including the geometrical configuration, information regarding the shape and conformation, and surface properties. Higher-order descriptors, such as 4D, have also been developed, based on reference grids and MD modelling to generate ensembles. 3D methods are usually based on constructing an initial representation of the molecular structure from experiments, e.g., X-ray crystallography or NMR spectroscopy, or from computational methods. 3D structures can be generated manually or numerically using, e.g., quantum or molecular mechanics. They are then refined through minimisation of the conformational energy using methods such as DFT, or, less accurately, molecular mechanics.

Quantum chemical and MD-based descriptors [86] can improve accuracy but require a large computational budget, unless preexisting databases are available. The Hartree-Fock semi-empirical methods described in Sect. 3.6.2 are less time-consuming but are generally of insufficient accuracy, with DFT being considered optimal. In *fragment-based* or group contribution QSAR, the molecule is split into its components (molecular fragments) and relationships are developed between the targets and properties related to the molecular fragments, such as Molecular ACCess System (or MDL) keys [89] and molecular fingerprints [90].

The descriptors, whether 1D, 2D, 3D or higher, can exist in high-dimensional spaces, which has motivated the use of feature extraction methods [91] to find reduced-dimensional representations that can be used as efficient and accurate proxies. These methods include hand crafting of features, linear and nonlinear dimension reduction (Sects. 6.14 and 6.15), graph-based methods, autoencoders (similar to the encoder-decoder models in Sect. 6.12.5) and many others. Clustering (Sect. 6.16) can also be used to group similar molecules in order to achieve a better (localised) fit.

6.19.2 Examples of Redox Potential and Solubility Estimation

The main area in which machine learning has been used for the prediction of molecular properties relates to organic species for ORFB [92]. Large-scale laboratory measurements of redox potentials and solubilities using voltammetry and solubility measurement methods is costly and time-consuming. With the availability of electronic-structure calculation tools, it is in principle possible to screen or even design new molecules computationally. These methods are, nevertheless, also extremely time-consuming. Capturing certain properties, especially in complex environments such as solvents may not be possible with sufficient accuracy using low levels of theory such as Hartree-Fock or DFT with a simple functional, necessitating even more costly calculations.

Applications of machine learning to make feasible large-scale screening using quantum-mechanical calculations have focused on three issues: the redox potential, solubility in aqueous solvents, and stability. Although the motivation was not flow batteries in most cases, the results are often highly relevant to ORFB. The solubility of organic species is vitally important, and one of the great barriers to realising higher energy densities.

In [93], Schroeter et al. developed various QSAR models for the buffer and native solubilities based on different machine learning methods and 1664 Dragon descriptors, a mix of 1D, 2D and 3D [94]. They used a combination of databases containing over 5000 measurements in total. The methods included were a GP model (Sect. 6.7), SVR (Sect. 6.9), a random forest (RF) model and Lasso ridge regression (Sect. 6.3), with the GP seemingly giving the best performance. The GP was used with a SEARD kernel (6.64) to lower the number of descriptors to 200 by choosing those with the highest correlation lengths θ_i.

Boobier et al. [95] considered a broader range of methods, including an ANN (Sect. 6.12.1), and they used 3D quantum molecular descriptors based on DFT. As has been seen in other work, the authors concluded that the molecular descriptors are more important than the machine learning method used. Klopman and Hao [96] instead used a group-based approach for a database of over 1100 organic molecules, also finding that the fragment level descriptors are the main determinant of accuracy.

Kim et al. [97] developed a method for high-throughput molecular screening specifically for ORFBs, which they termed a 'multiple descriptor multiple kernel' (MultiDK) method. The method relies on using different 1D and 2D descriptors in combination, including fingerprints and functional keys, along with other physico-chemical properties. They also used the same method to find a pH-dependent solubility relationship. The machine learning method used was kernel regression (Sect. 6.6), with a mixture of linear and different nonlinear kernels (Sect. 6.6). The prediction accuracy using this method is shown in Fig. 6.3, for 1676 organic molecules. The authors compared the method to a single descriptor model (SD), a multiple descriptor model (MD), and tried various descriptors, both binary and non-binary, leading to different versions of MD and MultiDK.

Another 2D QSAR method was used by Suleyman and co-workers [98, 99] for the screening of quinone-like anolytes for aqueous ORFBs (in terms of solubility). The authors used a total of 123 descriptors, rationalised using a Pearson correlation analysis and fed into MLP, random forest and extreme-gradient-boost algorithms. From a candidate set of 3257 redox pairs, 205 were found to have a higher solubility (and lower redox potential calculated from DFT studies) than AQDS, which is commonly used in ORFBs.

The redox potential has been the focus of most attention, since it determines the cell voltage (and power) and is readily calculated *via* a thermodynamic cycle

Fig. 6.3 Solubility prediction accuracy using MultiDK on 1676 organic molecules. Here SD refers to a single descriptor model and MD to multiple descriptors. The coefficients xy in MDxy and MultiDKxy represent the number of embodied binary descriptors and the number of embodied nonbinary descriptors, respectively. Reprinted (adapted) with permission from [97]. Copyright 2017 American Chemical Society

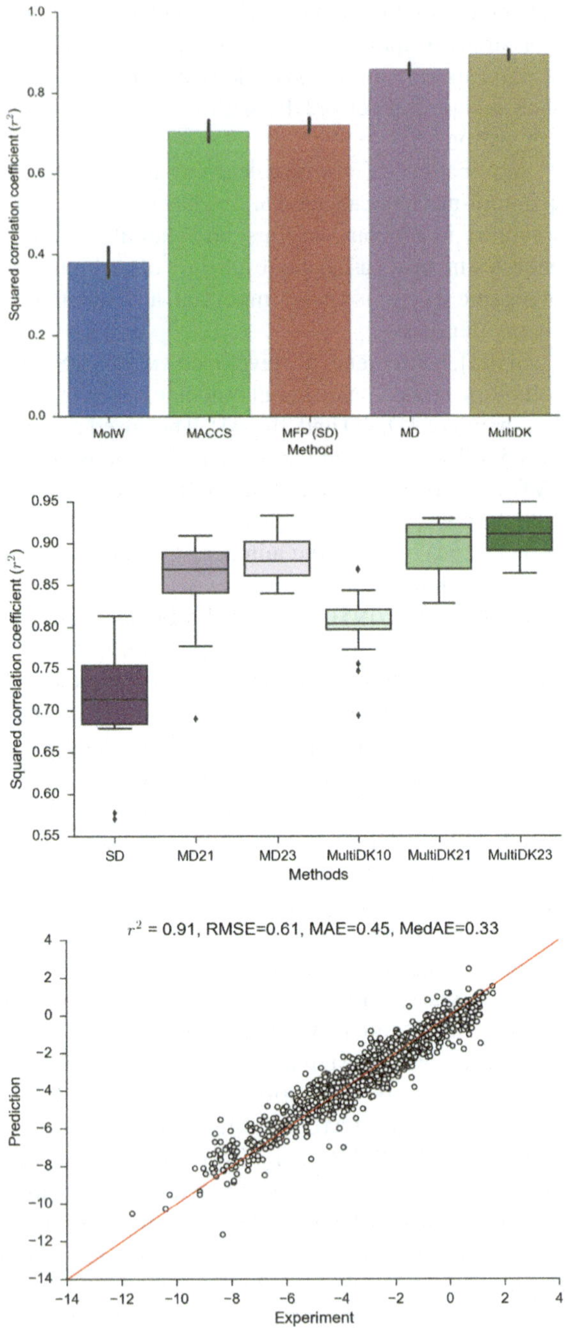

using ab-initio methods such as DFT. As discussed above, the high cost of these methods has motivated machine learning-assisted approaches to facilitate large-scale screening. Allam et al. [100] used a variety of machine learning methods (MLPs, kernel regression and gradient boosting) in combination with DFT to design novel organic electrode materials, motivated by Li-ion battery applications but applicable to ORFBs.

Electronic structure properties, including the adiabatic electron affinity (EA), HOMO and LUMO, and the HOMO-LUMO gap calculated from DFT, were used as descriptors. Additional descriptors were included, in the form of basic structural features such as the numbers of carbon, boron, oxygen, lithium and hydrogen atoms, as well as the number of aromatic rings. In one approach, these inputs were used in their original form, while an enhanced version added a penalty term (in the Manhattan norm) to the loss function in each method after expanding the number of inputs to include hand-crafted features. The enhanced version yielded the best performance in combination with kernel regression. It has to be noted that the inputs used in this case (HOMO and LUMO) are required either from experiments (cyclic voltammetry) or from electronic structure calculations, which means that large-scale screening would be time-consuming.

Doan et al. [101] used a GP model to predict the oxidation potential of homobenzylic ethers (HBEs), with the motivation being the formation of passivating films on the electrodes of nonaqueous ORFB, which reduce the cycle life. The oxidation potentials were calculated using DFT and the GP model was used as a direct replacement. The authors proposed a system in which redox-active cores are connected to a molecular scaffold *via* a cleavable tether. By altering the electrode potential, the passivating film could be removed by triggering a cleavage reaction, with HBEs chosen as the molecular scaffold for mesolytic cleavage.

A total of 49 inputs were used in the model, including the molecular weight, topological surface area, number of valence electrons, and number of aromatic rings. The inputs were transformed using PCA (Sect. 6.14.1), from which 30 features were selected based on the generalised variance (6.270) as the final set of descriptors. The authors further employed an *active learning* step to conduct Bayesian optimisation for sequentially selecting query points to add to the training data in order to perform an optimisation with the GP model.

Ghule et al. [102] investigated the use of four machine learning models for predicting the redox potentials of phenazine derivatives in dimethoxyethane, with the motivation being ORFBs. The models used a small data set generated from DFT calculations, relating only to phenazine derivatives with one type of functional group per molecule. The 'SelectKBest' function in the scikit-learn Python library was used to reduce the number of features for the machine learning models to 100. The authors then compared a GP model, SVR, kernel regression, and automatic relevance determination regression (ARDR), using the features as inputs. ARDR is essentially Bayesian linear regression (Sect. 6.5), with an elliptical prior

$$p(\mathbf{w} \mid \mathbf{A}) = \mathcal{N}(\mathbf{0}, \mathbf{A}^{-1}), \quad \text{diag}(\mathbf{A}) = (\lambda_1, \ldots, \lambda_M)^T \tag{6.427}$$

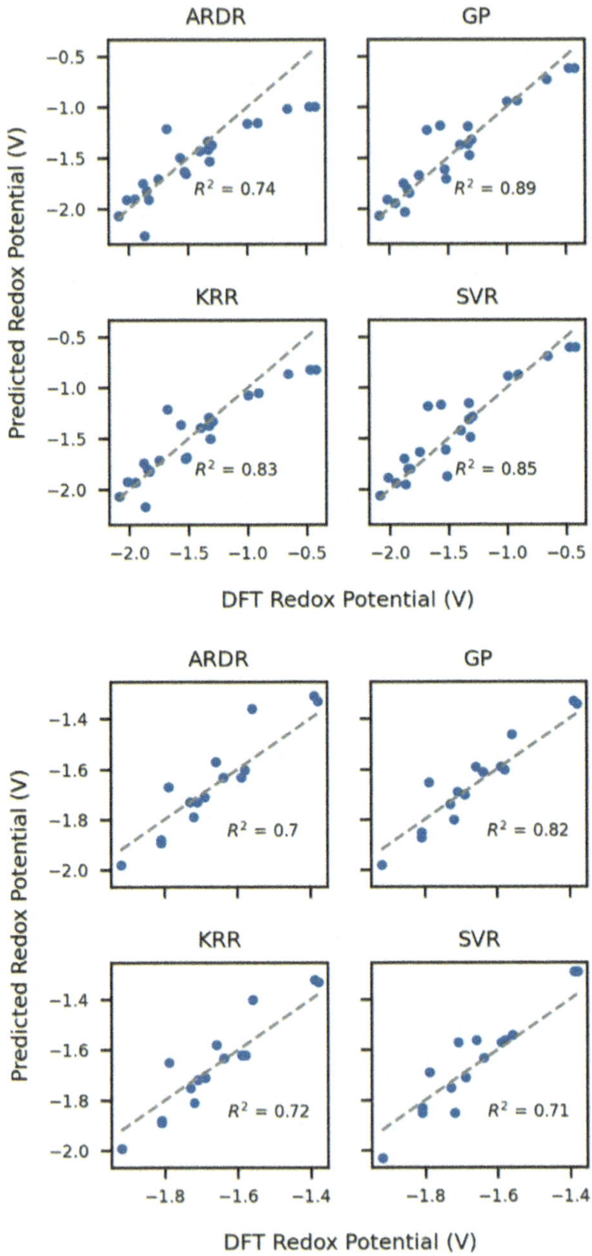

Fig. 6.4 Plots showing machine learning predictions on (top) a test set of 22 diverse phenazine derivatives with multiple types of functional groups, and (bottom) a two functional group test set of 15 phenazine derivatives. Reprinted with permission from [102] under the Creative Commons license https://creativecommons.org/licenses/by-nc-nd/4.0/

on the weights $\mathbf{w}$ rather than the spherical prior (6.33). This means that each component w_i of $\mathbf{w}$ has a zero mean and its own variance λ_i. The results on two different test sets are shown in Fig. 6.4, from which it can be seen that the GP model performs best overall.

Augmenting the training set with DFT results on 15 two-functional group derivatives was found to the accuracy on a three-functional group test set. Moreover, a descriptor related to molecular size and the partial charges was found to be the most important. From a physical point of view they concluded that the redox potential for derivatives with multiple functional groups was correlated with functional groups having either a strong electron-donating or a strong electron-withdrawing power.

6.20 Summary

In this chapter, we covered the basic principles of machine learning and introduced a number of advanced topics and methods. We presented a number of existing applications to flow batteries and outlined possible future applications. The potential for machine learning applications is vast, from systems level (which we did not cover) down to materials screening and design. Machine learning approaches are well established in the context of Li-ion batteries and fuel cells, but have received far less attention from the flow battery community thus far. The detailed descriptions in this chapter are intended to provide a clear guide to the implementation of machine learning to existing and future applications, for both newcomers and those experienced in machine learning.

Surrogate models, described in Chap. 3, can be used as replacements (or partial replacements) for complex physics-based simulations or experiments. The applications we described above are all, in one way or another, surrogate models. Particularly challenging are surrogate models for multi-variate outputs. We covered machine learning methods for this type of data in Sects. 6.10, 6.11.2 and 6.11.3. The neural networks in Sect. 6.12 can also be employed for multivariate inputs and outputs. There are, moreover, other approaches to multivariate-output surrogate modelling, and these approaches are covered in the next chapter, together with methods for time series or sequential data. Although such time-series methods are also data-driven, the nature of the data demands additional considerations.

References

1. C. Bishop, *Pattern Recognition and Machine Learning, Information Science and Statistics* (Springer, New York, 2006)
2. C.E. Rasmussen, C.K.I. Williams, *Gaussian Processes for Machine Learning* (MIT Press, Cambridge MA, USA, 2006)
3. M. Kennedy, A. O'Hagan, Predicting the output from a complex computer code when fast approximations are available. Biometrika **87**, 1–13 (2000)

4. H. Wackernagel, *Multivariate geostatistics: an introduction with applications*. (Springer Science & Business Media, 2013)
5. A.E. Gelfand, A.M. Schmidt, S. Banerjee, C.F. Sirmans, Nonstationary multivariate process modelling through spatially varying coregionalization. TEST **13**(2), 1–50 (2004)
6. S. Conti, A. O'Hagan, Bayesian emulation of complex multi-output and dynamic computer models. J. Statist. Plann. Inference **140**, 640–651 (2010)
7. T.E. Fricker, J.E. Oakley, N.M. Urban, Multivariate gaussian process emulators with nonseparable covariance structures. Technometrics **55**(1), 47–56 (2013)
8. D. Higdon, J. Gattiker, B. Williams, M. Rightley, Computer model calibration using high-dimensional output. J. Amer. Statist. Assoc. **103**, 570–583 (2008)
9. A. Narayan, C. Gittelson, D. Xiu, A stochastic collocation algorithm with multifidelity models. SIAM J. Sci. Comput. **36**(2), A495–A521 (2014)
10. M. Gerritsma, J.-B. van der Steen, P. Vos, G. Karniadakis, Time-dependent generalized polynomial chaos. J. Comput. Phys. **229**(22), 8333–8363 (2010)
11. Dongbin Xiu and George Em Karniadakis, The wiener-askey polynomial chaos for stochastic differential equations. SIAM J. Sci. Comput. **24**(2), 619–644 (2002)
12. D. Xiu, *Stochastic Collocation Methods: A Survey*. (Springer International Publishing, Cham, 2017), pp. 699–716
13. W.W. Xing, V. Triantafyllidis, A.A. Shah, P.B. Nair, N. Zabaras, Manifold learning for the emulation of spatial fields from computational models. J. Comput. Phys. **326**, 666–690 (2016)
14. L. Parussini, D. Venturi, P. Perdikaris, G.E. Karniadakis, Multi-fidelity Gaussian process regression for prediction of random fields. J. Comput. Phys. **336**(C), 36–50 (2017)
15. T.G. Kolda, B.W. Bader, Tensor decompositions and applications. SIAM Rev. **51**(3), 455–500 (2009)
16. S. Zhe, W. Xing, R.M. Kirby, Scalable high-order gaussian process regression, in *The 22nd International Conference on Artificial Intelligence and Statistics* (2019), pp. 2611–2620
17. H. Zhou, L. Li, H. Zhu, Tensor regression with applications in neuroimaging data analysis. J. Am. Stat. Assoc. **108**(502), 540–552 (2013)
18. X. Li, X. Da, H. Zhou, L. Li, Tucker tensor regression and neuroimaging analysis. Stat. Biosci. **10**(3), 520–545 (2018)
19. K. Lange, J. Chambers, W. Eddy, *Numerical Analysis for Statisticians*, vol. 2. (Springer, 1999)
20. I. Goodfellow, Y. Bengio, A. Courville, *Deep Learning*. (MIT Press, 2016). http://www.deeplearningbook.org
21. D.P. Kingma, J.B. Adam, A method for stochastic optimization (2014). arXiv:1412.6980
22. S. Hochreiter, J. Schmidhuber, Long short-term memory. Neural Comput. **9**(8), 1735–1780 (1997)
23. K. Cho, B. Van Merriënboer, D. Bahdanau, Y. Bengio, On the properties of neural machine translation: encoder-decoder approaches (2014). arXiv:1409.1259
24. P.J. Werbos, Generalization of backpropagation with application to a recurrent gas market model. Neural Netw. **1**(4), 339–356 (1988)
25. M. Schuster, K.K. Paliwal, Bidirectional recurrent neural networks. IEEE Trans. Signal Process. **45**(11), 2673–2681 (1997)
26. K. Cho, B. Van Merriënboer, C. Gulcehre, D. Bahdanau, F. Bougares, H. Schwenk, Y. Bengio, Learning phrase representations using RNN encoder-decoder for statistical machine translation (2014). arXiv:1406.1078
27. I. Sutskever, O. Vinyals, Q.V. Le, Sequence to sequence learning with neural networks, in *Advances in Neural Information Processing Systems*, vol. 27 (2014)
28. D. Bahdanau, K. Cho, Y. Bengio, Neural machine translation by jointly learning to align and translate (2014). arXiv:1409.0473
29. M.-T. Luong, H. Pham, C.D. Manning, Effective approaches to attention-based neural machine translation (2015). arXiv:1508.04025
30. A. Ruszczynski, *Nonlinear Optimization*. (Princeton University Press, 2011)
31. W.S. Torgerson, Multidimensional scaling: I. Theory and method. Psychometrika, **17**(4), 401–419 (1952)

32. B. Schölkopf, A. Smola, K.-R. Müller, Nonlinear component analysis as a kernel eigenvalue problem. Neural Comput. **10**(5), 1299–1319 (1998). (July)
33. S.T. Roweis, L.K. Saul, Nonlinear dimensionality reduction by locally linear embedding. Science **290**(5500), 2323–2326 (2000)
34. W. Xing, A.A. Shah, P.B. Nair, Reduced dimensional Gaussian process emulators of parametrized partial differential equations based on Isomap, in *Proceedings of the Royal Society of London A: Mathematical, Physical and Engineering Sciences*, vol. 471, no. 2174 (2014)
35. D. Donoho, C. Chui, R.R. Coifman, S. Lafon, Special issue: diffusion maps and wavelets diffusion maps. Appl. Comput. Harmon. Anal. **21**(1), 5–30 (2006)
36. J.B. Tenenbaum, V. De Silva, J.C. Langford, A global geometric framework for nonlinear dimensionality reduction. Science **290**(5500), 2319–2323 (2000)
37. C.K.I. Williams, On a connection between kernel PCA and metric multidimensional scaling. Mach. Learn. **46**, 11–19 (2002)
38. J. Ham, D.D. Lee, S. Mika, B. Schölkopf, A kernel view of the dimensionality reduction of manifolds, in *Proceedings of the Twenty-First International Conference on Machine Learning*. (ACM, 2004), pp. 47
39. H. Choi, S. Choi, Kernel isomap. Electron. Lett. **40**(25), 1612–1613 (2004)
40. R.R. Coifman, S. Lafon, A.B. Lee, M. Maggioni, B. Nadler, F. Warner, S.W. Zucker, Geometric diffusions as a tool for harmonic analysis and structure definition of data: diffusion maps. Proc. Natl. Acad. Sci. USA **102**(21), 7426–7431 (2005)
41. F.R.K. Chung, *Spectral Graph Theory*, vol. 92. (American Mathematical Soc., 1997)
42. R. Bellman, *Introduction to Matrix Analysis*, 2nd edn
43. B. Nadler, S. Lafon, R.R. Coifman, I.G. Kevrekidis, Diffusion maps, spectral clustering and eigenfunctions of Fokker–Planck operators, in *in Advances in Neural Information Processing Systems*, vol. 18 ed. by Y. Weiss, B. Schölkopf, J. Platt (MIT Press, Cambridge, MA, 2005), pp. 955–962
44. Z. Zhang, H. Zha, Principal manifolds and nonlinear dimensionality reduction via tangent space alignment. SIAM J. Sci. Comput. **26**(1), 313–338 (2004)
45. X. Zou, Q. Zhu, Adaptive neighborhood graph for ltsa learning algorithm without free-parameter. Int. J. Comput. Appl. **19**(4), 28–33 (2011)
46. Z. Zhang, J. Wang, H. Zha, Adaptive manifold learning. IEEE Trans. Pattern Anal. Mach. Intell. **34**(2), 253–265 (2011)
47. J. Wei, H. Peng, Y.-S. Lin, Z.-M. Huang, J.-B. Wang, Adaptive neighborhood selection for manifold learning, in *2008 International Conference on Machine Learning and Cybernetics*, vol. 1. (IEEE, 2008), pp. 380–384
48. Y. Zhan, J. Yin, Robust local tangent space alignment via iterative weighted pca. Neurocomputing **74**(11), 1985–1993 (2011)
49. H. Li, L. Teng, W. Chen, I.-F. Shen, Supervised learning on local tangent space, in *International Symposium on Neural Networks*. (Springer, 2005), pp. 546–551
50. P. Arias, G. Randall, G. Sapiro, Connecting the out-of-sample and pre-image problems in kernel methods, in *2007 IEEE Conference on Computer Vision and Pattern Recognition*. (2007), pp. 1–8. (June)
51. J.T.Y. Kwok, I.W.H. Tsang, The pre-image problem in kernel methods. IEEE Trans. Neural Netw. **15**(6), 1517–1525 (2004). (Nov)
52. S. Mika, B. Schölkopf, AJ. Smola, K.-R. Müller, M. Scholz, G. Rätsch. Kernel PCA and De-noising in feature spaces, in *Advances in Neural Information Processing Systems*, vol. 11. (Max-Planck-Gesellschaft, MIT Press, Cambridge, MA, USA, 1999), pp. 536–542. (June 1999)
53. P. Etyngier, F. Ségonne, R. Keriven, Shape priors using manifold learning techniques, in *IEEE 11th International Conference on Computer Vision, ICCV 2007*. (Rio de Janeiro, Brazil, 2007), pp. 1–8. (14–20 Oct 2007)
54. N. Thorstensen, F. Segonne, R. Keriven, *Pre-image as Karcher Mean Using Diffusion Maps: Application to Shape and Image Denoising*. (Springer, Berlin, 2009), pp. 721–732

55. X. Ma, N. Zabaras, Kernel principal component analysis for stochastic input model generation. J. Comput. Phys. **230**(19), 7311–7331 (2011)
56. B. Ganapathysubramanian, N. Zabaras, A non-linear dimension reduction methodology for generating data-driven stochastic input models. J. Comput. Phys. **227**(13), 6612–6637 (2008)
57. E.A. Nadaraya, On estimating regression. Theory of Probability & Its Applications **9**(1), 141–142 (1964)
58. C.K.I. Williams, On a connection between kernel PCA and metric multidimensional scaling. Mach. Learn. **46**(1), 11–19 (2002)
59. B. Nadler, S. Lafon, R.R. Coifman, I.G. Kevrekidis, Diffusion maps, spectral clustering and reaction coordinates of dynamical systems, in *Applied and Computational Harmonic Analysis*, vol. 21. (2006), pp. 113 – 127
60. U. von Luxburg, O. Bousquet, M. Belkin, On the convergence of spectral clustering on random samples: the normalized case, in *Learning Theory*, ed. by J. Shawe-Taylor, Y. Singer. Lecture Notes in Computer Science, vol. 3120 (Springer-Verlag, Berlin, 2004), pp.457–471
61. N. Lawrence, Probabilistic non-linear principal component analysis with gaussian process latent variable models. J. Mach. Learn. Res. **6**, 1783–1816 (2005)
62. M.E. Tipping, C.M. Bishop, Probabilistic principal component analysis. J. R. Stat. Soc. Ser. B (Stat. Methodol.) **61**(3), 611–622 (1999)
63. E.V. Bonilla, M.C. Kian, C. Williams. Multi-task gaussian process prediction, in *Advances in Neural Information Processing Systems*, vol. 20, ed. by J.C. Platt, D. Koller, Y. Singer, S.T. Roweis (2008), pp. 153–160
64. A.A. Shah, F. Yu, W.W. Xing, P.K. Leung, Machine learning for predicting fuel cell and battery polarisation and charge-discharge curves. Energy Rep. **8**, 4811–4821 (2022)
65. S. Wan, X. Liang, H. Jiang, J. Sun, N. Djilali, T. Zhao, A coupled machine learning and genetic algorithm approach to the design of porous electrodes for redox flow batteries. Appl. Energy **298**, 117177 (2021)
66. T. Li, C. Zhang, X. Li, Machine learning for flow batteries: opportunities and challenges. Chem. Sci. **13**, 4740–4752 (2022)
67. A.A. Shah, R. Tangirala, R. Singh, R.G.A. Wills, F.C. Walsh, A dynamic unit cell model for the all-vanadium flow battery. J. Electrochem. Soc. **158**(6), A671 (2011)
68. K. Hansen, G. Montavon, F. Biegler, S. Fazli, M. Rupp, M. Scheffler, O. Anatole Von Lilienfeld, A. Tkatchenko, K.-R. Muller, Assessment and validation of machine learning methods for predicting molecular atomization energies. J. Chem. Theory Comput. **9**(8), 3404–3419 (2013)
69. K. T Schütt, F. Arbabzadah, S. Chmiela, K.R. Müller, A. Tkatchenko, Quantum-chemical insights from deep tensor neural networks. Nat. Commun. **8**(1), 1–8 (2017)
70. S. Chmiela, A. Tkatchenko, H.E. Sauceda, I. Poltavsky, K.T. Schütt, K.-R. Müller, Machine learning of accurate energy-conserving molecular force fields. Sci. Adv. **3**(5), e1603015 (2017)
71. F. Brockherde, L. Vogt, L. Li, M.E. Tuckerman, K. Burke, K.-R. Müller, Bypassing the Kohn-Sham equations with machine learning. Nat. Commun. **8**(1), 1–10 (2017)
72. K. Ryczko, D.A. Strubbe, I. Tamblyn, Deep learning and density-functional theory. Phys. Rev. A **100**(2), 022512 (2019)
73. L. Li, J.C. Snyder, I.M. Pelaschier, J. Huang, U.-N. Niranjan, P. Duncan, M. Rupp, K.-R. Müller, K. Burke, Understanding machine-learned density functionals. Int. J. Quantum Chem. **116**(11), 819–833 (2016)
74. R. Nagai, R. Akashi, O. Sugino, Completing density functional theory by machine learning hidden messages from molecules. npj Comput. Mater. **6**(1), 1–8 (2020)
75. J.T. Margraf, K. Reuter, Pure non-local machine-learned density functional theory for electron correlation. Nat. Commun. **12**(1), 1–7 (2021)
76. J.A. Ellis, L. Fiedler, G.A. Popoola, N.A. Modine, J.A. Stephens, A.P. Thompson, A. Cangi, S. Rajamanickam, Accelerating finite-temperature Kohn-Sham density functional theory with deep neural networks. Phys. Rev. B **104**(3), 035120 (2021)

77. S. Dick, M. Fernandez-Serra, Machine learning accurate exchange and correlation functionals of the electronic density. Nat. Commun. **11**(1), 3509 (2020)
78. T.B. Blank, S.D. Brown, A.W. Calhoun, D.J. Doren, Neural network models of potential energy surfaces. J. Chem. Phys. **103**(10), 4129–4137 (1995)
79. S. Lorenz, A. Groß, M. Scheffler, Representing high-dimensional potential-energy surfaces for reactions at surfaces by neural networks. Chem. Phys. Lett. **395**(4–6), 210–215 (2004)
80. M. Rupp, A. Tkatchenko, K.-R. Müller, O.A. Von Lilienfeld, Fast and accurate modeling of molecular atomization energies with machine learning. Phys. Rev. Lett. **108**(5), 058301 (2012)
81. Sergei Manzhos and Tucker Carrington Jr, A random-sampling high dimensional model representation neural network for building potential energy surfaces. J. Chem. Phys. **125**(8), 084109 (2006)
82. J. Behler, M. Parrinello, Generalized neural-network representation of high-dimensional potential-energy surfaces. Phys. Rev. Lett. **98**(14), 146401 (2007)
83. X. Zheng, H. LiHong, X.J. Wang, G.H. Chen, A generalized exchange-correlation functional: the neural-networks approach. Chem. Phys. Lett. **390**(1–3), 186–192 (2004)
84. F.A. Faber, L. Hutchison, B. Huang, J. Gilmer, S.S. Schoenholz, G.E. Dahl, O. Vinyals, S. Kearnes, P.F. Riley, O.A. von Lilienfeld, Prediction errors of molecular machine learning models lower than hybrid dft error. J. Chem. Theory Comput. **13**(11), 5255–5264 (2017)
85. S. Raghunathan, U. Deva Priyakumar, Molecular representations for machine learning applications in chemistry. Int. J. Quantum Chem. **122**(7), e26870 (2022)
86. L. Wang, J. Ding, L. Pan, D. Cao, H. Jiang, X. Ding, Quantum chemical descriptors in quantitative structure-activity relationship models and their applications. Chemom. Intell. Lab. Syst. **217**, 104384 (2021)
87. W.M. Berhanu, G.G. Pillai, A.A. Oliferenko, A.R. Katritzky, Quantitative structure-activity/property relationships: the ubiquitous links between cause and effect. ChemPlusChem **77**(7), 507–517 (2012)
88. H. Hong, Q. Xie, W. Ge, F. Qian, H. Fang, L. Shi, S. Zhenqiang, R. Perkins, W. Tong, Mold2, molecular descriptors from 2d structures for chemoinformatics and toxicoinformatics. J. Chem. Inf. Model. **48**(7), 1337–1344 (2008)
89. J.L. Durant, B.A. Leland, D.R. Henry, J.G. Nourse, Reoptimization of mdl keys for use in drug discovery. J. Chem. Inf. Comput. Sci. **42**(6), 1273–1280 (2002)
90. G. Graziano, Fingerprints of molecular reactivity. Nat. Rev. Chem. **4**(5), 227–227 (2020)
91. S.A. Alsenan, I.M. Al-Turaiki, A.M. Hafez, Feature extraction methods in quantitative structure-activity relationship modeling: A comparative study. IEEE Access **8**, 78737–78752 (2020)
92. P. Leung, A.A. Shah, L. Sanz, C. Flox, J.R. Morante, Q. Xu, M.R. Mohamed, C. Ponce de León, F.C. Walsh, Recent developments in organic redox flow batteries: a critical review. J. Power Sources **360**, 243–283 (2017)
93. T.S. Schroeter, A. Schwaighofer, S. Mika, A. Ter Laak, D. Suelzle, U. Ganzer, N. Heinrich, K.-R. Müller, Estimating the domain of applicability for machine learning qsar models: a study on aqueous solubility of drug discovery molecules. J. Comput. Aided Mol. Des. **21**, 485–498 (2007)
94. A. Mauri, V. Consonni, M. Pavan, R. Todeschini, Dragon software: an easy approach to molecular descriptor calculations. Match **56**(2), 237–248 (2006)
95. S. Boobier, D.R.J. Hose, J. Blacker, B. Nguyen, Machine learning with physicochemical relationships: Solubility prediction in organic solvents and water. Nat. Commun. **11**, 11 (2020)
96. G. Klopman, H. Zhu, Estimation of the aqueous solubility of organic molecules by the group contribution approach. J. Chem. Inf. Comput. Sci. **41**, 439–45 (2001)
97. S. Kim, A. Jinich, A. Aspuru-Guzik, Multidk: a multiple descriptor multiple kernel approach for molecular discovery and its application to the discovery of organic flow battery electrolytes. J. Chem. Inf. Model. **57**, 06 (2016)
98. Q. Zhang, A. Khetan, E. Sorkun, F. Niu, A. Loss, I. Pucher, S. Er, Data-driven discovery of small electroactive molecules for energy storage in aqueous redox flow batteries. Energy Storage Mater. **47**, 167–177 (2022)

99. M.C. Sorkun, J.M. V.A. Koelman, S. Er, Pushing the limits of solubility prediction via quality-oriented data selection. iScience **24**(1), 101961 (2021)
100. O. Allam, R. Kuramshin, Z. Stoichev, B.W. Cho, S.W. Lee, S.S. Jang, Molecular structure-redox potential relationship for organic electrode materials: density functional theory-machine learning approach. Mater. Today Energy **17**, 100482 (2020)
101. H. Doan, G. Agarwal, H. Qian, M. Counihan, J. Rodriguez Lopez, J. Moore, R. Assary, Quantum chemistry-informed active learning to accelerate the design and discovery of sustainable energy storage materials. Chem. Mater. **32**(15), 6338–6346 (2020)
102. S. Ghule, S.R. Dash, S. Bagchi, K. Joshi, K. Vanka, Predicting the redox potentials of phenazine derivatives using dft-assisted machine learning. ACS Omega, **7**(14), 11742–11755 (2022)

Chapter 7
Time Series Methods and Alternative Surrogate Modelling Approaches

7.1 Introduction

As discussed in Chap. 3, one of the main applications of machine learning is in the development of surrogate models, which are computationally cheap compared to the original physics-based simulations. The most common way to develop a surrogate model is to use machine learning, but there are two main alternatives, namely *multi-fidelity* and *reduced-order* models. The first of these can rely heavily on machine learning, while the latter does not usually involve any machine learning although it does rely on data from the original model. Machine learning can be introduced when the problem is parameter dependent and/or nonlinear. Reduced-order models are considered *intrusive* in that modifications to the original model or numerical formulation are required.

In some cases, it might not be desirable to replace a model directly with machine learning. For example, if a process is well described by a set of equations, e.g., fluid flow and heat transfer within some component of an RFB, replacing the governing equations entirely with a data-driven model may lead to inaccuracies, especially when the whole flow or temperature field is required (in 1D, 2D or 3D). A particular process could be is of overriding importance in a certain applications, such as heat transport for the heat management of stacks, so rather than considering a full model incorporating all processes, focus can be placed on a well-defined sub-model.

In such cases, when the problem is relatively simple, it may be advantageous to modify the original numerical formulation in such a way as to reduce its size (what we mean by size will be discussed in Sect. 7.3). This leads to the concept of reduced-order models (ROMs) or *model order reduction*. The original application area for ROMs was linear time-invariant pure dynamical systems, but they have been extended to spatially-varying nonlinear dynamical systems by employing certain approximations. The advantages of this approach are that (a) it largely retains the original physical formulation and (b) it naturally handles spatial variations, without the need for considering multivariate targets are their intercorrelations. The

A. A. Shah et al., *New Paradigms in Flow Battery Modelling*,
Engineering Applications of Computational Methods 16,
https://doi.org/10.1007/978-981-99-2524-7_7

disadvantages are that it does not naturally handle parameter dependence or nonlinearities, so that we would only recommend its used in the case of linear or weakly-nonlinear problems for which a precise set of governing equations are known, although ROMs have been frequently applied to the quasilinear Navier-Stokes equations.

Multi-fidelity models leverage data from models of differing accuracy and associated computational burden. The term 'fidelity' can be understood to mean 'accuracy' or 'complexity'. For example, one may consider a model that incorporates all known phenomena as a *high-fidelity* model, and an alternative model that neglects or simplifies certain of these phenomena as a *low-fidelity* model. As another example, one can consider 3D and 1D models of the same process as high- and low-fidelity, respectively. Another route to developing models of different fidelity is to relax numerical settings, for example, to use larger time steps and fewer grid points. The number of fidelities is arbitrary and need not be fixed at 2.

Typically, the lower fidelity results are obtained much more rapidly. This suggests a strategy in which reliance is placed on these low-fidelity results, while the high-fidelity model is used sparingly. By combining the low- and high-fidelity data (often referred to as *data fusion*), it can be possible to make predictions at the highest fidelity for new test inputs. This involves creating surrogate models of the maps between successive fidelities. A surrogate for the lowest fidelity can also be constructed, although in some methods this is not the case, and to make predictions at a test input the original low-fidelity model result is required. Pure machine learning approaches work only with high-fidelity data, so that the *offline* process of generating data can be much more burdensome than a multi-fidelity approach, which requires (in principle) only a fraction of the same data at high-fidelity.

As also mentioned in Chap. 6, problems involving data that represents a sequence of values in time or some other ordered index are somewhat different from pure supervised machine learning problems. To explain this in detail, consider Fig. 7.1, in which two sets of data are plotted. Here, x is some index related to an ordering of the values of y. The most familiar example is $x = t$, i.e, time, which can take values either in a continuous or discrete interval. In the context of RFBs, y could be the capacity at the beginning of discharge and x could be the cycle number relating to repeated constant-current charge-discharge cycling. Another example of a time series would be words in a sentence. In this case y would be a vector equal to a one-hot encoding of a word, while x would signify the order in which the word appears in the sentence. Alternatively, in these problems we can consider x to be implicit and work only with the sequence of values y_i corresponding to some x_i, with $x_i - x_{i-1} =$ constant for all i.

In Fig. 7.1a, we have the classical setup for regression. With a good design, we can cover the entire design space (x region) of interest. The goal is to find the map underlying the data (red plus symbols). In a typical time series problem, as illustrated in Fig. 7.1b, we have data (blue plus symbols) only up the present time or present index x. The goal is the same, to find the map underlying the data, but the problem is now much more complicated since there is no indication of the shape of the curve beyond the data points, in contrast to the problem illustrated in Fig. 7.1a. The

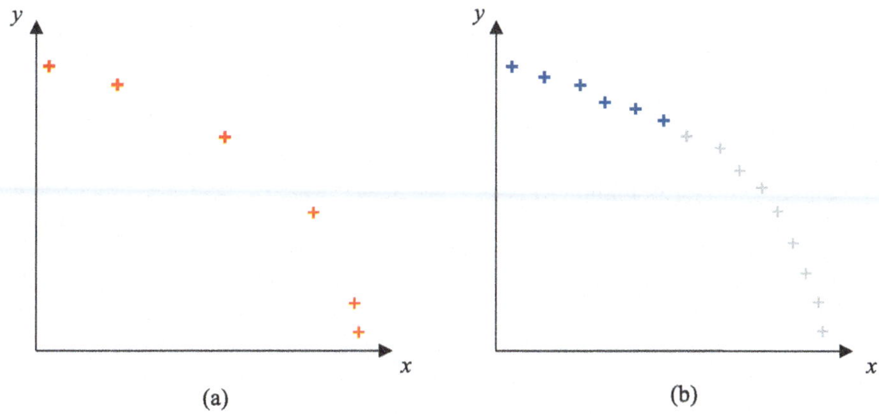

Fig. 7.1 Illustration of the difference between time series and regression analysis

difference is essentially between *interpolating* between data points, as in the first example, and *extrapolating* data, as in the second example. Particularly challenging are problems in which there is a marked change in the curve or trajectory outside the the range of x values for which data is available.

Time series problems can be tackled using classical supervised machine learning approaches, taking y values as targets and x values as inputs and assuming i.i.d. observations. In general, however, this is not recommended and a host of specialised methods for time series data have been developed. In particular, the simple supervised approach ignores the correlations between sequence values that are adjacent or close. Time series methods seek to exploit these correlations, with some sort of simplification—considering all correlations would lead to an intractable problem.

Although time series methods have not been used in flow battery modelling, as far as we are aware, they have been used extensively for Li-ion and fuel cell systems. In this chapter we introduce some of the main methods as a motivation for their adoption and application to important topics, such as RFB capacity degradation. The main omissions are Markov and hidden Markov models, along with some prominent state-space modelling approaches such as the Kalman filters.

In Sect. 7.2 we outline some prominent multi-fidelity methods, followed by reduced-order modelling in the form of proper orthogonal decomposition in Sect. 7.3. In Sect. 7.4 we introduce the time series methods and finally in Sect. 7.5 we present a case study in the use of multi-fidelity models for electrochemical systems.

7.2 Multi-fidelity Models

Multi-fidelity modelling essentially combines the results from models of different *fidelity*, loosely meaning of different complexity and, therefore, associated computational cost. The number of high-fidelity simulations should be kept to minimum, with a greater reliance on low-fidelity data, which is of lower accuracy and cheaper to obtain [1–4]. In this way, multi-fidelity models can be more time-efficient than pure machine learning approaches, which rely solely on the high-fidelity data. There can also be intermediate levels of fidelity, with a hierarchy in terms of accuracy and cost going from low to high.

The usual way to construct multi-fidelity models is to develop surrogate models for the mappings between successive fidelities [3, 5]. When predictions at new model inputs (parameters) are required, there are two choices for obtaining the output at the lowest fidelity. Either a surrogate model is developed for the mapping between the inputs and low-fidelity outputs, or the original low-fidelity model is used to generate outputs at the test inputs [6–9]. In the latter case, the multi-fidelity model is essentially a correction to the low-fidelity model.

In linear autoregression (LAR) and its variants [1, 10], it is assumed that the relationship between successive fidelities is linear. This approach was extended to a nonlinear autoregressive model (NARGP) [2] by placing GP priors over the mappings between fidelities. It can be applied to univariate as well as multivariate outputs [11]. Although NARGP increases model flexibility, it relies on the low-fidelity solution as an input to the high-fidelity GP, essentially leading to an expensive *deep GP* stucture [12]. It therefore requires sampling or other approximate Baye's techniques to perform inference.

Another approach uses a particular variant of stochastic collocation (SC) [13] based on a greedy procedure to select high-fidelity inputs in order to generate the high-fidelity outputs. A SC approximation of the high-fidelity outputs, which is basically an expansion in terms of a known basis, is obtained by using low-fidelity data corresponding to the same design points in order to approximate the basis-expansion coefficients. Low-fidelity simulations are performed each time a prediction is required, which can be a problem when the low-fidelity simulations are themselves expensive to obtain, as can often be the case.

The number of hyperparameters in the nonlinear GP-based approaches scales linearly with the dimension of the output space, which presents a problem for high-dimensional output spaces, e.g., field data. Greedy NAR [14] combines the NARGP and SC approaches using a generalised LAR model, in which the high-fidelity solution is given as linear map of the low-fidelity solution in a feature space. Although it does not rely on low-fidelity simulations to make predictions, it still suffers from the scaling problem encountered in NARGP for high-dimensional output spaces. Xing et al. [15] developed a scalable tractable alternative to NARGP for multivariate outputs, based on approximating the residuals between successive fidelities using a GP. This method does not rely on using the lower fidelity output as an input to the GP, so involves only l hyperparameters, in which l is the number of input parameters.

Below we discuss the nature of the data required for multi-fidelity models. We then outline a number of the methods mentioned above.

7.2.1 Multi-fidelity Data

Gathering the data is the first step in multi-fidelity modelling, having already decided on the definitions of the fidelities. Simulations are conducted using the original models at the different fidelities, $f = 1, \ldots, F$, with F denoting the highest fidelity. The inputs at each fidelity, which lie in some design space $\mathcal{X} \subset \mathbb{R}^l$, are labelled

$$\Xi^f = \{\boldsymbol{\xi}_n^f\}_{n=1}^{N_f} \subset \mathcal{X} \tag{7.1}$$

The corresponding outputs are labelled

$$\{\mathbf{y}_n^f\}_{n=1}^{N_f} \subset \mathbb{R}^d, \quad \{y_n^f\}_{n=1}^{N_f} \subset \mathbb{R} \tag{7.2}$$

in which $\mathbf{y}_n^f = \boldsymbol{\eta}^f(\boldsymbol{\xi}_n)$ and $y_n^f = \eta^f(\boldsymbol{\xi}_n)$, for latent functions $\boldsymbol{\eta}^f(\cdot)$ or $\eta^f(\cdot)$ in the multivariate and scalar case, respectively. Defining $\mathbf{Y}^f \in \mathbb{R}^{d \times N_f}$ ($d = 1$ for the scalar case) as a matrix containing the $\mathbf{y}_n^f$ as columns, a compact notation for the outputs is $\{\mathbf{Y}^f\}_{f=1}^F$. It is almost always assumed that the data is noiseless and that

$$\Xi^f \subset \Xi^{f-1} \tag{7.3}$$

This nested structure is essential for the majority of methods.

7.2.2 Autoregressive Models Based on Gaussian Processes

We present only the univariate case, with a relatively straightforward extension to multivariate data [11]. The general formulation for autoregressive multi-fidelity GP models is

$$\eta^f(\boldsymbol{\xi}) = g^f\left(\eta^{f-1}(\boldsymbol{\xi})\right) + \delta^f(\boldsymbol{\xi}) \tag{7.4}$$

in which $g^f(\cdot)$ is a mapping between fidelity f and $f - 1$, and $\delta^f(\cdot)$ is assumed to be a GP. A linear mapping

$$g^f(y) = cy \tag{7.5}$$

for some constant c and a GP prior over y, independent of $\delta^f(\cdot)$, leads to the *linear autoregressive* (LAR) model of Kennedy and O'Hagan [1]. This model can be implemented efficiently [10] by replacing the prior over $\eta^{f-1}(\boldsymbol{\xi})$ with the corresponding GP posterior $\eta_*^{f-1}(\boldsymbol{\xi})$, found at the previous level, assuming $\Xi^f \subset \Xi^{f-1}$. This leads to the scalar GP model with posterior (6.70) and maximum likelihood solution (6.73) in Sect. 6.7.

LAR was extended by Perdikaris et al. [2] to a nonlinear autoregressive GP (GPNAR) formulation by assuming a GP prior over g^f and absorbing $\delta^f(\boldsymbol{\xi})$ into g^f, assuming that the two GPs are independent

$$\eta^f(\boldsymbol{\xi}) = g^f\left(\boldsymbol{\xi}, \eta_*^{f-1}(\boldsymbol{\xi})\right) \tag{7.6}$$

With GPs over g^f and η^{f-1}, the posterior for $f \geq 2$ is not a GP, and neither is it tractable. Replacement of the prior by the posterior, on the other hand, relaxes the deep GP structure, and instead allows for sequential training from lower to higher fidelities. A fully deep approach developed by Cutajar et al. can be found in [16]. Since $\boldsymbol{\xi}$ and $\eta_*^{f-1}(\boldsymbol{\xi})$ belong to different spaces, a separable covariance k^f for the GP over $g^f(\boldsymbol{\xi}, \eta_*^{f-1}(\boldsymbol{\xi}))$ was assumed

$$\begin{aligned} &k^f\left(\boldsymbol{\xi}, \boldsymbol{\xi}', \eta_*^{f-1}(\boldsymbol{\xi}), \eta_*^{f-1}(\boldsymbol{\xi}')|\boldsymbol{\theta}_\xi^f, \boldsymbol{\theta}_y^f, \boldsymbol{\theta}_\delta^f\right) \\ &= k_\xi^f(\boldsymbol{\xi}, \boldsymbol{\xi}'|\boldsymbol{\theta}_\xi^f) \times k_y^f\left(\eta_*^{f-1}(\boldsymbol{\xi}), \eta_*^{f-1}(\boldsymbol{\xi}')|\boldsymbol{\theta}_y^f\right) + k_\delta^f(\boldsymbol{\xi}, \boldsymbol{\xi}'|\boldsymbol{\theta}_\delta^f) \end{aligned} \tag{7.7}$$

for covariance functions k_ξ^f, k_y^f and k_δ^f, each with their own hyperparameters $\boldsymbol{\theta}_j^f$.

Assuming $\Xi^f \subset \Xi^{f-1}$, a GP model is trained at each fidelity using Ξ^f, $\eta_*^{f-1}(\Xi^f)$ as inputs and $\mathbf{Y}^f$ as outputs, except at fidelity 1, in which case the GP model uses Ξ^1 and $\mathbf{Y}^1$ as inputs and outputs, respectively. In this notation, $\eta_*^{f-1}(\Xi^f)$ is the set of points $\{\eta_*^{f-1}(\boldsymbol{\xi}_n^f)\}_n$, which is given by the data $\mathbf{Y}^{f-1}$. For high-dimensional output spaces, the input space for these GPs when $f \geq 2$ is of dimension $d + l$, which compromises stability and accuracy when using an ARD kernel such as the SEARD in (6.64).

Making predictions requires sampling from the predictive posterior at each fidelity f for a test input $\boldsymbol{\xi}_*$

$$p(\eta^f(\boldsymbol{\xi}_*)) = \int p\left(\eta_*^f\left(\boldsymbol{\xi}_*, \eta_*^{f-1}(\boldsymbol{\xi}_*)\right)\right) p(\eta_*^{f-1}(\boldsymbol{\xi})_*) d\boldsymbol{\xi}_* \tag{7.8}$$

in which the integral can be approximated by Monte Carlo integration, and propagating the sampled outputs to the next fidelity as inputs, which is a very expensive procedure.

7.2.3 Residual Gaussian Process Model

The *residual Gaussian process* (ResGP) method [15] also assumes that $\Xi^f \subset \Xi^{f-1}$. Rather than the concatenating structure of GPNAR, the high-fidelity data is decomposed as

$$\eta^f(\boldsymbol{\xi}) = r^1(\boldsymbol{\xi}) + \cdots + r^f(\boldsymbol{\xi}) = \sum_{k=1}^{f} r^k(\boldsymbol{\xi}) \tag{7.9}$$

involving residuals $r^k(\boldsymbol{\xi}) = \eta^k(\boldsymbol{\xi}) - \eta^{k-1}(\boldsymbol{\xi})$ for fidelities $k = 2, \ldots, F$ and with $r^1(\boldsymbol{\xi}) = \eta^1(\boldsymbol{\xi})$. Independent GP priors

$$r^f(\boldsymbol{\xi}) \sim \mathcal{GP}(0, k^f(\boldsymbol{\xi}, \boldsymbol{\xi}'|\boldsymbol{\theta}^f) + (\sigma^f)^2) \tag{7.10}$$

are assumed for each residual function, so that

$$\eta^F(\boldsymbol{\xi}) \sim \mathcal{GP}\left(0, \sum_{f=1}^{F} k^f(\boldsymbol{\xi}, \boldsymbol{\xi}'|\boldsymbol{\theta}^f) + (\sigma^f)^2\right) \tag{7.11}$$

At fidelity 1, the data Ξ^1, $\mathbf{Y}^1$ is used to find the maximum likelihood solution (6.73) for the hyperparameters $\{\sigma^1, \boldsymbol{\theta}^1\}$, with predictive posterior (6.70) used for $r^1(\boldsymbol{\xi})$. For $f = 2, \ldots, F$, the data Ξ^f, $\mathbf{R}^f := \mathbf{Y}^f - \mathbf{Y}^{f-1}_{\mathbf{e}_f \cap \mathbf{e}_{f-1}}$ is then used to obtain estimates of $\{\sigma^f, \boldsymbol{\theta}^f\}$, which are placed in the predictive posterior (6.70) for $r^f(\boldsymbol{\xi})$. In this notation

$$\mathbf{Y}^{f-1}_{\mathbf{e}_f \cap \mathbf{e}_{f-1}} \tag{7.12}$$

denotes the outputs at fidelity $f-1$ corresponding to inputs $\Xi^f \subset \Xi^{f-1}$. The fidelity F posterior has the following tractable form

$$\begin{aligned}
\eta^F(\boldsymbol{\xi}) &\sim \mathcal{N}\left(\mu^F(\boldsymbol{\xi}), v^F(\boldsymbol{\xi})\right) \\
\mu^F(\boldsymbol{\xi}) &= \sum_{f=1}^{F} \mathbf{k}^f(\boldsymbol{\xi})^T(\mathbf{K}^f + (\sigma^f)^2\mathbf{I})^{-1}\mathbf{R}^f, \\
v^F(\boldsymbol{\xi}) &= \sum_{f=1}^{F} [k^f(\boldsymbol{\xi}, \boldsymbol{\xi}|\boldsymbol{\theta}^f) - (\mathbf{k}^f(\boldsymbol{\xi}))^T(\mathbf{K}^f + (\sigma^f)^2\mathbf{I})^{-1}\mathbf{k}^f(\boldsymbol{\xi})], \\
\mathbf{k}^f(\boldsymbol{\xi}) &= [k^f(\boldsymbol{\xi}, \boldsymbol{\xi}_1|\boldsymbol{\theta}^f), \ldots, k^f(\boldsymbol{\xi}, \boldsymbol{\xi}_{N_f}|\boldsymbol{\theta}^f)]^T,
\end{aligned} \tag{7.13}$$

in which $[\mathbf{K}^f]_{ij} = k^f(\boldsymbol{\xi}_i, \boldsymbol{\xi}_j|\boldsymbol{\theta}^f)$ is the covariance matrix at fidelity f and $\mathbf{k}^f(\boldsymbol{\xi})$ denotes the vector of covariances between $r^f(\boldsymbol{\xi})$ and points in $\mathbf{R}^f$.

7.2.4 *Stochastic Collocation for Multi-fidelity Modelling*

Consider a two-fidelity setting with models defined by mappings $\boldsymbol{\eta}^f : \mathcal{X} \to \mathbb{R}^d$, $f \in \{1, 2\}$. Here, we will present the stochastic collocation (SC) approach of [13] for the multivariate case since it involves no additional effort. The approach consists of using N data points $\mathbf{y}_n^2 = \boldsymbol{\eta}^2(\boldsymbol{\xi}_n)$ to find an approximating function $\boldsymbol{\eta}^2(\boldsymbol{\xi})$ of the high-fidelity mapping in the form

$$\boldsymbol{\eta}^2(\boldsymbol{\xi}) = \sum_{n=1}^{M} c_n(\boldsymbol{\xi})\mathbf{y}_n^2 \tag{7.14}$$

in which $c_n(\boldsymbol{\xi})$ are undetermined coefficients depending on $\boldsymbol{\xi}$. This is clearly an interpolation, with coefficients (usually assumed to be polynomials) approximated using data at carefully selected collocation points, e.g., Gauss-Lobatto-Legendre points [17]. Obtaining the high-fidelity data $\mathbf{y}_n^2$, on the other hand, is deemed to be too costly, so instead the fidelity-1 solutions $\mathbf{y}_n^1 = \boldsymbol{\eta}^1(\boldsymbol{\xi}_n)$ are used for approximating the $c_n(\boldsymbol{\xi})$. Given $\mathbf{y}^1 = \boldsymbol{\eta}^1(\boldsymbol{\xi})$ for an arbitrary $\boldsymbol{\xi}$, the $c_n(\boldsymbol{\xi})$ are determined from the following set of conditions

$$\sum_{n=1}^{N} c_n(\boldsymbol{\xi})(\mathbf{y}_n^1)^T\mathbf{y}_n^1 = (\mathbf{y}^1)^T\mathbf{y}_n^1, \quad \forall \mathbf{y}_n^1 \tag{7.15}$$

This can be interpreted as either a projection of $\boldsymbol{\eta}^1(\boldsymbol{\xi})$ onto the space of all linear combinations of the low-fidelity data, i.e., span$\{\mathbf{y}_n^1\}$, or as an interpolation. As demonstrated in [14], it is also equivalent to a GP model using a linear kernel, i.e., Bayesian linear regression.

A certain greedy point selection strategy was used in [13] in order to generate the high-fidelity observations $\mathbf{y}_n^2$. It relies on sampling inputs from $\mathcal{X}$ at random to obtain N fidelity-1 results $\mathbf{y}_n^1$ that are used to construct an $N \times N$ *Gramian matrix* $\mathbf{G}^1$

$$[G^1]_{l,n} = (\mathbf{y}_l^1)^T\mathbf{y}_n^L, \quad l, n = 1, \ldots, N \tag{7.16}$$

The collocation points are selected using a sampling refinement. An LU decomposition is performed on $\mathbf{G}^1$ with complete pivoting. The low-fidelity data is then ordered in a certain manner by calculating a matrix $\mathbf{Q}$ that satisfies

$$\mathbf{P}\mathbf{G}^L\mathbf{Q} = \mathbf{L}\mathbf{U} \tag{7.17}$$

in which $\mathbf{L}$ and $\mathbf{U}$ are lower and upper triangular matrices, and $\mathbf{P}$ and $\mathbf{Q}$ are reordering permutation matrices for the rows and columns of $\mathbf{G}$, respectively. This defines $m \ll N$ sampling points with associated indices $n_1, \ldots, n_m$ for calculating the $c_n(\boldsymbol{\xi})$ and for high-fidelity predictions at a test input $\boldsymbol{\xi} \in \mathcal{D}$. The latter are given by (7.14) at the sampling points

$$\mathbf{y}^H(\boldsymbol{\xi}) = \sum_{i=1}^{m} c_{n_i}(\boldsymbol{\xi})\mathbf{y}^H(\boldsymbol{\xi}_{n_i}) \tag{7.18}$$

The $c_i(\boldsymbol{\xi})$ are obtained from a least-squares fit by projection onto the low-fidelity data

$$\begin{pmatrix} G^L_{i_1,i_1} & \cdots & G^L_{i_1,i_m} \\ \vdots & \ddots & \vdots \\ G^L_{i_1,i_m} & \cdots & G^L_{i_m,i_m} \end{pmatrix} \begin{pmatrix} c_{n_1}(\boldsymbol{\xi}) \\ \vdots \\ c_{n_m}(\boldsymbol{\xi}) \end{pmatrix} = \begin{pmatrix} k\left(\mathbf{y}^L(\boldsymbol{\xi}), \mathbf{y}^L(\boldsymbol{\xi}_{i_1})\right) \\ \vdots \\ k\left(\mathbf{y}^L(\boldsymbol{\xi}), \mathbf{y}^L(\boldsymbol{\xi}_{i_m})\right) \end{pmatrix} \tag{7.19}$$

The inputs ξ_{n_i} in (7.18), which are used to generate the high-fidelity data, are chosen according to a *weak greedy* procedure. To explain the selection, we consider a general low-fidelity model

$$u^L(\boldsymbol{\xi}) : \mathcal{D} \to \mathcal{V}^L \tag{7.20}$$

in a low-fidelity Hilbert space $\mathcal{V}^L$ with inner product $\langle \cdot, \cdot \rangle^L$. The distance between a subspace $\mathcal{W} \subset \mathcal{V}^L$ and any function $v(\boldsymbol{\xi}) \in \mathcal{V}^L$ is taken to be the infimum (greatest lower bound) of the distance between v and $\mathcal{W}$

$$d^L(v, W) = \inf_{w \in \mathcal{W}} ||(I - P_{\mathcal{W}})v||^L \tag{7.21}$$

in which we use the metric induced by the inner product $\|\cdot\|^L = \langle \cdot, \cdot \rangle^L$. Here, $P_{\mathcal{W}}$ is the orthogonal projection operator onto $\mathcal{W}$, while I is an identity operator. For the case $u(\boldsymbol{\xi}) = \mathbf{y}^L(\boldsymbol{\xi})$ and $\mathcal{V}^L \subset \mathbb{R}^d$, the spaces are Euclidean, which are finite-dimensional Hilbert spaces using the standard inner product.

In a class of methods referred to as *radial basis methods* (RBMs), usually for single-fidelity problems involving an output $u(\boldsymbol{\xi}) : \mathcal{D} \to \mathcal{V}$, a popular greedy approach selects R query points $\boldsymbol{\xi}_1, \ldots, \boldsymbol{\xi}_R$ to generate a set of basis vectors $u(\boldsymbol{\xi}_1), \ldots, u(\boldsymbol{\xi}_R)$ from the functional manifold $\{u(\boldsymbol{\xi}) : \boldsymbol{\xi} \in \mathcal{D}\}$. This procedure is carried in such a way as to minimise the distance as we have defined it above between $\{u(\boldsymbol{\xi}) : \boldsymbol{\xi} \in \mathcal{D}\}$ and $\text{span}\{u(\boldsymbol{\xi}_1), \ldots, u(\boldsymbol{\xi}_R)\} \subset \mathcal{V}$.

The selection of the collocation points follows a similar route, but due to the cost of obtaining the high-fidelity data, the low-fidelity model is instead used to select these points. The selection proceeds iteratively, successively adding points $\boldsymbol{\xi}_*$ to a currently optimal set Γ_n at some n, by maximising the distance between $u^L(\boldsymbol{\xi}_*)$ and $\mathcal{U}^L(\Gamma_n) = \text{span}\{u(\boldsymbol{\xi}) \mid \boldsymbol{\xi} \in \Gamma_n\}$

$$\boldsymbol{\xi}_* = \text{argmax}_{\boldsymbol{\xi} \in \overline{\mathcal{D}}} d^L(u^L(\boldsymbol{\xi}), \mathcal{U}^L(\Gamma_n)), \quad \Gamma_{n+1} = \Gamma_n \cup \{\boldsymbol{\xi}_*\}, \quad \overline{\mathcal{D}} \subseteq \mathcal{D}, \quad \Gamma_0 = \emptyset \tag{7.22}$$

Carrying out this procedure in a continuous space $\overline{\mathcal{D}} = \mathcal{D}$ is not computationally practical by virtue of the potentially large volume of queries to u^L along with the absence of an expression for $\nabla_{\boldsymbol{\xi}} u^L$, which would lead to costly numerical approximations. On the other hand, it can be carried out for a discrete finite-cardinality set $\overline{\mathcal{D}} \subset \mathcal{D}$.

Narayan et al. [13] proved that the LU decomposition (7.17) of $\mathbf{G}^L$ to obtain $n_1, \ldots, n_m$, and $\Gamma_m = \{\boldsymbol{\xi}_{n_i}\}$ solves the optimisation problem (7.22) on the finite-cardinality set $\overline{\mathcal{D}} = \{\boldsymbol{\xi}_n\}_{n=1}^M$, assuming that $\overline{\mathcal{D}}$ is sufficiently dense that all important variations in the high-fidelity output are captured.

7.3 Reduced Order Models

Reduced order models (ROMs) involve the modification a system of differential equations (the model) or its numerical formulation, rather than replacing the entire model. In this sense, since they alter the original model, they are known as *intrusive* methods. The basic idea of ROMs is to reduce the 'size' of the original problem by restricting solutions to a *subspace* of the original solution space. Characterising this subspace *via* a basis is the key to applying the ROM method. The archetypal ROM is based on *proper orthogonal decomposition* (POD), which uses a set of solutions computed from the full model in order to construct the basis.

We first introduce some notation. $\mathcal{D} \subset \mathbb{R}^L$, $L = 1, 2, 3$, is a bounded, regular domain and $\mathbf{x} = (x_1, \ldots, x_L)$ denotes a point in $\mathcal{D}$. $t \in [0, T]$ is a time interval of interest and as usual $\boldsymbol{\xi} \in \mathcal{X} \subset \mathbb{R}^l$ is vector of parameters that are contained in the system of differential equations. $\mathcal{H}$ is used to denote a separable Hilbert space with corresponding inner product $(\cdot, \cdot)_{\mathcal{H}}$ and norm $|| \cdot ||_{\mathcal{H}}$ induced by the inner product. In most cases $\mathcal{H} = L^2(\mathcal{D})$, i.e., the space of square integrable (equivalence classes of) functions with the inner product

$$(v, v')_{L^2(\mathcal{D})} = \int_{\mathcal{D}} v(\mathbf{x})v'(\mathbf{x})d\mathbf{x} \quad || \cdot ||_{L^2(\mathcal{D})} = \sqrt{(v, v')_{L^2(\mathcal{D})}} \tag{7.23}$$

defined in Chap. 3 but reproduced here for convenience. The space $L^2(0, T; \mathcal{H})$ is defined by functions $u(\mathbf{x}, t)$ such that $t \mapsto u(\cdot, t; \boldsymbol{\xi})$ is a measurable map, from the interval $(0, T)$ to the space $\mathcal{H}$, with the norm

$$\|u\|_{L^2(0,T;\mathcal{H})} := \int_0^T ||u(\cdot, t; \boldsymbol{\xi})||_{\mathcal{H}} dt < \infty \tag{7.24}$$

7.3.1 Discretisations and Galerkin Projections onto a Subpsace

For the purposes of illustration, we restrict our attention to a parameter-dependent parabolic partial differential equation, in which the dependent variable is $u(\mathbf{x}, t; \boldsymbol{\xi})$

$$\begin{aligned} \partial_t u + \mathcal{L}(\boldsymbol{\xi})u + \mathcal{N}(\boldsymbol{\xi})u &= g(\mathbf{x}; \boldsymbol{\xi}) \quad (\mathbf{x}, t) \in \mathcal{D} \times (0, T] \\ u(\mathbf{x}, 0; \boldsymbol{\xi}) &= u_0(\mathbf{x}; \boldsymbol{\xi}) \quad \mathbf{x} \in \mathcal{D} \end{aligned} \tag{7.25}$$

The model is completed by a set of boundary conditions. $\mathcal{L}(\boldsymbol{\xi})$ and $\mathcal{N}(\boldsymbol{\xi})$ respectively denote linear and nonlinear operators that are dependent on the parameters. The full dependence on $\boldsymbol{\xi}$ is *via* these operators, and/or the source $g(\mathbf{x}; \boldsymbol{\xi})$ and/or the initial and boundary conditions.

For each ξ, it is assumed that $u \in L^2(0, T; \mathcal{H})$, so that $u(\cdot, t; \xi) \in \mathcal{H}$ for each $t \in (0, T)$. Using finite differences, finite volumes or nodal finite elements, spatial discretization of (7.25) leads to an ODE system (semi-discrete problem)

$$\dot{\mathbf{u}}(t; \xi) = \mathbf{A}(\xi)\mathbf{u}(t; \xi) + \mathbf{f}(\mathbf{u}(t; \xi); \xi), \quad \mathbf{u}(0; \xi) = \mathbf{u}_0(\xi) \tag{7.26}$$

in which $\mathbf{u}(t; \xi) = (u_1(t; \xi), \ldots, u_d(t; \xi))^T$, known as the *solution vector*. d represents the degrees of freedom in the numerical formulation, which corresponds to the number of grid points in a finite-difference approximation or the number of basis functions using a nodal finite-element basis. In (7.26), $\mathbf{A}(\xi) \in \mathbb{R}^{d\times d}$ is essentially a discretisation of the term $\mathcal{L}(\xi)u$, while the nonlinear term $\mathbf{f}(\mathbf{u}(t; \xi); \xi) \in \mathbb{R}^d$ is a discretisation of the terms $\mathcal{N}(\xi)u$, $g(\mathbf{x}; \xi)$, and possibly the conditions at the boundary. The form of $\mathbf{A}(\xi)$ is dictated by the manner in which $\mathcal{L}(\xi)$ depends on ξ. In simple cases we obtain a convenient affine form

$$\mathbf{A}(\xi) = \sum_i c_i(\xi)\mathbf{A}_i \tag{7.27}$$

with known functions $c_i(\xi)$ and constant matrices $\mathbf{A}_i$.

The exact correspondence between $\mathbf{u}(t; \xi)$ and $u(\mathbf{x}, t; \xi)$ as well as the forms of $\mathbf{A}(\xi)$ and $\mathbf{f}(\mathbf{u}; \xi)$ depend on the discretisation method, as does the incorporation of the boundary conditions. When using finite differences, the problem (7.25) is solved in its original form, with boundary conditions entering through $\mathbf{f}(\mathbf{u}; \xi)$. In a finite-element (FE) formulation, a weak formulation of (7.25) is instead solved, obtained by multiplying (7.25) by test functions $v \in \mathcal{H}$, or functions in a dense subspace $\mathcal{V}$ of $\mathcal{H}$. The boundary conditions in the finite-element case are incorporated in $\mathbf{f}$, or within the definition of $\mathcal{H}$.

In all discretisations (assuming a nodal finite-element basis), the coefficients $u_i(t; \xi)$ of $\mathbf{u}(t; \xi) \in \mathbb{R}^d$ are approximations to $u(\mathbf{x}^{(i)}, t; \xi)$, in which $\mathbf{x}^{(i)} \in \overline{\mathcal{D}}$, $i = 1, \ldots, d$, are spatial locations (that define the discretisation of the solution domain grid points, nodes or volume centres). Solution of the semi-discrete problem (7.26) yields solutions vectors

$$\mathbf{u}^{(i)}(\xi) := \mathbf{u}(t^{(i)}; \xi), \quad t^{(i)} = 1, \ldots, M \tag{7.28}$$

which are referred to as *snapshots*.

A *Galerkin projection* approximates (7.26) in a low-dimensional subspace $\mathcal{S}(\xi) \subset \mathbb{R}^d$ for each fixed $\xi \in \mathcal{X}$. To the subspace $\mathcal{S}(\xi)$ we assign an orthonormal basis $\mathbf{v}_j(\xi) \in \mathbb{R}^d$, $j = 1, \ldots, r$, with $\dim(\mathcal{S}(\xi)) = r \ll d$. It is important to note that the subspace is specific to the particular input. An approximation of $\mathbf{u}$ in the space $\text{span}(\mathbf{v}_1(\xi), \ldots, \mathbf{v}_r(\xi)) = \mathcal{S}$ is given by

$$\mathbf{u}_r(t; \xi) = \sum_{j=1}^{r} a_j(t; \xi)\mathbf{v}_j(\xi) = \mathbf{V}_r(\xi)\mathbf{a}(t; \xi) \tag{7.29}$$

in which $\mathbf{V}_r(\boldsymbol{\xi}) = [\mathbf{v}_1(\boldsymbol{\xi}) \ldots \mathbf{v}_r(\boldsymbol{\xi})]$ and $\mathbf{a} = (a_1(t;\boldsymbol{\xi}), \ldots, a_r(t;\boldsymbol{\xi}))^T$ is a new solution vector associated with the low-dimensional subspace. Replacing $\mathbf{u}$ with $\mathbf{u}_r$ in (7.26), a Galerkin projection onto $\{\mathbf{v}_i(\boldsymbol{\xi})\}$ yields

$$\dot{\mathbf{a}}(t;\boldsymbol{\xi}) = \mathbf{A}_r(\boldsymbol{\xi})\mathbf{a}(t;\boldsymbol{\xi}) + \mathbf{f}_r(\mathbf{a}(t;\boldsymbol{\xi});\boldsymbol{\xi}), \quad \mathbf{a}(0;\boldsymbol{\xi}) = \mathbf{V}_r(\boldsymbol{\xi})^T\mathbf{u}_0(\boldsymbol{\xi}) \tag{7.30}$$

in which

$$\mathbf{A}_r(\boldsymbol{\xi}) := \mathbf{V}_r(\boldsymbol{\xi})^T\mathbf{A}(\boldsymbol{\xi})\mathbf{V}_r(\boldsymbol{\xi}), \quad \mathbf{f}_r(\mathbf{a}(t;\boldsymbol{\xi});\boldsymbol{\xi}) := \mathbf{V}_r(\boldsymbol{\xi})^T\mathbf{f}(\mathbf{V}_r(\boldsymbol{\xi})\mathbf{a}(t;\boldsymbol{\xi});\boldsymbol{\xi}) \tag{7.31}$$

Thus, the original system (7.26) of d equations is replaced by the system (7.30) of r equations for coefficients $a_i(t;\boldsymbol{\xi})$. The remaining task is to construct the basis $\{\mathbf{v}_j(\boldsymbol{\xi})\}_{j=1}^r$. By far the most common method uses the snapshots $\{\mathbf{u}^{(i)}(\boldsymbol{\xi})\}_{i=1}^M$ and is called proper orthogonal decomposition (POD).

7.3.2 Proper Orthogonal Decomposition via Karhunen-Loeve Theory

POD is confusingly motivated and/or presented in different ways in the literature (under different names), including as an error minimisation and as a variance maximisation, in both a continuous and a discrete form [18]. In order to derive POD and explain these various approaches, let us start by assuming that $u(\mathbf{x}, t;\boldsymbol{\xi})$ is a sample path of a zero-mean random field indexed by $(\mathbf{x}, t)$ [19–21]. The underlying probability space is $(\Omega, \mathcal{A}, p)$, for a sample space Ω, an event space $\mathcal{A}$ and a probability measure p. The goal is to construct a basis for $u(\mathbf{x}, t;\boldsymbol{\xi})$, $(\mathbf{x}, t) \in \mathcal{D} \times [0, T]$ in some optimal sense, to be defined, for each fixed $\boldsymbol{\xi} \in \mathcal{X}$. This is achieved in POD by using the ensemble of continuous snapshots $\{u(\mathbf{x}; t^{(j)}, \boldsymbol{\xi})\}_{j=1}^M$, $\mathbf{x} \in \mathcal{D}$, which are analogues of the discrete-level snapshots $\mathbf{u}^{(i)}(\boldsymbol{\xi})$. The following assumptions are required to apply *Karhunen-Loéve* (KL) theory [22], which is central to constructing the basis

1. $u(\mathbf{x}, t;\boldsymbol{\xi})$ is continuous in a quadratic mean (q.m.) sense, with respect to $\mathbf{x}$
2. $u(\mathbf{x}, t;\boldsymbol{\xi})$ is stationary with respect to t

Assumption 2 implies that the spatial autocovariance function for $u(\mathbf{x}, t;\boldsymbol{\xi})$ takes on the following simplified form

$$\mathbb{E}\left[u(\mathbf{x}, t;\boldsymbol{\xi})u(\mathbf{x}', t;\boldsymbol{\xi})\right] = C(\mathbf{x}, \mathbf{x}';\boldsymbol{\xi}), \quad \mathbf{x}, \mathbf{x}' \in \mathcal{D} \tag{7.32}$$

At a fixed $t \in [0, T]$, $u(\mathbf{x}, t;\boldsymbol{\xi})$ can be regarded as a single-parameter random field, with index $\mathbf{x} \in \mathcal{D}$ [21]. Realisations or sample paths of this field, which are generated for a fixed $\omega \in \Omega$, are non-random (deterministic) functions $u(\cdot, t;\boldsymbol{\xi}) : \mathcal{D} \to \mathbb{R}$. Assumption 1 implies that $u(\cdot, t;\boldsymbol{\xi}) \in L^2(\mathcal{D})$, $\forall t \in [0, T]$, and therefore that $u(\mathbf{x}, t;\boldsymbol{\xi}) \in L^2(0, T; L^2(\mathcal{D}))$.

KL theory [22] states that for a fixed t, $u(\mathbf{x}, t; \boldsymbol{\xi})$ is the q.m. limit of the following sequence of partial sums for some set of undetermined coefficients $a_i(t; \boldsymbol{\xi})$

$$u(\mathbf{x}, t; \boldsymbol{\xi}) = \lim_{m\to\infty} \sum_{i=1}^{m} a_i(t; \boldsymbol{\xi}) v_i(\mathbf{x}; \boldsymbol{\xi}) \tag{7.33}$$

in which $v_i(\mathbf{x}; \boldsymbol{\xi})$ are deterministic functions that form an orthonormal basis for $L^2(\mathcal{D})$. By Markovs' inequality, a corollary of the KL theorem is that the partial sums in (7.33) also converge in probability to $u(\mathbf{x}, t; \boldsymbol{\xi})$. Moreover, convergence in quadratic mean (in the $L^2(\mathcal{D})$ norm) implies convergence in mean ($L^1(\mathcal{D})$ norm). It is important to note that if we adopt an approximation of $u(\mathbf{x}, t; \boldsymbol{\xi})$ based the partial sum defined (7.33), the randomness will enter only through time t.

The $v_i(\mathbf{x}; \boldsymbol{\xi})$, called *POD modes*, are given by the eigenfunctions of an integral operator with kernel $C(\mathbf{x}, \mathbf{x}'; \boldsymbol{\xi})$

$$\mathcal{C} v_i(\mathbf{x}; \boldsymbol{\xi}) := \int_{\mathcal{D}} C(\mathbf{x}, \mathbf{x}'; \boldsymbol{\xi}) v_i(\mathbf{x}'; \boldsymbol{\xi}) d\mathbf{x}' = \lambda_i'(\boldsymbol{\xi}) v_i(\mathbf{x}; \boldsymbol{\xi}) \quad i \in \mathbb{N} \tag{7.34}$$

with corresponding non-negative, real eigenvalues $\lambda_i'(\boldsymbol{\xi}) > \lambda_{i+1}'(\boldsymbol{\xi})$ $\forall i \in \mathbb{N}$. Moreover

$$\mathbb{E}[a_i(t; \boldsymbol{\xi})] = 0, \quad \mathbb{E}[a_i(t; \boldsymbol{\xi}) a_j(t; \boldsymbol{\xi})] = \lambda_i'(\boldsymbol{\xi}) \delta_{ij} \tag{7.35}$$

meaning that the $a_i(t; \boldsymbol{\xi})$ are mutually uncorrelated random processes indexed by t, since t is arbitrary. Expectations with respect to the underlying probability measure can be replaced by time averages, under the assumption of *ergodicity*

$$\mathbb{E}[X] = \int_{\Omega} X(\omega) p(d\omega) = \int_t X(t) dt \tag{7.36}$$

for any process X indexed by t.

The basis $\{v_i(\mathbf{x}; \boldsymbol{\xi})\}_{i\in\mathbb{N}}$ has two equivalent interpretations. For any orthonormal basis $\{\varphi_i\}_{i=1}^{\infty} \subset L^2(\mathcal{D})$, the $v_i(\mathbf{x}; \boldsymbol{\xi})$ satisfy the following variance maximisation property

$$\sum_{i=1}^{r} \mathbb{E}[(u, v_i)^2] = \sum_{i=1}^{r} \lambda_i'(\boldsymbol{\xi}) > \sum_{i=1}^{r} \mathbb{E}[(u, \varphi_i)^2], \quad \forall r \in \mathbb{N} \tag{7.37}$$

Equivalently, for any arbitrary orthonormal basis $\{\varphi_i\}_{i=1}^{\infty} \subset L^2(\mathcal{D})$, we have an error minimisation property

$$\mathbb{E}\left[\|u - \sum_{i=1}^{r} a_i v_i\|^2 \right] \leq \mathbb{E}\left[\|u - \sum_{i=1}^{r} a_i \varphi_i\|^2 \right], \quad \forall \{\varphi_i\}_{i=1}^{\infty} \subset L^2(\mathcal{D}) \tag{7.38}$$

with the expectations given by $\mathbb{E}[||\cdot||^2] = ||\cdot||^2_{L^2(0,T;L^2(\mathcal{D}))}$ by assuming ergodicity.

In practice, the continuous problem (7.34) requires a numerical approximation. We first define a sample-based approximation of the covariance function $C(\mathbf{x}, \mathbf{x}'; \boldsymbol{\xi})$ from a matrix of the solution vectors

$$\mathbf{U}(\boldsymbol{\xi}) := [\mathbf{u}^{(1)}(\boldsymbol{\xi}) \ldots \mathbf{u}^{(m)}(\boldsymbol{\xi})] \tag{7.39}$$

The sample (spatial) covariance matrix is then given by

$$\mathbf{C}(\boldsymbol{\xi}) = \mathbf{U}(\boldsymbol{\xi})\mathbf{U}(\boldsymbol{\xi})^T \approx \mathbb{E}[\mathbf{u}(t; \boldsymbol{\xi})\mathbf{u}(t; \boldsymbol{\xi})^T] \tag{7.40}$$

which is a discrete form of $C(\mathbf{x}, \mathbf{x}'; \boldsymbol{\xi})$.

Defining equally-spaced quadrature points $\{\mathbf{x}^{(i)}\}_{i=1}^d$ and $\{t^{(j)}\}_{j=1}^M$ for the integration, we can now employ a simple mid-point rule to approximate (7.34)

$$\mathbf{C}(\boldsymbol{\xi})\mathbf{v}_j(\boldsymbol{\xi}) = \lambda_j(\boldsymbol{\xi})\mathbf{v}_j(\boldsymbol{\xi}) \tag{7.41}$$

in which the POD modes $\mathbf{v}_j(\boldsymbol{\xi})$ are discrete-level analogues of the eigenfunctions $v_j(\mathbf{x}; \boldsymbol{\xi})$. Problem (7.41) is precisely a principal component analysis (PCA) (Sect. 6.14.1) for the eigenvectors $\mathbf{v}_j(\boldsymbol{\xi}) \in \mathbb{R}^d$ and corresponding eigenvalues, arranged according to $\lambda_j(\boldsymbol{\xi}) \geq \lambda_{j+1}(\boldsymbol{\xi})$, $j = 1, \ldots, d-1$. Thus, a PCA analysis on the covariance matrix formed from the snapshot matrix leads to the required basis $\mathbf{v}_j(\boldsymbol{\xi})$ in (7.29).

Alternative interpolations for (7.34) are possible. For example, in a FE formulation $u(\mathbf{x}, t; \boldsymbol{\xi})$ and $v_i(\mathbf{x}; \boldsymbol{\xi})$ are approximated in an FE basis, such as piecewise linear functions, with $\{\psi_i(\mathbf{x})\}_{i=1}^d \subset L^2(\mathcal{D})$. This leads to

$$\mathbf{C}(\boldsymbol{\xi})\mathbf{M}\mathbf{v}_i(\boldsymbol{\xi}) = \lambda_i(\boldsymbol{\xi})\mathbf{v}_i(\boldsymbol{\xi}) \tag{7.42}$$

in which $\mathbf{M}$ is the mass matrix, having components $M_{ij} = (\psi_i(\mathbf{x}), \psi_j(\mathbf{x}))$. Setting $\overline{\mathbf{v}}(\boldsymbol{\xi}) = \mathbf{M}^{1/2}\mathbf{v}(\boldsymbol{\xi})$ leads to

$$\mathbf{M}^{1/2}\mathbf{C}(\boldsymbol{\xi})\mathbf{M}^{1/2}\overline{\mathbf{v}}(\boldsymbol{\xi}) = \lambda(\boldsymbol{\xi})\overline{\mathbf{v}}(\boldsymbol{\xi}) \tag{7.43}$$

The eigenpairs of $\mathbf{M}^{1/2}\mathbf{C}(\boldsymbol{\xi})\mathbf{M}^{1/2}$, which we can label $\overline{\mathbf{v}}_i(\boldsymbol{\xi})$, $\lambda_i(\boldsymbol{\xi})$, $i = 1, \ldots, d$, then furnish the basis vectors for POD

$$\mathbf{v}_i(\boldsymbol{\xi}) = \mathbf{M}^{-1/2}\overline{\mathbf{v}}_i(\boldsymbol{\xi}) \tag{7.44}$$

arranged in the desired order.

7.3.3 Generalisations of POD Based on Alternative Hilbert Spaces

A generalisation of the POD problem consists of solving the problem

$$\min_{\{\varphi_i\}} \|u - \sum_{i=1}^{r} a_i \varphi_i\|^2_{L^2(0,T;\mathcal{H})} \tag{7.45}$$

over all $\mathcal{H}$-orthonormal bases $\{\varphi_i(\mathbf{x})\}_{i=1}^{\infty} \subset \mathcal{H}$, with the solution given by the eigenfunctions of the new operator

$$\mathcal{R}v := \mathbb{E}\left[u(u, v)_{\mathcal{H}}\right] = \int_0^T u(u, v)_{\mathcal{H}} dt \tag{7.46}$$

Since time and spatial averaging operators commute, $\mathcal{R} = \mathcal{C}$ when $\mathcal{H} = L^2(\mathcal{D})$.

7.3.4 Temporal Autocovariance Function and the Method of Snapshots

There are two alternative but equivalent versions of POD that are often implemented in practice, the first of which is called the *method of snapshots*, which is especially relevant for cases in which $M \ll d$. In this version, the focus is placed on the temporal autocovariance

$$K(t, t'; \boldsymbol{\xi}) = \int_{\mathcal{D}} u(\mathbf{x}, t; \boldsymbol{\xi}) u(\mathbf{x}, t'; \boldsymbol{\xi}) d\mathbf{x} \tag{7.47}$$

and the eigenvalue problem

$$\mathcal{K}a_i(t; \boldsymbol{\xi}) := \int_0^T K(t, t'; \boldsymbol{\xi}) a_i(t'; \boldsymbol{\xi}) dt' = \lambda_i'(\boldsymbol{\xi}) a_i(t; \boldsymbol{\xi}) \tag{7.48}$$

The eigenfunctions $a_i(t; \boldsymbol{\xi})$ are orthogonal and are equal to the POD coefficients defined earlier, while the eigenvalues are identical to those of the operator $\mathcal{C}$. Based on the known relationship

$$\mathbb{E}[a_i(t; \boldsymbol{\xi}) a_j(t; \boldsymbol{\xi})] = \lambda_i'(\boldsymbol{\xi}) \delta_{ij} \tag{7.49}$$

the basis provided by the POD modes can be obtained from

$$v_i(\mathbf{x}; \boldsymbol{\xi}) = \frac{1}{\lambda_i'(\boldsymbol{\xi})} \int_0^T u(\mathbf{x}, t; \boldsymbol{\xi}) a_i(t; \boldsymbol{\xi}) dt \tag{7.50}$$

Discretisation of (7.48) in space and time yields

$$\mathbf{K}(\boldsymbol{\xi})\mathbf{a}_i(\boldsymbol{\xi}) = \lambda_i \mathbf{a}_i(\boldsymbol{\xi}) \tag{7.51}$$

in which

$$\mathbf{K}(\boldsymbol{\xi}) := \mathbf{U}(\boldsymbol{\xi})^T \mathbf{U}(\boldsymbol{\xi}) \tag{7.52}$$

is a kernel matrix, the entries of which are given by $K_{ij} = \mathbf{u}^{(i)}(\boldsymbol{\xi})^T \mathbf{u}^{(j)}(\boldsymbol{\xi})$, which are the discrete forms of $K(t^{(i)}, t^{(j)}; \boldsymbol{\xi})$. An eigendecomposition of the kernel matrix yields

$$\mathbf{K}(\boldsymbol{\xi}) = \mathbf{A}(\boldsymbol{\xi})\boldsymbol{\Lambda}(\boldsymbol{\xi})\mathbf{A}(\boldsymbol{\xi})^T, \quad \boldsymbol{\Lambda}(\boldsymbol{\xi}) = \mathrm{diag}(\lambda_1(\boldsymbol{\xi}), \ldots, \lambda_M(\boldsymbol{\xi})) \tag{7.53}$$

with

$$\mathbf{A}(\boldsymbol{\xi}) = [\mathbf{a}_1(\boldsymbol{\xi}) \ldots \mathbf{a}_M(\boldsymbol{\xi})] \tag{7.54}$$

In this discretisation, the j-th component of $\mathbf{a}_i(\boldsymbol{\xi})$, labelled $a_{i,j}(\boldsymbol{\xi})$, is an approximation of $a_i(t^{(j)}; \boldsymbol{\xi})$, which in turn yields the following analogue of (7.50)

$$v_i(\mathbf{x}; \boldsymbol{\xi}) = \frac{1}{\lambda_i(\boldsymbol{\xi})} \sum_{j=1}^{M} u(\mathbf{x}, t^{(j)}; \boldsymbol{\xi}) a_{i,j}(\boldsymbol{\xi}) \tag{7.55}$$

This shows that the basis functions are in fact linear combinations of the snapshots. For discrete space and time we can use the normalisation $\mathbf{a}_i(\boldsymbol{\xi}) \mapsto \mathbf{a}_i'(\boldsymbol{\xi})/\sqrt{\lambda_i(\boldsymbol{\xi})}$ to obtain the modes

$$\mathbf{v}_i(\boldsymbol{\xi}) = \mathbf{U}(\boldsymbol{\xi})\mathbf{a}_i'(\boldsymbol{\xi})/\sqrt{\lambda_i(\boldsymbol{\xi})} \tag{7.56}$$

A singular value decomposition (SVD) is frequently used in *lieu* of both the PCA and method-of-snapshots variants. An SVD of $\mathbf{U}(\boldsymbol{\xi})$ yields $\mathbf{U}(\boldsymbol{\xi}) = \mathbf{A}'(\boldsymbol{\xi})\boldsymbol{\Lambda}(\boldsymbol{\xi})^{1/2}$ $\mathbf{V}(\boldsymbol{\xi})^T$, in which $\mathbf{V}(\boldsymbol{\xi}) = [\mathbf{v}_1(\boldsymbol{\xi}) \ldots \mathbf{v}_M(\boldsymbol{\xi})]$ and $\mathbf{A}'(\boldsymbol{\xi}) = [\mathbf{a}_1'(\boldsymbol{\xi}) \ldots \mathbf{a}_M'(\boldsymbol{\xi})]$. From the point of view of the SVD, the columns of $\mathbf{A}'(\boldsymbol{\xi})$ and $\mathbf{V}(\boldsymbol{\xi})$, which are equal to the eigenvectors of $\mathbf{K}(\boldsymbol{\xi})$ and $\mathbf{C}(\boldsymbol{\xi})$, are the left and right singular vectors. Since $\mathbf{v}_i(\boldsymbol{\xi}) = k\mathbf{U}(\boldsymbol{\xi})\mathbf{a}_i'(\boldsymbol{\xi})$ for any $k \in \mathbb{R}$, we can normalise the $\mathbf{v}_i(\boldsymbol{\xi})$ by setting $k = 1/\sqrt{\lambda_i(\boldsymbol{\xi})}$, which leads precisely to the previously obtained expressions for the POD modes.

7.3.5 *Parameter Dependence*

A major stumbling block in ROMs is constructing a basis (the POD modes for example) across the parameter space $\mathcal{X}$, since the basis derived in the previous section is valid only for a particular $\boldsymbol{\xi}$, which led to the snapshot matrix $\mathbf{U}(\boldsymbol{\xi})$. In theory, a new snapshot matrix is required for each new input, which presents a bottleneck for parameter-dependent problems. Here, machine learning can play a role in devising feasible approaches.

The most common approaches can be categorised as follows

1. A *global basis* is constructed across the input space
2. A *local basis* is constructed by interpolation
3. The *matrices* in problem (7.26) are interpolated

For certain types of problems, the matrices in (7.26) can be written (or approximated) as affine combinations $\mathbf{A}(\boldsymbol{\xi}) = \sum_i c_i(\boldsymbol{\xi})\mathbf{A}_i$, in which the $\mathbf{A}_i$ are constant matrices and the coefficients $c_i(\boldsymbol{\xi})$ are known. In this case, there is little effort required to assemble new reduced-order systems as the parameter values change [23–27].

So-called global basis methods [23, 28–30] use multiple local snapshot matrices $\mathbf{U}(\boldsymbol{\xi}^{(j)})$ for $\boldsymbol{\xi}^{(j)} \in \mathcal{X}$, $j = 1, \ldots, N$, and extract a single basis $\mathbf{v}_i(\boldsymbol{\xi})$ from a global snapshot matrix

$$[\mathbf{U}(\boldsymbol{\xi}^{(1)}), \ldots, \mathbf{U}(\boldsymbol{\xi}^{(n)})] \in \mathbb{R}^{d \times NM} \tag{7.57}$$

The drawbacks of this approach are that the POD optimality is violated, that the growth of the global snapshot matrix with the number of samples becomes unmanageable, and that the validity of the resulting basis can be limited to small windows of input space [31]. Interpolation of bases or matrices is an alternative approach, e.g., linear interpolation of a basis [31] or its interpolation in a tangent space, after mapping local bases to a tangent space of a Grassman manifold [32]. Similar methods can be applied to the matrices (7.26), which avoids having to compute these matrices at each new input [33].

7.3.6 Nonlinearity and the Discrete Empirical Interpolation Method

Incorporating strong (high-order polynomial or non-polynomial) nonlinearities $\mathbf{f}(\cdot; \boldsymbol{\xi}) \in \mathbb{R}^d$ in (7.26) is another issue that compromises the ROM approach, since the cost of computing the nonlinearity is dependent on d. Local polynomial expansions [34, 35] can be applied but the accuracy of this approach is dependent upon the strength of the nonlinearity and/or the region of state space under consideration. Additionally, the computational cost grows exponentially with the order of the polynomial used in the expansion.

Hyper-reduction methods were developed to overcome the deficiencies of linearisation methods. One of the best-known approaches is the empirical interpolation method (EIM), which essentially interpolates the nonlinearity at certain spatial points using a basis that is empirically derived [24, 25]. The *discrete empirical interpolation method* (DEIM), is a discrete form of EIM and is applied to the system resulting from the spatial discretisation [36, 37].

In both of these approaches, a subspace is derived for approximating the nonlinearity, with a greedy algorithm to select the points at which to interpolate. The same approach can further be applied to the approximation of non-affine versions of the matrices discussed earlier [38]. The *Gauss-Newton with approximated tensors* (GNAT) method is applied to the fully-discretised system (in space and time), and instead solves a residual-minimisation problem [39], leading to a Petrov-Galerkin problem.

In the DEIM method we seeks vectors $\mathbf{w}_i(\boldsymbol{\xi}) \in \mathbb{R}^d$, $i = 1, \ldots, d$, to form a subspace

$$\text{span}(\mathbf{w}_1(\boldsymbol{\xi}), \ldots, \mathbf{w}_s(\boldsymbol{\xi})) \subset \mathbb{R}^d \tag{7.58}$$

with some $s \ll d$, that provides a good approximation of $\mathbf{f}(\boldsymbol{u}(t; \boldsymbol{\xi}); \boldsymbol{\xi})$ for any arbitrary t. In other words, we look for an approximation

$$\mathbf{f}(\boldsymbol{u}(t; \boldsymbol{\xi}); \boldsymbol{\xi}) \approx \mathbf{W}(\boldsymbol{\xi})\mathbf{h}(t; \boldsymbol{\xi}), \quad \mathbf{W}(\boldsymbol{\xi}) = [\mathbf{w}_1(\boldsymbol{\xi}) \ldots \mathbf{w}_s(\boldsymbol{\xi})] \tag{7.59}$$

in which $\mathbf{h}(t; \boldsymbol{\xi}) \in \mathbb{R}^s$. To obtain the basis $\{\mathbf{w}_i(\boldsymbol{\xi})\}_{i=1}^d$ we use snapshots $\{\mathbf{f}^{(i)}(\boldsymbol{\xi})\}_{i=1}^M$, in which we define

$$\mathbf{f}^{(i)}(\boldsymbol{\xi}) = \mathbf{f}(\boldsymbol{u}^{(i)}(\boldsymbol{\xi}); \boldsymbol{\xi}) \tag{7.60}$$

We can form the matrix

$$\mathbf{F}(\boldsymbol{\xi}) = [\mathbf{f}^{(1)}(\boldsymbol{\xi}) \ldots \mathbf{f}^{(m)}(\boldsymbol{\xi})] \tag{7.61}$$

and perform a PCA on $\mathbf{F}(\boldsymbol{\xi})\mathbf{F}(\boldsymbol{\xi})^T$ or an SVD on $\mathbf{F}(\boldsymbol{\xi})$ in order to obtain the vectors $\{\mathbf{w}_i(\boldsymbol{\xi})\}_{i=1}^d$, arranged in the usual manner such that the associated eigenvalues decay.

The system $\mathbf{f}(\boldsymbol{u}(t; \boldsymbol{\xi}); \boldsymbol{\xi}) = \mathbf{W}(\boldsymbol{\xi})\mathbf{h}(t; \boldsymbol{\xi})$ is overdetermined, so that the DEIM method has to find an optimal solution across all s-dimensional subspaces, in some well-defined sense. We can define a matrix

$$\mathbf{P} = [\mathbf{e}_{p_1} \ldots \mathbf{e}_{p_s}] \in \mathbb{R}^{d \times s} \tag{7.62}$$

in which the $\mathbf{e}_{p_i} \in \mathbb{R}^d$ are the standard Euclidean basis vectors. Provided that $\mathbf{P}^T\mathbf{W}(\boldsymbol{\xi})$ is nonsingular, the following holds

$$\begin{aligned} \mathbf{f}_r\,(\mathbf{a}(t; \boldsymbol{\xi}); \boldsymbol{\xi}) &= \mathbf{V}_r(\boldsymbol{\xi})^T \mathbf{f}\,(\mathbf{V}_r(\boldsymbol{\xi})\mathbf{a}(t; \boldsymbol{\xi}); \boldsymbol{\xi}) \\ &\approx \mathbf{V}_r(\boldsymbol{\xi})^T \mathbf{W}(\boldsymbol{\xi})\mathbf{h}(t; \boldsymbol{\xi}) \\ &= \mathbf{V}_r(\boldsymbol{\xi})^T \mathbf{W}(\boldsymbol{\xi})(\mathbf{P}^T\mathbf{W}(\boldsymbol{\xi}))^{-1}\mathbf{P}^T\mathbf{f}(\boldsymbol{u}(t; \boldsymbol{\xi}); \boldsymbol{\xi}) \\ &= \mathbf{V}_r(\boldsymbol{\xi})^T \mathbf{W}(\boldsymbol{\xi})(\mathbf{P}^T\mathbf{W}(\boldsymbol{\xi}))^{-1}\mathbf{f}(\mathbf{P}^T\boldsymbol{u}(t; \boldsymbol{\xi}); \boldsymbol{\xi}) \end{aligned} \tag{7.63}$$

with the assumption that $\mathbf{f}\,(\cdot; \boldsymbol{\xi})$ acts point-wise. We than specify the indices $p_i \in \{1, 2, \ldots, d\}$, $i = 1, \ldots, s$, using a greedy procedure [36], for which, for a given value of s, we can bound the error as follows

$$\begin{aligned} ||\mathbf{f} - \widehat{\mathbf{f}}|| &\le ||(\mathbf{P}^T\mathbf{W}(\boldsymbol{\xi}))^{-1}||\;||(\mathbf{I} - \mathbf{W}(\boldsymbol{\xi})\mathbf{W}(\boldsymbol{\xi})^T)\mathbf{f}|| \\ \widehat{\mathbf{f}} &:= \mathbf{W}(\boldsymbol{\xi})(\mathbf{P}^T\mathbf{W}(\boldsymbol{\xi}))^{-1}\mathbf{P}^T\mathbf{f} \end{aligned} \tag{7.64}$$

with $\widehat{\mathbf{f}}$ being the approximation of $\mathbf{f}$ in the DEIM method. Considering $\mathbf{f}$ to be a function of t, the estimate above holds at each t by virtue of the factor $||(\mathbf{I} - \mathbf{W}(\boldsymbol{\xi})\mathbf{W}(\boldsymbol{\xi})^T)\mathbf{f}||$; this factor is the square error in the approximation of $\mathbf{f}$ using Range($\mathbf{W}(\boldsymbol{\xi})$). The SVD on $\mathbf{F}(\boldsymbol{\xi})$ yields an error bound that is approximately uniform in t as $M \to \infty$, which explains why this basis is chosen.

7.4 Time Series Methods

Time series methods are important for analysing certain types of data from batteries, in particular data that is related to degradation. For Li-ion batteries, a number of approaches have been developed to predict capacity fade based on cycling data available for single or multiple batteries. In this section, we provide a detailed introduction to several popular time-series methods, in the hope that we can motivate their application to flow batteries; we are not aware of any applications to date.

7.4.1 Basic Approaches and Data Embedding

A time series consists of a sequence of values y_n, $n = 1, \ldots, N$, that are ordered in some fashion. The usual goal is to predict values in the sequence for $n > N$. The sequence or series can be considered as a realisation (or sample path) of some random process, labelled $\{Y_n\}$ or Y_n, which is indexed by some parameter n.

The time series data $\{y_n\}$ comprises only one sample path from the entire set of sample paths (the *ensemble*), which necessitates an approximation with regards to the expectation operator $\mathbb{E}[\cdot]$. Usually, ergodicity is assumed, which allows $\mathbb{E}[\cdot]$ to be defined with respect to n rather than probability measure and sample space underlying the process. From this point onwards, we will not distinguish between the process Y_n and its realisations y_n in the notation.

Any random process $\{y_n\}$, $n \in \mathbb{N}$, is called *wide-sense stationary* if both its expectation and the autocovariance

$$\mu = \mathbb{E}[y_n], \quad k(n, n') = \mathbb{E}[(y_n - \mu)(y_{n'} - \mu')] \tag{7.65}$$

respectively, do not depend on n. This implies that the mean μ of the process is constant and that the autocovariance $k(n, n')$ has the special form $k(n - n')$. In a time series analysis, this form of stationarity is normally assumed, as opposed to the strong form. We will normally used the notation y_n rather than $\{y_n\}$ for a process, unless it is important to make a distinction between the process, the sequence values such as y_n, $n = 1, \ldots, N$, or a random variable generated at a fixed n.

In many time-series methods, the data set is *embedded* to define a modified set of inputs and outputs that can be used in a supervised machine learning approach. One way of distinguishing between models relates to the number of data points in the sequence that are used for predicting future values, as well as the number of values that are to be predicted. A *one-step-ahead* forecast relates to the prediction of a single y_n based on $m \geq 1$ past values *via* an *autoregressive map*

$$f : \mathbb{R}^m \to \mathbb{R} \tag{7.66}$$

The map can take the explicit form

$$y_{n+D} = f(y_{n-m+1}, \ldots, y_n) \tag{7.67}$$

Here, D is a *delay*, while m is called the *embedding lag* (or embedding order). The function f may be linear or nonlinear, and is used to estimate the true latent function underlying the relationship between y_n and the past values (if it exists). In the notation, we will only distinguish between the true and estimated values when it is important to do so.

In the case of *multi-step-ahead* forecasting, in which multiple values y_n are predicted, there are several options. In *iterated* or recursive strategies, the $f(\cdot)$ in (7.67) is applied repeatedly up to some window or *horizon* of time or other index. When the delay is 1, we may start with the estimate

$$\widehat{y}_{N+1} = f(y_{N-m+1}, \ldots, y_N) \tag{7.68}$$

This is based on the observed (known) values in the argument of the right-hand side. We then use this estimate to predict at the next index value

$$\widehat{y}_{N+2} = f(y_{N-m+2}, \ldots, \widehat{y}_{N+1}) \tag{7.69}$$

and repeat the process up to some horizon h, i.e., obtain estimates up to $\widehat{y}_{N+h}$. The drawback of this approach is that it suffers from error propagation, meaning that any errors are compounded as the iterations proceed.

On the other hand, we may take a *direct strategy*, which consists of developing h independent models $f_k : \mathbb{R}^m \to \mathbb{R}$

$$y_{n+k} = f_k(y_{n-m+1}, \ldots, y_n), \quad k = 1, \ldots, h \tag{7.70}$$

All of these models are required for forecasting the h future values, which obviously incurs a much higher computational cost. Moreover, it ignores any correlations between the predicted values. To overcome some of these issues, a *hybrid direct-recursive* strategy can be employed, which is again based on h models. In this case, however, the input window (number of values) is enlarged at each step by including the prior forecasts, that is

$$y_{n+k} = f_k(y_{n-m+1}, \ldots, y_{n+k-1}), \quad k = 1, \ldots, h \tag{7.71}$$

We can take as an example the forecast for the 3rd value

$$\widehat{y}_{N+3} = f_2(y_{N-m+1}, \ldots, \widehat{y}_{N+1}, \widehat{y}_{N+2}) \tag{7.72}$$

which uses the previous two forecasts $\widehat{y}_{N+1}$ and $\widehat{y}_{N+2}$. Neither the direct nor the hybrid strategies are desirable for large values of h, given the associated costs.

The final approach avoids the issues of the direct and hybrid approaches and is called *multi-input-multi-output* (MIMO) or joint. As the name suggests, it is based on a map

$$\mathbf{f} : \mathbb{R}^m \to \mathbb{R}^h \tag{7.73}$$

or in detail

$$(y_{n+1}, \ldots, y_{n+h})^T = \mathbf{f}(y_{n-m+1}, \ldots, y_n) \tag{7.74}$$

in which the next h values are predicted in a single step. This estimator can in principle be used recursively although the number of estimates $\widehat{y}_{n+j}$, $j = 1, \ldots, h$, used for the next forecast step must be decided.

To develop the mappings above, we first need to shape the data. With a delay of $D = 1$ (for the purposes of illustration), the data is rewritten as a set of inputs and outputs, with the inputs equal to the following vectors of values $\mathbf{x}_l$ used in the function $f(\cdot)$ or $\mathbf{f}(\cdot)$, and the outputs equal to the following vectors of values $\mathbf{y}_l$ to be predicted

$$\begin{aligned} \mathbf{x}_l &= (y_l, y_{l+1} \ldots, y_{l+m-1})^T \\ \mathbf{y}_l &= (y_{l+m}, y_{l+m+1} \ldots, y_{l+m+h-1})^T \\ l &= 1, \ldots, L = N - m - h + 1 \end{aligned} \tag{7.75}$$

Having defined input-output pairs, we can now use a statistical model of the form

$$\mathbf{y}_l = \mathbf{f}(\mathbf{x}_l) + \epsilon \tag{7.76}$$

with $\mathbf{f} : \mathbb{R}^m \to \mathbb{R}^h$. ϵ is an error term, which is usually assumed to be Gaussian or set to zero. The index l is not to be confused with n, which refers to the time series indexing. For the case $h = 1$, i.e., univariate outputs, we replace $\mathbf{y}_l$ and $\mathbf{f}$ with y_l and f. When additionally $m = 1$, the model is reduced to a first-order *Markov model*. We next introduce some classical approaches to time series analysis.

7.4.2 Autoregressive Integrated Moving Average Models

Autoregressive integrated moving average (ARIMA) models do not explicitly use an embedding, but an embedding is implicit in their design. They are linear examples of *state space* modelling, which is a unifying framework for a range of models that includes the Kalman filter and hidden Markov models, as well as the Gaussian process dynamical model (GPDM) introduced in Sect. 7.4.5 below.

ARIMA models take the form (7.76) with a forecast horizon $h = 1$ and a linear map f. A zero-mean stationary process (without deterministic components) $\{y_n\}$, $n \in \mathbb{N}$, can be written in a particular way, which we call the *infinite moving average* or MA(∞) form [40]

$$y_n = z_n - \sum_{i=1}^{\infty} \beta_i z_{n-i} = \beta(L) z_n, \quad \beta(L) = 1 - \sum_{i=1}^{\infty} \beta_i L^i \tag{7.77}$$

Stationarity implies that

$$\sum_{i=0}^{\infty} \beta_i^2 < \infty \tag{7.78}$$

In this expression, z_i is a white noise innovation process with zero-mean

$$\mathbb{E}[z_i z_{i+k}] = \sigma^2 \mathbf{1}_{\{k=0\}} \tag{7.79}$$

Here, $\mathbf{1}_\mathbf{A}$ denotes the indicator function on a set $\mathbf{A}$, meaning that it is zero outside $\mathbf{A}$ and equal to 1 on $\mathbf{A}$. Moreover

$$\sigma^2 = \mathrm{Var}(z_i) = \mathbb{E}[z_i^2] \tag{7.80}$$

Thus, the z_i are uncorrelated and have constant mean and variance. If they are normally distributed, they are therefore i.i.d. In (7.77), L is called a *lag operator* or a shift operator, for obvious reasons. The autocovariance of y_n is

$$k(\tau = n - n') = \begin{cases} \sigma^2 \sum_{i=0}^{\infty} \beta_i \beta_{i+\tau} & 0 \le \tau \le q \\ 0 & \text{otherwise} \end{cases} \tag{7.81}$$

If the MA(∞) form (7.77) is convergent, we call the process y_n a *causal* or linear process. A finite-dimensional version, termed the MA(q) approximation, confines the MA(∞) terms to the first q

$$y_n = \beta_q(L) z_n = (1 - \sum_{i=1}^{q} \beta_i L^i) z_n \tag{7.82}$$

If the model (7.77) is invertible, it can be written in an alternative form called the *infinite autoregressive* or AR(∞) form

$$y_n - \sum_{i=1}^{\infty} \gamma_i y_{n-i} = z_n \quad \text{or} \quad \gamma(L) y_n = z_n \tag{7.83}$$

Here

$$\gamma(L) = 1 - \sum_{i=1}^{\infty} \gamma_i L^i, \quad \sum_{i=1}^{\infty} \gamma_i^2 < \infty \tag{7.84}$$

The AR(∞) model (or process) essentially describes y_n in terms of lagged dependent variables, whereas the MA(∞) form describes y_n in terms of lagged errors. The equivalent finite-dimensional AR(p) approximation is

$$\gamma_p(L)y_n = (1 - \sum_{i=1}^{p} \gamma_i L^i)y_n = z_n \tag{7.85}$$

This is the form of model (7.67), with a delay $d = 1$ and p equivalent to m. We note for later discussion that if $\gamma_p(L)$ has an inverse, the process can be written in MA(∞) form

$$y_n = \gamma_p(L)^{-1} z_n \tag{7.86}$$

by virtue of the fact that since $\gamma_p(L)^{-1}$ is an infinite series. As an example, for an AR(1) processes, $(1 - \gamma L)^{-1} = \sum_{i=0}^{\infty} (\gamma L)^i$. This is easily seen by recursively applying $y_{n-i} = \gamma y_{n-i-1} + z_{n-i}$, $i = 1, \ldots$, to the RHS of the AR(1) model.

The process is guaranteed to be both causal and stationary if all of the roots of $\gamma_p(w)$, $w \in \mathbb{C}$, lie outside the unit circle $|w| = 1$. For example, for $\gamma_1(L) = 1 - \gamma L$, this requires $|\gamma| < 1$. For the AR(1) model, it holds that

$$k(\tau) = \gamma k(\tau - 1) \tag{7.87}$$

which is recursive, whereupon $k(\tau) = \gamma^\tau k(0)$; since $|\gamma| < 1$, there is a geometric decay in the autocovariance, which is a property shared by all AR processes.

Based on combining the AR and MA models, we may develop a so-called ARMA(p, q) model, defined as follows

$$y_n - \sum_{i=1}^{p} \gamma_i y_{n-i} = z_n - \sum_{i=1}^{q} \beta_i z_{n-i} \quad \text{or} \quad \gamma_p(L)y_n = \beta_q(L)z_n \tag{7.88}$$

In order to be stationary, causal and invertible, now both of $\gamma_p(w)$ and $\beta_q(w)$, $w \in \mathbb{C}$, must have roots that lie outside the unit circle, and, furthermore, must not possess any common roots. ARMA models can be converted to an MA(∞) form

$$y_n = [\gamma_p(L)^{-1}\beta_q(L)]z_n \tag{7.89}$$

so that the autocovariance decays geometrically. Formally, these models are only applicable to stationary processes. For non-stationary processes y_n we use the lag operator to create a stationary process, namely, we define a d-th differenced process $(1 - L)^d y_n$. We may then use an ARMA model for the differenced process, provided it is stationary. The original series is termed *differenced of order d* and we use the notation I(d) for such a process. The ARMA model in this case is called an ARIMA(p, d, q) process (the I stands for 'integrated'), and it satisfies

$$\gamma_p(L)\Delta^d y_n = \beta_q(L)z_n, \quad \Delta = 1 - L \tag{7.90}$$

The innovation process is almost always assumed to be Gaussian

$$z_n \sim \mathcal{GP}(0, \sigma^2\delta(n, n')) \tag{7.91}$$

in which $\mathcal{GP}(\cdot, \cdot)$ is a GP as defined in Sect. 6.7 and $\delta(n, n')$ is the Kronecker-delta function. For each fixed n, the random variables z_n are therefore i.i.d. with a mean of 0 and a variance of σ^2. The full set of parameters can be collected in a vector

$$\boldsymbol{\theta} = (\gamma_1, \ldots, \gamma_p, \beta_1, \ldots, \beta_q, \sigma^2)^T \tag{7.92}$$

and the goal is to estimate these parameters, usually by maximising the likelihood (Sect. 6.2), i.e., the joint density $p(\mathbf{Y} \mid \boldsymbol{\theta})$ of the data $\mathbf{Y} = (y_1, \ldots, y_N)^T$, conditioned on $\boldsymbol{\theta}$, or the density of a differenced series when $d > 1$. With the GP assumption on z_n, the y_n are i.i.d. Gaussian, conditioned on the past observations, which leads to the following log likelihood solution

$$\begin{aligned}&\text{argmin}_{\boldsymbol{\theta}}\left(-\ln p(\mathbf{Y} \mid \boldsymbol{\theta}) = \frac{N}{2}\ln(2\pi) + \frac{N}{2}\ln|\Gamma(\boldsymbol{\theta})| + \frac{1}{2}\mathbf{Y}^T\Gamma(\boldsymbol{\theta})^{-1}\mathbf{Y}\right)\\ &\Gamma(\boldsymbol{\theta}) = \mathbb{E}(\mathbf{Y}\mathbf{Y}^T) = [k(n - n')]_{n,n'=1,\ldots,N}\end{aligned} \tag{7.93}$$

Γ is the covariance matrix of $\mathbf{Y}$ (a random vector), with a parametric dependence on $\boldsymbol{\theta}$ *via* (7.90). We note that there are various different maximum likelihood implementations for ARIMA models, which we will not discuss in this book.

Once the parameters are determined, predictions can be made by writing the ARIMA(p, d, q) model as

$$\psi(L)y_n = \beta_q(L)z_n, \quad \psi(L) = \gamma_p(L)(1 - L)^d \tag{7.94}$$

in which the parameters ψ_i, $i = 1, \ldots, p + d$, are dependent on γ_i. This does not mean that it is an ARMA($p + d, q$) process, since the coefficients do not satisfy the stationarity conditions. We can alternatively write

$$y_n - \sum_{i=1}^{p+d} \psi_i y_{n-i} = z_n - \sum_{j=1}^{q} \beta_j z_{n-j} \tag{7.95}$$

It can be shown that the conditional expectation $\mathbb{E}[y_{n+1} \mid \mathbf{Y}]$ is the least-squares optimal linear predictor of y_{n+1}. In fact, by the assumption on the innovation process, y_n is a GP, and so $\mathbb{E}[y_{n+1} \mid \mathbf{Y}]$ is optimal among all predictors. Therefore, we may take expectations conditioned on $\mathbf{Y}$ to recursively obtain an h horizon prediction

$$\widehat{y}_N(h) = \sum_{i=1}^{p+d} \psi_i \widehat{y}_N(h-i) + \widehat{z}_N(h) + \sum_{j=1}^{q} \beta_j \widehat{z}_N(h-j) \tag{7.96}$$

Here we use the notation

$$\widehat{y}_N(j) = \mathbb{E}[y_{N+j} \mid \mathbf{Y}], \quad \widehat{z}_N(j) = \mathbb{E}[z_{N+j} \mid \mathbf{Y}] \tag{7.97}$$

For $j \leq 0$, we replace the estimate $\widehat{y}_N(j)$ with the data point, and $\widehat{z}_N(j)$ is obtained by recursion. When $j \geq 0$, $\widehat{y}_N(j)$ represents the predicted value and

$$\widehat{z}_N(j) = \mathbb{E}[z_{N+j} \mid \mathbf{Y}] = 0 \tag{7.98}$$

by virtue of the fact that z_{N+j} is independent of $\mathbf{Y}$.

7.4.3 Nonlinear Univariate Gaussian Process Autoregression

A second popular approach is Gaussian process autoregression (GPNAR), which conforms to the model (7.76), and is used with data of the form (7.75). GPNAR was discussed in Sect. 7.2.2 in the context of multi-fidelity models, in which the fidelity plays the role of time. In this section we provide full details of its application to time series using a simple method for prediction that relies only on the predictive mean from the previous time step (a so-called *mean-prediction*). GPNAR assumes a prior GP over $f : \mathbb{R}^m \to \mathbb{R}$ as well as $\epsilon : \mathbb{R}^m \to \mathbb{R}$

$$\begin{aligned} f(\mathbf{x}_l) &\sim \mathcal{GP}\left(\mathbf{h}(\mathbf{x}_l)^T \boldsymbol{\beta}, k(\mathbf{x}_l, \mathbf{x}_{l'} | \boldsymbol{\theta})\right) \\ \epsilon(\mathbf{x}_l) &\sim \mathcal{GP}\left(0, \sigma^2 \delta(\mathbf{x}_l, \mathbf{x}_{l'})\right) \end{aligned} \tag{7.99}$$

with covariance function $k(\mathbf{x}_l, \mathbf{x}_{l'} | \boldsymbol{\theta})$ depending on hyperparameters $\boldsymbol{\theta}$ (see Sect. 6.7). The noise is i.i.d. Gaussian with zero-mean and variance σ^2.

Here we include a non-zero-mean function in the form of a linear combination of known basis functions

$$\mathbf{h}(\mathbf{x}_l) = (h_1(\mathbf{x}_l), \ldots, h_M(\mathbf{x}_l))^T \tag{7.100}$$

with coefficients $\boldsymbol{\beta}$ that are unknown. This model is equivalent to $y_l = \mathbf{h}(\mathbf{x}_l)^T \boldsymbol{\beta} + g(\mathbf{x}_l)$, in which $g(\mathbf{x}_l) \sim \mathcal{GP}\left(0, k(\mathbf{x}_l, \mathbf{x}_{l'} | \boldsymbol{\theta}) + \delta(\mathbf{x}_l, \mathbf{x}_{l'}) \sigma^2\right)$ is a zero-mean GP. We place a prior over the coefficients $\boldsymbol{\beta}$ as follows

$$\boldsymbol{\beta} \sim \mathcal{N}(\boldsymbol{\beta} \mid \mathbf{b}, \mathbf{B}) \tag{7.101}$$

in which $\mathbf{b}$ and $\mathbf{B}$ are hyperparameters to be determined. In [41] it is shown that $\boldsymbol{\beta}$ can be integrated out, leading to

$$y_l \sim \mathcal{GP}\left(\mathbf{h}(\mathbf{x}_l)^T\mathbf{b}, k(\mathbf{x}_l, \mathbf{x}_{l'}|\boldsymbol{\theta}) + \delta(\mathbf{x}_l, \mathbf{x}_{l'})\tau + \mathbf{h}(\mathbf{x}_l)^T\mathbf{B}\mathbf{h}(\mathbf{x}_l)\right) \tag{7.102}$$

Letting $\mathbf{Y}_m = (y_1, \ldots, y_L)^T$, and using the notation $\Theta = \{\boldsymbol{\theta}, \mathbf{b}, \mathbf{B}, \tau\}$, the predictive posterior can be derived using standard rules for Gaussians

$$\begin{aligned} f(\mathbf{x}_l) \mid \Theta &\sim \mathcal{N}\left(\mu(\mathbf{x}_l), \sigma^2(\mathbf{x}_l)\right) \\ \mu(\mathbf{x}_l) &= \mathbf{k}(\mathbf{x}_l)^T(\mathbf{K}+\tau\mathbf{I})^{-1}\mathbf{Y}_m + \mathbf{R}^T\bar{\beta} \\ \sigma^2(\mathbf{x}_l) &= k(\mathbf{x}_l, \mathbf{x}_l|\boldsymbol{\theta}) - \mathbf{k}(\mathbf{x}_l)^T(\mathbf{K}+\tau\mathbf{I})^{-1}\mathbf{k}(\mathbf{x}_l) \\ \mathbf{R} &= \mathbf{h}(\mathbf{x}_l) - \mathbf{H}(\mathbf{K}+\tau\mathbf{I})^{-1}\mathbf{k}(\mathbf{x}_l) \\ \bar{\beta} &= (\mathbf{B}^{-1} + \mathbf{H}(\mathbf{K}+\tau\mathbf{I})^{-1}\mathbf{H}^T)^{-1}(\mathbf{H}(\mathbf{K}+\tau\mathbf{I})^{-1}\mathbf{Y}_m + \mathbf{B}^{-1}\mathbf{b}) \end{aligned} \tag{7.103}$$

in which

$$\begin{aligned} \mathbf{k}(\mathbf{x}_l) &= (k(\mathbf{x}_l, \mathbf{x}_1|\boldsymbol{\theta}), \ldots, k(\mathbf{x}_l, \mathbf{x}_{N-m+1}|\boldsymbol{\theta}))^T \\ \mathbf{H} &= [\mathbf{h}(\mathbf{x}_1)\ldots\mathbf{h}(\mathbf{x}_{N-m+1})] \\ [\mathbf{K}]_{ll'} &= k(\mathbf{x}_l, \mathbf{x}_{l'}|\boldsymbol{\theta}), \quad l, l' = 1, \ldots, N-m+1 \end{aligned} \tag{7.104}$$

The hyperparameters Θ are usually obtained by a log-likelihood maximisation, namely,

$$\text{argmax}_{\Theta} - \frac{1}{2}\ln|\mathbf{K}+\tau\mathbf{I}+\mathbf{H}^T\mathbf{B}\mathbf{H}| - \frac{1}{2}\left(\mathbf{H}^T\mathbf{b} - \Delta\right)^T\left(\mathbf{K}+\tau\mathbf{I}+\mathbf{H}^T\mathbf{B}\mathbf{H}\right)^{-1}\left(\mathbf{H}^T\mathbf{b} - \Delta\right) \tag{7.105}$$

Equations (7.103)–(7.105) can applied iteratively using the posterior mean $\mu(\mathbf{x}_l)$, $l = L+1, \ldots$, which approximates $y_{N+1}, \ldots$, at each step. We note, however, that while the predictive variance in the first estimate is exact, subsequent variances underestimate the true variance since they are based on the GP mean. Overcoming this limitation requires expensive sampling methods, as discussed in Sect. 7.2.2.

7.4.4 Autoregression Networks

The recurrent networks (RNNs) in Sect. 6.12.3 and the encoder-decoder models in Sect. 6.12.5 are designed for sequence prediction and are essentially of the form

$$\mathbf{y}_l = \mathbf{f}(\mathbf{x}_l) \tag{7.106}$$

with a nonlinear map $\mathbf{f}$ and data (7.75). They can be used iteratively to forecast up to an arbitrary horizon. Note also that $\boldsymbol{\epsilon} = \mathbf{0}$ in the vanilla case but an error can be included explicitly using Bayesian neural networks or implicitly using a regularisation term. The map $\mathbf{f} : \mathbb{R}^m \to \mathbb{R}^h$ can also be a multi-layer perceptron (Sect. 6.12.1), a convolutional network (Sect. 6.12.2), a bi-directional RNN (Sect. 6.12.4), or indeed any other network (including hybrid networks such as the combination of a CNN with an LSTM).

When $h = 1$, y_l represents the next step in the sequence $\mathbf{x}_l$ and an iterative implementation is straightforward. When $h > 1$, $\mathbf{y}_l$ represents the next h values in the sequence $\mathbf{x}_l$ and a choice has to be made in terms of the number of estimates in $\mathbf{y}_l$ to use at the next iteration. This is a free choice, although the potential for error propagation suggest that a minimal number estimates should be used.

7.4.5 Gaussian Process Dynamical Models

The Gaussian process dynamical model (GPDM) is an unsupervised methods that can find low-dimensional embeddings of time series data and/or can be used to predict future states. It is based on a mapping from a latent space to an observation space, and a model for the dynamics of the latent variables [42]. It is one of a number of nonlinear state-space models, with hidden Markov models (HMMs) and extended Kalman filtering being two other popular methods. We consider a case of vector-valued time series data $\mathbf{Y} = [\mathbf{y}_1 \ldots \mathbf{y}_T]^T \in \mathbb{R}^{T\times D}$, with T observations $\mathbf{y}_n \in \mathbb{R}^D$. Let $\mathbf{X} = [\mathbf{x}_1 \ldots \mathbf{x}_T]^T \in \mathbb{R}^{T\times Q}$ to be a matrix of corresponding latent variables $\mathbf{x}_n \in \mathbb{R}^Q$, in which we can demand (if desired) that $Q \ll D$. GPDM is based on the following first-order Markov model

$$\begin{aligned}\mathbf{x}_n &= f(\mathbf{x}_{n-1}; \mathbf{A}) + \mathbf{n}_{x,n} = \sum_{i=1}^{K} \mathbf{a}_i \phi_i(\mathbf{x}_{n-1}) + \mathbf{n}_{x,n} := \mathbf{A}^T \phi(\mathbf{x}_{n-1}) + \mathbf{n}_{x,n} \\ \mathbf{y}_n &= g(\mathbf{x}_n; \mathbf{B}) + \mathbf{n}_{y,n} = \sum_{j=1}^{M} \mathbf{b}_j \psi_j(\mathbf{x}_n) + \mathbf{n}_{y,n} := \mathbf{B}^T \psi(\mathbf{x}_n) + \mathbf{n}_{y,n}\end{aligned} \tag{7.107}$$

Here, the rows of $\mathbf{A} = [\mathbf{a}_1 \ldots \mathbf{a}_K]^T \in \mathbb{R}^{M\times D}$ and $\mathbf{B} = [\mathbf{b}_1 \ldots \mathbf{b}_M]^T \in \mathbb{R}^{K\times Q}$ are unknown weights and $\{\phi_i(\mathbf{x})\}$, $\{\psi_j(\mathbf{x})\}$ are sets of basis functions. The noise terms $\mathbf{n}_{y,n}$, $\mathbf{n}_{x,n}$ are zero-mean GPs.

The basis functions and weights can be eliminated using kernel substitution, marginalising out both $\mathbf{A}$ and $\mathbf{B}$ using Gaussian prior distributions. We start by placing independent priors over the columns $\widehat{\mathbf{b}}_d \in \mathbb{R}^M$, $d = 1, \ldots, D$, of $\mathbf{B}$

$$p(\widehat{\mathbf{b}}_d \mid w_d) = \mathcal{N}\left(\mathbf{0}, w_d^{-2}\mathbf{I}\right) \tag{7.108}$$

This model is equivalent to

$$\widehat{\mathbf{y}}_d = \boldsymbol{\Psi}\widehat{\mathbf{b}}_d + \mathbf{n}_{y,d} \tag{7.109}$$

for the columns $\widehat{\mathbf{y}}_d$ of $\mathbf{Y}$, in which $\boldsymbol{\Psi} = [\psi(\mathbf{x}_1) \ldots \psi(\mathbf{x}_T)]^T$ and $\mathbf{n}_{y,d} \sim \mathcal{N}(\mathbf{0}, w_d^{-2}\sigma_Y^2\mathbf{I})$. In this original formulation, the error variance $w_d^{-2}\sigma_Y^2$ contains the scale factor w_d^{-2} [42]. Actually, this is not necessary, and the formulation can be generalised as in Sect. 6.15.6 of Chap. 6. The assumptions above lead to

$$p(\widehat{\mathbf{y}}_d \mid \mathbf{X}, \widehat{\mathbf{b}}_d, w_d, \sigma_Y) = \mathcal{N}\left(\boldsymbol{\Psi}\widehat{\mathbf{b}}_d, w_d^{-2}\sigma_Y^2\mathbf{I}\right) \tag{7.110}$$

Applying standard conditioning results to the distributions $p(\widehat{\mathbf{y}}_d \mid \mathbf{X}, \widehat{\mathbf{b}}_d, w_d, \sigma_Y)$ and $p(\widehat{\mathbf{b}}_d \mid w_d)$ leads to the elimination of $\widehat{\mathbf{b}}_d$

$$p(\widehat{\mathbf{y}}_d \mid \mathbf{X}, w_d, \sigma_Y) = \mathcal{N}\left(\mathbf{0}, w_d^{-2}\mathbf{K}_Y\right), \quad \mathbf{K}_Y = \boldsymbol{\Psi}\boldsymbol{\Psi}^T + \sigma_Y^2\mathbf{I} \tag{7.111}$$

in which $\mathbf{K}_Y$ is a kernel matrix that is unscaled.

Denoting the components of $\widehat{\mathbf{y}}_d$ by $\widehat{\mathbf{y}}_{d,n}$, we see that it is possible to use kernel substitution to replace the covariances between these components by

$$\text{cov}(\mathbf{y}_{d,n}, \mathbf{y}_{d,n'}) = w_d^{-2}k_Y(\mathbf{x}_n, \mathbf{x}_{n'}|\boldsymbol{\theta}_Y) \tag{7.112}$$

in which $k_Y(\mathbf{x}_n, \mathbf{x}_{n'}|\boldsymbol{\theta}_Y)$ is an equivalent kernel with associated hyperparameters $\boldsymbol{\theta}_Y$. For example, we can use the (by now familiar) squared exponential

$$k_Y(\mathbf{x}_n, \mathbf{x}_{n'}|\boldsymbol{\theta}_Y) = \theta_{Y,1}\exp\left(-\frac{\theta_{Y,2}}{2}\|\mathbf{x}_n - \mathbf{x}_{n'}\|^2\right) + \theta_{Y,3}^{-1}\delta(\mathbf{x}_n - \mathbf{x}_{n'}) \tag{7.113}$$

with $\boldsymbol{\theta}_Y = \left\{\theta_{Y,1}, \theta_{Y,2}, \theta_{Y,3}\right\}$, $\theta_{Y,3}^{-1} = \sigma_Y^2$. The likelihood can be written as

$$\begin{aligned} p(\mathbf{Y} \mid \mathbf{X}, \boldsymbol{\theta}_Y, \mathbf{W}) &= \prod_{n=1}^{T} p(\widehat{\mathbf{y}}_d \mid \mathbf{X}, \boldsymbol{\theta}_Y, w_d) \\ &= \frac{|\mathbf{W}|^T}{\sqrt{(2\pi)^{TD}|\mathbf{K}_Y|^D}}\exp\left\{-\frac{1}{2}\text{tr}\left(\mathbf{K}_Y^{-1}\mathbf{Y}\mathbf{W}^2\mathbf{Y}^\top\right)\right\} \end{aligned} \tag{7.114}$$

with $\mathbf{W} = \text{diag}\,(w_1 \ldots w_D)$.

For the latent mapping the situation is more complicated due to the Markov assumption, which uses $\mathbf{x}_{n-1}$ as the input for $\mathbf{x}_n$. We place independent (isotropic) priors over the columns $\widehat{\mathbf{a}}_q \in \mathbb{R}^K$, $q = 1, \ldots, Q$, of $\mathbf{A}$, now employing standard normals (without the w_i)

$$p(\widehat{\mathbf{a}}_q) = \mathcal{N}(\mathbf{0}, \mathbf{I}) \tag{7.115}$$

The probability of $\mathbf{x}_1$ is taken into account separately to fully specify the model, since $\mathbf{x}_0$ is unknown. Excluding the first column of $\mathbf{X}$, the model for each of the resulting columns $\widehat{\mathbf{x}}_{q\sim 1}$ is

$$\widehat{\mathbf{x}}_{q\sim 1} = \boldsymbol{\Phi}_{\sim T}\widehat{\mathbf{a}}_q + \mathbf{n}_{x,q} \tag{7.116}$$

in which $\boldsymbol{\Phi}_{\sim T} = [\phi(\mathbf{x}_1)\ldots\phi(\mathbf{x}_{T-1})]^T$ and $\mathbf{n}_{x,q} \sim \mathcal{N}(\mathbf{0}, \sigma_X^2\mathbf{I})$ is i.i.d. noise, with variance σ_X^2. Therefore

$$p(\widehat{\mathbf{x}}_{q\sim 1} \mid \widehat{\mathbf{a}}_q) = \mathcal{N}\left(\boldsymbol{\Phi}_{\sim T}\widehat{\mathbf{a}}_q, \sigma_X^2\mathbf{I}\right) \tag{7.117}$$

and integrating out $\mathbf{a}_q$ yields

$$p(\widehat{\mathbf{x}}_{q\sim 1} \mid \sigma_X) = \mathcal{N}(\mathbf{0}, \mathbf{K}_{X\sim T}), \quad \mathbf{K}_{X\sim T} = \mathbf{\Phi}_{\sim T}\mathbf{\Phi}_{\sim T}^T + \sigma_X^2\mathbf{I} \tag{7.118}$$

where $\mathbf{K}_{X\sim T}$ can again be replaced by an equivalent kernel $k_X(\mathbf{x}_n, \mathbf{x}_{n'}|\boldsymbol{\theta}_X)$. For sequence or time series problems, we often combine common kernels with a linear kernel, e.g.

$$k_X(\mathbf{x}_n, \mathbf{x}_{n'}|\boldsymbol{\theta}_X) = \theta_{1,X}\exp\left(-\frac{\theta_{2,X}}{2}\left\|\mathbf{x}-\mathbf{x}'\right\|^2\right) + \theta_{3,X}\mathbf{x}^T\mathbf{x}' + \theta_{4,X}^{-1}\delta(\mathbf{x}_n - \mathbf{x}_{n'}) \tag{7.119}$$

in which $\boldsymbol{\theta}_X = \{\theta_{1,X}, \theta_{2,X}, \theta_{3,X}, \theta_{4,X}\}$, $\theta_{4,X}^{-1} = \sigma_X^2$. The joint (non-Gaussian) likelihood is

$$\begin{aligned} p(\mathbf{X} \mid \boldsymbol{\theta}_X) &= p(\mathbf{x}_1)\prod_{n=2}^{T} p(\widehat{\mathbf{x}}_{q\sim 1} \mid \boldsymbol{\theta}_X) \\ &= \frac{p(\mathbf{x}_1)}{\sqrt{(2\pi)^{(T-1)Q}|\mathbf{K}_{X\sim T}|^Q}}\exp\left\{-\frac{1}{2}\mathrm{tr}\left(\mathbf{K}_{X\sim T}^{-1}\mathbf{X}_{\sim 1}\mathbf{X}_{\sim 1}^{\top}\right)\right\} \end{aligned} \tag{7.120}$$

in which $\mathbf{X}_{\sim 1}$ excludes the first row from $\mathbf{X}$. As mentioned earlier, $\mathbf{x}_0$ is unknown, so that a specified distribution is placed over $\mathbf{x}_1$ using, for example, and isotropic Gaussian.

To estimate the latent variables and all hyperparameters we can use a maximum likelihood approach. In order to ameliorate overfitting, inverse priors can be placed over the hyperparameters

$$p(\boldsymbol{\theta}_X) \propto \prod_i \theta_{i,X}^{-1},\ p(\boldsymbol{\theta}_Y) \propto \prod_j \theta_{j,Y}^{-1}, \quad p(\mathbf{W}) = \prod_d w_d^{-1} \tag{7.121}$$

with a posterior satisfying

$$p(\mathbf{X}, \boldsymbol{\theta}_X, \boldsymbol{\theta}_Y, \mathbf{W} \mid \mathbf{Y}) \propto p(\mathbf{Y} \mid \mathbf{X}, \boldsymbol{\theta}_Y, \mathbf{W})p(\mathbf{X} \mid \boldsymbol{\theta}_X)p(\boldsymbol{\theta}_X)p(\boldsymbol{\theta}_Y)p(\mathbf{W}) \tag{7.122}$$

Minimising $-\mathcal{L} = -\ln p(\mathbf{X}, \boldsymbol{\theta}_X, \boldsymbol{\theta}_Y, \mathbf{W} \mid \mathbf{Y})$ (ignoring constants) then provides the solution

$$\begin{aligned} \mathbf{X}^*, \boldsymbol{\theta}_X^*, \boldsymbol{\theta}_Y^*, \mathbf{W}^* &= \underset{\mathbf{X}, \boldsymbol{\theta}_X \boldsymbol{\theta}_Y, \mathbf{W}}{\mathrm{argmin}} -\mathcal{L}(\mathbf{X}, \boldsymbol{\theta}_X, \boldsymbol{\theta}_Y, \mathbf{W}) \\ \mathcal{L}(\mathbf{X}, \boldsymbol{\theta}_X, \boldsymbol{\theta}_Y, \mathbf{W}) &= -\frac{Q}{2}\ln|\mathbf{K}_{X\sim T}| - \frac{1}{2}\mathrm{tr}\left(\mathbf{K}_{X\sim T}^{-1}\mathbf{X}_{\sim 1}\mathbf{X}_{\sim 1}^{\top}\right) + T\ln|\mathbf{W}| + \frac{D}{2}\ln|\mathbf{K}_Y| \\ &\quad - \frac{1}{2}\mathrm{tr}\left(\mathbf{K}_Y^{-1}\mathbf{Y}\mathbf{W}^2\mathbf{Y}^{\top}\right) - \sum_i \ln\theta_{i,X} - \sum_j \ln\theta_{j,X} - \sum_d \ln w_d \end{aligned} \tag{7.123}$$

To make predictions, the simplest methods is *mean-prediction* (as in the autoregressive GP method of Sect. 7.4.3). That is, only the expected value from the GP posterior is used, which neglects the variances (or at least makes them unreliable). $\mathbf{x}_t$ conditioned on $\mathbf{x}_{n-1}$ is predicted as follows

$$\begin{aligned} p(\mathbf{x}_n \mid \mathbf{x}_{n-1}, \boldsymbol{\theta}_X) &= \mathcal{N}\left(\boldsymbol{\mu}_X(\mathbf{x}_{n-1}), v_X(\mathbf{x}_{n-1})\mathbf{I}\right) \\ \boldsymbol{\mu}_X(\mathbf{x}) &= \mathbf{X}_{\sim 1}^{\top}\mathbf{K}_{X\sim T}^{-1}\mathbf{k}_{X\sim T}(\mathbf{x}) \\ v_X(\mathbf{x}) &= k_X(\mathbf{x},\mathbf{x}|\boldsymbol{\theta}_X) - \mathbf{k}_{X\sim T}(\mathbf{x})^{\top}K_{X\sim T}^{-1}\mathbf{k}_{X\sim T}(\mathbf{x}) \\ \mathbf{k}_{X\sim T}(\mathbf{x}) &= (k_X(\mathbf{x},\mathbf{x}_1|\boldsymbol{\theta}_X),\ldots,k_X(\mathbf{x},\mathbf{x}_{T-1}|\boldsymbol{\theta}_X))^T \end{aligned} \tag{7.124}$$

To be precise, this means that $\mathbf{x}_{n-1} = \boldsymbol{\mu}_X(\mathbf{x}_{n-2})$ is used to predict $\mathbf{x}_n$ from Eq. (7.124). In the same way, $\mathbf{y}_n$ can be predicted from the following posterior using the estimated value of $\mathbf{x}_n$

$$\begin{aligned} p(\mathbf{y}_n \mid \mathbf{x}_n, \boldsymbol{\theta}_Y, \mathbf{W}) &= \mathcal{N}\left(\boldsymbol{\mu}_Y(\mathbf{x}_n), v_Y(\mathbf{x}_n)\mathbf{W}^{-2}\right) \\ \boldsymbol{\mu}_Y(\mathbf{x}) &= \mathbf{Y}^{\top}\mathbf{K}_Y^{-1}\mathbf{k}_Y(\mathbf{x}) \\ v_Y(\mathbf{x}) &= k_Y(\mathbf{x},\mathbf{x}|\boldsymbol{\theta}_Y) - \mathbf{k}_Y(\mathbf{x})^{\top}K_Y^{-1}\mathbf{k}_Y(\mathbf{x}) \\ \mathbf{k}_Y(\mathbf{x}) &= (k_Y(\mathbf{x},\mathbf{x}_1|\boldsymbol{\theta}_Y),\ldots,k_Y(\mathbf{x},\mathbf{x}_T|\boldsymbol{\theta}_Y))^T \end{aligned} \tag{7.125}$$

7.4.6 Adjusting for Deterministic Trends and Seasonality

A time series can often be decomposed into various sub-components, which helps to simplify the analysis and to select a suitable method, often by transforming it into a stationary process. Underlying the series or process may be a deterministic part such as a linear decay. Moreover, there may be a *seasonal* trend, i.e., some fluctuations occurring at regular intervals. There are two main ways to deal with such effects, namely *additive* and *multiplicative* models. In the first case, we may write the series or process in the form

$$y_n = d_n + \pi_n + f_n \tag{7.126}$$

in which d_n is the deterministic part (non-random), π_n is a deterministic seasonal part and f_n is the *remainder* or *residual* or *irregular* process. Any deterministic part can be removed by defining a new process: $y_n \mapsto y_n - d_n$. This of course relies on full knowledge of d_n, which is often not available. We assume from hereon, that any *known* deterministic trend has been removed, so that $y_n = \pi_n + f_n$.

If it is constant, the seasonal process satisfies

$$\pi_n = \pi_{n+ip'}, \quad i = \pm 1, \pm 2, \ldots \tag{7.127}$$

for some period p'. The process y_n can then be turned into a stationary process (assuming that the residual process is stationary) using a so-called *seasonal differencing operator*

$$\Delta_{p'} = 1 - L^{p'}, \quad \Delta_{p'} y_n = y_n - y_{n-p'} \tag{7.128}$$

We can apply this operator to y_n to yield

$$\Delta_{p'} y_n = \Delta_{p'}(\pi_n + f_n) = \Delta_{p'} f_n \tag{7.129}$$

by virtue of the fact that $\pi_n = \pi_{n+ip'}$. The result is therefore stationary, under the assumption on f_n, because the property of stationarity is retained by a process following the application of a differencing operation. This leads to a *multiplicative seasonal* ARIMA (SARIMA) model

$$\begin{aligned} \Gamma_P(L^{p'})\gamma_p(L)\Delta_{p'}^D y_n &= \beta_q(L)\Theta_Q(L^{p'})z_n \\ \Delta_{p'}^D &= (1 - L^{p'})^D \\ \Gamma_P(L^{p'}) &= 1 - \Gamma_1 L^{p'} - \ldots - \Gamma_P(L^{p'})^P \\ \Theta_Q(L^{p'}) &= 1 - \Theta_1 L^{p'} - \ldots - \Theta_Q(L^{p'})^Q \end{aligned} \tag{7.130}$$

Here we define a seasonal differencing operator $\Delta_{p'}^D$, a seasonal AR operator $\Gamma_P(L^{p'})$ and a seasonal MA operator $\Theta_Q(L^{p'})$. We call such a model the SARIMA $(P, D, Q)_{p'} \times (p, 1, q)$ model or process, with usually $D = 1$.

7.4.7 Tests for Stationarity

When deciding on an appropriate model for a time series (selecting embedding orders, lags, (S)ARIMA models) it is useful to first conduct a series of tests (alongside visual inspection), rather than use a brute force approach that cycles through different model choices. The main tests include those for stationarity, which is particularly important for the application of ARIMA methods, and inspections of the *sample autocorrelation* and *partial autocorrelation* functions, which helps in deciding the model choices with respect to the AR and MA components.

Unit root tests are used to test for stationarity; they are one-tailed hypothesis tests with a null hypothesis being the existence of a unit root. From the preceding discussions on ARIMA models, this would indicate a non-stationary process. The contrary hypothesis is that there is a root of magnitude exceeding 1, suggesting a stationary I(0) process. The most often employed test is called the augmented Dicky-Fuller (ADF) test, which is based on the process obeying the following ARMA model

$$y_n = \phi y_n + \sum_{i=1}^{p} \psi_i \Delta y_{n-i} + z_n \tag{7.131}$$

The p lagged terms Δs_{n-i} are used to approximate an ARMA model. Denoting by $\widehat{\phi}$ the maximum likelihood or least-squares estimator of ϕ, we use the test statistic

$$t = \frac{\widehat{\phi} - 1}{SE(\widehat{\phi})} \tag{7.132}$$

where $SE(\widehat{\phi})$ denotes the standard error in the estimator.

To select a value of p we can use information criteria, with the Bayesian or Schwarz criterion (BIC) and the Akaike criterion (AIC) [43] being the most commonly employed. For an ARMA model involving k parameters $\boldsymbol{\theta}$, and a least-squares (or other) estimate $\widehat{\boldsymbol{\theta}}$ based on N data points, these criteria are defined as follows

$$\begin{aligned} \text{BIC} &= -2 \ln L(\widehat{\boldsymbol{\theta}}) + k \ln N \\ \text{AIC} &= -2 \ln L(\widehat{\boldsymbol{\theta}}) + 2k \end{aligned} \tag{7.133}$$

in which $L(\widehat{\boldsymbol{\theta}})$ denotes the likelihood function. If the series is not stationary, successively higher order differencing can be applied until the differenced series is stationary. The order of the differencing required for stationarity set the value of d in an ARIMA(p, d, q) model.

7.4.8 Autocorrelation and Partial Autocorrelation Analyses

Alternative and in fact complimentary methods for selecting values of p and q in ARMA models, as well as to inform embedding choices, are analyses of the autocorrelation function (ACF) and the partial autocorrelation function (PACF). The former is approximated from the time series *via* the sample autocorrelation

$$\begin{aligned} \rho(\tau) &= \frac{k(\tau)}{k(0)} \\ k(\tau) &= \frac{1}{N} \sum_{n=\tau}^{N-1} (y_n - \bar{y})(y_{n-\tau} - \bar{y}), \quad \tau = 0, \ldots, N-1 \end{aligned} \tag{7.134}$$

in which the sample mean of the time series is denoted $\bar{y}$. The ACF is defined as the function $\rho(\tau)$ for a lag τ, meaning the number of time units n between the values for which the correlation is estimated. A plot of this function can be used to decide the order q of the MA part of the model, as well as to test for stationarity. A slow decay in the ACF would indicate a near-unit-root process, as opposed to a geometric decay that is indicative of a stationary process, as illustrated for the non-stationary process in Fig. 7.2.

On the other hand, the simulated MA(2) process in Fig. 7.2 exhibits a cut-off after the second lag (the values within the blue lines are statistically insignificant with 95%

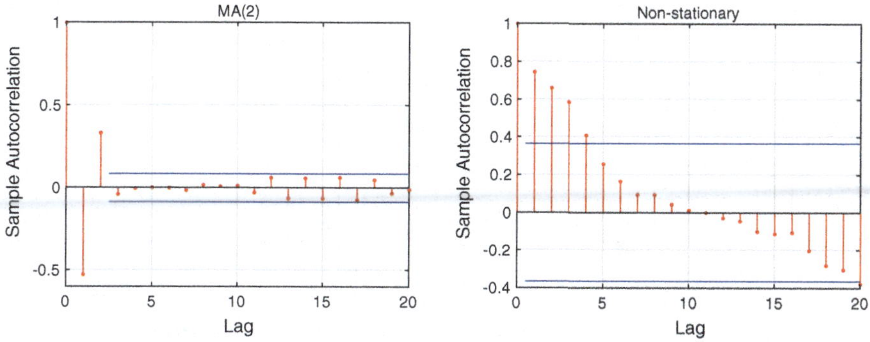

Fig. 7.2 Illustration of the ACF for an MA(2) and a non-stationary process

confidence). This behaviour is indicative of an MA(2) process. A gradual geometric decay in the ACF, on the other hand, would indicate an AR process.

The PACF, denoted $\phi_{\tau\tau}$, where τ is the lag, is defined as follows

$$\begin{aligned} \phi_{11} &= \mathrm{corr}(y_{n+1}, y_n) = \rho(1) \\ \phi_{\tau\tau} &= \mathrm{corr}(y_{n+\tau} - \widehat{y}_{n+\tau}, y_n - \widehat{y}_n) \end{aligned} \tag{7.135}$$

in which

$$\begin{aligned} \widehat{y}_{n+\tau} &= \mathbb{E}[y_{n+\tau} \mid y_{n+\tau-1}, \ldots, y_{n+1}] \\ \widehat{y}_n &= \mathbb{E}[y_n \mid y_{n+\tau-1}, \ldots, y_{n+1}] \end{aligned} \tag{7.136}$$

are optimal linear predictors of $y_{n+\tau}$ and y_n conditioned on the data. The $\phi_{\tau\tau}$ estimate $\mathrm{corr}(y_{n+\tau}, y_n)$ by eliminating the intermediate relationship with the samples $y_{n+\tau-1}, \ldots, y_{n+1}$. They can be approximated using the sample autocorrelation based on recursive procedures. In an AR(p) process, theoretically $\phi_{\tau\tau} = 0$ for $\tau > p$, so that inspection of the PACF can determine p. Figure 7.3 shows the ACF and PACF for an AR(4) process, in which it can be seen that the ACF decays slowly, while the PACF tails off after the 4th lag (note that lag 0 corresponds to the correlation of y_n with itself so that we expect it to have a value of 1).

For an ARIMA process we would expect to see a gradual decay in both the PACF and ACF, while for a seasonal process, values of both the ACF and PACF for lags in the vicinity of each period would be statistically significant. It has to be noted, however, that visual inspection of the ACF and PACF will not yield any definitive information for many complicated time series.

7.5 Multi-fidelity Modelling for Electrochemical Systems

We are not aware of any existing multi-fidelity or reduced-order modelling applications to flow batteries, but there have been a small number of attempts to use

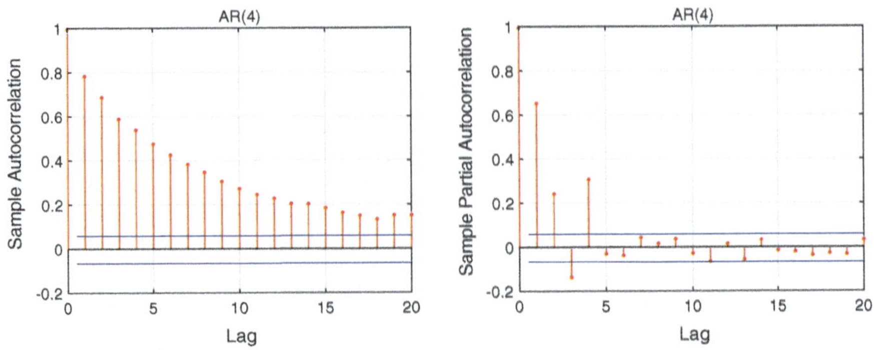

Fig. 7.3 Illustration of the ACF and PACF for an AR(4) process

multi-fidelity models for fuel cells [15] and reduced-order models for Li-ion batteries [44, 45]. Here we describe the ResGP model of Xing et al. [15], which was used in the study of a solid-oxide fuel cell, comparing it to GreedyNAR, GPNAR and stochastic collation, all of which were described in Sect. 7.2.

The authors considered a steady-state 3D solid-oxide fuel model, incorporating charge balances in the electron- and ion-conducting phases based on Ohm's law (4.16), the flow in the gas channels based on Navier-Stoke's equations (3.14), the flow in the porous electrodes based on Brinkman's equation (3.30) and mass balances (3.26) of the species in the porous electrodes and channels. A Maxwell-Stefan model was employed for diffusion and convection of the species. The charge-transfer kinetics were described using the Butler-Volmer law (4.25). The reactions in an SOFC are

$$\begin{aligned} H_2 + O^{2-} &\rightarrow H_2O + 2e^- \\ O_2 + 4e^- &\rightarrow 2O^{2-} \end{aligned} \tag{7.137}$$

in the anode and cathode, respectively. It was assumed that the cell operated in potentiostatic mode, so that the cell voltage was varied and the current density was calculated.

The authors chose as inputs

1. The electrode porosities $\epsilon \in [0.4, 0.85]$
2. The cell voltage $E_c \in [0.2, 0.85]$ V
3. The temperature $T \in [973, 1273]$ K
4. The channel pressures $P \in [0.5, 2.5]$ atm.

Using a Sobol sequence (Sect. 3.7.2), 60 points were selected to conduct low- and high-fidelity simulations using the finite-element method, with a further 40 points generated at high-fidelity for testing. The low-fidelity (F1) model was defined by 3164 mapped elements and a relative error tolerance of 0.1, and the high-fidelity (F2) model was defined by 37064 elements and a relative error tolerance of 0.001. The outputs used for the multi-fidelity models were the electrolyte current density

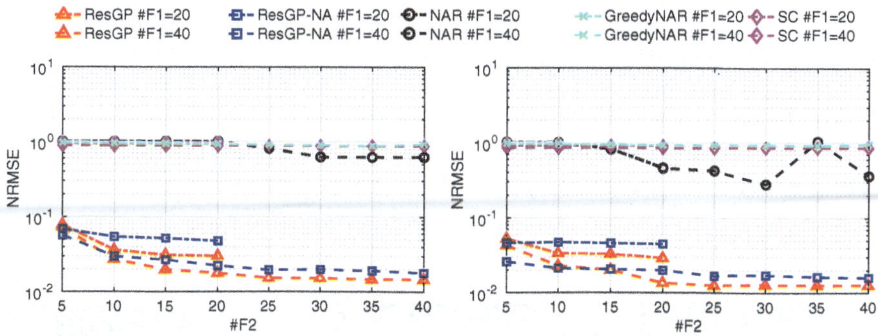

Fig. 7.4 NRMSE for all methods. The left-hand figure corresponds the current density and the right-hand figure to the electrolyte potential. Reproduced from [15] with permission. Copyright Elsevier

(A m^{-2}) and the ionic potential (V) in a plane located at the channel centres. Both outputs were vectorised values of these quantities at 5000 spatial locations.

The accuracy was assessed using a normalised root mean square error

$$NRMSE = \sqrt{\frac{\sum_{n=1}^{N}\sum_{j=1}^{d}(y_{nj} - \hat{y}_{nj})^2}{\sum_{n=1}^{N}\sum_{j=1}^{d}(\hat{y}_{nj})^2}} \tag{7.138}$$

in which y_{jd} ($\hat{y}_{jd}$) is the j-th coefficient of the n-th prediction (real value or 'ground truth'). The number of test points was $N = 40$.

The NRMSE values obtained using a five-fold cross validation for the cases of 20 and 40 F1 training points are shown in Fig. 7.4, for both the ionic conductivity and the current density. In these figures, the number of F2 training points is gradually increased. There are two cases for ResGP. The one labelled ResGP-NA employed active learning to select the high-fidelity training points based on maximising the information gain [15]. The other version ResGP did not use active learning.

As can be seen from Fig. 7.4, NARGP and Greedy NAR perform poorly, which is not unexpected because the dimensionality of the output space ($d = 5000$) is high. 5001 hyperparameters are required since the low-fidelity output is used as an input for the high-fidelity mapping. Learning this many hyperparameters with a maximum likelihood is highly challenging for any optimisation algorithm. In the case of stochastic collocation, without access to the F1 data for testing, the performance is also poor. ResGP exhibits a steady decline in the NRMSE with an increasing number of F2 training points. With 40 F1 and 10 F2 training points, the NRMSE is 97% lower than that for the next best method on the current density data, when active learning is used. Similarly, for the ionic potential, the RMSE is around 90% lower than that for the next best method, for all numbers of F2 training points.

ResGP (without active learning) predictions of the current density and ionic potential using 40 F1 and 20 F2 training points are shown in Figs. 7.5 and 7.6, together with the actual test cases and point-wise absolute differences between the predictions and tests. In both of these figures, the predictions with the lowest and highest error are shown, alongside a prediction near the median error. It is evident that even in the case of the highest errors, the accuracy of ResGP is both qualitatively and quantitatively high.

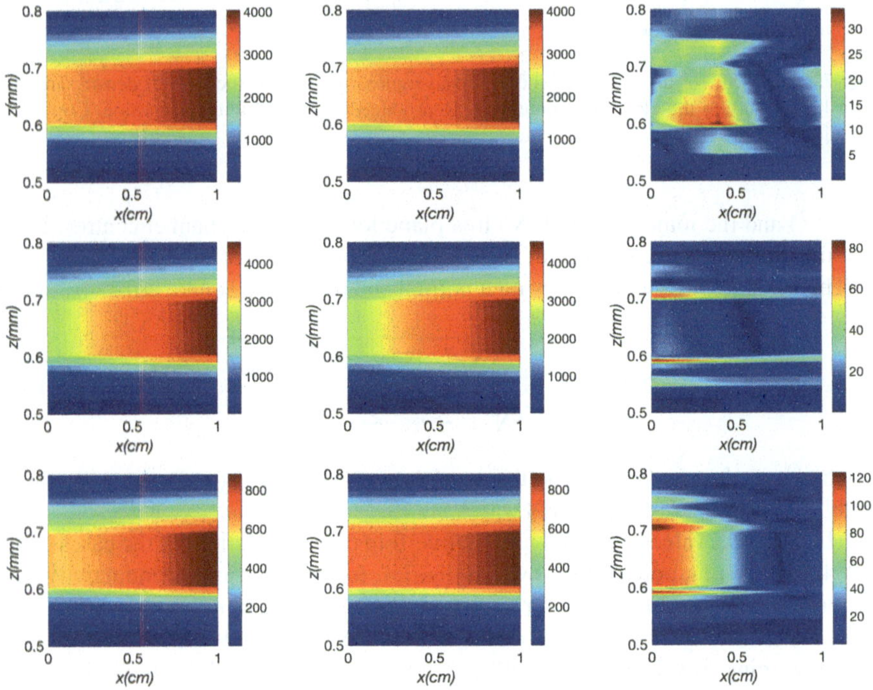

Fig. 7.5 Predictions of the electrolyte current density (A m^{-2}) for 40 F1 and 20 F2 training points. From the top to the bottom the predictions correspond to the lowest error, (near) median and highest errors. The first column is the prediction, the second is the ground truth and the third is the point-wise absolute differences. Reproduced from [15] with permission. Copyright Elsevier

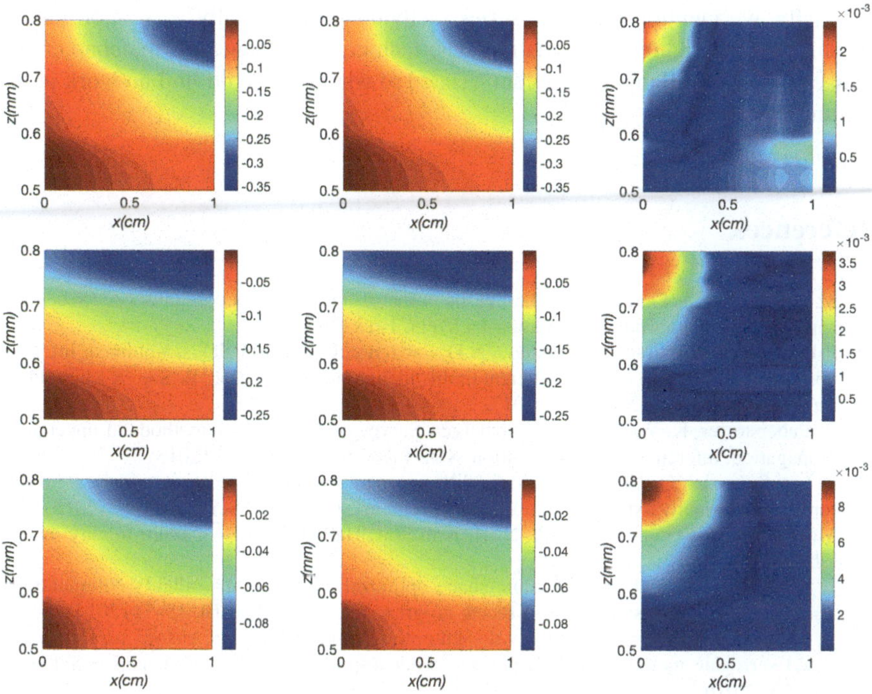

Fig. 7.6 Predictions of the ionic potential (V) for 40 F1 and 20 F2 training points. From the top to the bottom the predictions correspond to the lowest error, (near) median and highest errors. The first column is the prediction, the second is the ground truth and the third is the point-wise absolute differences. Reproduced from [15] with permission. Copyright Elsevier

7.6 Summary

In this chapter we introduced two alternatives to the machine-learning surrogate modeling approach described in Chaps. 3 and 6. We hope that we have inspired readers that are familiar with the latter approach to consider these alternatives, especially multi-fidelity models, which in some cases can be more powerful than pure machine learning.

There is a scenario in particular in which multi-fidelity models could gain traction, and that is when it is not necessarily better to 'throw out' the physics-based model entirely. When accuracy is paramount, as opposed to simulation effort and time, it may well be appropriate to use a low-fidelity model and find a correction to this model through machine learning. We imagine that this could be the case for electronic-structure calculations, in which, e.g., DFT could be considered low-fidelity and coupled cluster (CC) theory considered high-fidelity. We could in principle build a multi-fidelity model that takes as input the raw DFT result, which is comparatively cheap to obtain, and provides CC-level accuracy.

Finally, we mention that time series methods are potentially very important to the study of flow battery degradation. The techniques presented in the chapter can be used for this challenging problem, related to which we could find no work in the existing literature.

References

1. M.C. Kennedy, A. O'Hagan, Predicting the output from a complex computer code when fast approximations are available. Biometrika **87**(1), 1–13 (2000)
2. P. Perdikaris, M. Raissi, A. Damianou, N.D. Lawrence, G.E. Karniadakis, Nonlinear information fusion algorithms for data-efficient multi-fidelity modelling. Proc. R. Soc. A Math. Phys. Eng. Sci. **473**(2198), 20160751 (2017)
3. B. Peherstorfer, K. Willcox, M. Gunzburger, Survey of multifidelity methods in uncertainty propagation, inference, and optimization. SIAM Rev. **60**(3), 550–591 (2018)
4. B. Liu, S. Koziel, Q. Zhang, A multi-fidelity surrogate-model-assisted evolutionary algorithm for computationally expensive optimization problems. J. Comput. Sci. **12**, 28–37 (2016)
5. M.G. Fernandez-Godino, C. Park, N.-H. Kim, R.T. Haftka, Review of multi-fidelity models (2016)
6. L. Leifsson, S. Koziel, Aerodynamic shape optimization by variable-fidelity computational fluid dynamics models: a review of recent progress. J. Comput. Sci. **10**, 45–54 (2015)
7. L. Leifsson, S. Kozie, Multi-fidelity design optimization of transonic airfoils using physics-based surrogate modeling and shape-preserving response prediction. J. Comput. Sci. **1**(2), 98–106 (2010)
8. G. Venter, R.T. Haftka, J.H. Starnes, Construction of response surface approximations for design optimization. AIAA J. **36**(12), 2242–2249 (1998)
9. M. Eldred, A. Giunta, S. Collis, Second-order corrections for surrogate-based optimization with model hierarchies, in *AIAA Paper: Multidisciplinary Analysis and Optimization Conference, 2004* (2004)
10. L. Le Gratiet, *Multi-fidelity Gaussian process regression for computer experiments*. PhD thesis, Université Paris-Diderot-Paris VII, 2013
11. L. Parussini, D. Venturi, P. Perdikaris, G.E. Karniadakis, Multi-fidelity Gaussian process regression for prediction of random fields. J. Comput. Phys. **336**(C), 36–50 (2017)
12. A. Damianou, N. Lawrence, Deep gaussian processes, in *Artificial Intelligence and Statistics*. (2013), pp. 207–215
13. A. Narayan, C. Gittelson, D. Xiu, A stochastic collocation algorithm with multifidelity models. SIAM J. Sci. Comput. **36**(2), A495–A521 (2014)
14. W. Xing, M. Razi, R.M. Kirby, K. Sun, A.A. Shah, Greedy nonlinear autoregression for multifidelity computer models at different scales. Energy and AI **1**, 100012 (2020)
15. W.W. Xing, A.A. Shah, P. Wang, S. Zhe, Q. Fu, R.M. Kirby, Residual gaussian process: A tractable nonparametric Bayesian emulator for multi-fidelity simulations. Appl. Math. Model. **97**, 36–56 (2021)
16. K. Cutajar, M. Pullin, A. Damianou, N. Lawrence, J. González, Deep gaussian processes for multi-fidelity modeling (2019). arXiv:1903.07320
17. A. Lunardi, *Interpolation Theory*, vol. 9. (Springer, 2009)
18. A.A. Shah, W.W. Xing, V. Triantafyllidis, Reduced-order modelling of parameter-dependent, linear and nonlinear dynamic partial differential equation models. Proc. R. Soc. A Math. Phys. Eng. Sci. **473**(2200), 20160809 (2017)
19. L. Sirovich, Turbulence and the dynamics of coherent structures: part i: Coherent structures. Quarterly Appl. Math. **45**, 561–571 (1987)
20. G. Berkooz, P. Holmes, J.L. Lumley, The proper orthogonal decomposition in the analysis of turbulent flows. Ann. Rev. Fluid Mech. **25**, 539–575 (1993)

21. A.J. Newman, Model reduction via the Karhunen-Loeve expansion part i: an exposition. *Technical Report T.R.96-32*, University of Maryland, College Park, MD., 1996
22. E. Wong, *Stochastic Processes in Information and Dynamical Systems*. (McGraw-Hill, 1971)
23. T. Bui-Thanh, K. Willcox, O. Ghattas, Model reduction for large-scale systems with high-dimensional parametric input space. SIAM J. Sci. Comput. **30**, 3270–3288 (2008)
24. M.A. Grepl, A.T. Patera, A posteriori error bounds for reduced-basis approximations of parametrized parabolic partial differential equations. ESAIM: M2AN **39**(1), 157–181 (2005)
25. M.A. Grepl, Y. Maday, N.C. Nguyen, A.T. Patera, Efficient reduced-basis treatment of nonaffine and nonlinear partial differential equations. ESAIM: M2AN **41**(3), 575–605 (2007)
26. T. Bui-Thanh, K. Willcox, O. Ghattas, Parametric reduced-order models for probabilistic analysis of unsteady aerodynamic applications. AIAA J. **46**(10), 2520–2529 (2008)
27. M. Barrault, Y. Maday, N.C. Nguyen, A.T. Patera, An "empirical interpolation" method. C. R. Acad. Sci. Paris Ser. I Math **339**, 667–672 (2004)
28. J.A. Taylor. Dynamics of large scale structures in turbulent shear layers. (Department of Mechanical & Aeronautical Engineering, Clarkson University, NY, Rept. MAE-354, 2001)
29. J.A. Taylor, M.N. Glauser, Towards practical flow sensing and control via pod and lse based low-dimensional tools. J. Fluids Eng. **126**(3), 337–345 (2004)
30. J. Degroote, J. Vierendeels, K. Willcox, Interpolation among reduced-order matrices to obtain parameterized models for design, optimization and probabilistic analysis. Int. J. Numer. Meth. Fluids **63**(2), 207–230 (2010)
31. T. Lieu, C. Farhat, M. Lesoinne, Reduced-order fluid/structure modeling of a complete aircraft configuration. Comput. Methods Appl. Mech. Eng. **195**(41), 5730–5742 (2006)
32. D. Amsallem, C. Farhat, Interpolation method for adapting reduced-order models and application to aeroelasticity. AIAA J. **46**(7), 1803–1813 (2008)
33. D. Amsallem, J. Cortial, K. Carlberg, C. Farhat, A method for interpolating on manifolds structural dynamics reduced-order models. Int. J. Numer. Meth. Eng. **80**(9), 1241–1258 (2009)
34. Y. Chen, Model order reduction for nonlinear systems. Master's thesis, MIT, Cambridge, MA, 1999
35. Z. Bai, Krylov subspace techniques for reduced-order modeling of large-scale dynamical systems. Appl. Numer. Math. **43**(1), 9–44 (2002)
36. S. Chaturantabut, D.C. Sorensen, Nonlinear model reduction via discrete empirical interpolation. SIAM J. Sci. Comput. **32**(5), 2737–2764 (2010)
37. S. Chaturantabut, D.C. Sorensen, A state space error estimate for pod-deim nonlinear model reduction. SIAM J. Numer. Anal. **50**(1), 46–63 (2012)
38. P. Benner, S. Gugercin, K. Willcox, A survey of projection-based model reduction methods for parametric dynamical systems. SIAM Rev. **57**(4), 483–531 (2015)
39. K. Carlberg, C. Farhat, J. Cortial, D. Amsallem, The GNAT method for nonlinear model reduction: effective implementation and application to computational fluid dynamics and turbulent flows. J. Comput. Phys. **242**, 623–647 (2013)
40. C. Chatfield, *The Analysis of Time Series: An Introduction*. (Chapman and Hall/CRC, 2003)
41. A. O'Hagan, Curve fitting and optimal design for prediction. J. R. Stat. Soc. Ser. B (Methodol.) **40**(1), 1–42 (1978)
42. J. Wang, A. Hertzmann, D.J. Fleet, Gaussian process dynamical models, in *Advances in Neural Information Processing Systems*, vol. 18 (2005)
43. A. Chakrabarti, J.K. Ghosh, Aic, bic and recent advances in model selection, in *Philosophy of Statistics*, ed. by P.S. Bandyopadhyay, M.R. Forster. Handbook of the Philosophy of Science, vol. 7 (North-Holland, Amsterdam, 2011), pp.583–605
44. L. Cai, R.E. White, An efficient electrochemical-thermal model for a lithium-ion cell by using the proper orthogonal decomposition method. J. Electrochem. Soc. **157**, A1188–A1195 (2010)
45. L. Cai, R.E. White, Reduction of model order based on proper orthogonal decomposition for lithium-ion battery simulation. J. Electrochem. Soc. **156**, A154–A161 (2009)

Chapter 8
Summary and Outlook

Redox flow batteries (RFBs) are likely to be a key technology in the drive towards decarbonising grids, which relies heavily on intermittent power-generating technologies, and therefore requires energy storage capabilities. Demand and smart management can go some way towards off-setting the need for energy storage, but can in no way act as a replacement. Of the energy storage solutions available, redox flow batteries is the only one that is simultaneously site-independent, scaleable, energy-efficient and zero-carbon.

Modelling and simulation can play important roles in further developing, optimising and operating RFBs, their components and their materials. In this book, we have introduced the reader to a wide variety of methods that are appropriate for the modelling of redox flow batteries, both physics-based and data-driven, as well as methods that combine both of these approaches. There are indeed many that we did not cover, and their omission does not diminish their importance; we have strived not to stray too far outside our own domains of expertise and understanding.

Traditional macroscopic models based on the continuum theory of matter have been successfully applied to the study of a broad range of electrochemical energy conversion technologies. The limitations of these models were discussed in Chap. 3, chiefly their inability to resolve details at small scales. With the advent of widely available computational resources, powerful multi-core processors, parallelisation and graphical processing unit (GPU) programming, together with advances in other modelling approaches, there are ample opportunities to make new contributions to the area of flow-battery modelling and simulation. These future contributions could go some way towards realising the potential of flow batteries.

It has long been hoped that electronic-structure and molecular-dynamics simulations can facilitate the high-throughput screening and design of new materials, including energy materials. Currently, however, such an approach is not feasible

A. A. Shah et al., *New Paradigms in Flow Battery Modelling*,
Engineering Applications of Computational Methods 16,
https://doi.org/10.1007/978-981-99-2524-7_8

given the long simulation times involved, especially given that standard methods such as DFT and MD are frequently of insufficient accuracy, necessitating higher levels of theory such as *ab-initio* MD. Here, machine learning can play a role, accelerating simulations and making possible multi-scale modelling within reasonable timescales. This sort of fusion, on the other hand, is heavily reliant on the quantity and quality of data, and on expressive numerical descriptors of the molecules under consideration. Although much progress has been made in this area, there are still many challenges to overcome.

Above all, we hope that this book has inspired readers to adopt some of the approaches that we have presented, or has provided new ideas for those already familiar with one or more of these approaches.

Appendix A
Solving Linear Systems

A.1 Linear Systems

We consider linear systems

$$\mathbf{A}\mathbf{x} = \mathbf{b} \tag{A.1}$$

in which we look for a solution $\mathbf{x} \in \mathbb{R}^n$ given $\mathbf{b} \in \mathbb{R}^n$ and some matrix $\mathbf{A} \in \mathbb{R}^{n\times n}$. This equation is linear in the sense that the operator or mapping $\mathbf{A}$ is linear

$$\begin{aligned} \mathbf{A}(\mathbf{x}+\mathbf{y}) &= \mathbf{A}\mathbf{x} + \mathbf{A}\mathbf{y} \\ \mathbf{A}(a\mathbf{x}) &= a\mathbf{A}\mathbf{x} \end{aligned} \tag{A.2}$$

for any two vectors $\mathbf{x} \in \mathbb{R}^n$ and $\mathbf{y} \in \mathbb{R}^n$, and any scalar $a \in \mathbb{R}$. These sorts of equations arise in many settings in the physical sciences, engineering, computer science and other areas. A great many numerical methods have been developed to solve such equations efficiently and accurately.

We can write the matrix as $\mathbf{A} = [a_{ij}]$ in which $a_{ij}, i, j = 1, \ldots, n$, are the elements of the matrix. Some special matrices are the diagonal and upper-diagonal matrices

$$\begin{bmatrix} a_{11} & 0 & \ldots & 0 \\ 0 & a_{22} & \ldots & 0 \\ \vdots & \vdots & \ddots & \vdots \\ 0 & 0 & \ldots & a_{nn} \end{bmatrix}, \quad \begin{bmatrix} a_{11} & a_{12} & \ldots & a_{1n} \\ 0 & a_{22} & \ldots & a_{2n} \\ \vdots & \vdots & \ddots & \vdots \\ 0 & 0 & \ldots & a_{nn} \end{bmatrix} \tag{A.3}$$

respectively, with lower-diagonal matrices similarly defined. Tri-diagonal matrices take the form

A. A. Shah et al., *New Paradigms in Flow Battery Modelling*, Engineering Applications of Computational Methods 16, https://doi.org/10.1007/978-981-99-2524-7

$$\begin{bmatrix} a_{11} & a_{12} & 0 & 0 & \dots & 0 \\ a_{21} & a_{22} & a_{23} & 0 & \dots & 0 \\ 0 & a_{32} & a_{33} & a_{34} & \dots & 0 \\ 0 & 0 & \ddots & \ddots & \ddots & \vdots \\ \vdots & \vdots & \ddots & \ddots & \ddots & a_{(n-1)n} \\ 0 & 0 & \dots & 0 & a_{n(n-1)} & a_{nn} \end{bmatrix} \tag{A.4}$$

and are often encountered in the numerical solution to partial differential equations.

The inverse of a square matrix $\mathbf{A} \in \mathbb{R}^{n\times n}$ is another matrix $\mathbf{A}^{-1} \in \mathbb{R}^{n\times n}$ defined by $\mathbf{A}^{-1}\mathbf{A} = \mathbf{A}\mathbf{A}^{-1} = \mathbf{I}$, in which $\mathbf{I}$ is the identity matrix. The inverse can be used to solve the linear system (A.1), namely, $\mathbf{x} = \mathbf{A}^{-1}\mathbf{b}$. However, this is not a very efficient solution in most computational tasks; the inverse is costly to compute and could be ill-conditioned (see later). In practice, algorithms use other methods, some of which we outline below.

A.2 Gauss Elimination

Gauss elimination is based on a substitution procedure, which involves eliminating certain of the x_i from the equations, solving for x_n and back substituting to successively calculate the remaining unknowns. Let us return to the linear system, which can be written explicitly as

$$\begin{aligned} a_{11}x_1 + \ldots + a_{1n}x_n &= b_1 \\ a_{21}x_1 + \ldots + a_{2n}x_n &= b_2 \\ a_{31}x_1 + \ldots + a_{3n}x_n &= b_3 \\ &\vdots \\ a_{n1}x_1 + \ldots + a_{nn}x_n &= b_n \end{aligned} \tag{A.5}$$

in which x_i and b_i are the components of $\mathbf{x}$ and $\mathbf{b}$, respectively. The first part of the algorithm is called forward elimination and it reduces the system to one that is upper-triangular.

We eliminate the first unknown x_1 from all but the first of Eqs. (A.5) by multiplying the first equation with a_{21}/a_{11}

$$a_{21}x_1 + \frac{a_{21}}{a_{11}}a_{12}x_2 + \ldots + \frac{a_{21}}{a_{11}}a_{1n}x_n = \frac{a_{21}}{a_{11}}b_1 \tag{A.6}$$

We now subtract this equation from the second of Eqs. (A.5) to obtain

$$\left(a_{22} - \frac{a_{21}}{a_{11}}a_{12}\right)x_2 + \ldots + \left(a_{2n} - \frac{a_{21}}{a_{11}}a_{1n}\right)x_n = b_2 - \frac{a_{21}}{a_{11}}b_1 \tag{A.7}$$

which can be written as

$$a_{22}^{(1)}x_2 + \ldots + a_{2n}^{(1)}x_n = b_2^{(1)} \tag{A.8}$$

with $a_{22}^{(1)} = \left(a_{22} - \frac{a_{21}}{a_{11}}a_{12}\right)$, etc. We repeat this procedure for the remaining equations, e.g., in the next step we multiply the first equation by a_{31}/a_{11} and subtract the result from the third of Eqs. (A.5), and so on. The end result is

$$\begin{aligned} a_{11}x_1 + a_{12}x_2 + a_{13}x_3 + \ldots + a_{1n}x_n &= b_1 \\ a_{22}^{(1)}x_2 + a_{23}^{(1)}x_3 + \ldots + a_{2n}^{(1)}x_n &= b_2^{(1)} \\ a_{32}^{(1)}x_2 + a_{33}^{(1)}x_3 + \ldots + a_{3n}^{(1)}x_n &= b_3^{(1)} \\ &\vdots \\ a_{n2}^{(1)}x_2 + a_{n3}^{(1)}x_3 + \ldots + a_{nn}^{(1)}x_n &= b_n^{(1)} \end{aligned} \tag{A.9}$$

In this procedure, the first of Eqs. (A.5) is called the pivot equation and a_{11} is called the pivot coefficient. In the next part of the algorithm, the procedure above is repeated to eliminate x_2 from the third to the last of Eqs. (A.9). In the first step we multiply the second of Eqs. (A.9) by $a_{32}^{(1)}/a_{22}^{(1)}$ and subtract the result from the third equation, resulting in

$$\left(a_{33}^{(1)} - \frac{a_{32}^{(1)}}{a_{22}^{(1)}}a_{23}^{(1)}\right)x_3 + \ldots + \left(a_{3n}^{(1)} - \frac{a_{32}^{(1)}}{a_{22}^{(1)}}a_{2n}^{(1)}\right)x_n = b_3^{(1)} - \frac{a_{32}^{(1)}}{a_{22}^{(1)}}b_2^{(1)} \tag{A.10}$$

which can be written as

$$a_{33}^{(2)}x_3 + \ldots + a_{3n}^{(2)}x_n = b_3^{(2)} \tag{A.11}$$

We again repeat this to eliminate x_2 from the fourth to the last of (A.9) to obtain

$$\begin{aligned} a_{11}x_1 + a_{12}x_2 + a_{13}x_3 + \ldots + a_{1n}x_n &= b_1 \\ a_{22}^{(1)}x_2 + a_{23}^{(1)}x_3 + \ldots + a_{2n}^{(1)}x_n &= b_2^{(1)} \\ a_{33}^{(2)}x_3 + \ldots + a_{3n}^{(2)}x_n &= b_3^{(2)} \\ &\vdots \\ a_{n3}^{(2)}x_3 + \ldots + a_{nn}^{(2)}x_n &= b_n^{(2)} \end{aligned} \tag{A.12}$$

In this case the second equation in (A.9) was the pivot equation. We proceed in this fashion, now eliminating x_3 from the fourth to the last of Eqs. (A.12), with the third equation becoming the pivot equation, and then moving on to the elimination of x_4, and so on. The final set of equations is

$$\begin{aligned} a_{11}x_1 + a_{12}x_2 + a_{13}x_3 + \ldots + a_{1n}x_n &= b_1 \quad \text{(a)} \\ a_{22}^{(1)}x_2 + a_{23}^{(1)}x_3 + \ldots + a_{2n}^{(1)}x_n &= b_2^{(1)} \\ a_{33}^{(2)}x_3 + \ldots + a_{3n}^{(2)}x_n &= b_3^{(2)} \\ &\vdots \\ a_{nn}^{(n-1)}x_n &= b_n^{(n-1)} \end{aligned} \tag{A.13}$$

which is easily solved, since

$$x_n = \frac{b_n^{(n-1)}}{a_{nn}^{(n-1)}} \tag{A.14}$$

(A.14) can be substituted into the equation $(n-1)$ to obtain x_{n-1}. This result and x_n are then used in equation $(n-2)$ to obtain x_{n-2}, and so on. This part of the algorithm is called back-substitution, with the general formula

$$x_i = \frac{b_i^{(i-1)} - \sum_{j=i+1}^{n} a_{ij}^{(i-1)} x_j}{a_{ii}^{(i-1)}} \tag{A.15}$$

A.3 Ill-Conditioned Systems and Pivoting

Let us consider a simple example

$$\begin{aligned} x_1 + 2x_2 &= 10 \\ 1.1x_1 + 2x_2 &= 10.4 \end{aligned} \tag{A.16}$$

which is easy to solve by hand

$$\begin{aligned} x_1 &= \frac{2 \cdot 10 - 2 \cdot 10.4}{1 \cdot 2 - 2 \cdot 1.1} = 4 \\ x_2 &= \frac{1 \cdot 10.4 - 1.1 \cdot 10}{1 \cdot 2 - 2 \cdot 1.1} = 3 \end{aligned} \tag{A.17}$$

Now consider a similar looking example where we make a small change in one coefficient

$$\begin{aligned} x_1 + 2x_2 &= 10 \\ 1.05x_1 + 2x_2 &= 10.4 \end{aligned} \tag{A.18}$$

the solution to which is

$$\begin{aligned} x_1 &= \frac{2 \cdot 10 - 2 \cdot 10.4}{1 \cdot 2 - 2 \cdot 1.05} = 8 \\ x_2 &= \frac{1 \cdot 10.4 - 1.05 \cdot 10}{1 \cdot 2 - 2 \cdot 1.05} = 1 \end{aligned} \tag{A.19}$$

Notice that a small change in one coefficient leads to a quite different solution. The issue here is that the denominators $(1 \cdot 2 - 2 \cdot 1.1)$ and $(1 \cdot 2 - 2 \cdot 1.05)$ are close to zero, so dividing by these numbers can change the result dramatically if we change one of the coefficients.

Let us try to make this observation more precise and devise a general rule that tells us when we will encounter this type of behaviour. Consider the general form

$$\begin{aligned} a_{11}x_1 + a_{12}x_2 &= b_1 \\ a_{21}x_1 + a_{22}x_2 &= b_2 \end{aligned} \tag{A.20}$$

with solution

$$\begin{aligned} x_1 &= \frac{a_{22} \cdot b_1 - a_{12} \cdot b_2}{a_{11} \cdot a_{22} - a_{21} \cdot a_{12}} \\ x_2 &= \frac{a_{21} \cdot b_1 - a_{11} \cdot b_2}{a_{11} \cdot a_{22} - a_{21} \cdot a_{12}} \end{aligned} \tag{A.21}$$

We can also write the system as

$$\underbrace{\begin{bmatrix} a_{11} \; a_{12} \\ a_{21} \; a_{22} \end{bmatrix}}_{\mathbf{A}} \begin{bmatrix} x_1 \\ x_2 \end{bmatrix} = \begin{bmatrix} b_1 \\ b_2 \end{bmatrix} \tag{A.22}$$

Notice that the determinant of the matrix $\mathbf{A}$ is the denominator $|\mathbf{A}| = a_{11} \cdot a_{22} - a_{21} \cdot a_{12}$ in (A.21). The inverse of $\mathbf{A}$ is

$$\mathbf{A}^{-1} = \frac{1}{|\mathbf{A}|} \begin{bmatrix} a_{22} & -a_{12} \\ -a_{21} & a_{11} \end{bmatrix} \tag{A.23}$$

When $|\mathbf{A}|$ is close to zero, $\mathbf{A}^{-1}$ changes dramatically when small alterations are made to the coefficients of $\mathbf{A}$; that the solution to the linear system is $\mathbf{x} = \mathbf{A}^{-1}\mathbf{b}$. The problem for general systems is the same: the size of $|\mathbf{A}|$ determines how stable the solutions are in some sense. We can define ill-conditioned systems as those for which a small change in one or more coefficients results in a large change in the solutions.

Well-conditioned systems, on the other hand, are those for which a small change in one or more coefficients results in a correspondingly small change in the solutions. It would be convenient to quantify the condition of a system using the value of the determinant but this is not entirely possible because we can change the determinant by multiplying one or more of the equations by scalars without changing the solution. One way to partially circumvent this difficulty is to scale the equations so that the maximum value of the coefficients in every equation is equal to 1. There are several techniques for improving the accuracy of solutions to ill-conditioned systems. The aforementioned scaling is one such technique, and we will now look at another.

One strategy for checking the condition of the matrix when using Gauss elimination is to calculate the determinant at the end of the process. Recall that the determinant is the product of the diagonals $a_{11}a_{22}^{(1)} \dots a_{nn}^{(n-1)}$ because the final system corresponds to an upper-triangular matrix. Partial pivoting is a way to deal with ill-conditioning and consists of scaling and rearranging the equations. Recall that we divide by the pivot element, which is a problem if it is small. We therefore locate the largest coefficient in the column containing the pivot element, excluding those entries above the pivot, and swap the row containing this coefficient with the row that defines the pivot equation. We can then (optionally) divide the new pivot equation by the pivot element.

Full pivoting involves switching columns but is rarely used because it swaps the x_i and leads to high complexity. When implementing partial pivoting a modification is required to find the determinant. If we change a row, the determinant changes sign. If we call p the number of times a row is swapped (pivoted) then the determinant is

$$(-1)^p a_{11}a_{22}^{(1)} \dots a_{nn}^{(n-1)} \tag{A.24}$$

This does not alter the size, only the sign.

A.4 Gauss-Jordan Method

The Gauss-Jordan method is a variation of Gauss elimination, with the major difference being that when an unknown is eliminated in the Gauss-Jordan method, it is eliminated from all of the equations except one. In addition, all rows are normalised by their pivot elements, so that the elimination step results in an identity matrix (rather than a triangular matrix). The method is best illustrated by an example

$$\begin{aligned} 3x_1 - 0.1x_2 - 0.2x_3 &= 7.85 \\ 0.1x_1 + 7x_2 - 0.3x_3 &= -19.3 \\ 0.3x_1 - 0.2x_2 + 10x_3 &= 71.4 \end{aligned} \tag{A.25}$$

We can write this in a so-called augmented matrix form, in which the right-hand sides are placed in the last column

$$\left[\begin{array}{ccc|c} 3 & -0.1 & -0.2 & 7.85 \\ 0.1 & 7 & -0.3 & -19.3 \\ 0.3 & -0.2 & 10 & 71.4 \end{array}\right] \tag{A.26}$$

We normalise the first row by dividing it by the pivot element (3)

$$\left[\begin{array}{ccc|c} 1 & -0.0333 & -0.0666 & 2.61667 \\ 0.1 & 7 & -0.3 & -19.3 \\ 0.3 & -0.2 & 10 & 71.4 \end{array}\right] \tag{A.27}$$

The x_1 term can be eliminated from the second row by subtracting 0.1 times the first row from the second row. Similarly, subtracting 0.3 times the first row from the third row will eliminate the x_1 term from the third row. The result is

$$\left[\begin{array}{ccc|c} 1 & -0.0333 & -0.0666 & 2.6167 \\ 0 & 7.0033 & -0.2933 & -19.5617 \\ 0 & -0.1900 & 10.0200 & 70.6150 \end{array}\right] \tag{A.28}$$

Next, we normalise the second row by dividing it by 7.00333

$$\left[\begin{array}{ccc|c} 1 & -0.0333 & -0.0666 & 2.6167 \\ 0 & 1 & -0.0418 & -2.7932 \\ 0 & -0.1900 & 10.0200 & 70.6150 \end{array}\right] \tag{A.29}$$

Removing the x_2 terms from the first and third equations yields

$$\left[\begin{array}{ccc|c} 1 & 0 & -0.0681 & 2.5236 \\ 0 & 1 & -0.0418 & -2.7932 \\ 0 & 0 & 10.0120 & 70.0843 \end{array}\right] \tag{A.30}$$

The third row is then normalised by dividing through with 10.0120

$$\left[\begin{array}{ccc|c} 1 & 0 & -0.0681 & 2.5236 \\ 0 & 1 & -0.0418 & -2.7932 \\ 0 & 0 & 1 & 70.0000 \end{array}\right] \tag{A.31}$$

and finally the x_3 terms can be eliminated from the first and the second rows

$$\left[\begin{array}{ccc|c} 1 & 0 & 0 & 3.0000 \\ 0 & 1 & 0 & -2.5000 \\ 0 & 0 & 1 & 7.0000 \end{array}\right] \tag{A.32}$$

The flop count for Gauss-Jordan elimination is n^3, which is higher than the $2n^3/3$ count for Gauss elimination, so it is not as frequently used but remains a popular method in engineering.

A.5 LU Decomposition

We now look at an alternative elimination method, based on LU decomposition. The LU decomposition method is appealing because the time-consuming elimination step in Gauss elimination can be formulated so that it involves only operations on the matrix of coefficients. Thus, it is well suited for those situations in which we wish to solve the system with many right-hand side vectors **b**. Another motive for

LU decomposition is that it provides an efficient means for computing the matrix inverse (the inverse has a number of valuable applications in engineering practice). It also provides a means for evaluating the system condition.

Gauss elimination becomes inefficient when solving equations with the same coefficients (the entries of **A**), but with different right-hand sides **b**. Recall that Gauss elimination involves two steps: forward elimination and back-substitution. Of these two, the forward-elimination step comprises the bulk of the computational effort, particularly for large systems. LU decomposition methods separate the time-consuming elimination of the coefficients from the manipulations of **b**. Thus, once **A** has been decomposed, systems with different values of **b** can be solved in an efficient manner.

To illustrate LU decomposition we look at a 3-equation system—extension to larger systems is straightforward. We can rearrange the system as follows

$$\mathbf{Ax} - \mathbf{b} = \mathbf{0} \tag{A.33}$$

Suppose that we can find an upper-diagonal matrix **U** and a vector **d** such that **x** also satisfies

$$\begin{bmatrix} u_{11} & u_{12} & u_{13} \\ 0 & u_{22} & u_{23} \\ 0 & 0 & u_{33} \end{bmatrix} \begin{bmatrix} x_1 \\ x_2 \\ x_3 \end{bmatrix} = \begin{bmatrix} d_1 \\ d_2 \\ d_3 \end{bmatrix} = \mathbf{d} \tag{A.34}$$

This is similar to the manipulation that occurs in the first step of Gauss elimination. That is, we use elimination to reduce the system to upper-triangular form. (A.34) is the same as

$$\mathbf{Ux} - \mathbf{d} = \mathbf{0} \tag{A.35}$$

Now suppose that we can find a lower-diagonal matrix **L** such that

$$\mathbf{L}\left(\mathbf{Ux} - \mathbf{d}\right) = \mathbf{Ax} - \mathbf{b} \tag{A.36}$$

with the additional requirement that all entries of **L** along the diagonal are equal to 1

$$\begin{bmatrix} 1 & 0 & 0 \\ f_{21} & 1 & 0 \\ f_{31} & f_{32} & 1 \end{bmatrix} \tag{A.37}$$

If Eq. (A.36) holds, it follows that

$$\mathbf{LU} = \mathbf{A} \quad \text{and} \quad \mathbf{Ld} = \mathbf{b} \tag{A.38}$$

We can use a two-step strategy to solve the linear system (A.1)

1. LU decomposition step. **A** is factored or 'decomposed' into lower **L** and upper **U** triangular matrices

2. Substitution step. **L** and **U** are used to determine a solution **x** for a right-hand side **b**

The second step itself consists of two steps:

1. First, $\mathbf{Ld} = \mathbf{b}$ is used to generate an intermediate vector **d** by forward substitution (starting from the first element).
2. Then, the result is substituted into $\mathbf{Ux} - \mathbf{d} = \mathbf{0}$, which can be solved by back-substitution (starting from the last element) for **x**.

We will see how this is implemented in the context of Gauss elimination.

Although it might appear at face value to be unrelated to LU decomposition, Gauss elimination can be used to decompose **A** into **L** and **U**. This can be seen easily for the **U** part, which follows from forward elimination; recall that the forward-elimination step reduces the original coefficient matrix **A** to the form

$$\begin{bmatrix} u_{11} & u_{12} & u_{13} \\ 0 & u_{22} & u_{23} \\ 0 & 0 & u_{33} \end{bmatrix} = \mathbf{U} \tag{A.39}$$

which is in the desired upper-triangular format. In fact, the matrix **L** is also produced during this step. Let us see why this is the case with a three-equation system.

$$\begin{bmatrix} a_{11} & a_{12} & a_{13} \\ a_{21} & a_{22} & a_{23} \\ a_{31} & a_{32} & a_{33} \end{bmatrix} \begin{bmatrix} x_1 \\ x_2 \\ x_3 \end{bmatrix} = \begin{bmatrix} b_1 \\ b_2 \\ b_3 \end{bmatrix} \tag{A.40}$$

The first step in Gauss elimination is to multiply row 1 by the factor $f_{21} = a_{21}/a_{11}$ and subtract the result from the second row to eliminate the term $a_{21}x_1$. Similarly, row 1 is multiplied by $f_{31} = a_{31}/a_{11}$ and the result is subtracted from the third row to eliminate $a_{31}x_1$. The final step is to multiply the modified second row by $f_{32} = a_{32}^{(1)}/a_{22}^{(1)}$ and subtract the result from the third row to eliminate $a_{32}^{(1)}x_2$.

Let us define the following matrix

$$\mathbf{L} = \begin{bmatrix} 1 & 0 & 0 \\ f_{21} & 1 & 0 \\ f_{31} & f_{32} & 1 \end{bmatrix} \tag{A.41}$$

With the definitions of **L** and **U** above, it turns out that $\mathbf{A} = \mathbf{LU}$. In the general case (for a system of arbitrary size), we can use the Doolittle algorithm to obtain **L** and **U**. The Doolittle algorithm obtains **U** by eliminating all of the entries below the main diagonal column-by-column, starting from the left. It does this by creating a series of lower-diagonal matrices based on the factors f_{ij}.

Define $\mathbf{A}^{(0)} = \mathbf{A}$, i.e., the original matrix. We will generate a series of matrices $\mathbf{A}^{(0)}, \mathbf{A}^{(1)}, \ldots, \mathbf{A}^{(n-1)}$ by multiplication with a series of lower-triangular matrices $\mathbf{L}_1, \mathbf{L}_2, \ldots, \mathbf{L}_{n-1}$. In the k-th step, eliminate the matrix elements below the main diagonal in the k-th column of $\mathbf{A}^{(k-1)}$ by adding to the i-th row of this matrix the k-th row multiplied by

$$f_{ik} := \frac{a_{ik}^{(k-1)}}{a_{kk}^{(k-1)}}, \quad i = k+1, k+2, \ldots, n \tag{A.42}$$

So, in the first step ($k = 1$), we want to eliminate all entries below a_{11} in column 1; that is all entries a_{2i}, for $i = 2, 3, \ldots, n$. In the second step ($k = 2$), we want to eliminate all entries below the modified element $a_{22}^{(1)}$ in column 2; that is all entries $a_{2i}^{(1)}$, for $i = 3, 4, \ldots, n$. At each step k we transform $\mathbf{A}^{(k-1)}$ to $\mathbf{A}^{(k)}$ by performing these operations.

We can perform the elimination at step k by multiplying $\mathbf{A}^{(k-1)}$ to the left with the lower-triangular matrix

$$\mathbf{L}_k = \begin{pmatrix} 1 & 0 & & \cdots & & 0 \\ 0 & \ddots & \ddots & & & \\ & & 1 & & & \\ \vdots & & f_{(k+1)k} & & & \vdots \\ & & \vdots & & \ddots & 0 \\ 0 & & f_{nk} & & & 1 \end{pmatrix} \tag{A.43}$$

This means that

$$\mathbf{A}^{(k)} = \mathbf{L}_k \mathbf{A}^{(k-1)} \tag{A.44}$$

After $n-1$ steps, we will have eliminated all of the matrix elements below the main diagonal, so we obtain an upper-triangular matrix $\mathbf{A}^{(n-1)} = \mathbf{U}$. We can then surmise that

$$\begin{aligned} \mathbf{A} &= \mathbf{L}_1^{-1}\mathbf{L}_1\mathbf{A}^{(0)} \\ &= \mathbf{L}_1^{-1}\mathbf{A}^{(1)} \\ &= \mathbf{L}_1^{-1}\mathbf{L}_2^{-1}\mathbf{L}_2\mathbf{A}^{(1)} \\ &= \mathbf{L}_1^{-1}\mathbf{L}_2^{-1}\mathbf{A}^{(2)} \\ &= \ldots \\ &= \mathbf{L}_1^{-1}\ldots\mathbf{L}_{n-1}^{-1}\mathbf{A}^{(n-1)} \\ &= \left(\mathbf{L}_1^{-1}\ldots\mathbf{L}_{n-1}^{-1}\right)\mathbf{U} \end{aligned} \tag{A.45}$$

We can now define a matrix $\mathbf{L}$ as follows

$$\mathbf{L} = \mathbf{L}_1^{-1}\ldots\mathbf{L}_{n-1}^{-1} \tag{A.46}$$

The inverse of a lower-triangular matrix $\mathbf{L}_k$ is again a lower-triangular matrix and the multiplication of two lower-triangular matrices is again a lower-triangular matrix. Therefore $\mathbf{L}$ is the required lower-triangular matrix and is it is given explicitly by

$$
\mathbf{L} = \begin{pmatrix} 1 & 0 & \dots & & 0 \\ f_{21} & \ddots & \ddots & & \\ & & 1 & & \\ \vdots & & f_{(k+1)k} & \ddots & \vdots \\ & & \vdots & 1 & 0 \\ f_{n1} & & f_{nk} & f_{n(n-1)} & 1 \end{pmatrix} \tag{A.47}
$$

A.6 Solving Linear Systems with LU Decomposition

When solving linear systems using LU decomposition, we may want to use pivoting to ensure stability. This involves swapping rows so the final result resembles $\mathbf{PA} = \mathbf{LU}$ for some matrix $\mathbf{P}$ that defines the pivoting. Our main goal is to solve

$$
\mathbf{Ax} = \mathbf{b} \tag{A.48}
$$

Suppose that we have performed pivoting and subsequently obtained a decomposition $\mathbf{PA} = \mathbf{LU}$; then

$$
\mathbf{LUx} = \mathbf{PAx} = \mathbf{Pb} \tag{A.49}
$$

The solution proceeds in two steps

1. First we define $\mathbf{Ux} = \mathbf{d}$ and solve the equation $\mathbf{Ld} = \mathbf{Pb}$ for $\mathbf{d}$.
2. Then we solve the equation $\mathbf{Ux} = \mathbf{d}$ for $\mathbf{x}$.

In both steps we are dealing with triangular matrices, so we can solve by forward or backward substitution.

In the first step, for a 3×3 system we have

$$
\begin{bmatrix} 1 & 0 & 0 \\ f_{21} & 1 & 0 \\ f_{31} & f_{32} & 1 \end{bmatrix} \begin{bmatrix} d_1 \\ d_2 \\ d_3 \end{bmatrix} = \begin{bmatrix} \mathbf{Pb}_1 \\ \mathbf{Pb}_2 \\ \mathbf{Pb}_3 \end{bmatrix} \tag{A.50}
$$

in which $\mathbf{Pb}_i$ is the i-th element of $\mathbf{Pb}$. We calculate d_1, then d_2 then d_3: this is called forward substitution for obvious reasons. In the second step

$$
= \begin{bmatrix} u_{11} & u_{12} & u_{13} \\ 0 & u_{22} & u_{23} \\ 0 & 0 & u_{33} \end{bmatrix} \begin{bmatrix} x_1 \\ x_2 \\ x_3 \end{bmatrix} = \begin{bmatrix} d_1 \\ d_2 \\ d_3 \end{bmatrix} \tag{A.51}
$$

We then calculate x_3, then x_2 then x_1, which is called backward substitution. The solution to a larger system proceeds in the same manner.

A.7 Iterative Methods

So far we have looked only at elimination methods, which involve algebraic manipulations of the original linear system. In contrast, *iterative methods* successively refine proposed solutions to the linear system, starting with some initial guess. The Gauss-Seidel method is the most commonly used iterative method. To illustrate the idea, let us once more return to a 3×3 system. We can write x_1 in terms of x_2 and x_3

$$x_1 = \frac{b_1 - a_{12}x_2 - a_{13}x_3}{a_{11}} \tag{A.52}$$

Similarly

$$x_2 = \frac{b_2 - a_{31}x_1 - a_{32}x_3}{a_{22}}, \quad x_2 = \frac{b_3 - a_{31}x_1 - a_{32}x_2}{a_{33}} \tag{A.53}$$

In Gauss-Seidel we start with initial guesses x_1^0, x_2^0 and x_3^0, e.g., all 0, and substitute the values into the first equation to find a new x_1^1. We substitute this value in the second equation along with x_3^0 to find a new x_2^1. Finally, we use the new values of x_1^1 and x_2^1 in the last equation to obtain a new x_3^1. We then repeat the process from the beginning, with the first equation used to obtain a new value x_1^2 for x_1, and so on. This iterative procedure is continued until the solution converges according to some criterion, e.g.

$$\left| \frac{x_i^j - x_i^{j-1}}{x_i^j} \right| < \epsilon, \qquad \forall i \tag{A.54}$$

There is a variant of Gauss-Seidel called Jacobi iteration, which does not use the updated x_1, x_2 and x_3 at the current iteration. Instead, the updates are calculated entirely on the basis of the current values and this procedure is again repeated until convergence.

In the general case, for systems of arbitrary size, we follow the same procedure. We can first rewrite the system of equations as follows

$$\begin{aligned}
x_1 &= \frac{1}{a_{11}}(b_1 - a_{12}x_2 - a_{13}x_3 - \ldots - a_{1n}x_n) \\
x_2 &= \frac{1}{a_{22}}(b_2 - a_{21}x_1 - a_{23}x_3 - \ldots - a_{2n}x_n) \\
&\vdots \\
x_n &= \frac{1}{a_{nn}}(b_n - a_{n1}x_1 - a_{n2}x_2 - \ldots - a_{n(n-1)}x_{n-1})
\end{aligned} \tag{A.55}$$

and make an initial guess for each x_i, then proceed with either Gauss-Seidel or the Jacobi method.

Appendix B
Solving Ordinary Differential Equations

B.1 Ordinary Differential Equations

Ordinary differential equations are those that contain only ordinary derivatives. They are ubiquitous in science and engineering, e.g., mechanical systems and population dynamics. A famous example is Newton's second law

$$m\frac{dv}{dt} = mg - cv \tag{B.1}$$

for the velocity of an object of mass m falling in the Earth's atmosphere with drag coefficient c. This equation is accompanied by an initial condition, e.g., $v(t = 0) = v_0$. Another version of Newton's second law is for a mass-spring system, with displacement x

$$m\frac{d^2x}{dt^2} + c\frac{dx}{dt} + kx = 0 \tag{B.2}$$

in which c is a damping coefficient and k is the spring constant. In this case, two initial conditions are required, e.g., $x(0) = x_0$ and $dx/dt(0) = a$.

(B.1) is called a first-order equation and (B.2) is called a second-order equation; the order of the equation is defined by highest order derivative. Both of these examples are linear equations, meaning that if we write the equations as $\mathcal{L}v = 0$ or $\mathcal{L}x = 0$ for a differential operator $\mathcal{L}$, it necessarily holds that $\mathcal{L}(v_1 + v_2) = \mathcal{L}v_1 + \mathcal{L}v_2$ or $\mathcal{L}(x_1 + x_2) = \mathcal{L}x_1 + \mathcal{L}x_2$ for functions v_1 and v_2 or x_1 and x_2. For nonlinear equations, this property is not satisfied. An example of a nonlinear equation is

$$m\frac{d^2x}{dt^2} + c\frac{dx}{dt} + kx^3 = 0 \tag{B.3}$$

A. A. Shah et al., *New Paradigms in Flow Battery Modelling*,
Engineering Applications of Computational Methods 16,
https://doi.org/10.1007/978-981-99-2524-7

in which the nonlinearity appears in the source term kx^3. Nonlinearities can also appear in the differential part of the operator through the coefficients, e.g., $v^2 dv/dt$ or the argument of the derivative, e.g., $d(v^2)/dt$. We note that the independent variable t need not be time. It could be distance or any other quantity. When the dependent variable is time, we refer to the equation and associated initial conditions as an initial-value problem, otherwise it is called a boundary-value problem.

The number of boundary or initial conditions (for unique solutions) is equal to the order of the equation, since each integration of the equation introduces a constant that must be specified. For initial-value problems, the conditions are one-sided, meaning that they specify information at some initial time. In the boundary-value problems, values of the dependent variable and/or its derivatives are specified at the boundaries.

We can also have systems of ODEs in which there is more than one dependent variable, e.g.

$$\begin{aligned} m\frac{d^2x_1}{dt^2} + \frac{dx_1}{dt} + k_1x_1 + l_1x_2 &= 0 \\ m\frac{d^2x_2}{dt^2} + \frac{dx_2}{dt} + k_2x_1^2 - l_2\ln x_2 &= 0 \end{aligned} \tag{B.4}$$

Examples include population dynamics, molecular dynamics, traffic control, mechanical systems, and many others. There are analytical methods for solving ODEs, e.g., characteristic solutions and particular solutions for second-order equations, or integrating factors for first-order equations. These only apply to linear equations or nonlinear equations that have a particular structure. In general we must resort to numerical methods.

There is a vast array of numerical methods for solving ODEs. For some problems, more care is required, e.g., stiff system problems in which one or more of the dependent variables change on a much smaller or larger time scale than the others. To solve an ODE numerically we first develop a discrete-form approximation of the equation using discrete values $t_i, i = 1, 2, \ldots,$ of the independent variable, and then apply an iterative scheme to propagate the solution forwards in time or to reach some convergence criterion if the independent variable is not related to time. The latter case (boundary value problems) is somewhat more involved. In *shooting methods*, for example, we transform the problem into an initial-value problem, which is solved for different initial conditions to find a solution that it consistent with the boundary conditions in the original problem. In this appendix we will only cover first-order initial-value problems. Higher order problems can be solved using similar methods or in some cases by transforming the problem into a first-order problem.

The discrete time values t_i define *time steps* i and *step sizes* $\Delta t_i = t_i = t_{i-1}$, which are not necessarily equal for all i. Simple schemes use information only from the current time step (one-step). More sophisticated schemes use corrections, adaptive step sizes, more terms, or information at multiple previous time steps (multi-step).

To see how this works we first consider the simplest method for solving the following ODE

$$\frac{dy}{dt} = f(t, y) \tag{B.5}$$

If $f(t, y)$ is linear in y, the equation is linear, otherwise it is nonlinear. The first step is to approximate the derivative dy/dt. Let us suppose that we have the numerical solution at some time t_i, i.e., we know $y_i \approx y(t_i)$. We can then attempt to find the value of y at some time t_{i+1}, which is close to t_i. This requires that $\Delta t = t_{i+1} - t_i$ is small, otherwise the errors will be large. For example, we could use the approximation

$$\frac{dy}{dt} \approx \frac{y_{i+1} - y_i}{t_{i+1} - t_i} = \frac{\Delta y}{\Delta t} \tag{B.6}$$

in which $y_{i+1} \approx y(t_{i+1})$ and $\Delta y = y_{i+1} - y_i$.

We next have to approximate $f(t, y)$ in the interval $[t_i, t_{i+1}]$. We could set $f(t, y)$ to be equal to a constant $f(t_i, y_i)$ in this interval and write

$$\frac{y_{i+1} - y_i}{t_{i+1} - t_i} = \frac{y_{i+1} - y_i}{\Delta t} = f(t_i, y_i) \tag{B.7}$$

as an approximation. This leads us to

$$y_{i+1} = y_i + \Delta t f(t_i, y_i) \tag{B.8}$$

which is called the explicit Euler method (it gives an explicit solution at t_{i+1}). We can then use the same approach to find $y_{i+2}, y_{i+3}, \ldots$, which defines an iterative scheme that we call a time stepping scheme.

Another way to obtain the same result is to use a Taylor series expansion about t_i to yield y_{i+1}

$$y_{i+1} = y_i + \Delta t \frac{dy}{dt}(t_i) + O(\Delta t^2) = y_i + \Delta t f(t_i, y_i) + \mathcal{O}(\Delta t^2) \tag{B.9}$$

We can instead use a Taylor expansion to obtain y_i from information at t_{i+1}

$$y_i = y_{i+1} - \Delta t \frac{dy}{dt}(t_{i+1}) + O(\Delta t^2) = y_{i+1} + \Delta t f(t_{i+1}, y_{i+1}) + O(\Delta t^2) \tag{B.10}$$

Then

$$y_{i+1} = y_i + \Delta t f(t_{i+1}, y_{i+1}) \tag{B.11}$$

This equation has no explicit solution if $f(y, t)$ is not linear in y, and is called the implicit Euler method.

We can also view these method in terms of finite differences. In the explicit Euler method, we set the derivative at t_i to be equal to $f(t_i, y_i)$

$$\frac{dy}{dt}(t_i) = \frac{y_{i+1} - y_i}{\Delta t} = f(t_i, y_i) \tag{B.12}$$

The approximation $\frac{y_{i+1}-y_i}{\Delta t}$ is called a forward difference (quotient) since it approximates $\frac{dy}{dt}(t_i)$ using forward information at t_{i+1}. Explicit Euler is also therefore also called the forward Euler method. In the implicit Euler method, we set the derivative at t_{i+1} to be equal to $f(t_{i+1}, y_{i+1})$

$$\frac{dy}{dt}(t_{i+1}) = \frac{y_{i+1} - y_i}{\Delta t} = f(t_{i+1}, y_{i+1}) \tag{B.13}$$

In this case $\frac{y_{i+1}-y_i}{\Delta t}$ is called a backward difference since it approximates $\frac{dy}{dt}(t_{i+1})$ using backward information at t_i.

B.2 Error Analysis

Error analysis is concerned with characterising the error of a numerical scheme. Consider the following Taylor expansion

$$y(t_{i+1}) = y(t_i) + \Delta t f(t_i, y(t_i)) + \frac{1}{2}\Delta t^2 \frac{d^2 y}{dt^2}(\tau) \tag{B.14}$$

for some $\tau \in [t_i, t_{i+1}]$. The forward Euler scheme is

$$y_{i+1} = y_i + \Delta t f(t_i, y_i) \tag{B.15}$$

We can find some η such that

$$f(t_i, y(t_i)) = f(t_i, y_i) - f_y(t_i, \eta)(y(t_i) - y_i) = f(t_i, y_i) - f_y(t_i, \eta)E_i \tag{B.16}$$

in which E_i is called the global truncation error and $f_y = \partial f/\partial y$ (we assume that f, f_y and f_t are continuous and bounded for all values of t and y). We now have

$$y(t_{i+1}) = y(t_i) + \Delta t f(t_i, y_i) - \Delta t f_y(t_i, \eta)E_i + \frac{1}{2}\Delta t^2 \frac{d^2 y}{dt^2}(\tau) \tag{B.17}$$

Subtracting this from the forward Euler equation yields

$$y_{i+1} - y(t_{i+1}) = y_i - y(t_i) + \Delta t f_y(t_i, \eta)E_i - \frac{1}{2}\Delta t^2 \frac{d^2 y}{dt^2}(\tau) \tag{B.18}$$

or

$$E_{i+1} = E_i(1 + \Delta t f_y(t_i, \eta)) - \frac{1}{2}\Delta t^2 \frac{d^2 y}{dt^2}(\tau) \tag{B.19}$$

This result tells us about the relationship between the errors at successive steps. The local truncation error is defined to be the error in step $i + 1$ when there is no error in step i. Hence, the local truncation error for forward Euler is

$$-\frac{1}{2}\Delta t^2 \frac{d^2 y}{dt^2}(\tau) \tag{B.20}$$

and is $\mathcal{O}(\Delta t^2)$. The quantity $E_i(1 + \Delta t f_y(t_i, \eta))$ represents the error at step $i + 1$ caused by the error at step i. This propagated error is larger than E_i if $f_y > 0$ and smaller than E_i if $f_y < 0$. Having $f_y < 0$ is generally desirable because it causes truncation errors to diminish as they propagate. Without proof, the global error at time t can be shown to be bounded by

$$|E| \leq K(e^{kt} - 1)\Delta t \tag{B.21}$$

in which K and k are constants.

There are two important conclusions to this analysis. The first is that the error vanishes as $\Delta t \to 0$ and thus the truncation error can be made arbitrarily small by reducing the step size. The second conclusion is that the error in the method is approximately proportional to Δt. The order of a numerical method is the number of factors of Δt in the global truncation error estimate. Euler's method is therefore a first-order method. Halving the step size in a first-order method reduces the error by approximately a factor of 2. As with Euler's method, the order of most methods is 1 less than the power of Δt in the local truncation error.

The implicit Euler problem is

$$y_{i+1} = y_i + \Delta t f(t_{i+1}, y_{i+1}) \tag{B.22}$$

which we cannot solve explicitly, requiring us to use a root finding method. Despite the associated cost, implicit methods such as implicit Euler provide more stable solutions for certain types of problems (especially stiff problems). Explicit methods can fail or require very small time steps in such cases. We can write implicit Euler as

$$g(Y) = y_i + \Delta t f(t_{i+1}, Y) - Y = 0 \tag{B.23}$$

in which $Y = y_{i+1}$, and we wish to solve $g(Y) = 0$ for Y. We could, for example, use the Newton-Raphson method

$$Y_{k+1} = Y_k - \left(\frac{\partial g}{\partial Y}(Y_k)\right)^{-1} g(Y_k) \tag{B.24}$$

starting with some initial guess $Y = Y_0$, which could be $Y_0 = y_i$. Once we find Y, we set $y_{i+1} = Y$ and repeat the process.

Generally speaking, we can use larger time steps Δt in implicit Euler in order to achieve the same level of accuracy as explicit Euler. In particular, implicit Euler is much better for solving stiff problems. Consider a solution to an ODE of the form $y = e^{-t} + e^{-1000t}$: the second part decays very rapidly while the first part decays slowly. We need very small time steps to capture the first part, which can easily be obscured in the solution. This type of problem is called stiff.

B.3 Predictor-Corrector Methods

To improve upon explicit Euler, we can use a so-called *corrector step*. In a *predictor step*, explicit Euler yields an intermediate solution y_{i+1}^0

$$y_{i+1}^0 = y_i + \Delta t f(t_i, y_i) \tag{B.25}$$

The slope at the end of the interval $[t_i, t_{i+1}]$ can be estimated as $f(t_{i+1}, y_{i+1}^0)$, while the slope at the beginning of the interval $[t_i, t_{i+1}]$ can be estimated as $f(t_i, y_i)$. These two slopes can be combined to obtain an average slope for the interval

$$\overline{\frac{dy}{dt}} = \frac{f(t_i, y_i) + f(t_{i+1}, y_{i+1}^0)}{2} \tag{B.26}$$

This is called the corrector equation. The Heun method is a multi-step predictor-corrector approach that does the following

$$\begin{aligned} y_{i+1}^0 &= y_i + \Delta t f(t_i, y_i) \\ y_{i+1} = y_i + \Delta t \overline{\frac{dy}{dt}} &= y_i + \Delta t \left(\frac{f(t_i, y_i) + f(t_{i+1}, y_{i+1}^0)}{2} \right) \end{aligned} \tag{B.27}$$

It is called multi-step because information at more than one time is involved, in contrast to the Euler method. The corrector equation can be treated as implicit and applied successively until iteration. Next we will look at a powerful set of methods called Runge-Kutta, of which there are many implementations.

B.4 Runge-Kutta Methods

Runge-Kutta (RK) methods achieve the accuracy of a Taylor series approach without requiring the calculation of higher-order derivatives. Many variations exist but all can be cast in the generalised form

$$y_{i+1} = y_i + \Delta t \phi(t_i, y_i, \Delta t) \tag{B.28}$$

in which $\phi(t_i, y_i, \Delta t)$ is called an increment function. The increment function can be interpreted as a representative slope over the interval $[t_i, t_{i+1}]$. It can be written in the following general form

$$\phi = a_1 k_1 + a_2 k_2 + \ldots + a_n k_n \tag{B.29}$$

with the k_i defined by

$$
\begin{aligned}
k_1 &= f(t_i, y_i) \\
k_2 &= f(t_i + p_1 \Delta t, y_i + q_{11} k_1 \Delta t) \\
&\dots \\
k_n &= f(t_i + p_{n-1} \Delta t, y_i + q_{(n-1)1} k_1 \Delta t + q_{(n-1)2} k_2 \Delta t \dots + q_{(n-1)(n-1)} k_{n-1} \Delta t)
\end{aligned} \tag{B.30}
$$

in which the a_i, p_i and q_{ij} are constants.

B.5 Second-Order Runge-Kutta Methods

Various types of Runge-Kutta methods can be devised by employing different numbers of terms n in the increment function, and the first-order RK method ($n = 1$) is equivalent to Euler's method. Once n is chosen, values for a_i, p_i and q_{ij} are evaluated by comparing $y_{i+1} = y_i + \Delta t \phi(t_i, y_i, \Delta t)$ to a Taylor series expansion. Thus, at least for the lower-order versions, the number of terms n usually represents the order of the approach. For example, second-order RK methods have a local truncation error that is $\mathcal{O}(\Delta t^3)$ and a global error that is $\mathcal{O}(\Delta t^2)$. They take the form

$$
y_{i+1} = y_i + \Delta t (a_1 k_1 + a_2 k_2) \quad \begin{cases} k_1 = f(t_i, y_i) \\ k_2 = f(t_i + p_1 \Delta t, y_i + q_{11} k_1 \Delta t) \end{cases} \tag{B.31}
$$

The derivation of the values of a_i, p_i and q_{ij} using a Taylor series expansion is straightforward but messy, even for order-2. We simply state the result

$$
a_1 + a_2 = 1, \quad a_2 p_1 = \frac{1}{2}, \quad a_2 q_{11} = \frac{1}{2} \tag{B.32}
$$

Since we have three equations with four unknowns, we must assume a value of one of the unknowns, e.g., a_2. Given that we can choose from an infinite number of values for a_2, there are an infinite number of second-order RK methods. Every version would yield exactly the same results if the solution to the ODE was quadratic, linear or a constant. However, in general they yield different results.

We will look at three of the most commonly used and preferred versions.

1. If $a_2 = 1/2$ we obtain $a_1 = 1/2$ and $p_1 = q_{11} = 1$

$$
i + 1 = y_i + \Delta t \left(\frac{k_1 + k_2}{2} \right) \quad \begin{cases} k_1 = f(t_i, y_i) \\ k_2 = f(t_i + \Delta t, y_i + k_1 \Delta t) \end{cases} \tag{B.33}
$$

   which is precisely Heun's method.

2. If $a_2 = 1$ we obtain $a_1 = 0$ and $p_1 = q_{11} = 1/2$

$$
y_{i+1} = y_i + \Delta t k_2 \quad \begin{cases} k_1 = f(t_i, y_i) \\ k_2 = f(t_i + \Delta t/2, y_i + k_1 \Delta t/2) \end{cases} \tag{B.34}
$$

   called the midpoint method.

3. If $a_2 = 2/3$ we obtain $a_1 = 1/3$ and $p_1 = q_{11} = 3/4$

$$y_{i+1} = y_i + \Delta t\left(\frac{1}{3}k_1 + \frac{2}{3}k_2\right) \quad \begin{cases} k_1 = f(t_i, y_i) \\ k_2 = f(t_i + 3\Delta t/4, y_i + 3k_1\Delta t/4) \end{cases} \tag{B.35}$$

called Ralston's method.

Ralston's method provides a bound on the minimum of the truncation error for the second-order RK algorithms.

B.6 Third-Order and Higher Runge-Kutta Methods

For $n = 3$ we can derive a similar result using a Taylor series, yielding six equations with eight unknowns; we specify two unknowns and determine the rest. A common version is

$$y_{i+1} = y_i + \frac{1}{6}\Delta t(k_1 + k_2 + k_3), \quad \begin{cases} k_1 = f(t_i, y_i) \\ k_2 = f(t_i + \Delta t/2, y_i + k_1\Delta t/2) \\ k_3 = f(t_i + \Delta t, y_i - k_1\Delta t + 2k_2\Delta t) \end{cases} \tag{B.36}$$

The local truncation error in this case is $\mathcal{O}(\Delta t^4)$ and the global error is $\mathcal{O}(\Delta t^3)$.

The most popular RK methods are fourth order and the most commonly used form is

$$y_{i+1} = y_i + \frac{1}{6}\Delta t(k_1 + 2k_2 + 2k_3 + k_4), \quad \begin{cases} k_1 = f(t_i, y_i) \\ k_2 = f(t_i + \Delta t/2, y_i + k_1\Delta t/2) \\ k_3 = f(t_i + \Delta t/2, y_i + k_2\Delta t/2) \\ k_4 = f(t_i + \Delta t, y_i + k_3\Delta t) \end{cases} \tag{B.37}$$

We can also use higher-order methods but any gain in accuracy is usually offset by the increased computational effort.

B.7 Adaptive Runge-Kutta

A constant step size can be a serious limitation. In some problems, for most of the time range the solution changes gradually, suggesting a large step size should be used. For some regions, however, the solution undergoes an abrupt change, requiring very small step sizes. If the step size is too small the solution time can become long, while if it is too large, the results may be inaccurate.

One way to adapt the step size while running the algorithm is to estimate the local truncation error. For example, at step i we could employ both a fourth- and a fifth-order RK method, with the difference between the results used to estimate the local truncation error. Since the function evaluations are similar for both methods, this is an efficient scheme.

Once we have the error estimate, say $\Delta = \mathcal{O}(\Delta t^5)$, we can use it to adjust the step size. In general, the strategy is to increase the step size if the error is small and decrease it if the error is large. We can use a criterion such as

$$\Delta t_0 = \Delta t \times \left|\frac{\Delta_0}{\Delta}\right|^{1/5} \tag{B.38}$$

which calculates the step Δt_0 that would have given a desired tolerance of Δ_0. If $|\Delta| > |\Delta_0|$ the equation tells us by how much to decrease the step size; we then retry the present (failed) step. If $|\Delta| < |\Delta_0|$ the equation tells by how much we can safely increase the step size for the next step.

B.8 Multi-step Methods

In multi-step methods we use information at more than just the current time t_i to propagate the solution forwards. We start by discussing the Adams-Bashforth formulae. From the fundamental theorem of calculus

$$y(t_{i+1}) = y(t_i) + \int_{t_i}^{t_{i+1}} y'(t)\, dt = y(t_i) + \int_{t_i}^{t_{i+1}} f(t, y(t))dt \tag{B.39}$$

We define

$$A = \int_{t_i}^{t_{i+1}} f(t, y(t))dt = \int_{t_i}^{t_{i+1}} F(t)dt \tag{B.40}$$

To obtain a value for A we can use an interpolating polynomial $P(t)$, which is an approximation of $f(t, y(t))$. Here $F(t) = f(t, y(t))$ is considered as a function of t.

A Lagrange interpolating polynomial of order k uses function values $P(t)$ at specified locations t_j and polynomials of a certain form $L_j(t)$ to approximate $F(t)$

$$F(t) \approx P(t)= \sum_{j=0}^{k} F(t_j)L_j(t) \tag{B.41}$$

in which

$$L_j(t) := \prod_{\substack{0\leq m\leq k \\ m\neq j}} \frac{t - t_m}{t_j - t_m} \tag{B.42}$$

Let us use a linear interpolation for $F(t) = f(t, y(t))$, requiring the points t_i and t_{i-1}

$$P(t) = f(t_i, y_i)\frac{t - t_{i-1}}{t_i - t_{i-1}} + f(t_{i-1}, y_{i-1})\frac{t - t_i}{t_{i-1} - t_i} \tag{B.43}$$

We can then approximate A as follows

$$\begin{aligned} A &= \int_{t_i}^{t_{n+1}} f(t, y(t))\, dt \approx \int_{t_i}^{t_{n+1}} P(t)\, dt \\ &= \int_{t_i}^{t_{i+1}} \left[f(t_i, y_i)\frac{t - t_{i-1}}{t_i - t_{i-1}} + f(t_{i-1}, y_{i-1})\frac{t - t_i}{t_{i-1} - t_i} \right] dt \end{aligned} \tag{B.44}$$

leading to

$$A = \frac{3\Delta t}{2} f(t_i, y_i) - \frac{\Delta t}{2} f(t_{i-1}, y_{i-1}) \tag{B.45}$$

From (B.39) and (B.40) we then obtain an iterative scheme as follows (with $y(t_{i+1}) \approx y_{i+1}$, etc.)

$$y_{i+1} = y_i + \frac{3\Delta t}{2} f(t_i, y_i) - \frac{\Delta t}{2} f(t_{i-1}, y_{i-1}) \tag{B.46}$$

called the two-step Adams-Bashforth formula, with local truncation error $\mathcal{O}(\Delta t^3)$. We can instead use a quadratic interpolation at points t_i, t_{i-1} and t_{i-2} to obtain

$$y_{i+1} = y_i + \Delta t \left(\frac{23}{12} f(t_i, y_i) - \frac{4}{3} f(t_{i-1}, y_{i-1}) + \frac{5}{12} f(t_{i-2}, y_{i-2}) \right) \tag{B.47}$$

called the three-step Adams-Bashforth formula, with local truncation error $\mathcal{O}(\Delta t^4)$. We can go further to obtain higher step formulae. We note that these are explicit methods.

The Adams-Moulton methods are similar to the Adams-Bashforth methods but are implicit methods. They are derived in the same way but the interpolating polynomials also use information at t_{i+1}. The 3 and 4 step methods are

$$\begin{aligned} y_{i+1} &= y_i + \Delta t \left(\frac{5}{12} f(t_{i+1}, y_{i+1}) + \frac{2}{3} f(t_i, y_i) - \frac{1}{12} f(t_{i-1}, y_{i-1}) \right) \\ y_{i+1} &= y_i + \Delta t \left(\frac{9}{24} f(t_{i+1}, y_{i+1}) + \frac{19}{24} f(t_i, y_i) - \frac{5}{24} f(t_{i-1}, y_{i-1}) + \frac{1}{24} f(t_{i-2}, y_{i-2}) \right) \end{aligned} \tag{B.48}$$

respectively. As before, a root finding method is used to find the solution. One issue is how to obtain information at t_{i-1}, t_{i-2}, etc. when $i = 0$. This is resolved by starting at $i = 0$ with Runge-Kutta or Euler and proceeding with very small time steps for a few iterations, before using the formulae above.

B.9 Predictor-Corrector Methods Based on Adams-Bashforth and Adams-Moulton

We can combine the Adams-Bashforth and Adams-Moulton methods using the first as a predictor and the second as a corrector. For example, the prediction step could be a four-step Adams-Bashforth

$$y_{i+1} = y_i + \frac{\Delta t}{24}\left(55 f(t_i, y_i) - 59 f(t_{i-1}, y_{i-1}) + 37 f(t_{i-2}, y_{i-2}) - 9 f(t_{i-3}, y_{i-3})\right) \tag{B.49}$$

We then use the four-step Adams-Moulton method as the corrector step

$$y_{i+1} = y_i + \frac{\Delta t}{24}\left(9 f(t_{i+1}, y_{i+1}) + 19 f(t_i, y_i) - 5 f(t_{i-1}, y_{i-1}) + f(t_{i-2}, y_{i-2})\right) \tag{B.50}$$

The corrector equation can be treated as implicit and applied successively until the iterations converge.

Appendix C
Solving Partial Differential Equations

C.1 Partial Differential Equations

Partial differential equations (PDEs) contain partial derivatives and, therefore, more than one independent variable. A famous example is the heat conduction equation

$$\frac{\partial T}{\partial t} - \frac{\lambda}{\rho C_p}\frac{\partial^2 T}{\partial x^2} = S(T, t, x) \tag{C.1}$$

for the temperature T of an object with thermal conductivity λ, density ρ and specific heat capacity C_p. If the source term S is zero we say that the equation is homogeneous. If S depends nonlinearly on T, e.g., $S = Ae^{-E/RT}$, we say that the equation is nonlinear, otherwise it is linear. In general, linearity and nonlinearity are defined as in the ODE case in Appendix B, in terms the differential operator that defines the equation. For example, in (C.1),

$$\mathcal{L} = \frac{\partial}{\partial t} - \frac{\lambda}{\rho C_p}\frac{\partial^2}{\partial x^2} - S(T, t, x) \tag{C.2}$$

For linearity, we require that $\mathcal{L}(f_1 + f_2) = \mathcal{L}f_1 + \mathcal{L}f_2$ for any two functions f_1 and f_2, and otherwise the equation is nonlinear. This equation is 1D (in x) and time dependent. If the time derivative is zero we say that the equation is steady state or time independent.

A steady-state heat conduction equation in 2D is

$$\nabla^2 T = \frac{\partial^2 T}{\partial x^2} + \frac{\partial^2 T}{\partial y^2} = 0 \tag{C.3}$$

A. A. Shah et al., *New Paradigms in Flow Battery Modelling*,
Engineering Applications of Computational Methods 16,
https://doi.org/10.1007/978-981-99-2524-7

and is also called Laplace's equation. This and the preceding PDE are second order, meaning that the highest order of the derivatives is two. Equation (C.1) is called a parabolic PDE while the steady state-form is an example of an elliptic PDE.

We can also have first-order PDEs, e.g.

$$a(t, x, u)\frac{\partial u}{\partial t} + b(t, x, u)\frac{\partial u}{\partial x} = c(t, x, u) \tag{C.4}$$

which is a hyperbolic PDE. A famous example is the inviscid Burger's equation

$$\frac{\partial u}{\partial t} + u\frac{\partial u}{\partial x} = 0 \tag{C.5}$$

which can describe shock formation in a gas-dynamic system (e.g., air in a tube) or the propagation of waves in general, e.g., in acoustics. Another hyperbolic equation is the second-order wave equation, which takes the form (in 1D)

$$\frac{\partial^2 T}{\partial t^2} = c\frac{\partial^2 T}{\partial x^2} \tag{C.6}$$

In order to solve any of these equations we have to specify: (1) the domain and (b) the initial and boundary conditions (initial conditions only for time-dependent PDEs). The domain is the spatial region and time interval of interest. For example, in the 1D heat conduction equation we could consider a rod of length 10 m and define the domain as $x \in [0, 10]$ with $t \in [0, t_f]$ for some final time t_f. For the 2D heat conduction equation we may define the domain as $(x, y) \in [0, l] \times [0, l]$, which is a box of length l.

We can denote the interior of the spatial domain by Ω, which we will assume is bounded. We will also assume for simplicity that the solutions and their derivatives are smooth up to at least the orders in which they appear in the equation and that the problem is *well-posed*, which informally means that a solution exists, it is unique and it depends continuously on the initial-boundary conditions and source term. The interior Ω does not include the boundary, which we would normally denote $\partial\Omega$. The PDE is only valid inside Ω, while on the boundary $\partial\Omega$ we must specify boundary conditions, and at the initial time we must specify initial conditions in Ω. These conditions are independent of the PDE, but both taken together must be consistent (lead to a well-posed problem). Problems involving boundary conditions alone are called boundary-value problems (as in the ODE case). Those involving both initial and boundary conditions are called initial-boundary value problems.

The number of initial and boundary conditions required to obtain unique solutions depends on the type of PDE, primarily the order of the highest order derivative. In general, the nature of the initial and boundary conditions (for unique solutions that satisfy required smoothness properties) depends on the type of PDE (elliptic, hyperbolic, hyperbolic) and the nature of the domain and boundary, and is more complicated than was the case with ODEs. We will specify the conditions for selected important examples. Burger's equation requires one initial and one boundary condition because it is time dependent and contains a first-order spatial derivative. The dynamic 1D heat conduction equation requires one initial and two boundary conditions because the spatial derivative is second order. The steady-state 2D heat conduction equation requires four boundary conditions, two for each second-order spatial derivative. The 1D wave equation requires two initial conditions and two boundary conditions.

Boundary conditions can be of three types:

1. Dirichlet conditions specify the value of the dependent variable at a boundary point or portion of the boundary. For example, for the 1D heat conduction equation, we could set $T = 100$ at $x = 0$ and for the 2D steady heat conduction equation, we could set $T = 100$ at $x = 0$, $y \in [0, l]$.
2. Neumann conditions specify the value of the derivative of the dependent variable at a boundary point or portion of the boundary. For example, in the case of the 1D heat conduction equation, we could set

$$-\lambda \frac{\partial T}{\partial x}\bigg|_{x=0} = \alpha \tag{C.7}$$

which is the heat flux at $x = 0$.
3. A third type is called a mixed or Robin boundary condition and takes the form

$$\beta T + \gamma \left(\lambda \frac{\partial T}{\partial x} \right)\bigg|_{x=0} = \alpha \tag{C.8}$$

for some constants (or functions of x and/or t) β, α, γ.

In most cases of practical interest, analytical solutions are not possible to find, especially for systems of PDEs that are coupled, and we must resort to numerical solutions. We now discuss how to obtain such numerical solutions.

C.2 Finite Difference Method

In the finite difference method (FDM) we use a discretisation of the derivatives based on finite differences, which we saw in Appendix B for pure time-dependent problems. A discretisation is first performed on the spatial domain, which means splitting up the domain into intervals in 1D or into cells in 2D or 3D. The cells or intervals need not be of the same size, and in the case of 2D and 3D problems, we can split the domain into polygons, with often triangles or tetrahedra used in finite-element and finite-volume methods. The resulting collection of these simplices is called a triangulation. To keep matters simple, we will only look at regular sized meshes, and will not consider anything more complicated than quadrilaterals.

Let us start by considering a PDE with dependent variable $u(x, t)$. From now on, we will use the more compact notation

$$u_t = \frac{\partial u}{\partial t}, \quad u_x = \frac{\partial u}{\partial x}, \quad u_{xx} = \frac{\partial^2 u}{\partial x^2} \tag{C.9}$$

We can define equi-distant points between $[0, l]$ with separation h, i.e., $x_i - x_{i-1} = h$. A Taylor expansion of $u(x, t)$ about x_i is

$$u(x_i + h, t) = u(x_i, t) + hu_x(x_i, t) + \frac{h^2 u_{xx}(x_i, t)}{2!} + \dots \tag{C.10}$$

so that

$$u_x(x_i, t) = \frac{u(x_i + h, t) - u(x_i, t)}{h} + \mathcal{O}(h) \tag{C.11}$$

This is called a forward difference approximation of $u_x(x_i, t)$. If instead we use the Taylor expansion

$$u(x_i - h, t) = u(x_i, t) - hu_x(x_i, t) + \frac{h^2 u_{xx}(x_i, t)}{2!} + \dots \tag{C.12}$$

we obtain a backward difference approximation of $u_x(x_i, t)$

$$u_x(x_i, t) = \frac{u(x_i, t) - u(x_{i-1}, t)}{h} + \mathcal{O}(h) \tag{C.13}$$

Both of these are first-order approximations, meaning that the errors are $\mathcal{O}(h)$. If we subtract the second Taylor expansion from the first we obtain

$$u_x(x_i, t) = \frac{u(x_{i+1}, t) - u(x_{i-1}, t)}{2h} + \mathcal{O}(h^2) \tag{C.14}$$

which is called a central difference approximation of $u_x(x_i, t)$. Central differences are second-order approximations, meaning that the errors are $\mathcal{O}(h^2)$.

If instead we add the two Taylor series expansions we obtain

$$u(x_{-1}, t) + u(x_{i+1}, t) = 2u(x_i, t) + h^2 u_{xx}(x_i, t) + \mathcal{O}(h^4) \tag{C.15}$$

because all odd-order derivatives hu_x, $h^3 u_{xxx}$, etc. will cancel. Therefore we obtain the following second-order symmetric FD approximation to u_{xx}

$$u_{xx}(x_i, t) = \frac{u(x_i + h, t) - 2u(x_i, t) + u(x_i - h, t)}{h^2} + \mathcal{O}(h^2) \tag{C.16}$$

Note that we can also derive this formula from a finite difference approximation of u_x.

C.3 Finite Difference Method for a 1D Hyperbolic Equation

Now let us suppose that $u(x, t)$ satisfies a first-order hyperbolic PDE

$$u_t + cu_x = 0, \quad c > 0 \tag{C.17}$$

For the time derivative, we can use forward Euler (Appendix B) with a constant time step Δt. Introducing the notation $u_i^n = u(n\Delta t, x_i)$, i.e., the solution at x_i at time step n, forward Euler is expressed as

$$u_t(x_i, n\Delta t) = \frac{u_i^{n+1} - u_i^n}{\Delta t} + \mathcal{O}(\Delta t) \tag{C.18}$$

We can now try to solve the equation using a forward-Euler in time and backward-difference in space scheme

$$\frac{u_i^{n+1} - u_i^n}{\Delta t} + c\frac{u_i^n - u_{i-1}^n}{h} = 0 \tag{C.19}$$

Rearranging this equation we obtain

$$u_i^{n+1} = u_i^n - \frac{c\Delta t}{h}(u_i^n - u_{i-1}^n) = u_i^n - \nu(u_i^n - u_{i-1}^n) \tag{C.20}$$

in which $\nu = \frac{c\Delta t}{h}$ is called the Courant-Friedrichs-Lewy (CFL) number.

Using instead a forward difference for space we obtain

$$u_i^{n+1} = u_i^n - \nu(u_{i+1}^n - u_i^n) \tag{C.21}$$

and using a central difference for space (called forward time-centred space (FTCS)) yields

$$u_i^{n+1} = u_i^n - \frac{1}{2}\nu(u_{i+1}^n - u_{i-1}^n) \tag{C.22}$$

C.4 von Neumann Stability Analysis

At this point we must not proceed blindly. To decide on the best scheme we can conduct a von Neumann stability analysis by examining the growth of the error, which is assumed to be of the form

$$e(x, t) = \sum_{\beta} A_\beta e^{\gamma t} e^{j\beta x} \tag{C.23}$$

which is a sum of Fourier modes. γ is in general complex and depends on β. $e_i^n = u(x_i, t_i) - u_i^n$ corresponds to $t = n\Delta t$, $x = ih$, so let us consider a single mode

$$e_i^n = A_\beta e^{\gamma n \Delta t} e^{j\beta i h} \tag{C.24}$$

Substituting this into the backward-space formula yields, after dividing through by $A_\beta e^{\gamma n \Delta t} e^{j\beta i h}$

$$e^{\gamma \Delta t} = 1 - \nu + \nu e^{-j\beta h} = [1 - \nu + \nu \cos(\beta h)] - j\nu \sin(\beta h) \tag{C.25}$$

the magnitude of which is

$$\begin{aligned} |e^{\gamma \Delta t}| &= [1 - \nu + \nu \cos(\beta h)]^2 - \nu^2 \sin^2(\beta h) \\ &= (1 - \nu)^2 + 2(1 - \nu)\nu \cos(\beta h) + \nu^2(\cos^2(\beta h) + \sin^2(\beta h)) \end{aligned} \tag{C.26}$$

Using $1 - \cos(b) = 2\sin^2(b/2)$ this becomes

$$\begin{aligned} |e^{\gamma \Delta t}| &= (1 - \nu)^2 + 2(1 - \nu)\nu(1 - 2\sin^2(\beta h/2)) + \nu^2 \\ &= 1 - 4(1 - \nu)\nu \sin^2(\beta h/2)) \end{aligned} \tag{C.27}$$

We want the magnitude to be less the 1, so that the scheme is stable. Since $\sin^2(\beta h/2) \geq 0$ this means that we must have $(1 - \nu)\nu > 0$, and since $\nu > 0$ we obtain

$$\nu = \frac{c\Delta t}{h} \leq 1 \tag{C.28}$$

which is the called the CFL condition. The CFL condition places a restriction on the step size, in order to obtain accurate and stable solutions. If $c < 0$ we need to use the forward spatial difference, otherwise we use the backward difference.

C.5 The Lax-Friedrichs and Leapfrog Methods

If we repeat the stability analysis with the central difference scheme, we will find that it is always unstable (the error magnitude is always > 1). This can be corrected by the Lax-Friedrichs method, which replaces u_i^n with a neighbour average

$$u_i^{n+1} = \frac{1}{2}(u_{i-1}^n + u_{i+1}^n) - \frac{1}{2}\nu(u_{i+1}^n - u_{i-1}^n) \tag{C.29}$$

leading to stability under the CFL condition. This method is equivalent to replacing the original equation with

$$u_t + cu_x = \frac{h^2}{2\Delta t} u_{xx} \tag{C.30}$$

which corresponds to adding a regularisation or dissipation term on the right-hand side. Much in the same way as regularisation in machine learning (Chap. 6), this introduces some error but stabilises the numerical scheme.

Both FTCS and Lax-Friedrichs are one-level schemes with first-order approximations for the time derivative and second-order approximations for the spatial derivative. Under these circumstances $c\Delta t$ should be much smaller than h in practice. Second-order accuracy in time can be achieved if we set

$$u_t(x_i, n\Delta t) = \frac{u_i^{n+1} - u_i^{n-1}}{2\Delta t} + O(\Delta t^2) \tag{C.31}$$

which is a central difference in time. Combined with FTCS we obtain

$$u_i^{n+1} = u_i^{n-1} - \nu(u_{i+1}^n - u_{i-1}^n) \tag{C.32}$$

called the Leapfrog scheme.

The Leapfrog scheme is 2-step, requiring information at n and $n-1$. It still requires the CFL condition to be satisfied but larger Δt are possible in practise. To obtain the initial values (recall that we require information at $n-1$), we use forward Euler or an alternative scheme. Higher order time-stepping methods of the type discussed in Appendix B can also be used to solve this problem. Implicit schemes are usually avoided because larger time steps that violate the CFL condition lead to inaccuracies, so there is no advantage gained.

The CFL condition is important because it is a physical constraint, i.e., the distance moved by the wave $c\Delta t = dx/dt \times \Delta t$ must be less than the spacing between points, otherwise information is lost. Similar CFL conditions exist for solutions to other types of equations.

C.6 Finite Difference Method for Parabolic Equations

We now move onto the 1D heat conduction equation in the domain $0 \leq x \leq l, t \geq 0$.

$$\begin{aligned} u_t &= \lambda u_{xx}, \quad \lambda > 0 \\ u(x, 0) = u_0(x), \quad u(0, t) &= g(t), \quad u(l, t) = f(t) \end{aligned} \tag{C.33}$$

We could attempt a forward in time, second-order symmetric spatial FD approximation as follows

$$\frac{u_i^{n+1} - u_i^n}{\Delta t} = \lambda \frac{u_{i+1}^n - 2u_i^n + u_{i-1}^n}{h^2}, \quad i = 1, \ldots, N-1 \tag{C.34}$$

or

$$u_i^{n+1} = u_i^n (1 - 2\mu) + \mu(u_{i+1}^n + u_{i-1}^n) \tag{C.35}$$

in which $\mu = \lambda\frac{\Delta t}{h^2}$. The initial condition gives

$$u_i^0 = u_0(x_i), \quad \text{for all } i = 1, \ldots, N \tag{C.36}$$

and the boundary conditions yield

$$u_0^{n+1} = g((n+1)\Delta t), \quad u_N^{n+1} = f((n+1)\Delta t) \quad \text{for all } n = 0, 1, 2, \ldots \tag{C.37}$$

We now have an explicit set of equations to solve the problem for each $i = 1, \ldots, N-1$, starting at $n = 0$. It can be shown that this explicit method is both convergent and stable if $\mu \leq 1/2$ or

$$\Delta t \leq \frac{1}{2}\frac{h^2}{\lambda} \tag{C.38}$$

In reality, it can easily suffer from instabilities, so we may want to use an implicit method. To develop such a method we can use the forward time t_{i+1} to approximate the second derivative

$$u_{xx}(x_i, t_{i+1}) = \frac{u_{i+1}^{n+1} - 2u_i^{n+1} + u_{i-1}^{n+1}}{h^2} \tag{C.39}$$

We then use backward (implicit) Euler for $u_t(x_i, t_{i+1})$

$$\frac{u_i^{n+1} - u_i^n}{\Delta t} = \lambda\frac{u_{i+1}^{n+1} - 2u_i^{n+1} + u_{i-1}^{n+1}}{h^2}, \quad i = 1, \ldots, N-1 \tag{C.40}$$

This results in the equations

$$-\mu u_{i+1}^{n+1} + (1+2\mu)u_i^{n+1} - \mu u_{i-1}^{n+1} = u_i^n \tag{C.41}$$

in which $\mu = \lambda\frac{\Delta t}{h^2}$. The boundary conditions are as in (C.37). In contrast to the explicit method, we now have a system of linear equations that we must solve together.

C.7 The θ Method

The θ method combines the implicit and explicit schemes as follows

$$\frac{u_i^{n+1} - u_i^n}{\Delta t} = \theta\lambda\frac{u_{i+1}^{n+1} - 2u_i^{n+1} + u_{i-1}^{n+1}}{h^2} + (1-\theta)\lambda\frac{u_{i+1}^n - 2u_i^n + u_{i-1}^n}{h^2} \tag{C.42}$$

for some number θ, which results in the equations

$$-\mu\theta u_{i+1}^{n+1} + (1+2\mu\theta)u_i^{n+1} - \mu\theta u_{i-1}^{n+1} = u_i^n + \mu(1-\theta)(u_{i+1}^n - 2u_i^n + u_{i-1}^n) \tag{C.43}$$

with boundary conditions (C.37) and $\mu = \lambda\frac{\Delta t}{h^2}$. Again have a system of linear equations that we must solve together.

We can write the system as

$$(\mathbf{I} - \mu\theta\mathbf{A})\mathbf{u}^{n+1} = (\mathbf{I}' + \mu(1-\theta)\mathbf{A})\mathbf{u}^n + \mathbf{f}^{n+1} \tag{C.44}$$

in which

$$\mathbf{u}^{n+1} = (u_0^{n+1}, \ldots, u_N^{n+1})^T \in \mathbb{R}^{N+1} \tag{C.45}$$

while $\mathbf{I} \in \mathbb{R}^{(N+1)\times(N+1)}$ is the identity and $\mathbf{A}$ is a tri-diagonal matrix with zeros along the top and bottom rows

$$\mathbf{A} = \begin{bmatrix} 0 & 0 & 0 & & & 0 \\ 1 & -2 & 1 & & & \\ & 1 & -2 & 1 & & \\ & & \ddots & \ddots & \ddots & \\ & & & 1 & -2 & 1 \\ 0 & & & 0 & 0 & 0 \end{bmatrix} \tag{C.46}$$

The vector $\mathbf{f}^{n+1}$ is defined by

$$\mathbf{f}^{n+1} = (g((n+1)\Delta t), 0, \ldots, 0, f((n+1)\Delta t))^T \in \mathbb{R}^{N+1} \tag{C.47}$$

The matrix $\mathbf{I}' \in \mathbb{R}^{(N+1)\times(N+1)}$ is equal to the identity matrix with the entries at $(1, 1)$ and $(N+1, N+1)$ set to zero. When $\theta = 1/2$, we have a special case called the Crank-Nicolson (CN) scheme, which is the only case in which the θ scheme is second-order accurate in time.

We essentially use a centred difference approximation of $u_{xx}(x_i, t_{n+1/2})$, i.e., an average of $u_{xx}(x_i, t_{n+1})$ and $u_{xx}(x_i, t_{n+1})$. We can then interpret the forward Euler as a central difference approximation at $t_{n+1/2}$

$$u_t(x_i, t_{n+1/2}) = \frac{u_i^{n+1} - u_i^n}{\Delta t} + \mathcal{O}(\Delta t^2) \tag{C.48}$$

We also need knowledge regarding the stability of the θ scheme in order to select the time steps, for which we can perform a stability analysis as before. We substitute a Fourier mode $u_i^n = A(\beta)^n e^{j\beta x_i}$ into the numerical scheme, but this time we look directly at the solution and use $A(\beta)^n$ rather than $(e^{\gamma(\beta)\Delta t})^n$, requiring that $|A(\beta)| < 1$. Substituting this mode into the numerical scheme yields

$$A(\beta) = \frac{1 - 4\mu(1-\theta)\sin^2(\beta h/2)}{1 + 4\mu\theta\sin^2(\beta h/2)} \tag{C.49}$$

for which we used the same trigonometric identities as before. Since A is real, we require that $-1 \le A \le 1$. Clearly, $A < 1$, so that we only need $A \ge -1$, for which

$$2\mu \sin^2 (\beta h/2)\,(1 - 2\theta) \le 1 \tag{C.50}$$

Since $\sin^2 (\beta h/2) \le 1$, this is assured if

$$2\mu(1 - 2\theta) \le 1 \tag{C.51}$$

which is always satisfied if $1/2 \le \theta \le 1$. For $0 \le \theta \le 1/2$, the scheme is conditionally stable, requiring

$$\Delta t \le \frac{h^2}{2\lambda(1 - 2\theta)} \tag{C.52}$$

C.8 Other Boundary Conditions

Next we investigate the implementation of other types of boundary conditions. The general case is a Robin condition, of which the Neumann condition is a special case. Let us therefore consider

$$\alpha u(0, t) + \beta u_x(0, t) = \gamma \tag{C.53}$$

There are different ways to implement a condition such as this, and we start with a simple forward difference. At $x_0 = 0$ we may write

$$u_x(0, n\Delta t) = \frac{u_1^n - u_0^n}{h} + \mathcal{O}(h) \tag{C.54}$$

We then approximate the Robin condition as follows

$$\alpha u_0^{n+1} + \beta \frac{u_1^{n+1} - u_0^{n+1}}{h} = \gamma \tag{C.55}$$

If we are using the forward Euler scheme

$$u_i^{n+1} = u_i^n\,(1 - 2\mu) + \mu(u_{i+1}^n + u_{i-1}^n) \tag{C.56}$$

and we can set $i = 1$ to obtain

$$u_1^{n+1} = u_1^n\,(1 - 2\mu) + \mu(u_2^n + u_0^n) \tag{C.57}$$

Equations (C.55) and (C.57) are simultaneous equations for u_0^{n+1} and u_1^{n+1}, which we can easily solve. If we use Eq. (C.55) to approximate the mixed boundary condition with the θ method, we need to adapt the system to solve for u_0^{n+1}

$$(\mathbf{I}-\mu\theta\mathbf{A})\mathbf{u}^{n+1}=(\mathbf{I}'+\mu(1-\theta)\mathbf{A})\mathbf{u}^{n}+\mathbf{f}^{n+1} \tag{C.58}$$

The vector $\mathbf{f}^{n+1}$ is now

$$\mathbf{f}^{n+1}=(\gamma h,0,\ldots,0,f((n+1)\Delta t))^{T}\in\mathbb{R}^{N+1} \tag{C.59}$$

and the matrix $\mathbf{A}$ is unchanged but we need to modify $(\mathbf{I}-\mu\theta\mathbf{A})$ as follows

$$\mathbf{I}-\mu\theta\mathbf{A}=\begin{bmatrix} \alpha h-\beta & \beta & 0 & & & 0 \\ -\mu\theta & 1+2\mu\theta & -\mu\theta & & & \\ & & \ddots & \ddots & \ddots & \\ & & & -\mu\theta & 1+2\mu\theta & -\mu\theta \\ 0 & & & 0 & 0 & 1 \end{bmatrix} \tag{C.60}$$

Unfortunately, this method is first order in space, and so is not very accurate. An alternative method for applying the boundary condition is to use a central difference

$$u_x(0,n\Delta t)=\frac{u_1^n-u_{-1}^n}{2h}+\mathcal{O}(h^2) \tag{C.61}$$

in which we introduce a fictitious (ghost) node $x_{-1}=x_0-h$ to the left of x_0. This means that we may approximate the Robin boundary condition using (at t_n this time)

$$\alpha u_0^n+\beta\frac{u_1^n-u_{-1}^n}{2h}=\gamma \tag{C.62}$$

If we are using the forward Euler scheme we then obtain

$$u_0^{n+1}=u_0^n(1-2\mu)+\mu(u_1^n+u_{-1}^n) \tag{C.63}$$

From Eq. (C.62) we have that

$$u_{-1}^n=u_1^n+\frac{2h\alpha}{\beta}u_0^n-\frac{2h\gamma}{\beta} \tag{C.64}$$

from which the update for $u(x_0,t)$ is

$$u_0^{n+1}=u_0^n(1-2\mu)+\mu\left(2u_1^n+\frac{2h\alpha}{\beta}u_0^n-\frac{2h\gamma}{\beta}\right) \tag{C.65}$$

and in the θ method we use

$$\alpha u_0^{n+1}+\beta\frac{u_1^{n+1}-u_{-1}^{n+1}}{2h}=\gamma,\quad \alpha u_0^n+\beta\frac{u_1^n-u_{-1}^n}{2h}=\gamma \tag{C.66}$$

We now apply the θ equation for $i = 0$ to obtain

$$-\mu\theta u_1^{n+1} + (1 + 2\mu\theta)u_0^{n+1} - \mu\theta u_{-1}^{n+1} = u_0^n + \mu(1-\theta)(u_1^n - 2u_0^n + u_{-1}^n) \tag{C.67}$$

We can use Eqs. (C.66) to replace u_{-1}^{n+1} and u_{-1}^n in Eq. (C.67) in order to obtain a new equation that we include in the linear system.

C.9 Solving the Linear System

The linear systems obtained in the parabolic PDE problem can be solved in different ways, e.g., Gauss elimination (Appendix A). If the system matrix is diagonally dominant, more efficient methods such as the Thomas algorithm for tri-diagonal systems can be used. Variants of the Thomas algorithm exist for matrices that are almost tri-diagonal, such as the one we are considering. The Thomas algorithm uses forward elimination to obtain an upper-triangular matrix but elimination is only required for the diagonal entries below the main diagonal so that it is $\mathcal{O}(N)$ rather than $\mathcal{O}(N^3)$, which is the cost of Gauss elimination. It is also a form of LU decomposition.

Consider the system

$$\begin{bmatrix} b_1 & c_1 & & & 0 \\ a_2 & b_2 & c_2 & & \\ & a_3 & b_3 & \ddots & \\ & & \ddots & \ddots & c_{n-1} \\ 0 & & & a_n & b_n \end{bmatrix} \begin{bmatrix} x_1 \\ x_2 \\ x_3 \\ \vdots \\ x_n \end{bmatrix} = \begin{bmatrix} d_1 \\ d_2 \\ d_3 \\ \vdots \\ d_n \end{bmatrix} \tag{C.68}$$

for $\mathbf{x} = (x_1, \ldots, x_n)^T$. As before (see Appendix A) we eliminate the terms below the main diagonal, but now we only need to eliminate the a_i terms

$$a_i x_{i-1} + b_i x_i + c_i x_{i+1} = d_i \quad (a_1 = c_n = 0) \tag{C.69}$$

We perform forward elimination, e.g., the second row

$$(\text{equation } 2) \cdot b_1 - (\text{equation } 1) \cdot a_2 \tag{C.70}$$

which eliminates x_1, and continuing this process we obtain new coefficients

$$c'_i = \begin{cases} \dfrac{c_i}{b_i} & ; i = 1 \\ \dfrac{c_i}{b_i - a_i c'_{i-1}} & ; i = 2, 3, \ldots, n-1 \end{cases}$$
$$d'_i = \begin{cases} \dfrac{d_i}{b_i} & ; i = 1 \\ \dfrac{d_i - a_i d'_{i-1}}{b_i - a_i c'_{i-1}} & ; i = 2, 3, \ldots, n. \end{cases} \tag{C.71}$$

We then solve by back-substitution using

$$\begin{aligned} x_n &= d'_n \\ x_i = d'_i - c'_i x_{i+1} \qquad i &= n-1, n-2, \ldots, 1 \end{aligned} \tag{C.72}$$

The system is not tri-diagonal, but it can be made tri-diagonal by removing the values u_0^n and u_N^n from $\mathbf{u}^n$ and solving instead for $\mathbf{u}_r^n = (u_1^n, \ldots, u_{N-1}^n)^T$

$$\mathbf{B}\mathbf{u}_r^{n+1} = \mathbf{r}^{n+1} \tag{C.73}$$

The matrix **B** will be tri-diagonal and the right-hand side $\mathbf{r}^{n+1}$ will contain u_0^n and u_N^n in the first and last entries, respectively. For example, for $i = 1$

$$-\mu\theta u_2^{n+1} + (1 + 2\mu\theta)u_1^{n+1} - \mu\theta u_0^{n+1} = u_1^n + \mu(1-\theta)(u_2^n - 2u_1^n + u_0^n) \tag{C.74}$$

which can be written as

$$-\mu\theta u_2^{n+1} + (1 + 2\mu\theta)u_1^{n+1} = u_1^n + \mu(1-\theta)(u_2^n - 2u_1^n + u_0^n) + \mu\theta u_0^{n+1} \tag{C.75}$$

We can now place the value of u_0^n into this formula and do the same for $i = N$. The resulting matrix $\mathbf{B} = \mathbf{I} - \mu\theta\mathbf{A}$ will be tri-diagonal and of size $(N-1) \times (N-1)$.

We can also use iterative methods to solve the linear system, such as the Gauss-Seidel method in Appendix A. One variant of Gauss-Seidel is called successive over-relaxation and we will see it later when we look at the solution of elliptic equations, for which we also have a sparse matrix but its shape is block-diagonal, since the problem is 2D.

C.10 Finite Difference Method for Elliptic PDEs

Consider now an elliptic PDE in 2D

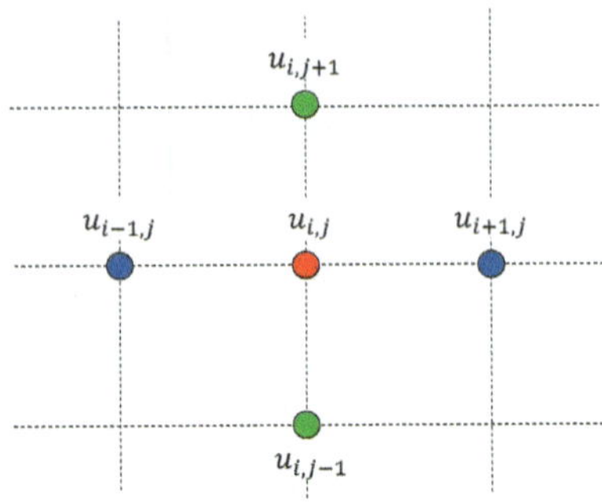

Fig. C.1 Finite difference stencil for an elliptic equation in 2D

$$\begin{aligned} u_{xx} + u_{yy} &= 0 \quad x \in \Omega \\ u &= g(x, y) \quad x \in \partial\Omega \end{aligned} \tag{C.76}$$

which is Laplace's equation, or Poisson's equation if there is a source term

$$u_{xx} + u_{yy} = f(x, y) \quad x \in \Omega \tag{C.77}$$

Assume a rectangular domain $0 \leq x \leq l_x, 0 \leq y \leq l_y$, which we discretise as before to generate grid points $x_i, i = 0, \ldots, N, y_j, j = 0, \ldots, M$, where $x_i = ih, y_j = jk$, and $k = l_y/M$. We use the label $u_{i,j}$ for the FD approximation of $u(x_i, y_j)$. The FD approximations of the partial derivatives are the same as before, but now we approximate in two directions (see Fig. C.1) to obtain

$$\frac{u_{i+1,j} - 2u_{i,j} + u_{i-1,j}}{h^2} + \frac{u_{i,j+1} - 2u_{i,j} + u_{i,j-1}}{k^2} = f(x_i, y_i) \tag{C.78}$$

and for simplicity we set $k = h$ so that

$$-u_{i+1,j} - u_{i-1,j} - u_{i,j+1} - u_{i,j-1} + 4u_{i,j} = -h^2 f(x_i, y_i) \tag{C.79}$$

Equation (C.79) is valid for $i = 1, \ldots, N-1$, $j = 1, \ldots, M-1$, while at the boundary (see Fig. C.2)

$$u_{0,j} = g(0, y_j), \quad u_{N,j} = g(l_x, y_j), \quad u_{i,0} = g(x_i, 0), \quad u_{i,M} = g(x_i, l_y) \tag{C.80}$$

Define a vector of values at the interior grid points or nodes as follows

$$\mathbf{u} = (u_{1,1}, u_{2,1}, \ldots, u_{N-1,0}, u_{1,2}, \ldots, u_{N-1,2}, \ldots, u_{1,M-1}, \ldots u_{N-1,M-1})^T \tag{C.81}$$

We can write the equations for the interior nodes as a linear system $\mathbf{Au} = \mathbf{b}$ with a block matrix $\mathbf{A}$ of size $(N-1)(M-1) \times (N-1)(M-1)$

Fig. C.2 Boundary nodes for an elliptic equation in 2D

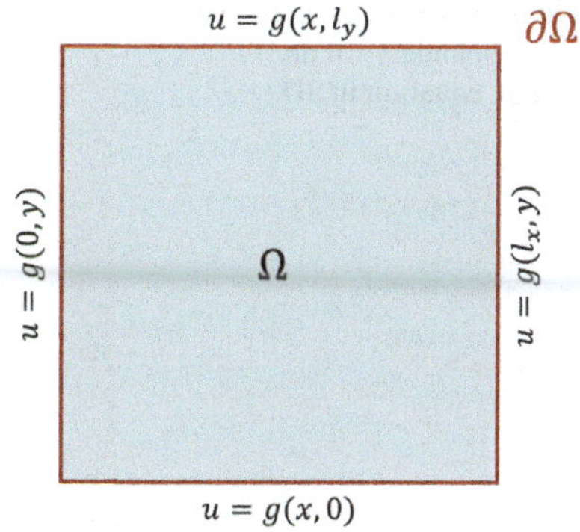

$$
\mathbf{A} = \begin{bmatrix} \mathbf{D} & -\mathbf{I} & 0 & 0 & 0 & \dots & 0 \\ -\mathbf{I} & \mathbf{D} & -\mathbf{I} & 0 & 0 & \dots & 0 \\ 0 & -\mathbf{I} & \mathbf{D} & -\mathbf{I} & 0 & \dots & 0 \\ \vdots & \ddots & \ddots & \ddots & \ddots & \ddots & \vdots \\ 0 & \dots & 0 & -\mathbf{I} & \mathbf{D} & -\mathbf{I} & 0 \\ 0 & \dots & \dots & 0 & -\mathbf{I} & \mathbf{D} & -\mathbf{I} \\ 0 & \dots & \dots & \dots & 0 & -\mathbf{I} & \mathbf{D} \end{bmatrix} \tag{C.82}
$$

in which $\mathbf{D} \in \mathbb{R}^{(N-1)\times(N-1)}$ is given by

$$
\mathbf{D} = \begin{bmatrix} 4 & -1 & 0 & 0 & 0 & \dots & 0 \\ -1 & 4 & -1 & 0 & 0 & \dots & 0 \\ 0 & -1 & 4 & -1 & 0 & \dots & 0 \\ \vdots & \ddots & \ddots & \ddots & \ddots & \ddots & \vdots \\ 0 & \dots & 0 & -1 & 4 & -1 & 0 \\ 0 & \dots & \dots & 0 & -1 & 4 & -1 \\ 0 & \dots & \dots & \dots & 0 & -1 & 4 \end{bmatrix} \tag{C.83}
$$

The vector $\mathbf{b}$ contains the values $f_{i,j} = f(x_i, y_j)$.

At the interior nodes next to the boundary ($i = 1, i = N-1, j = 1, j = M-1$) we also have to include the boundary values. Below we give examples of points next to the boundary and of how we can define the value of the corresponding component of $\mathbf{b}$. For the point $i = 1, j = 1$, i.e., (x_1, y_1) (see Fig. C.3)

$$
\begin{aligned}
-u_{2,1} - u_{0,1} - u_{1,2} - u_{1,0} + 4u_{1,1} &= -h^2 f_{1,1} \\
-u_{2,1} - g(0, y_1) - u_{1,2} - g(x_1, 0) + 4u_{1,1} &= -h^2 f_{1,1} \\
-u_{2,1} - u_{1,2} + 4u_{1,1} &= -h^2 f_{1,1} + g(0, y_1) + g(x_1, 0)
\end{aligned} \tag{C.84}
$$

and for the point $i = 2, j = 1$, i.e., (x_2, y_1)

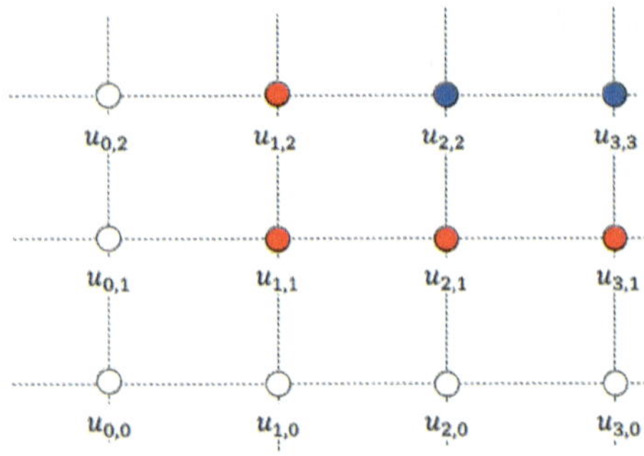

Fig. C.3 Interior nodes next to the boundary for an elliptic equation in 2D

$$\begin{aligned} -u_{3,1} - u_{1,1} - u_{2,2} - u_{2,0} + 4u_{2,1} &= -h^2 f_{2,1} \\ -u_{3,1} - u_{1,1} - u_{2,2} + 4u_{2,1} &= -h^2 f_{2,1} + g(x_2, 0). \end{aligned} \tag{C.85}$$

C.11 Successive Over-Relaxation

A solution to the system $\mathbf{Au} = \mathbf{b}$ can be obtained using a variety of methods, with Jacobi and Gauss-Seidel iteration (Appendix A) being two methods that are commonly used. We can rearrange then FD formulae and start with an initial guess $u^0_{i,j}$ for all values of $u_{i,j}$, then iterate as follows for $k = 1, 2, \ldots$

$$u^{k+1}_{i,j} = \frac{1}{4}\left(u^k_{i+1,j} + u^k_{i-1,j} + u^k_{i,j+1} + u^k_{i,j-1} - h^2 f(x_i, y_i)\right) \tag{C.86}$$

until convergence. This is Jacobi iteration, while for Gauss-Seidel we immediately use the new values $u^{k+1}_{i,j}$ as we increment i and j. A variant of this method called successive over-relaxation works by decomposing the matrix $\mathbf{A}$ into a diagonal component $\mathbf{D}$, and strictly lower- and upper-triangular components $\mathbf{L}$ and $\mathbf{U}$. We can write $\mathbf{A}$ as

$$\mathbf{A} = \begin{bmatrix} a_{11} & a_{12} & \cdots & a_{1n} \\ a_{21} & a_{22} & \cdots & a_{2n} \\ \vdots & \vdots & \ddots & \vdots \\ a_{n1} & a_{n2} & \cdots & a_{nn} \end{bmatrix} \tag{C.87}$$

Then $\mathbf{A} = \mathbf{D} + \mathbf{L} + \mathbf{U}$ with

$$\mathbf{D} = \begin{bmatrix} a_{11} & 0 & \cdots & 0 \\ 0 & a_{22} & \cdots & 0 \\ \vdots & \vdots & \ddots & \vdots \\ 0 & 0 & \cdots & a_{nn} \end{bmatrix}, \mathbf{L} = \begin{bmatrix} 0 & 0 & \cdots & 0 \\ a_{21} & 0 & \cdots & 0 \\ \vdots & \vdots & \ddots & \vdots \\ a_{n1} & a_{n2} & \cdots & 0 \end{bmatrix}, \mathbf{U} = \begin{bmatrix} 0 & a_{12} & \cdots & a_{1n} \\ 0 & 0 & \cdots & a_{2n} \\ \vdots & \vdots & \ddots & \vdots \\ 0 & 0 & \cdots & 0 \end{bmatrix} \tag{C.88}$$

The linear system can be written as

$$(\mathbf{D} + \omega \mathbf{L})\mathbf{u} = \omega \mathbf{b} - [\omega \mathbf{U} + (\omega - 1)\mathbf{D}]\mathbf{u} \tag{C.89}$$

in which $\omega > 1$ is a constant called the relaxation factor. Successive over-relaxation is an iterative technique that performs the following updates

$$\mathbf{u}^{(k+1)} = (\mathbf{D} + \omega \mathbf{L})^{-1}\left(\omega \mathbf{b} - [\omega \mathbf{U} + (\omega - 1)\mathbf{D}]\mathbf{u}^{(k)}\right) = \mathbf{L}_{w}\mathbf{u}^{(k)} + \mathbf{c} \tag{C.90}$$

Note that with $\omega = 1$ we recover the Gauss-Seidel method. Taking advantage of the triangular form of $(\mathbf{D} + \omega \mathbf{L})$, the elements $u_i^{(k+1)}$ of $\mathbf{u}^{(k+1)}$ can be computed sequentially using forward substitution

$$u_i^{(k+1)} = (1 - \omega)u_i^{(k)} + \omega \frac{1}{a_{ii}} \left(b_i - \sum_{j<i} a_{ij} u_j^{(k+1)} - \sum_{j>i} a_{ij} u_j^{(k)} \right) \tag{C.91}$$

for $i - 1, 2, \ldots, n$.

In this form we see that successive over-relaxation uses a weighted average of the previous iterate $u_i^{(k)}$ and the computed Gauss-Seidel iterate. Methods that use $\omega < 1$ are called under-relaxation methods and methods that use $\omega > 1$ are called over-relaxation methods. For stability, we strictly require that $\omega < 2$.

C.12 The Finite-Volume Method

The finite-volume method (FVM) is another discretisation technique, especially appropriate for conservation laws and fluid flows given its natural incorporation of boundary conditions, simplicity compared to the finite-element method (which does not naturally preserve mass fluxes across elements) and stability. The basis of the finite-volume method is the integral form of a conservation law. We divide the domain into control volumes (or cells) and approximate the integral conservation law in each of the control volumes.

Consider again the heat equation in $[x_0, x_l]$

$$\frac{\partial u}{\partial t} = \frac{\partial}{\partial x}\left(\lambda \frac{\partial u}{\partial x}\right) \tag{C.92}$$

The heat flux is $-\lambda u_x$ and the integral form of this equation is

$$\int_{x_0}^{x_l} \frac{\partial u}{\partial t} dx = \int_{x_0}^{x_l} \frac{\partial}{\partial x}\left(\lambda \frac{\partial u}{\partial x}\right) dx = \lambda \left.\frac{\partial u}{\partial x}\right|_{x_0}^{x_l} \tag{C.93}$$

Let us divide $[x_0, x_l]$ into N intervals of equal length h. Points are defined as $x_0, \ldots, x_N$, $x_i = ih$. We then define the cells/volumes as intervals in which x_i are the midpoints, $i = 1, \ldots, N-1$. The points x_0 and $x_{N+1} = x_l$ sit in half-cells, as illustrated in Fig. C.4. The end points of interval i are labelled $x_{i+1/2}$ and $x_{i-1/2}$ and we integrate the equation in the volume i

$$\int_{x_{i-1/2}}^{x_{i+1/2}} \frac{\partial u}{\partial t} dx = \lambda \frac{\partial u}{\partial x}\bigg|_{x_{i-1/2}}^{x_{i+1/2}} \tag{C.94}$$

We can use central differences for the end-point fluxes as follows

$$\lambda \frac{\partial u}{\partial x}\bigg|_{x_{i-1/2}} = \lambda \frac{u_i^n - u_{i-1}^n}{h} + \mathcal{O}(h^2), \quad \lambda \frac{\partial u}{\partial x}\bigg|_{x_{i+1/2}} = \lambda \frac{u_{i+1}^n - u_i^n}{h} + \mathcal{O}(h^2) \tag{C.95}$$

in which u_i^n is defined as the spatial average of $u(x, t_n)$ in cell i

$$u_i^n = \frac{1}{h} \int_{x_{i-1/2}}^{x_{i+1/2}} u(x, t_n) dx \tag{C.96}$$

We approximate the left-hand side of Eq. (C.94) as follows:

$$\int_{x_{i-1/2}}^{x_{i+1/2}} \frac{\partial u}{\partial t}(x, t_n) dx = \frac{\partial}{\partial t} \int_{x_{i-1/2}}^{x_{i+1/2}} u(x, t_n) dx = h \frac{\partial u_i^n}{\partial t} \tag{C.97}$$

We can now use any time-stepping scheme for the time derivative, e.g., forward Euler, to obtain

$$\frac{u_i^{n+1} - u_i^n}{\Delta t} h = \lambda \frac{u_{i+1}^n - u_i^n}{h} - \lambda \frac{u_i^n - u_{i-1}^n}{h} \tag{C.98}$$

which we can rewrite as

$$u_i^{n+1} = u_i^n + \frac{\lambda \Delta t}{h^2} \left(u_{i+1}^n + u_{i-1}^n - 2u_i^n \right), \quad i = 1, \ldots, N-1 \tag{C.99}$$

The stability criterion is the same as that for the forward Euler FD scheme. The treatment of Dirichlet boundary conditions is very simple, e.g., on the left we have the following equation for cell 1

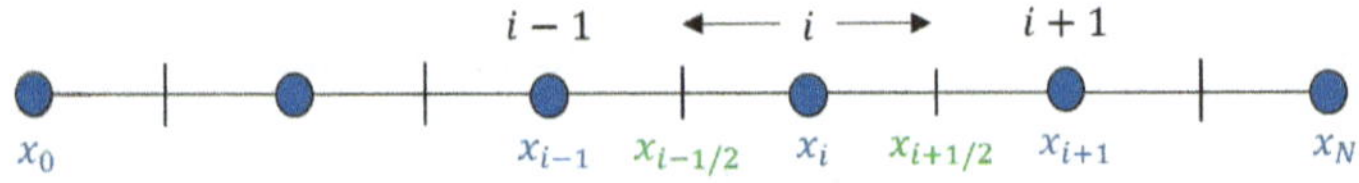

Fig. C.4 1D finite-volume cells including the boundary

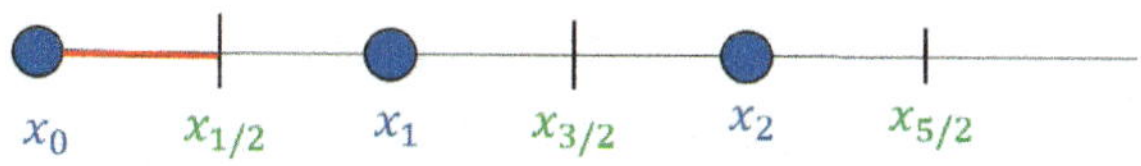

Fig. C.5 Treatment of a Dirichlet condition for a 1D parabolic problem in the finite-volume method

$$u_1^{n+1} = u_1^n + \frac{\lambda \Delta t}{h^2}\left(u_2^n + u_0^n - 2u_1^n\right) \tag{C.100}$$

for which we simply specify u_0^n (see Fig. C.5).

For a Neumann condition at x_0 we integrate the equation over the half-cell in which x_0 sits

$$\frac{u_0^{n+1} - u_0^n}{\Delta t}\frac{h}{2} = \lambda \frac{\partial u}{\partial x}\bigg|_{x_{1/2}} - \lambda \frac{\partial u}{\partial x}\bigg|_{x_0} \tag{C.101}$$

The flux at x_0 is specified, say α, and we use central differencing for the other derivative

$$\begin{aligned} \frac{u_0^{n+1} - u_0^n}{\Delta t}\frac{h}{2} &= \lambda \frac{u_1^n - u_0^n}{h} - \alpha \\ u_0^{n+1} = u_0^n &= \frac{2\lambda \Delta t}{h^2}\left(u_1^n - u_0^n\right) - \frac{2\Delta t}{h}\alpha \end{aligned} \tag{C.102}$$

Appendix D
Gradient-Based Methods for Optimisation

D.1 Outline of Optimisation

Optimisation is an important tool in many branches of engineering. It also lies at the heart of most machine learning tasks, whether maximising a likelihood or minimising a loss function. Optimisation is concerned with finding the *best* outcome for a particular problem or process in engineering [1]. What we mean by 'best' depends on what we decide is the *objective*. We can try to optimise with a *single objective* or *multiple objectives*, e.g., find a balance between the cost and efficiency of a flow battery. The second case is much more complicated, involving subjective elements, and will not be covered here.

We start any optimisation process by defining the objective, which is the quantity we wish to minimise or maximise. It is a function of some inputs (arguments) x_i, which we can collect together as a vector $\mathbf{x} = (x_1, \ldots, x_n)^T$. The goal is then to find the value of $\mathbf{x}$ that minimises or maximises an *objective function* $f(\mathbf{x})$, which expresses the objective mathematically.

There are a number of ways to go about this task, some algebraic but mostly based on derivatives (of $f(\mathbf{x})$. The latter approach is almost always iterative, since an analytical solution to the optimisation problem cannot be found. The iterations start at some value of $\mathbf{x}$ and continue until convergence to a terminal value, at which the objective function attains an optimum value. In any method, we must try to find the *global extremum*, as opposed to a *local extremum*. Avoiding local extremum can be achieved by starting at many different values of $\mathbf{x}$ and selecting the terminal value that yields the highest or lowest value of the objective function. Alternatively, we can perturb a terminal value to see if we return to the same point. There is, however, no guarantee of reaching a global extremum.

A. A. Shah et al., *New Paradigms in Flow Battery Modelling*,
Engineering Applications of Computational Methods 16,
https://doi.org/10.1007/978-981-99-2524-7

We mention that there is a whole class of non-gradient methods that we shall not cover, namely, evolutionary optimisation methods. These are heuristic rule-based methods that are geared towards finding global extrema, but involve a sacrifice in terms of accuracy. Gradient-based methods are by far the most reliable, efficient and widely used in machine learning.

It is also important to note that optimization problems are frequently subject to constraints, which can take the form of one or more equalities, $g(\mathbf{x} = 0$ for some function $g(\cdot)$, or inequalities, $g(\mathbf{x} \leq 0$. Such problems are referred to as *constrained optimization* problems, as opposed to unconstrained problems that do not feature such constraints. Solutions to constrained problems can be constructed by transforming the problem into an unconstrained problem, which involves directly including the constraint in the objective function. The new objective function is called a *Lagrangian* and it involves *Lagrange multipliers*, which are unknown constants multiplying the constraint term in the Lagrangian (see for example Sect. 6.13 on support vector machines). We will not cover such problems in this appendix. Also not covered are certain advanced topics, notably dual and primal-dual formulations of optimization problems (some details and explanations can be found in Sect. 6.13).

D.2 Golden Section Search

We first look at a bracketing method called the *golden section search* [1], in which we seek a maximum or minimum of an objective function in an interval $[x_l, x_u] \subset \mathbb{R}$, assuming for simplicity that this extremum is unique. We then choose another point x_1 in the same interval, along with a fourth point x_2. If $f(x_1) > f(x_2)$ we eliminate $[x_l, x_2]$ because it does not contain the extremum. For this case, x_2 becomes the new x_l. If $f(x_2) > f(x_1)$ we eliminate $[x_1, x_u]$ and x_1 becomes the new x_u. As we iterate this procedure, the interval containing the extremum is reduced rapidly and the extremum is chosen from amongst $f(x_u)$, $f(x_1)$, $f(x_2)$ and $f(x_l)$. The only remaining question is how to choose the interior points x_1 and x_2: they are chosen according to the so-called *golden ratio*

$$x_1 = x_l + d, \quad x_2 = x_u - d, \quad d = \frac{\sqrt{5}-1}{2}(x_u - x_l) \tag{D.1}$$

In each iteration the interval is reduced by a factor of the golden ratio (around 61.8%). Because the original x_1 and x_2 were chosen using the golden ratio, we do not have to recalculate all the function values for the next iteration. For example, if $f(x_1) > f(x_2)$ the old x_1 becomes the new x_2 so we already have $f(x_2)$. We now only need to determine the new x_1 as before

$$x_1 = x_l + d = x_l + \frac{\sqrt{5}-1}{2}(x_u - x_l). \tag{D.2}$$

D.3 Newton's Method

Newton's method is an example of a *gradient-based algorithm*, meaning that we use the derivative of the objective function to make updates. In 1D, the extremum obviously occurs where $f'(x) = 0$. Setting $g(x) = f'(x)$, a Taylor series expansion about a point x_i is

$$g(x) = g(x_i) + (x - x_i)g'(x_i) + \dots \tag{D.3}$$

For $g(x) = 0$ we must have

$$x = x_i - \frac{g(x_i)}{g'(x_i)} \quad \text{or} \quad x = x_i - \frac{f'(x_i)}{f''(x_i)} \tag{D.4}$$

This formula holds if x is close to x_i. It provides a method for finding the roots of $f(x)$, starting with an initial guess x_0 for the location of the extremum and making a small change

$$x_1 = x_0 - \frac{f'(x_0)}{f''(x_0)} \tag{D.5}$$

The process is repeated with

$$x_{i+1} = x_i - \frac{f'(x_i)}{f''(x_i)}, \quad i = 2, 3, \dots \tag{D.6}$$

until some convergence criterion is reached.

Newton's method works well but is impractical for cases in which the derivatives cannot be evaluated analytically. For such cases, other approaches that do not involve derivative evaluations are available. For example, a secant-like version of Newton's method can be developed by using finite difference approximations for the derivatives. A bigger problem is that it may diverge (not reach a solution or oscillate) based on the nature of the function and the quality of the initial guess. It is usually employed only when we are close to the extremum. We can use hybrid approaches based on polynomial approximations of $f(x)$ or derivative information and bracketing to improve the convergence and stability.

D.4 General Gradient-Based Methods

Any numerical solution to an optimisation problem starts with an initial guess $\mathbf{x}_* = \mathbf{x}_0$, and proceeds to iteratively improve the guess based on some rules until some form of convergence criterion is met. The most widely used methods are *line search descent methods*, which perform updates as follows

$$\mathbf{x}_{i+1} = \mathbf{x}_i + \alpha_i \mathbf{d}_i, \quad i = 1, 2, \dots \tag{D.7}$$

in which $\mathbf{d}_i$ is the *search direction* and α_i is the *step size* or step length. What distinguishes different methods are the forms of the search direction and step size.

Gradient-based methods use the gradient of the function $\nabla f(\mathbf{x})$ to find a search direction $\mathbf{d}_i$. If $\mathbf{n}$ is a unit vector, then $\mathbf{n} \cdot \nabla f(\mathbf{x})$ is called the *directional derivative* of $f(\mathbf{x})$ in the direction $\mathbf{n}$ (another equivalent definition will be seen later). The directional derivative is the rate of change of $f(\mathbf{x})$ in the direction $\mathbf{n}$ at the point $\mathbf{x}$. Two useful properties of the directional derivative and gradient are

1. The *maximal directional derivative* of a function $f(\mathbf{x})$ at the point $\mathbf{x}$ is in the direction of the gradient vector $\nabla f(\mathbf{x})$.
2. If a surface is given by $f(\mathbf{x}) = c$ (isocontours), where c is a constant, then the normals to the surface are the vectors $\pm\nabla f(\mathbf{x})$.

The first is the most useful for our purpose, and will be proven below. It tells us that if we wish to maximise or minimise a function $f(\mathbf{x})$ and our current estimate is $\mathbf{x}_i$, we can move $\mathbf{x}_i$ in the direction $\nabla f(\mathbf{x})$ $(-\nabla f(\mathbf{x}))$, in which the function is increasing (decreasing) most rapidly, to obtain an improved estimate $\mathbf{x}_{i+1}$. Consider minimisation: in general, we wish to move $\mathbf{x}_i$ in a direction $\mathbf{d}_i$ in which $f(\mathbf{x})$ is decreasing at the point $\mathbf{x}_i$. We require $\mathbf{d}_i$ to be a *descent direction*, and fix it to have unit length. To check whether our $\mathbf{d}_i$ is valid we use the definition of a directional derivative of $f(\mathbf{x})$ in the direction $\mathbf{d}_i$ at the point $\mathbf{x}_i$, i.e., we impose

$$\mathbf{d}_i \cdot \nabla f(\mathbf{x}_i) = \mathbf{d}_i^T \nabla f(\mathbf{x}_i) < 0 \tag{D.8}$$

A Taylor expansion with a small α yields

$$f(\mathbf{x}_i + \alpha\mathbf{d}_i) - f(\mathbf{x}_i) = \alpha\mathbf{d}_i^T \nabla f(\mathbf{x}_i) + \mathcal{O}(\alpha^2) \tag{D.9}$$

For $f(\mathbf{x}_i + \alpha\mathbf{d}_i) - f(\mathbf{x}_i) < 0$ we require $\mathbf{d}_i^T \nabla f(\mathbf{x}_i) < 0$. The first property of directional derivatives above tells us that the maximum rate of decrease occurs when $\mathbf{d}_i = -\nabla f(\mathbf{x}_i)$. If, on the other hand, we want to maximise the objective function, the maximum rate of increase occurs when $\mathbf{d}_i = \nabla f(\mathbf{x}_i)$. The *gradient descent (ascent)* method corresponds to $\mathbf{d}_i = -\nabla f(\mathbf{x}_i)$ ($\mathbf{d}_i = \nabla f(\mathbf{x}_i)$). We are not, however, limited to these choices.

Once we have chosen a descent (or ascent) direction we need to choose the step length α_i, for which we can use a *line search* [1]. In an *exact line search* we find the α_i that gives the actual minimum at a given iteration, while in an *inexact search* we use a value that gives *sufficient* decrease (e.g., some constant).

D.5 Steepest Descent

In steepest descent we can consider minimising the following function

$$\psi_i(\alpha) = f(\mathbf{x}_i + \alpha\mathbf{d}_i) = f(\mathbf{x}_i - \alpha\nabla f(\mathbf{x}_i)) \tag{D.10}$$

$\psi_i(\alpha)$ is a function of one variable α and if we minimise it, we are finding the minimum of the function $f(\mathbf{x})$ for values of $\mathbf{x}$ starting at $\mathbf{x}_i$ and moving in the direction $-\nabla f(\mathbf{x}_i)$. Note also that $d\psi_i/d\alpha(0)$ is the rate of change of $f(\mathbf{x})$ in the direction $\mathbf{d}_i$ at the point $\mathbf{x}_i$ for any general $\mathbf{d}_i$. Therefore, $d\psi_i/d\alpha(0)$ is the directional derivative. Using the chain rule, we first differentiate w.r.t. $(\mathbf{x}_i + \alpha\mathbf{d}_i)$ and then differentiate $(\mathbf{x}_i + \alpha\mathbf{d}_i)$ w.r.t. α

$$\frac{d\psi_i}{d\alpha}(0) = \frac{d}{d\alpha} f(\mathbf{x}_i + \alpha\mathbf{d}_i)\Big|_{\alpha=0} = \nabla f(\mathbf{x}_i + \alpha\mathbf{d}_i)\frac{d}{d\alpha}(\mathbf{x}_i + \alpha\mathbf{d}_i)\Big|_{\alpha=0} \tag{D.11}$$

leading to

$$\frac{d\psi_i}{d\alpha}(0) = \nabla f(\mathbf{x}_i) \cdot \mathbf{d}_i \tag{D.12}$$

which was our previous definition of the directional derivative. By the Cauchy-Schwartz inequality

$$-\|\nabla f(\mathbf{x}_i)\|\|\mathbf{d}_i\| \leq \nabla f(\mathbf{x}_i) \cdot \mathbf{d}_i \leq \|\nabla f(\mathbf{x}_i)\|\|\mathbf{d}_i\| \tag{D.13}$$

and since $\|\mathbf{d}_i\| = 1$

$$-\|\nabla f(\mathbf{x}_i)\| \leq \nabla f(\mathbf{x}_i) \cdot \mathbf{d}_i \leq \|\nabla f(\mathbf{x}_i)\| \tag{D.14}$$

Thus, the maximum rate of increase of the function (maximum directional derivative) is the upper bound $\|\nabla f(\mathbf{x}_i)\|$ and the maximum rate of decrease of the function (minimum directional derivative) is the lower bound $-\|\nabla f(\mathbf{x}_i)\|$, proving the property stated earlier.

Returning to the line search, the value of α that yields the minimum of

$$\psi_i(\alpha) = f(\mathbf{x}_i + \alpha\mathbf{d}_i) = f(\mathbf{x}_i - \alpha\nabla f(\mathbf{x}_i)) \tag{D.15}$$

can be used to define the step length. We can then set

$$\mathbf{x}_{i+1} = \mathbf{x}_i - \alpha_i \nabla f(\mathbf{x}_i) \tag{D.16}$$

and iterate until the values of $\mathbf{x}_i$ converge. This method, using $\mathbf{d}_\mathbf{i} = -\nabla f(\mathbf{x}_i)$ and the exact line search above, is called the *steepest descent* method, a special case of gradient descent. For maximisation we would use $\mathbf{d}_\mathbf{i} = \nabla f(\mathbf{x}_i)$ and call it steepest ascent.

Consider a quadratic function

$$f(\mathbf{x}) = \frac{1}{2}\mathbf{x}^T\mathbf{Q}\mathbf{x} - \mathbf{b}^T\mathbf{x} \tag{D.17}$$

for some symmetric matrix $\mathbf{Q}$ and some vector $\mathbf{b}$. The gradient is given by $g(\mathbf{x}) = \nabla f(\mathbf{x}) = \mathbf{Q}\mathbf{x} - \mathbf{b}$ and in steepest descent we would use

$$\mathbf{x}_{i+1} = \mathbf{x}_i - \alpha_i \nabla f(\mathbf{x}_i) = \mathbf{x}_i - \alpha_i(\mathbf{Q}\mathbf{x}_i - \mathbf{b}) \tag{D.18}$$

We choose the step length that minimises $\psi(\alpha) = f(\mathbf{x}_i - \alpha\nabla f(\mathbf{x}_i))$, namely,

$$\psi(\alpha) = \frac{1}{2}(\mathbf{x}_i - \alpha g(\mathbf{x}_i))^T \mathbf{Q} (\mathbf{x}_i - \alpha g(\mathbf{x}_i)) - \mathbf{b}^T (\mathbf{x}_i - \alpha g(\mathbf{x}_i)) \tag{D.19}$$

Differentiating w.r.t. α and solving yields

$$\alpha g(\mathbf{x}_i)^T \mathbf{Q} g(\mathbf{x}_i) = g(\mathbf{x}_i)^T g(\mathbf{x}_i) \tag{D.20}$$

in which $g(\mathbf{x}) = \nabla f(\mathbf{x}) = \mathbf{Q}\mathbf{x} - \mathbf{b}$, so that

$$\alpha_i = \frac{g(\mathbf{x}_i)^T g(\mathbf{x}_i)}{g(\mathbf{x}_i)^T \mathbf{Q} g(\mathbf{x}_i)} \tag{D.21}$$

yielding the update rule

$$\mathbf{x}_{i+1} = \mathbf{x}_i - \frac{g(\mathbf{x}_i)^T g(\mathbf{x}_i)}{g(\mathbf{x}_i)^T \mathbf{Q} g(\mathbf{x}_i)} g(\mathbf{x}_i) \tag{D.22}$$

in which $g(\mathbf{x}_i) = \mathbf{Q}\mathbf{x}_i - \mathbf{b}$. In the general non-quadratic case, this method can be used with local approximations of the function using the quadratic form (D.17) at each iteration. The gradient can be approximated using finite differences if the function is not explicitly known. The method outlined above has a connection to the linear *conjugate gradient* method, but we shall not cover this class of methods.

D.6 Barzilai and Borwein Method

We can choose the step length in other ways, e.g., a fixed constant for all iterations, although this can lead to algorithms that diverge close to the extremum or else take too long to converge. Another exact line search method is from *Barzilai and Borwein* [2], in which we minimise

$$\psi_i(\alpha) = \|\Delta\mathbf{x} - \alpha\Delta f\|^2 \tag{D.23}$$

where $\Delta\mathbf{x} = \mathbf{x}_i - \mathbf{x}_{i-1}$ and $\Delta f = \nabla f(\mathbf{x}_i) - \nabla f(\mathbf{x}_{i-1})$. Differentiating w.r.t. α

$$\frac{d}{d\alpha}(\Delta\mathbf{x} - \alpha\Delta f)^T(\Delta\mathbf{x} - \alpha\Delta f) = 2\Delta f^T(\Delta\mathbf{x} - \alpha\Delta f) = 0 \tag{D.24}$$

with solution

$$\alpha = \frac{\Delta f^T \Delta \mathbf{x}}{\Delta f^T \Delta f} \tag{D.25}$$

which defines the step length α_i. The Barzilai and Borwein method is a quasi-Newton method, which we will outline later.

D.7 Backtracking

Another popular method is called *backtracking* [1], which is an inexact line search method. In backtracking we start with some α and successively reduce it by $\lambda\alpha$ for some $\lambda \in (0, 1)$ until we satisfy the *Wolfe conditions*

$$\alpha \to \lambda\alpha \to \lambda^2\alpha \to \lambda^3\alpha \ldots \tag{D.26}$$

This yields an α_i giving a 'sufficient' decrease in $f(\mathbf{x})$ at the point $\mathbf{x}_i$ without giving the greatest decrease in the direction $\nabla f(\mathbf{x}_i)$, i.e., it will not solve $\min_\alpha f(\mathbf{x}_i - \alpha\nabla f(\mathbf{x}_i))$. Backtracking can be applied with any search direction $\mathbf{d}_i$ and the Wolfe conditions are (without proof)

$$\begin{aligned} f(\mathbf{x}_i + \alpha_i \mathbf{d}_i) &\leq f(\mathbf{x}_i) + c_1 \alpha_i \mathbf{d}_i^T \nabla f(\mathbf{x}_i) \\ -\mathbf{d}_i^T \nabla f(\mathbf{x}_i + \alpha_i \mathbf{d}_i) &\leq -c_2 \mathbf{d}_i^T \nabla f(\mathbf{x}_i) \end{aligned} \tag{D.27}$$

with $c_1 \in (0, 1)$, and c_2 much larger, the values of which depend on the search direction used. The first condition is called the *Armijo rule* and the second is called the *curvature condition*. If we modify the curvature condition to the following

$$|\mathbf{d}_i^T \nabla f(\mathbf{x}_i + \alpha_i \mathbf{d}_i)| \leq c_2 |\mathbf{d}_i^T \nabla f(\mathbf{x}_i)| \tag{D.28}$$

we obtain the *strong Wolfe conditions*, which force α_i to yield a point close to a minimum of $\psi_i(\alpha)$. Typically we do not use the strong conditions because they increase computational complexity. The Wolfe conditions ensure convergence of the gradient to zero, with an additional motivation in the case of quasi-Newton methods explained later.

D.8 Newton-Raphson and Damped Newton Methods

A frequently used method in many applications is the *Newton-Raphson method*, which employs second-order derivatives, so that it is more computationally expensive but simultaneously more accurate. It is essentially an extension of the 1D Newton method, based on a quadratic approximation to $f(\mathbf{x})$ about a point $\mathbf{x}_i$

$$f(\mathbf{x}) \approx \widehat{f}(\mathbf{x}) = f(\mathbf{x}_i) + \nabla f(\mathbf{x}_i)(\mathbf{x} - \mathbf{x}_i)^T + \frac{1}{2}(\mathbf{x} - \mathbf{x}_i)^T \nabla^2 f(\mathbf{x}_i)(\mathbf{x} - \mathbf{x}_i) \quad \text{(D.29)}$$

$\mathbf{x}$ is chosen such that $\widehat{f}(\mathbf{x})$ attains a minimum in the region of $\mathbf{x}_i$ and the process is repeated until convergence, i.e., until $\|\nabla f(\mathbf{x}_i)\|$ or $\|\mathbf{x}_{i+1} - \mathbf{x}_i\|$ is sufficiently small.

We call $\nabla^2 f(\mathbf{x}_i)$ the *Hessian* matrix (evaluated at $\mathbf{x}_i$) and we will use the notation $H(\mathbf{x}_i) = \nabla^2 f(\mathbf{x}_i)$. It is a matrix of the second-order partial derivatives of $f(\mathbf{x})$ w.r.t. the inputs $x_1, \ldots, x_n$, and is symmetric matrix since $\partial^2 f/\partial x_k \partial x_l = \partial^2 f/\partial x_l \partial x_k$. At the current update $\mathbf{x} = \mathbf{x}_i$, a quadratic approximation of $f(\mathbf{x})$ around the point $\mathbf{x}_i$ is given by $\widehat{f}(\mathbf{x})$. Minimisation of $\widehat{f}(\mathbf{x})$ is a much simpler problem, with the minimum attained at $\mathbf{x} = \mathbf{x}_{i+1}$, where

$$\mathbf{x}_{i+1} = \mathbf{x}_i - H(\mathbf{x}_i)^{-1} \nabla f(\mathbf{x}_i) \quad \text{(D.30)}$$

Again, the iterations are continued until convergence. We may observe that this is a line search descent method with descent direction $\mathbf{d}_i = -H(\mathbf{x}_i)^{-1}\nabla f(\mathbf{x}_i)$. The Hessian has to be positive definite since

$$\nabla f(\mathbf{x}_i)^T \mathbf{d}_i = -\nabla f(\mathbf{x}_i)^T H(\mathbf{x}_i)^{-1} \nabla f(\mathbf{x}_i) \quad \text{(D.31)}$$

has to be negative. A matrix $\mathbf{A}$ is positive definite (p.d.) if $\mathbf{x}^T \mathbf{A} \mathbf{x} > 0$ for any $\mathbf{x}$. If $H(\mathbf{x}_i)$ is p.d. then $H(\mathbf{x}_i)^{-1}$ is p.d.

The Newton-Raphson method converges much faster than gradient descent, therefore requiring fewer iterations, but the Hessian is expensive to invert and may not be positive definite. There is a large class of methods devoted to modifying Newton's method to avoid computing the Hessian, called *quasi-Newton methods*. Before we look at these methods, we can consider a simple modification using a line search, which leads to the *damped Newton's method*

$$\mathbf{x}_{i+1} = \mathbf{x}_i - \alpha H(\mathbf{x}_i)^{-1} \nabla f(\mathbf{x}_i) \quad \text{(D.32)}$$

If $\alpha = 1$ we recover Newton-Raphson. Step sizes are typically chosen by backtracking: set $\mathbf{d}_i = H(\mathbf{x}_i)^{-1}\nabla f(\mathbf{x}_i)$ and at each iteration, we start with $\alpha = 1$ and while

$$f(\mathbf{x}_i + \alpha \mathbf{d}_i) > f(\mathbf{x}_i) + c_1 \alpha \mathbf{d}_i^T \nabla f(\mathbf{x}_i) \quad \text{(D.33)}$$

we set $\alpha \mapsto \beta\alpha$ for $\beta \in (0, 1)$ and $c_1 \in (0, 0.5)$ (the Wolfe condition). Damping can help to provide better convergence although of course it will be slower because smaller steps are taken. With this method we still require the Hessian, so next we will look at alternative methods that avoid Hessian computations altogether.

D.9 Quasi-Newton Methods

The Newton-Raphson method involves two main steps: (1) first compute the Hessian $H(\mathbf{x}_i) = \nabla^2 f(\mathbf{x}_i)$; (2) then solve the system $H(\mathbf{x}_i)\Delta\mathbf{x}_i = -\nabla f(\mathbf{x}_i)$ to obtain the increment $\Delta\mathbf{x}_i$ for $\mathbf{x}_{i+1} = \mathbf{x}_i + \Delta\mathbf{x}_i$. In *Quasi-Newton* methods we perform the update

$$\mathbf{x}_{i+1} = \mathbf{x}_i + \alpha_i \Delta\mathbf{x}_i \tag{D.34}$$

in which the search direction $\Delta\mathbf{x}_i$ is given by

$$\mathbf{B}\Delta\mathbf{x}_i = -\nabla f(\mathbf{x}_i) \tag{D.35}$$

for some *approximation* $\mathbf{B}$ of $H(\mathbf{x}_i)$. We require that $\mathbf{B}$ is easy to compute from $\mathbf{B}\Delta\mathbf{x}_i = -\nabla f(\mathbf{x}_i)$. Initialising $\mathbf{x} = \mathbf{x}_0$ and $\mathbf{B} = \mathbf{B}_0$ (p.d. and symmetric), we solve $\mathbf{B}_i\Delta\mathbf{x}_i = -\nabla f(\mathbf{x}_i)$ and update according to

$$\mathbf{x}_{i+1} = \mathbf{x}_i + \alpha_i \Delta\mathbf{x}_i \tag{D.36}$$

computing $\mathbf{B}_{i+1}$ from $\mathbf{B}_i$, and continuing until convergence.

Different quasi-Newton methods implement the step of computing $\mathbf{B}_{i+1}$ from $\mathbf{B}_i$ in different ways [3]. Since $\mathbf{B}_i$ already contains information about the Hessian, we can use a suitable matrix update to find $\mathbf{B}_{i+1}$. A reasonable requirement for $\mathbf{B}_{i+1}$ is

$$\nabla f(\mathbf{x}_{i+1}) = \nabla f(\mathbf{x}_i) + \mathbf{B}_{i+1}(\mathbf{x}_{i+1} - \mathbf{x}_i), \tag{D.37}$$

i.e., a first-order Taylor expansion. We can set

$$\begin{aligned} \mathbf{y} &= \nabla f(\mathbf{x}_{i+1}) - \nabla f(\mathbf{x}_i) \\ \mathbf{s} &= \mathbf{x}_{i+1} - \mathbf{x}_i \end{aligned} \tag{D.38}$$

to obtain the *secant equation*

$$\mathbf{B}_{i+1}\mathbf{s} = \mathbf{y} \tag{D.39}$$

We require $\mathbf{B}_{i+1}$ to be p.d. and symmetric, so we could attempt an approximation of the form

$$\mathbf{B}_{i+1} = \mathbf{B}_i + a\mathbf{u}\mathbf{u}^T \tag{D.40}$$

for some vector $\mathbf{u}$ and some constant a. Note that $\mathbf{u}\mathbf{u}^T = \mathbf{u} \circ \mathbf{u} = \mathbf{u} \otimes \mathbf{u}^T$ is an outer product defined in (6.159). As explained in Sect. 6.11.1, a matrix (second-order tensor or multi-array) of the form $a\mathbf{u}\mathbf{u}^T$ is of CP-rank 1. For this reason we call (D.40) a *symmetric rank-one (SR1) update*. Using the secant equation

$$(a\mathbf{u}^T\mathbf{s})\mathbf{u} = \mathbf{y} - \mathbf{B}_i\mathbf{s} \tag{D.41}$$

which only holds if $\mathbf{u}$ is a multiple of $\mathbf{y} - \mathbf{B}_i\mathbf{s}$. Putting $\mathbf{u} = \mathbf{y} - \mathbf{B}_i\mathbf{s}$ in the secant equation yields

$$a(\mathbf{y} - \mathbf{B}_i\mathbf{s})^T\mathbf{s}(\mathbf{y} - \mathbf{B}_i\mathbf{s}) = \mathbf{y} - \mathbf{B}_i\mathbf{s} \tag{D.42}$$

and we can right multiply both sides by $\mathbf{y} - \mathbf{B}_i\mathbf{s}$ to obtain

$$\begin{aligned} a &= \frac{1}{(\mathbf{y} - \mathbf{B}_i\mathbf{s})^T\mathbf{s}} \\ \mathbf{B}_{i+1} &= \mathbf{B}_i + \frac{(\mathbf{y} - \mathbf{B}_i\mathbf{s})(\mathbf{y} - \mathbf{B}_i\mathbf{s})^T}{(\mathbf{y} - \mathbf{B}_i\mathbf{s})^T\mathbf{s}} \end{aligned} \tag{D.43}$$

In the next step we will need to solve $\mathbf{B}_{i+1}\Delta\mathbf{x}_{i+1} = -\nabla f(\mathbf{x}_{i+1})$. As well as generating $\mathbf{B}_{i+1}$ from $\mathbf{B}_i$, we can also directly obtain $\mathbf{B}_{i+1}^{-1}$ from $\mathbf{B}_i^{-1}$. To do this we use the Sherman-Morrison formula

$$(\mathbf{A} + \mathbf{u}\mathbf{v}^T)^{-1} = \mathbf{A}^{-1} - \frac{\mathbf{A}^{-1}\mathbf{u}\mathbf{v}^T\mathbf{A}^{-1}}{1 + \mathbf{v}^T\mathbf{A}^{-1}\mathbf{u}} \tag{D.44}$$

for a matrix $\mathbf{A}$ and vectors $\mathbf{u}$, $\mathbf{v}$. Since $\mathbf{B}_{i+1} = \mathbf{B}_i + a\mathbf{u}\mathbf{u}^T$, we obtain

$$\mathbf{B}_{i+1}^{-1} = \mathbf{B}_i^{-1} + \frac{(\mathbf{s} - \mathbf{B}_i^{-1}\mathbf{y})(\mathbf{s} - \mathbf{B}_i^{-1}\mathbf{y})^T}{(\mathbf{s} - \mathbf{B}_i^{-1}\mathbf{y})^T\mathbf{y}}. \tag{D.45}$$

D.10 Symmetric Rank 2 Updates

The SR1 update is simple and cheap, but has a key shortcoming: it does not preserve positive definiteness. Rather than a rank-one update to $\mathbf{B}_i$, we can instead make a *rank-two update*

$$\mathbf{B}_{i+1} = \mathbf{B}_i + a\mathbf{u}\mathbf{u}^T + b\mathbf{v}\mathbf{v}^T \tag{D.46}$$

Following the same procedure as above would lead to the *Broyden-Fletcher-Goldfarb-Shanno (BFGS) update*, one of the most widely used of the quasi-Newton methods because it preserves positive definiteness

$$\mathbf{B}_{i+1} = \mathbf{B}_i - \frac{\mathbf{B}_i\mathbf{s}\mathbf{s}^T\mathbf{B}_i}{\mathbf{s}^T\mathbf{B}_i\mathbf{s}} + \frac{\mathbf{y}\mathbf{y}^T}{\mathbf{y}^T\mathbf{s}} \tag{D.47}$$

We can also generate $\mathbf{B}_{i+1}^{-1}$ cheaply as before using the Woodbury formula (a generalisation of the Sherman-Morrison formula). Another popular method directly computes a rank-two update of the inverse

$$\mathbf{B}_{i+1}^{-1} = \mathbf{B}_i^{-1} + a\mathbf{u}\mathbf{u}^T + b\mathbf{v}\mathbf{v}^T \tag{D.48}$$

leading to

$$\mathbf{B}_{i+1}^{-1} = \mathbf{B}_i^{-1} - \frac{\mathbf{B}_i^{-1}\mathbf{y}\mathbf{y}^T\mathbf{B}_i^{-1}}{\mathbf{y}^T\mathbf{B}_i^{-1}\mathbf{y}} + \frac{\mathbf{s}\mathbf{s}^T}{\mathbf{y}^T\mathbf{s}} \tag{D.49}$$

which we call the *Davidon-Fletcher-Powell (DFP) update*.

The *Broyden class* of updates combines SR1, BFGS and DFP by using a weighted average of two of these methods. There are numerous other methods in this category, but those above are by far the most popular in use and suffice for the vast majority of problems.

References

1. E.K.P. Chong, S.H. Zak, *An Introduction to Optimization*, vol. 75. (Wiley, 2013)
2. R. Fletcher, On the barzilai-borwein method, in *Optimization and Control with Applications*. (Springer, 2005), pp. 235–256
3. S. Wright, J. Nocedal et al., Numerical optimization. Springer Sci. **35**(67–68), 7 (1999)

The manufacturer's authorised representative in the EU is Springer Nature Customer Service Centre GmbH, Europaplatz 3, 69115 Heidelberg, Germany. If you have any concerns regarding our products, please contact ProductSafety@springernature.com

Printed and bound by CPI Group (UK) Ltd, Croydon, CR0 4YY
30/09/2026
02225314-0011